Lecture Notes in Energy

Volume 74

Lecture Notes in Energy (LNE) is a series that reports on new developments in the study of energy: from science and engineering to the analysis of energy policy. The series' scope includes but is not limited to, renewable and green energy, nuclear, fossil fuels and carbon capture, energy systems, energy storage and harvesting, batteries and fuel cells, power systems, energy efficiency, energy in buildings, energy policy, as well as energy-related topics in economics, management and transportation. Books published in LNE are original and timely and bridge between advanced textbooks and the forefront of research. Readers of LNE include postgraduate students and non-specialist researchers wishing to gain an accessible introduction to a field of research as well as professionals and researchers with a need for an up-to-date reference book on a well-defined topic. The series publishes single- and multi-authored volumes as well as advanced textbooks.

Indexed in Scopus and EI Compendex The Springer Energy board welcomes your book proposal. Please get in touch with the series via Anthony Doyle, Executive Editor, Springer (anthony.doyle@springer.com).

More information about this series at http://www.springer.com/series/8874

Tanay Sıdkı Uyar
Editor

Accelerating the Transition to a 100% Renewable Energy Era

Springer

Editor
Tanay Sıdkı Uyar
Energy Section
Department of Mechanical Engineering
Marmara University
Istanbul, Turkey

ISSN 2195-1284 ISSN 2195-1292 (electronic)
Lecture Notes in Energy
ISBN 978-3-030-40737-7 ISBN 978-3-030-40738-4 (eBook)
https://doi.org/10.1007/978-3-030-40738-4

This Springer imprint is published by the registered company Springer Nature Switzerland AG
The registered company address is: Gewerbestrasse 11, 6330 Cham, Switzerland

Preface

Coronavirus, COVID-1, is a virus that is too small to be seen. In a few months, it almost stopped all life on Earth. Not only did it stop the daily life, but in a short period of time it crossed all borders and oceans drawn by human beings quickly, reaching and infecting millions of people in 185 countries in almost over the world.

Its circulation on Earth continues, and it is not yet clear up to where it will continue and when it will stop. 343,608 people have died from the coronavirus, COVID-19, outbreak as of May 24, 2020. Worse still, nobody knows what will happen tomorrow and beyond. No one can say anything concrete.

Mankind has not been able to stop this epidemic for months with all its historical knowledge, and economic and technological capabilities. Far from stopping, it tries to reduce losses and increase the number of lives it can save.

Governments are racing in the fight against this virus crisis by undertaking the task of designing social experiences, economic powers, management approaches and priorities of the social relations they rely on, and measures and recovery packages with the slogan of "PROBLEM GLOBAL, SOLUTION NATIONAL". Despite the unknown future, these efforts and these packages have already reached a scale that will shape and restructure societies and economies in the coming years.

However, mankind is not a stranger to this type of epidemics. Humanity has faced many epidemic diseases and disasters throughout its history. When it became unsustainable due to the suffering caused by local or regional crises, it evolved to new social forms or disappeared locally, depending on the level of uneven development and productive forces and the severity of pain.

The fact that epidemics throughout the history of humanity have reached global dimensions in our era and are not limited to nations, local geographies, or even continents clearly reveals that the solution of the crisis today requires a global solidarity.

The ongoing crisis has made global connections more visible, strengthened the vision of a more flexible society at national and regional levels, as well as once again clearly revealed the great differences between countries' conditions and capacities.

For this reason, international cooperation is required to overcome the crisis and the struggle against the crisis should be at the level of global common dependency and responsibility. It is an imperative rather than a necessity to direct global investments wherever they are needed, including the most vulnerable countries and communities.

The extent and progress of the crisis require solidarity on a global scale, but it requires the bold implementation of the measures to determine the causes of the crisis without delay and eliminate them without wasting time.

Today, we cannot dissociate the cause of the coronavirus epidemic, which threatens the entire planet, and took and continues to take hundreds of thousands of lives, from the consequences of the climate crisis.

Global climate change caused by the plundering and looting of the natural values at the level of an uncontrolled war against nature in the second half of the twentieth century and afterward in an attempt to overcome its challenges by using advanced technology with an inexhaustible ambition for profit led to the global crises such as coronavirus. And it seems that crises caused by corona and similar viruses are neither the first nor the last.

The climate crisis and its consequences affect the entire planet, causing unlimited losses, especially loss of life, economies collapse, and epidemic diseases that cannot be dealt with and threaten the future of humanity.

Extreme atmospheric events, hurricanes, tornados, irregularity of precipitation regimes, floods, drought and desertification, melting glaciers and rising seas, large forest fires that could not be extinguished, destroyed ecosystems, drying lakes, lost water resources, polluted air, water, soil, all occur more frequently as the phenomena dragging the humanity to its end.

Air pollution caused by fossil fuels, especially coal-fired power plants, spreads over hundreds of kilometers. People living here easily get lung and other chronic diseases and are the first target in epidemics.

The use of coal, oil, and natural gas in the transportation, industry and housing sectors, both changes the global climate and the air pollution it creates causes the death of millions of people every year.

Coronavirus, on the other hand, kills patients suffering from chronic diseases (cancer, tuberculosis, heart, and lung diseases) caused by using fossil fuel, without the opportunity to use medication.

It is obvious that vaccination studies will last at least 1 year, due to inadequate number of tests in the initial process early diagnosis hasn't been possible and in the countries where preventive medicine is ignored the fight against the virus will extend the time and increase the losses.

If the necessary lessons are not taken and the causes cannot be eliminated by doing what is necessary, such crises will continue until they destroy the living species in the universe with an increasing severity.

Increasing loss of life is disruptive, and pressure on communities and economies will require comprehensive strategies to eliminate causes. A broader perspective is needed that sees energy, society, economy, and the environment as part of a unique, integrated system.

The steps taken within the scope of the fight against the crisis, the investments made, and the public resources spent should provide more than a rescue operation for the existing socio-economic structures.

Now, more than ever, public policies and investment decisions must be compatible with a vision of sustainable and fair future, free from political concerns.

In the context of combating the epidemic, incentives and improvement measures should encourage economic development and job creation, promote social equality and prosperity, and put the world on a climate-safe path.

The governments should resort to a renewable-energy-based energy transformation to bring a series of solutions at this difficult moment.

Decentralized technologies enable citizens and communities to participate more in energy decisions with transformative social implications. More importantly, it offers a proven approach to remote healthcare in energy-poor communities and adds a key element to the crisis response process.

Burning fossil fuels (coal, natural gas, and petroleum) in the atmosphere, which are the main causes of chronic diseases, should be stopped and transition to 100% renewable energy should take place as a permanent solution for the elimination of epidemics and for keeping people and other living creatures free from outbreaks. The energy transformation process should be accelerated to help revitalize renewable technologies and industries and create new jobs.

What should be done to stop the carbon emissions, which has become a vital necessity for humans and nature, should be applied immediately without any excuses and "but's".

Air is a nature element that makes life possible and sustainable.

Water that makes life possible is a product that cannot be obtained in any way other than what nature offers.

Soil, which gives plants life and can be cultivated, is the product of a process corresponding to tens of thousands of years.

Forests that enable the universe to breathe and provide the rain to the universe as a natural event are the products of hundreds of thousands of years. Fossil forests are world heritage formations that enable us to discover natural history. The formation of fossils is a process that takes millions of years and makes it possible through the scientific studies to have information about the climate and plant species related to the geological times.

These natural values should never be consumed for any reason. They should continue their existence in their natural environment and should be used based on sustainability of life in the way the nature offers.

We, as the promoters of transition to 100% renewable energy who worked together to produce this book, are trying to define the problems of conventional energy production and consumption, to find solutions that can be implemented and to develop related technologies and methodologies for effective long-term mitigation of the problems and protection of our unique common living space on Earth.

Permanent solution to prevent people from catching epidemics is to stop combustion of fossil fuels (coal, natural gas, and oil) in the atmosphere, which is the main cause of chronic diseases.

We need to develop and implement Local Green Deal's in cities and National Green Deal's in countries to make Global Green Deal a reality.

Istanbul, Turkey Tanay Sıdkı Uyar

Contents

Nomenclature, Acronyms and Abbreviations

Nomenclature

A	Area of wind turbine (m^2)
A	Area (m^2)
A	Solar panel irradiation area (m^2)
A	Surface area (m^2)
a, b	Constants of the heat pump *COP* versus temperature function (Eq. 37)
ALT	Residence time in the atmosphere years
C	Installed Capacity (kW)
c	Specific heat capacity (J/kg °C)
c	Unit emissions factor of a fuel (kg CO_2/kW-h)
c, d, e, g, h	Constant in Eqs. 33–35
C_{eq}	Equipment Life-Cycle Cost-Design Temperature Difference Factor (€·K/kW-h)
C_{hp}	Heat Pump Life-Cycle Cost-Design Temperature Difference Factor (€/K·kW-h)
CO_2	Carbon dioxide emission (kg of CO_2)
CO_{2base}	Base (Reference) emission (kg of CO_2)
Comp.	Compound
COP	Coefficient of performance
$COPEX$	Exergy-based coefficient of performance
C_R	Composite Rationality Index
D	District Pipe Inner Diameter (m)
$\dot{E}$	Energy flow rate (kW)
E	Actual voltage of cell (V)
E	Electrical energy (load) (kW-h)
E	Energy (kJ)
$E_{act,a}$	Activation overpotential at anode (V)
$E_{act,c}$	Activation overpotential at cathode (V)

E_{conc}	Concentration overpotential (V)
EDR	Ratio of carbon CO_2 emissions difference to the base emission, dimensionless
ELC	Exergy-Levelized Unit Cost (€/m^2 based on REMM)
EM	Composite embodiment cost (€)
EMR	Exergy embodiment recovery, years (See Fig. 22)
E_{ohmic}	Ohmic overpotential (V)
ex	Specific exergy (kJ/kg)
E_x	Exergy (kW)
$\dot{Ex}$	Exergy rate (kW)
$\dot{Ex}_{dest}$	Exergy destruction (kW)
$\dot{Ex}_D$	Exergy destruction rate (kW)
$\dot{Ex}_d$	Exergy destruction rate (kW)
$\dot{\mathrm{Ex}}$	Exergy rate (kW)
Ex	Specific exergy (kJ/kg)
F	Faraday constant (C/mol)
FC	Selling price of the solar panel (€)
GWP	Global warming potential
h	Enthalpy
H	Specific enthalpy (kJ/kg)
HSDI	Human (Sustainable) Development Index
I	Solar radiation (W/m^2)
I_d	Direct normal irradiation (W/m^2)
I_n	Net solar insolation normal to solar receiver surface (W/m^2)
J	Current density (A/m^2)
J_{oa}	Anodic exchange current density (A/m^2)
J_{oc}	Cathodic exchange current density (A/m^2)
L	One-way District circuit distance (m)
LH	Vlower heating value (kJ/kg)
m	Mass (kg)
$\dot{\mathrm{m}}$	Mass flow (kg/s)
$\dot{m}$	Mass flow rate (kg/s)
N	Molar flow rate (mole/s)
N	Number
n	Number of moles
ODI	Composite Ozone Depletion Index
ODP	Ozone Depleting Potential
P	Pressure (kPa)
PC	Unit power cost of a solar energy system (€/kW_{peak})
PEF	Primary energy factor
PER	Primary energy ratio (Reciprocal of *PEF*)
PEXR	Exergy-based primary energy ratio
P_s	Power demand for pump stations (kW)
$\dot{\mathrm{Q}}$	Heat (w)

$\dot{Q}$	Heat rate (kW)
Q, Q_H	Thermal energy (load) (kW-h)
R	Gas constant (kJ/kmol K)
r	Radius solar pond (m)
RCO_2	Reduction Potential Ratio
Re	Reynolds number
rms	Root-mean-square (electricity)
s	Specific entropy (kJ/kg K)
s, t, u	Coefficients in Eq. 2
S_c	Solar constant (1366 W/m^2)
$\dot{S}_{gen}$	Entropy generation rate (kW/K)
T	Temperature (°C or K)
TI	Transformation index
U	Heat transfer coefficient (W/m^{2}K)
V	Velocity (m/s)
V	Volumetric Flow (m^2/h)
V	Wind speed (m/s)
w	Correction factor for *COPEX* in Eq. 4, $0.85 \leq w \leq 1.0$
W	Power (W)
W	Weight of a panel (kg)
$\dot{W}$	Work rate (kW)
x	Length (m)
X	Power Split
Y	The ratio of the temperature changes of the Stratosphere and Troposphere
ΔP_S	Power demand of the pump stations per unit pipe length (kW/m)

Greek Letters

$\alpha, \beta, \delta, \gamma$	Atoms of C, H, N and O
α	Symmetry factor
η	Efficiency
η	Energy efficiency
η_I	First-law efficiency
η_{II}	Second-law efficiency
η_T	Power transmission and distribution efficiency
η_W	Wind turbine-to-electricity first-law efficiency
ρ	Density
ρ	Density of air
λ	Stoichiometric constant
λ_{mem}	Membrane water content
σ_{mem}	Membrane conductivity (1/Ω cm)

Δ	Difference
φ	Betz limit
ϕ	Golden ratio
ϕ	Fuel exergy to energy ratio
ψ	Exergy efficiency
ψ_R	Rational exergy management efficiency
ΔCO_2	Avoidable CO_2 emissions (kg CO_2/kW-h heat)
ΣCO_2	Total CO_2 emissions (Direct and avoidable) (kg CO_2/kW-h heat)
E	Unit exergy (kW/kW)

Subscripts

0	Ambient conditions
0	Reference state
a	Anode
a	Indoor air
act	Activation
app	Useful application (Temperature)
ar	Air
base	Base
BESS	Battery Energy Storage System
bt	Bottom
c	Cathode
c	Cooling
ch	Chemical
comp	Compressor
con	Condenser
D	Destruction
dem	Demand
des	Destroyed
e	Break-even
E	Electric
el	Electrolyser
en	Energy
ev	Evaporator
EVTC	Evacuated tube solar collectors
ex	Exergy
exc	Exchanger
f	Energy source, fuel
F	Fan
f	Fuel
FC	Fuel cell
G	Generator

g	Geothermal
H	Thermal (Heat)
HSZ	Heat storage zone
in	Inlet
in, out	Inlet and outlet connections of a hydronic circuit
int	Integrated
l, m	Local power plant, distant power plant, respectively
lm	Logarithmic
loss	Thermal losses
min, max	Minimum, maximum
net	Net power
o	Air
o	Standard state
o, ref	Reference
opt	Optimum
p	Product
P	Pump
PEM	Proton exchange membrane
PEV	Plug-in Electrical Vehicle
ph	Physical
pump	Pumping
PV	Photovoltaic
R	Reservoir, return
rcv	Receiver
ret, sup	Return, supply
s	Solar
s	Sun
SOC	State of Charge
sp	Solar pond
st	Storage
sur	Surface
sys	System
T	Power transmission, total, overall
T	Turbine
tot	Total
u	Useful
up	Upper
w	Water
w, wt	Wind, Wind Turbine
wf	Working fluid
WT	Wind turbine
X, EX	Exergy, exergetic

Acronyms and Abbreviations

Abs	Absorber
ABS, ADS	Absorption, Adsorption Cycle
AC, DC	Alternating Current, Direct Current
BESS	Battery Energy Storage System
CHP	Combined heat and power (Cogeneration)
Con	Condenser
COP	Coefficient of performance
Cu–Cl	Copper chlorine
DE, 4DE	District Energy System, 4th Generation District Energy System
DHW	Domestic Hot Water
ECBCS	Energy Conservation in Buildings and Community Systems Program
EEA	European Environment Agency
EES	Engineering equation solver
EIA	Energy Information Administration (US)
EPA	Environmental Protection Agency
EU	European Union
EVTC	Evacuated tube solar collector
EXT	Heat loss extraction
FC	Fuel cell
Gen	Generator
GSHP	Ground-source heat pump
HE	Heat exchanger, or Hydrogen economy
HEX	Heat exchanger
HPT	High pressure turbine
HRV	Heat-recovery ventilation
HSZ	Heat storage zone
HVAC	Heating, ventilating and air-conditioning (of buildings)
IEA	International Energy Agency
LHV	Lower heating value
LNG	Liquified natural gas
LowEx	Low exergy (building)
LPG	Liquified petroleum gas
LPT	Low pressure turbine
LVDC	Low-voltage DC power
mCHP	Micro CHP (Electrical power capacity <50 kW)
NCZ	Non-convective zone
nZEXB	Net-zero exergy building
OECD	The Organization for Economic Co-operation and Development
ORC	Organic Rankine cycle
OTEC	Ocean thermal energy conversion
P	Pump
PEM	Proton exchanger membrane

PEMFC	Proton exchange membrane fuel cell
PEV	Plug-in electrical vehicle
PV	Photovoltaic
PV, PVT	Photo voltaic, Photo-voltaic-heat
REMM	Rational exergy management model
SEAC	Single effect absorption cooling
SHF	Solar heliostat field
SOC	State of charge
SOE	Solid oxide electrolyser
SP	Solar pond
SPT	Solar power tower
ST	Solar tower
SWWCP	Solar-wind-wave composite platform
TES	Thermal energy storage
TP	Temperature-peaking
UCZ	Upper convective zone
US	United States
WEF	World Economic Forum
WGSR	Water gas shift reactor

List of Figures

Accelerating the Transition to 100% Renewable Era. But How? Exergy Rationality in the Built Environment

Role of IRENA for Global Transition to 100% Renewable Energy

The Renewable City: The Future of Low-Carbon Living

Development and Thermodynamic Analysis of a 100% Renewable Energy Driven Electrical Vehicle Charging Station with Sustainable Energy Storage

Community Wind Under the Auctions Model: A Critical Appraisal

100% Renewable Energy Generation with Integrated Solar Energy Systems

The Role of Hydrogen in Global Transition to 100% Renewable Energy

Solar Hydrogen's Role for a Sustainable Future

Design and Analysis of a New Environmentally Benign Ammonia-Based Solar Thermochemical Integrated Plant

An Overview of Hydrogen Production from Biogas

Underground Large-Scale Hydrogen Storage

Biomass and Its Thermochemical Conversion: Can It Be a Road Map for Transition to 100% Renewable Energy?

Role of Energy Storage in 100% Renewable Urban Areas

Efficient Use of Energy in Buildings—New Smart Trends ....

List of Boxes

Role of IRENA for Global Transition to 100% Renewable Energy

Accelerating the Transition to 100% Renewable Era. But How? Exergy Rationality in the Built Environment

Birol Kılkış

1 Introduction

According to the 2018 IEA Key World Energy Statistics Report, the share of buildings in energy consumption has lately exceeded the individual shares of industry and transport (IEA 2018).

IEA Statistics are all based on the *quantity* balances of energy and do not consider the *quality* balances of energy consumption in buildings with the quality of energy used in industry and transport. Their energy balance flow sheets, energy atlases, and main trade gas flow charts are based on the energy quantity based on the First Law of Thermodynamics. On the other hand, according to the Second Law of Thermodynamics, renewable energy resources generally have lower energy quality (Exergy) compared to fossil fuels and thereby present a better fit to the quality demands of the built environment, which also demand low quality of energy like for heating and cooling. This establishes a better-quality match and balance between renewable energy sources and the demands on the built environment side. Such a better-quality balance results in a much lesser amount of exergy destruction, where exergy destruction leads to additional but avoidable CO_2 emissions. Unless international energy statistics transform to the Second Law, renewables may not be able to prove their competence, especially in the built environment. Therefore, in the quest of increasing the share of renewables towards 100% for decarbonization, deceleration of global warming, and minimizing ozone depletion, buildings are both the prime concern and the success potential for decarbonization with renewables.

The new Horizon Europe (HE) program for the financing period 2021–2027 represents an important opportunity for the energy sector (RHC 2019). Committing 15 billion EUR for the Climate, Energy, and Transport Cluster, renewable heating and

B. Kılkış (✉)
European Technology and Innovation Platform Renewable Heating and Cooling (RHC), Brussels, Belgium
e-mail: birolkilkis@hotmail.com

T. S. Uyar (ed.), *Accelerating the Transition to a 100% Renewable Energy Era*, Lecture Notes in Energy 74, https://doi.org/10.1007/978-3-030-40738-4_1

cooling is an essential part of the energy transition, while heating and cooling represent 47% of the EU's energy consumption. Yet, only 18% of heating in Europe is currently based on renewable energy sources. Figure 2 shows the current contribution breakdown of renewables (EU 2016). The rest primarily belongs to gas, coal, and oil. In terms of fossil fuel use, almost half of the EU's buildings have individual boilers installed before 1992, with an efficiency of 60% or less. 22% of individual gas boilers, 34% of direct electric heaters, 47% of oil boilers and 58% of coal boilers are older than their technical lifetime. This finding points to an important and timely opportunity from the techno-economical point of view such that transition to renewables may be coupled with the transition from conventional heating and cooling equipment to renewable-compatible and renewable-friendly systems and equipment for buildings for the better-quality match. In Fig. 2, nuclear has been repositioned by the Author from renewables, because nuclear is neither renewable nor fossil in the classical meaning, but may be considered to be a low-carbon fuel, if large amounts of CO_2 embodiment taking place during their construction, nuclear fuel production, and nuclear waste disposal and storage are entirely ignored-which is often the case in the public domain.

Extending the scope to a global scale, renewable energy supplied approximately 10.3% of total global energy consumption for heating in 2015. Another 16.4% was supplied by traditional biomass, predominantly for cooking and heating in the developing world (REN 2019). Despite additional bio-heat, geothermal direct use, and solar thermal capacities were added, growth has been quite slow. Energy demand for cooling is also growing rapidly worldwide. Figure 3 shows that commercial and residential buildings in total have been a major consumer also in the USA since 1950, in excess of transportation and eventually their consumption exceeded industry slightly.

Although the carbon footprint and global warming are proportioned to buildings in almost the same ratio given in Figs. 1 and 2, the main mechanism leading to the current global crisis is not well understood. Without understanding the root causes of the global crisis, the utilization of renewable energy sources directly with conventional equipment and systems in the built environment will not be successful. Therefore, it is a prerequisite for understanding the deeply interconnected global warming and ozone depletion for a sustainable transition to 100% renewables (Fig. 3).

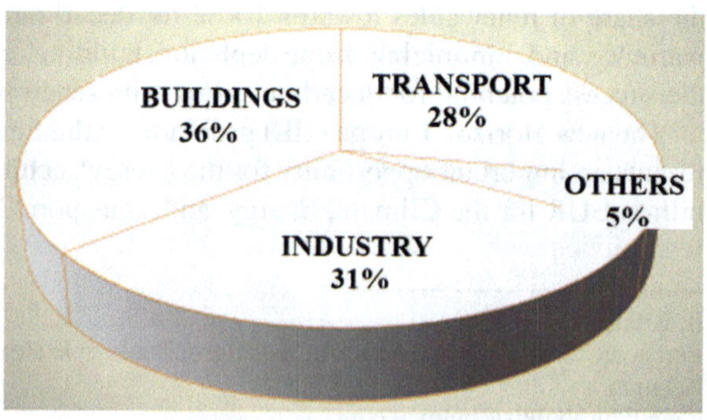

Fig. 1 Global energy consumption by sector (IEA 2018). (Building sector includes construction phase)

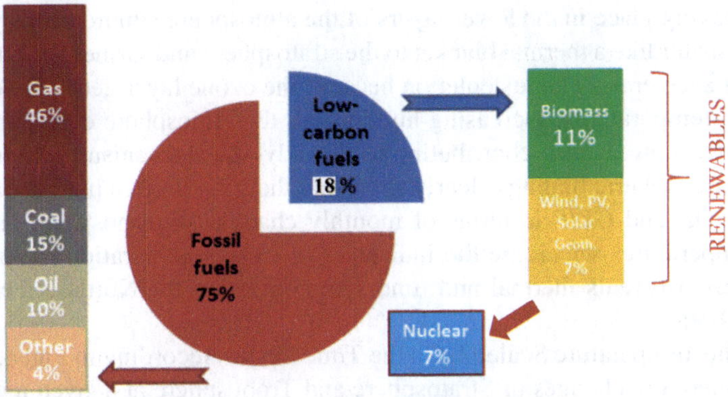

Fig. 2 Renewables for heating and cooling in Europe (EU 2016)

Fig. 3 US total energy consumption by sector, 1950–2017. Vertical axis shows consumption in quadrillion btu (EIA 2018)

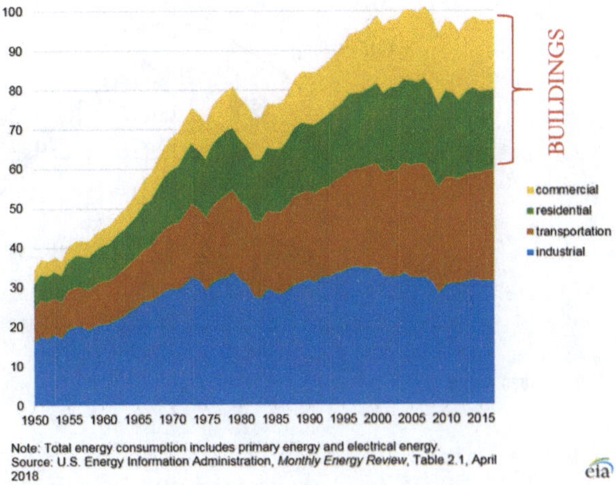

2 The Global Crisis-*What Is Happening in the Atmosphere?*

Today, the standard practice of evaluation and rating of 'high performance', 'sustainable', and 'green' buildings simply rely on the First Law of Thermodynamics and simple economy. Although their HVAC systems seem to be quite 'efficient' and green, new metrics, which are based on the Second-Law, namely exergy rationality metrics, show that these heating, ventilating, air-conditioning, and refrigeration practices are not rational and environmentally benign unless they are redesigned and re-equipped accordingly. Otherwise, ozone depletion and global warming will remain to accelerate. First, it must be understood what is happening in the atmosphere and look much deeper into the global warming issue than what the industry is claiming. Until today, ozone depletion potential (*ODP*) and global warming potential (*GWP*) of the refrigerants used in many applications were treated separately. In fact, global warming taking place in the lower layers of the atmosphere simultaneously reduces the heat transfer like a thermal blanket to the stratosphere and further cools the ozone layer. This accelerates ozone depletion because the ozone layer gets less dense with decreasing temperatures. Increasing humidity in the atmosphere due to man-made cooling towers etc. is also contributing to this adverse mechanism. In this respect, the latest atmospheric findings clearly show that there is almost a perfect correlation between *ODP* and *GWP* in terms of monthly changes in troposphere and stratosphere temperatures which are the indicators of CO_2 concentration and the ozone hole. Figure 4 reveals thermal and time symmetries for the Northern hemisphere (NASA 2019):

1. On the Temperature Scale, 2. On the Time Scale. According to Fig. 4, the ratio of the temperature changes of Stratosphere and Troposphere at a given month, *Y* is equal to π (Kilkis 2019a):

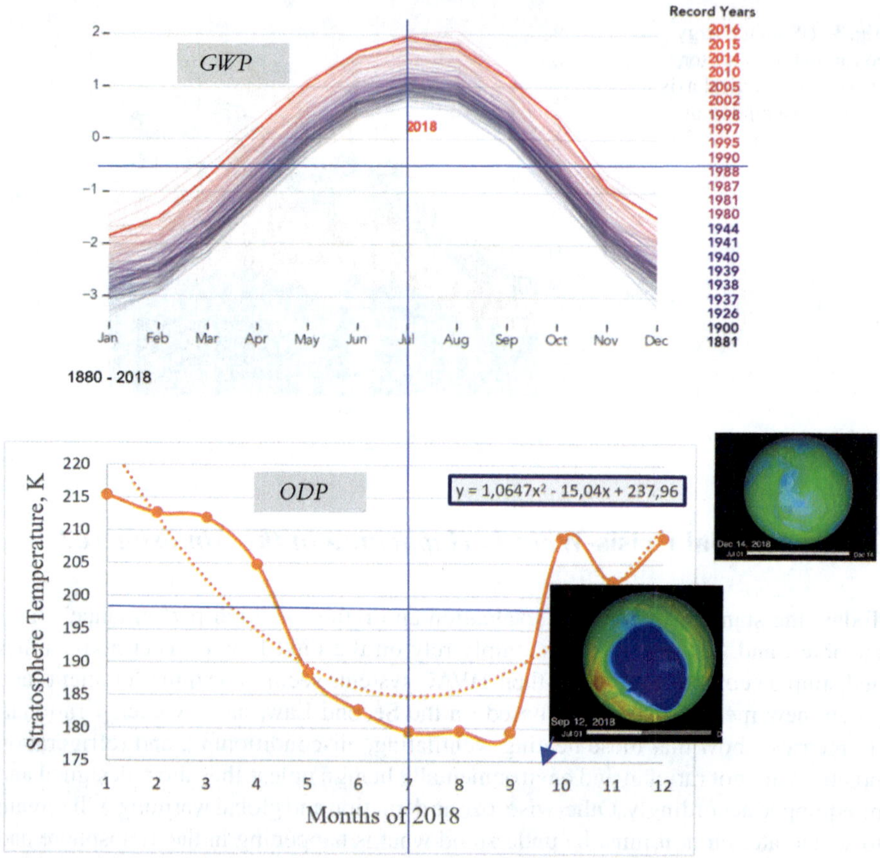

Fig. 4 Correlation between stratosphere and troposphere temperatures, 2018 (Kilkis 2019a)

$$Y = \frac{|\%\text{ Temperature Change in Stratosphere}|}{|\%\text{ Temperature Change in Troposphere}|} \approx \pi \quad (1)$$

The Y ratio gives the first clue about the direct dependence of ozone depletion on CO_2 concentration, or in other words, the *GWP* changes: on an annual basis while global temperature increases, ozone layer temperature decreases in a mirror image fashion. For example, the maximum global temperature occurs on the same day of July when the ozone layer is the coldest. Following this occurrence, the ozone hole becomes the largest after following a thermal lag of about two months (September). When the stratosphere temperature rises again in December, the ozone hole diameter becomes minimum (almost nil). Consequently, the maximum CO_2 concentration over the months in a year (2018), which is shown in Fig. 5 also coincides with the *GWP* curve given in Fig. 4. These findings lead to a new *ODP* and *GWP* correlation, including CO_2 emissions behavior. By a series of mathematical elaborations, the following composite formula was developed for refrigerants only, which transforms *ODP* into *ODI* in terms of the *GWP* value of a given refrigerant. The atmospheric residence time, *ALT* in years is also factored in (Kilkis 2019b):

$$ODI = \frac{sGWP^t}{(1 - ODP)} \times \left(\frac{ALT}{1}\right)^u \; \{s = 0.1, t = 0.03, u = 0.01, ODP < 1\} \quad (2)$$

Table 1 shows that even the best available less-ozone-harmful refrigerant, namely CO_2 with 0 *ODP* and 1 *GWP* has a non-zero *ODI*, because of its long residence time in the atmosphere. F gas has zero *ODP*, very high *GWP*, and less residence time. Therefore its *ODI* is higher than CO_2 gas.

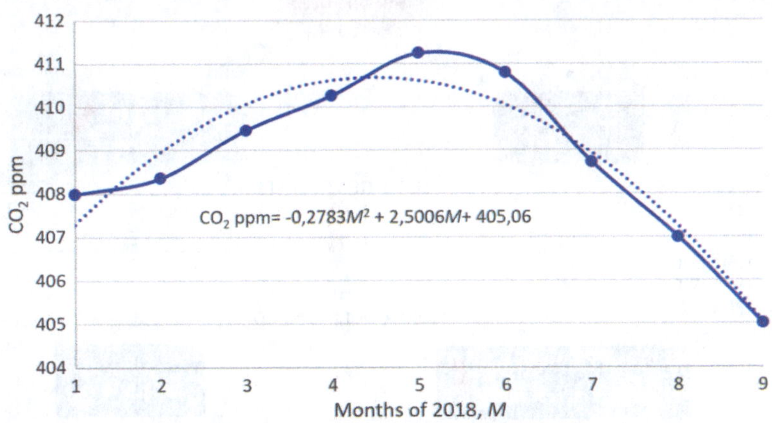

Fig. 5 Variation of CO_2 concentration during the nine months (Kilkis 2019a)

Table 1 Actual ozone depletion effect of F-Gas and CO_2 Gas (Kilkis 2019a)

Refrigerant	*ODP*	*GWP*	*ALT*	*ODI*
R744 (CO_2)	0	1	120 yr.	0.115
R227ea (F Gas)	0	3500	30 yr.	0.132

An Irreversible Cycle

Figure 6 shows a closed-loop cycle about the effects of building heating and cooling on global warming and ozone depletion. Today most of the heating is achieved by heating systems using fossil fuels. These systems emit CO_2, which causes global warming. CO_2 triggers more ozone depletion. Larger ozone hole permits more harmful rays such that global warming is accelerated and ice sheets are melted. More water vapor resulting from this melting process causes more global warming due to its greenhouse effect.

Global warming means more cooling demand, by which cooling equipment currently relies on electric power generated mainly by fossil fuels, causing more CO_2. The only positive impact of global warming with CO_2 emissions is less heating demand. However, because overall cooling loads involve latent cooling need (Humidity), their increasing impact on global warming and ozone depletion surpasses the reduced demand for heating. In cooling, *ODI*, CO_2, outdoor temperature, electricity demand increases while the chiller performance, *COP* decrease. This vicious cycle

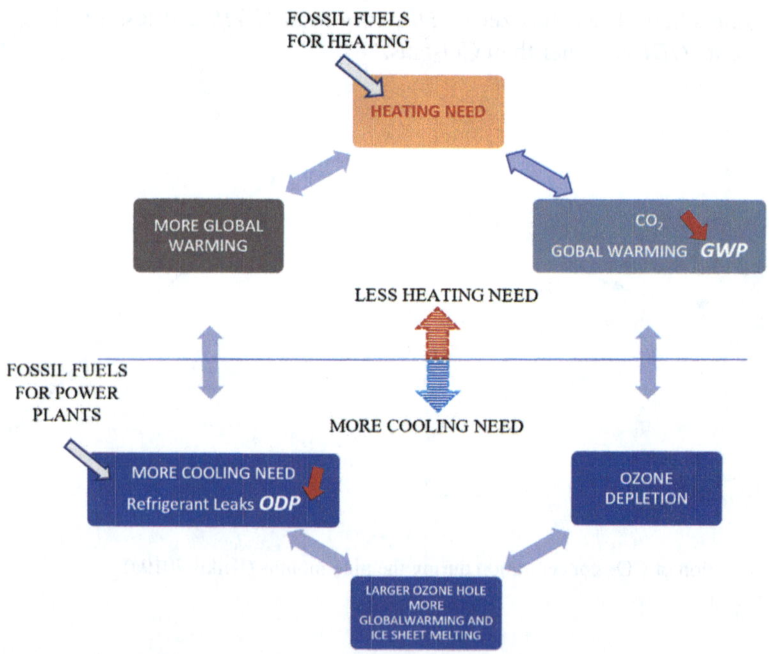

Fig. 6 Irreversible cycle of ozone depletion and global warming from the built environment

may only be broken by renewables on-site. The irreversibility circle for heating in Fig. 6 exemplifies how renewables may help to reduce the global crisis:

- Replace fossil fuels from building heating by various waste and renewable energy sources,
- This replacement will also support the ozone layer.
- Eliminate refrigerant leaks and/or eliminate harmful refrigerants by using captured CO_2 or by using hydrogen fuel cells, generated by water electrolysis with renewable energy sources,
- Eliminate fossil fuels from electric power plants by with solar and wind power plants,
- Reduce heating and cooling needs and transform to nearly-zero exergy buildings such that a better exergy balance is established between renewables and buildings. This will minimize exergy destructions and therefore avoidable CO_2 emissions.

On the other hand, Fig. 7 (Cowan et al. 2009) prooves that the cycle shown in Fig. 6 is indeed irreversible, because ozone-depleting refrigerant leakage ratios, *LR* are very slowly decreasing, according to a wide data scatter. This rate is even slower according to the least-squares fit by the Author of this Chapter. The small decrease in leakage rates is overwhelmed by the ever-increasing cooling loads.

Figure 8 helps to explain the most recent climatic and environmental disasters in terms of the difference between the unit exergy changes, ε between the stratosphere and troposphere (by a factor of about 7), which form a thermal shear between both layers. This temperature shear is coming closer to the earth, resulting in severe floods, and more frequent and stronger tornadoes, which happen today at locations of previously unheard of around the globe.

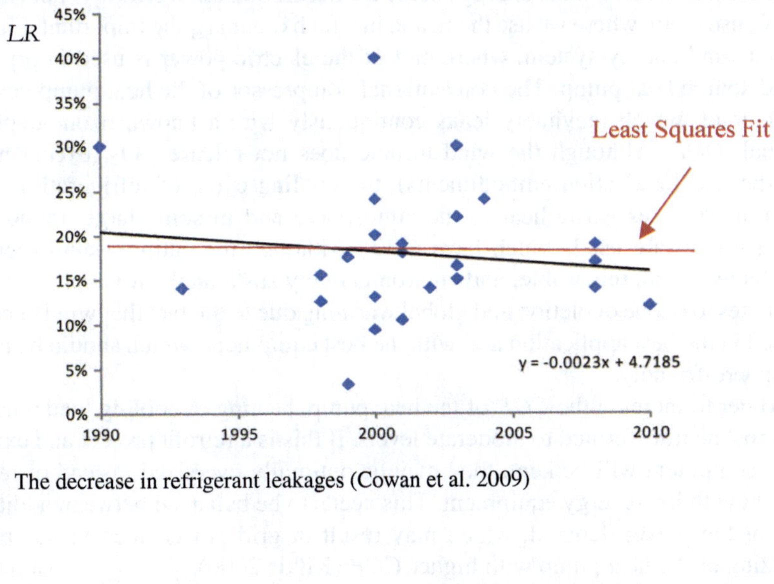

Fig. 7 The decrease in refrigerant leakages (Cowan et al. 2009)

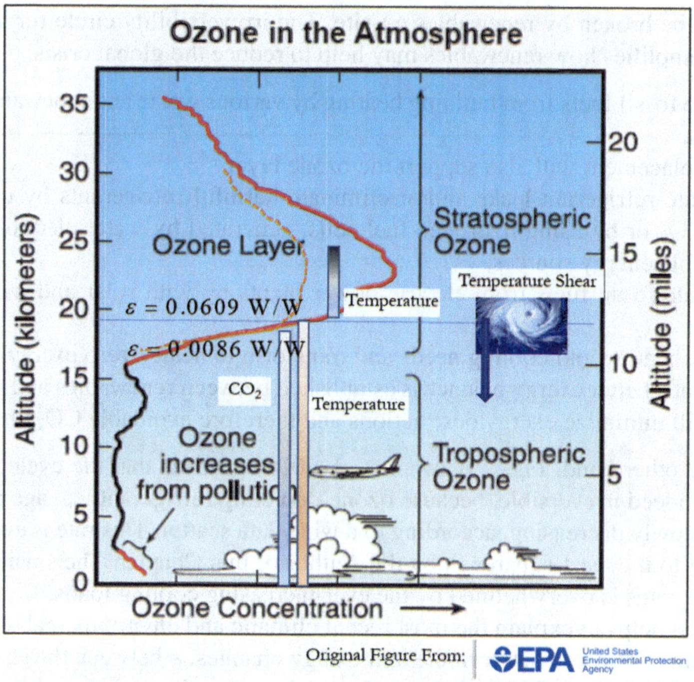

Fig. 8 Retraction of ozone layer and atmospheric temperatures (EPA 2019)

How Renewable Energy Sources are Innocent?

Just to use renewables at the energy source for decarbonization efforts is not enough. How we use them, where we use them, and in which order is quite important. Figure 9 shows a wind energy system, where part of the electric power is used in driving a ground-source heat pump. The conventional compressor of the heat pump contains a refrigerant, which inevitably leaks continuously with a known ozone-depletion potential, *ODP*. Although the wind turbine does not release CO_2 (except manufacturing and installation embodiments), the cooling cycle of refrigeration at the same time releases waste heat to the atmosphere and in some large applications cooling towers are used, which have *GWP*. Although the entire system seems to be perfectly green, renewable, and environmentally sustainable, it is not, because it contributes to ozone depletion and global warming due to the fact that wind energy is not used in the best application and with the best equipment, which should be totally atmosphere-friendly.

In order to increase the *COP* of the heat pump, heating or cooling load temperatures must be transformed to moderate levels. If this is a retrofit project and existing HVAC equipment will be kept, they may be optimally oversized instead of replacing them with low-exergy equipment. This needs to be balanced between additional pump or fan power demand, which may result in grid power need versus partial oversizing of the heat pump with higher *COP* (Kilkis 2000).

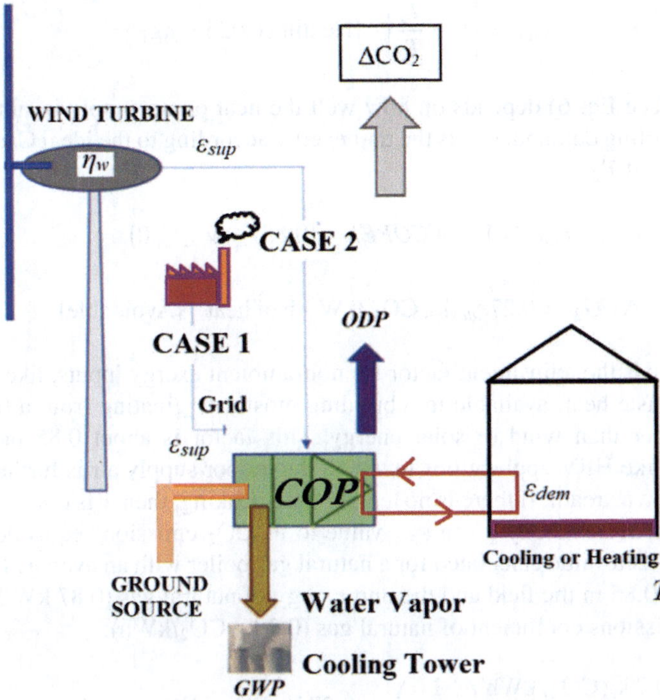

Fig. 9 Wind turbine operated heat pump has *GWP*, *ODP*, and ΔCO_2 emissions (Kilkis 2019; Kilkis et al. 2017)

This example shows that for 100% renewable targets, the following measures need to be taken:

- Track the renewable source (s) along the chain of emissions, demand, and energy transfer,
- Develop atmosphere-friendly energy conversion systems and equipment,
- Implement more renewable energy compatible applications like nearly-zero Exergy, or LowExergy (LowEx) buildings and similar applications,
- Observe exergy balance between pumping and heat (and or cold) transferred in district energy (DE) system,
- Minimize unit exergy destructions in the complete chain from supply to demand and waste. In this respect minimize the number of transformations from one form of power to another form of heat or vice versa. The same also holds for energy storage. Store energy in its original.

Unit exergy destructions, ε_{des} due to mismatches cause avoidable CO_2 emissions, which are related to the exergy-based *COP* of the ground-source heat pump, namely *COPEX*.

$$\varepsilon_{des} = \left|1 - \frac{T_1}{T_2}\right| \quad \{\text{Heating or Cooling}\} \tag{3}$$

COPEX (see Eq. 6) depends on how well the heat pump supply is suited to the heating or cooling demand. ε_{des} is the unit exergy according to the ideal Carnot cycle between T_1 and T_2.

$$\varepsilon_{des} = 1 - wCOPEX \quad \{0.85 \leq w \leq 1.0\} \tag{4}$$

$$\Delta CO_2 = 0.27\varepsilon_{des} \text{kg CO}_2/\text{kW - h of heat} \quad \{\text{Avoidable}\} \tag{5}$$

In Eq. 4, w is the adjustment factor for non-ambient exergy inputs, like a limited amount of waste heat, available in a building mostly originating from a fossil fuel expense, rather than wind or solar energy. This factor is about 0.85 in building applications like HRV applications in which the indoor supply air is further peaked by a boiler downstream. If there is no temperature peaking, then w is one. Equation 5 is based on converting any given ε_{des} value to its CO_2 emissions equivalent based on a standardized value calculated for a natural gas boiler with an average First-Law efficiency of 0.85 in the field and the unit exergy of natural gas (0.87 kW/kW), and unit CO_2 emissions coefficient of natural gas (0.2 kg CO_2/kWh):

$$\frac{0.2\,\text{kg CO}_2/\text{kWh}}{0.85\,\text{kW/kW}}\left(\frac{1}{0.87}\right) = 0.27\,\text{kg CO}_2/\text{kWh per unit}\,\varepsilon_{des}$$

$$COPEX = \psi_R \tag{6}$$

$$\psi_R = \frac{\varepsilon_{dem}}{\varepsilon_{\text{sup}}} \quad \{\text{Exergy destroyed upstream}\} \tag{7}$$

$$\psi_R = 1 - \frac{\sum \varepsilon_{dss}}{\varepsilon_{\text{strp}}} \quad \{\text{Exergy destroyed downstream}\} \tag{8}$$

In Eqs. 7 and 8, ψ_R is the Rational Exergy Management Model (REMM) efficiency, which watches the balance between the quality of energy among supply and demand points in the built environment. It must approach 1 in exergy-rational applications for minimum CO_2 emissions. For any heat pump application alone, with $COP > 1$, ψ_R needs to be multiplied by *COP*.

The EU target for electrification with heat pumps for heating and cooling is not exergy-rational if a natural gas-fired thermal power plant combusting natural gas at $T_f = 2235$ K and the grid electricity supplied is used by a heat pump with a *COP* of 3 for radiant floor heating at supply and return temperatures of 330 K and 320 K, respectively. For CASE 1 shown in Fig. 9 corresponds to grid power input from a thermal power plant *COPEX* is far less than one due to large exergy destructions (see Fig. 10). In the same token, ψ_R, which is responsible for CO_2 emissions is quite small. In a fully exergy-rational ($COPEX \rightarrow 1$), composite renewable energy system

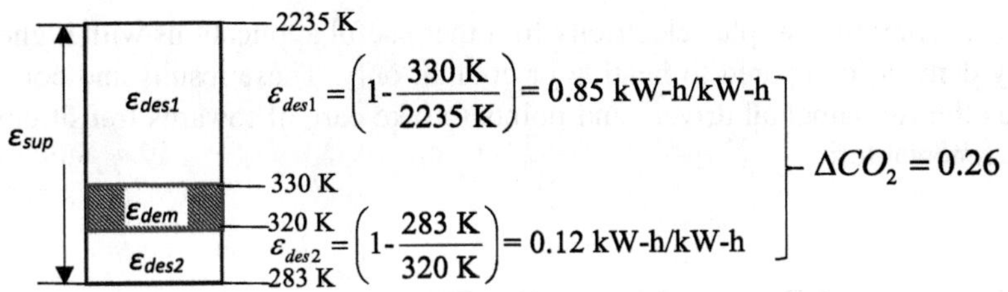

Fig. 10 Exergy flow diagram for the grid -operated heat pump (CASE 1) in heating mode, which is shown in Fig. 9

for Fig. 9, *COP* must approach to $(1/\psi_R) = 28.6$, quite unpractical. 283 K is the reference temperature, T_{ref}.

$$\psi_R = \frac{(1-320/330)}{(1-283/2235))} \times 3 = 0.1 \ \{\text{see Eq. 7, } COP \text{ is } 3\}$$
$$COPEX = \psi_R = 0.1 \qquad \{\text{see Eq. 6}\}$$

The virtual, Carnot-Cycle based source temperatures for solar and wind energies, T_f are determined by Eqs. 6 and 9. η_w is the efficiency of the wind turbine.

$$\frac{I_n}{S_c} = \frac{\left(1-\frac{T_{ref}}{T_f}\right)}{0.95} \tag{9}$$

0.95 W/W corresponds to the unit exergy at the surface of the sun (5778 K) relative to a reference temperature of 283 K on Earth if there was not any atmosphere. If I_n is equal to the solar constant (outside the atmosphere) T_f is 5778 K (Radiation from the sun taking place in a medium of nearly vacuum). In the real case of $I_n < S_c$, Eq. 9 applies for T_f. (In the atmosphere).

$$T_f = \frac{T_{ref}}{(1-\eta_w)} \tag{10}$$

From Eq. 10, the magnitude of the unit exergy, ε_{sup} of the wind turbine becomes equal to η_w. Then, CASE 2, which is wind turbine-operated heat pump shown in Fig. 9, may be solved, if the mechanical efficiency of the wind- to-electricity efficiency of the wind turbine, η_w is known. It is assumed that wind turbine-generated electricity is used to drive the heat pump. If η_w is 0.4, for example, then T_f from Eq. 10 is 472 K. After replacing 2235 K with this value in Fig. 9, ψ_R and *COPEX* become 0.23. ΔCO_2 (due to unit exergy unbalance between the kinetic energy of wind and heat (or cold) supplied by the heat pump) decreases to 0.1 kg CO_2/kWh of heat because ε_{des1} reduces to 0.3 kWh/kWh. This is a 2.6 times reduction from CASE 1. It may seem awkward to have a CO_2 emission responsibility. In fact, this is an avoidable emissions responsibility, that is avoidable to a large extent by selecting a larger-capacity wind turbine, with higher efficiency, a higher *COP* of the heat pump and thus generating surplus electricity for other useful applications with higher unit exergy demand compared to heating or cooling only. These results and comments give us the fundamental drivers and points to take care of towards transitioning to 100% renewables.

3 Exergy and Renewables to the Rescue

Decoupling economic development from CO_2 emissions is the most critical challenge of the global crisis (Fig. 11). While both power and heat are required for economic growth and wealthy urbanization, CO_2 emissions follow a parallel trend. 20 + 20 + 20 goals of EU, namely a 20% increase in energy savings, utilization of renewable energy sources, and efficiency, respectively each may reduce the increasing rate of CO_2 emissions but cannot decouple it from economic growth. HSDI is the Human (Sustainable) Development Index defined by UNDP. Economic growth turns into Sustainable Development once decoupling is achieved by exergy rationality allowing natural sinking to take over.

The compound CO_2 emissions, including the effect of exergy destruction, is given in Eqs. 11a and 11b. There are two components, namely direct emissions, and avoidable emissions, of which the latter can be avoided by reducing exergy destructions (Kilkis 2011).

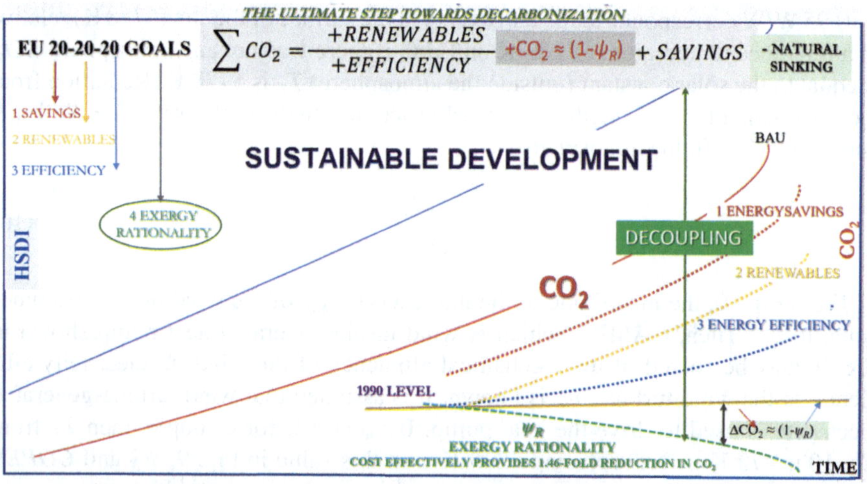

Fig. 11 Coupling between emissions and development (Kilkis 2019)

$$\underbrace{\sum CO_2}_{\text{COMPOUND}} = \underbrace{CO_2}_{\text{DIRECT}} + \underbrace{\Delta CO_2}_{\text{AVOIDABLE}} \tag{11a}$$

$$\sum CO_2 = \left[\frac{c_l}{\eta_l} + \overbrace{\frac{c_m}{\eta_m \eta_T}(1 - \psi_R)}^{\text{AVOIDABLE}}\right] Q_H + \frac{c_m}{\eta_m \eta_T} E \tag{11b}$$

With an increasing trend of transitioning to renewables, biogas, biofuel add-ons, and other technologies are already serving the second 20% target but at a slow pace. The third 20% target, namely efficiency, is also improving. For example, combined heat and power (CHP), condensing boilers, etc. are already approaching their theoretical limits, leaving not much room for improvement (Fig. 12).

The first 20% target is also on the right track, with smart grids, DC underground lines, and energy-saving measures. Yet, the proposed fourth 20% parameter, namely ψ_R remains unresolved and unknown, even though it has a large room for improvement. With existing technology and by simply changing the mindset from the First-Law to the Second-Law, this value may improve up to 80% and even more simply by

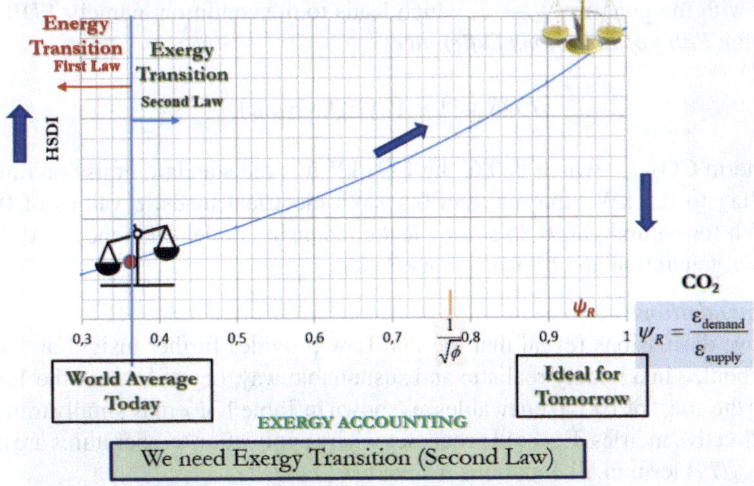

Fig. 12 Triple improvement of HSDI, CO_2 emission reduction, and ψ_R (Kilkis 2019)

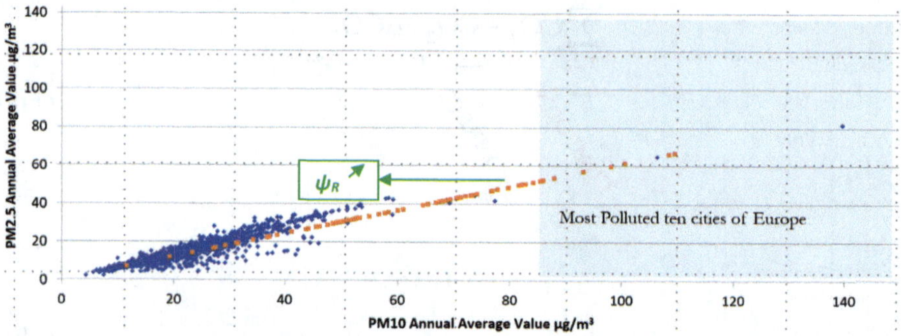

Fig. 13 The need for an integrated approach for cleaner cities. Derived from: (EEA 2018)

innovative combinations of equipment in a circular economy approach. For example, the most polluted ten cities in Europe may be improved by increasing ψ_R (see Fig. 13).

An increase in ψ_R, not only reduces the (1-ψ_R) term but, due to a decrease in exergy destructions, thermal and electrical power demands decrease as well as CO_2 emissions. Thus, triple decarbonization is obtained (Fig. 12). In Eq. 11b and Fig. 11, three decarbonization goals need to be supplemented hereby with a fourth goal, namely exergy rationality, ψ_R. Equation 11b clearly shows that CO_2 emissions may be substantially reduced if ψ_R is increased from 0.35 to a technically achievable value above 0.80. Even if the target for 100% renewables is achieved by solar heat, wind energy, solar PV farms, etc., such that the first term in the square brackets and the very last term are eliminated, ΣCO_2 may not be zero unless the renewables are utilized with the goal of $\psi_R \rightarrow 1$, which leads to the condition, namely $EDR \rightarrow 1$. *EDR* is the *Ratio of Emissions Difference*:

$$EDR = 1 - [CO_2/CO_{2\,baxe}] \tag{12}$$

The term CO_{2base}, which is 0.63 kg CO_2/kWh is the standard emission rate corresponding to 0.5 kWh thermal and 0.5 kWh electrical loads, *c* values of 0.2 kg CO_2/kWh for natural gas. 0.85 is a typical condensing boiler efficiency, and 0.35 is the power generation and transmission efficiency.

Recommendations

The above discussions reveal that the 2nd Law provides further insight and ability to decarbonize in a widely realistic and sustainable way, beyond where the 1st Law stops. In the quest of 100% renewables, as shown in Table 1, ψ_R may simply transform all EU directive metrics. For cold processes where application temperatures are below T_{ref}, (T_{ref}/T_s) term in all equations is inverted (Table 2).

$$PES_{EX} = \left(1 - \frac{1}{\left[\frac{CHPE\eta}{\mathrm{Re}fE\eta} + \frac{CHPH\eta}{REFH\eta}\right] \times \left(\frac{2-\psi_{R\,ref}}{2-\psi_R}\right)}\right) \times 100 \tag{13}$$

Table 2 Sample transfer functions for EU directives (Kilkis 2017)

Sample EU terms	1st Law	2nd Law	Comments
Performance coefficient	*COP*	COP_{EX}	Multiply *COP* by ψ_R
Primary energy ratio	*PER*	*PEXR*	(Inverse of *PEF*) Multiply *PER* by ψ_R
Primary energy factor	*PEF*	*PEFX*	PEF/ψ_R. Apply separately to electric and heat
Primary energy savings	*PES*	PES^{a}_{EX}	Cogeneration applications. Equation 13. ψ_{Rref} is 0.2
Ton-oil equivalent	Mtoe	MtoEX	Multiply Mtoe by ψ_{RF}

[a] $CHPE\eta$ and $CHPH\eta$ are the partial electric and thermal power efficiencies, respectively. Their denominators are their reference values

Exergy-Based Mtoe: Do not compare solar heat and oil:

$$MtoEX = \frac{\left(1 - \frac{T_{ref}}{T_f}\right)}{0.881} \times \text{Mtoe} = \psi_{RF} \times \text{Mtoe} \tag{14}$$

Here, ψ_{RF} is indexed to the unit exergy of crude oil, ε_F, namely 0.881 W/W.

4 Case Study

A solar PVT plant serves a Fourth-Generation district energy system (4DE). PVT plant generates both power and heat. Power is partially used to drive the circulation pumps in the district piping. Thermal power may be converted to cold by individual absorption (ABS) and or adsorption (ADS) units on customer site, on-demand. The rest of the generated power is distributed in the grid for electrical demand of different types and purposes, including mass transit. Figure 14 shows the basics of the system. The common mistake in the design and operation of such systems is the ignorance of the unit exergy difference between electrical and thermal powers. Among all ancillaries, which demand power circulation pumps need to be carefully optimized such that thermal exergy provided to the district, E_{XH} in Eq. 15 must exceed the exergy demand of the pumps, ignoring other parasitic losses and ancillary demand.

$$E_{XH} > E_{XP}(D, L) \tag{15}$$

Fig. 14 District energy system with solar PVT plant (Kilkis 2019)

The term E_{XP} is a function of the pipe diameter, D and the distance between the PVT plant and the district, L. For the limiting case of Eq. 15 and the given installed thermal capacity, C of the PVT plant, providing heat to the district between 330 and 320 K, serving radiant panels for heating and at the same time serving heat at 60 °C for DHW against Legionella risk.

$$E_{XH} = C\left(1 - \frac{320\,\text{K}}{340\,\text{K}}\right) = 0.059C$$

If for example, the power demand of the installed pump stations, is P_s is 15% of the thermal capacity, C, COP of the district energy system between the plant supply and district demand points looks quite favorable:

$$COP = \frac{C}{P_s} = \frac{C}{0.15C} = 6.7 \tag{16}$$

But $COPEX$ tells a different story if the average PV efficiency is 0.15:

$$COPEX = \frac{0.059C[\text{W/W}]}{0.15\text{C}[0.95\text{W/W}]} = 0.41$$

0.95 W/W is the unit exergy of electricity. From Eq. 7, ψ_R is 0.061 and the corresponding, avoidable ΔCO_2 emissions responsibility may be calculated from Eqs. 4, 5, and 6, which is 0.16 kg CO_2/kWh of heat delivered to the district. Therefore, although the solar energy source is 100% renewable, the CO_2 emission responsibility is not zero and this solar system is not 100% renewable. In order to improve the exergy performance of the system, pipe diameter may be increased to reduce pumping power demand per meter of pipe, ΔP_S at an expense of more cost and embodied CO_2 for the pipe material. The maximum one-way distance between the plant and the district L, namely L_{max} is related to exergy (Kilkis 2019). Consequently, the PVT systems should ideally be located within the district. However, this condition concludes that every building, in this case, should have their own PVT panel on their roofs and facades. This eliminates district thermal power distribution exergy destructions and solar energy utilization rate shall approach 100%.

$$L\text{max} < \frac{E_{XH}}{\Delta P_S(0.95)} \tag{17}$$

Topics presented so far quantifies the need for the transition to the Second-Law of Thermodynamics to effectively investigate the means and measures of transition to 100% renewables. In this respect, renewables should be utilized by paying attention to the following:

- The right quality of renewable supply,
- To the right (exergy-balanced) application(s),
- At the right order of utilization in the right cascade,

- At the right time,
- At the right location, and
- With proper storage without changing the form of energy, i.e. power to heat.

5 Barriers to 100% Renewables

Despite the need and the great potential of renewables, especially in the built environment for decarbonization and replacing the use of fossil fuels, the progress in renewable energy uptake, especially in heating and cooling, is quite slow against several barriers. The main barriers that yet to be removed for faster market penetration and globally wider use are:

1. ***Mistakes in the Name of 100% Renewables and Decarbonization***

Following the renewable chain from the supply point, many 'renewable' applications involve conventional systems and equipment, which are not exergy-compatible. Some of them need supplementary electric power from the grid, which is responsible for CO_2 emissions and even *ODI*. Many systems and equipment also involve *GWP*, like in Fig. 15. Equation 18 may approximate the equivalent CO_2 emission of the equipment, if their refrigerant types (GWP), annual charge amounts, *R* and annual leakage ratios, *LR* are known.

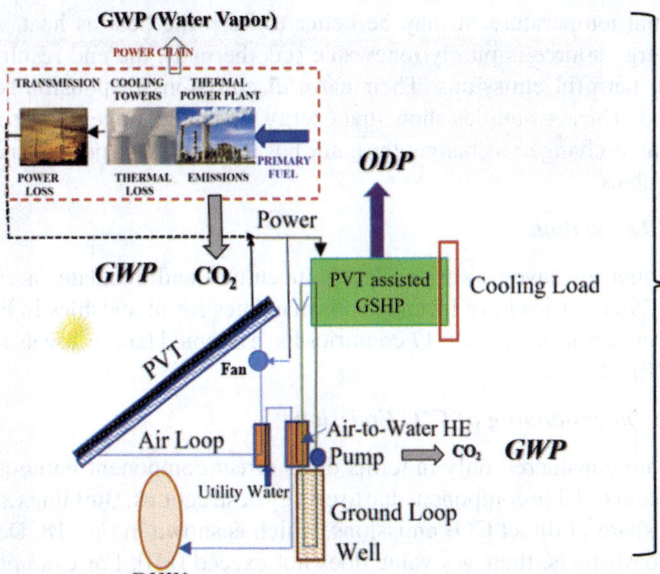

Fig. 15 Solar energy is not a complete solution alone. It depends on how we use it

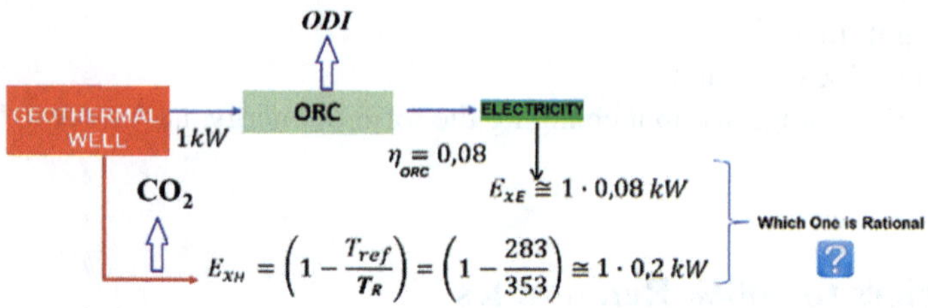

Fig. 16 Exergy rationale in geothermal energy (Kilkis 2019a)

$$CO_2 \cong \left(\frac{GWP}{1}\right)\left(\frac{LR}{100} \times R\right) \quad \{\text{kg } CO_2/\text{year indexed to } CO_2\text{refrigerant}\} \tag{18}$$

Equation 18 assumes that new refrigerants have nearly zero *ODP*. If the refrigerant properties are, $R = 0.4$ kg, $LR = 10\%$, $GWP = 1200$, then, CO_2 emission equivalency is approximately 48 kg CO_2/year, although the energy source is purely solar. This figure does not include additional emissions when supplementary grid power is used.

When electricity generation from low-exergy heat is considered, because power has a higher unit exergy, this idea will not work unless the 1st Law is also accounted for. Figure 16 shows two alternatives: 1-power generation with Organic Rankine Cycle (ORC) with 8% 1st Law efficiency (η_{ORC}), 2. direct utilization of heat at 80 °C (533 K). Alternative 1 has *ODI* responsibility and Alternative 2 releases CO_2 emissions unless re-injected to the ground. In Alternative 1, the exergy output, E_{XE} is 0.08 kW/kW. In Alternative 2, E_{XH} is higher: 0.2 kW/kW. Therefore, depending on the geothermal temperature, it may be better to keep the heat as heat. Again, although the energy source is purely renewable (geothermal), the end result is not 100% free from harmful emissions. Their harmful emissions responsibility must also be compared. These examples show that exergy is a game maker despite many describe it as a game-changer: it changes the game but also shows the proper strategies for 100% renewables.

2. ***Insufficient Legislation***

Although 128 countries have power regulatory incentives and mandates according to Fig. 17, only 29 countries have incentives or mandates for renewables in heating and cooling policies. Similarly, only 17 countries have adopted targets towards 2050 for zero CO_2 (REN 2019).

3. ***Insufficient Understanding of CO_2 Emissions***

CO_2 emissions are considered only in terms of its direct component without compounding with the avoidable component due to exergy destructions. Buildings already have the largest share of direct CO_2 emissions, which is shown in Fig. 18. Depending mainly on fossil fuels, their ψ_R value does not exceed 0.10. For example, if a natural-gas condensing boiler is used only for space heating, ψ_R is just about 0.06. The rest of the available exergy is destroyed. The result is a high avoidable CO_2 emission rate, even more than direct emission. When compounded with the already

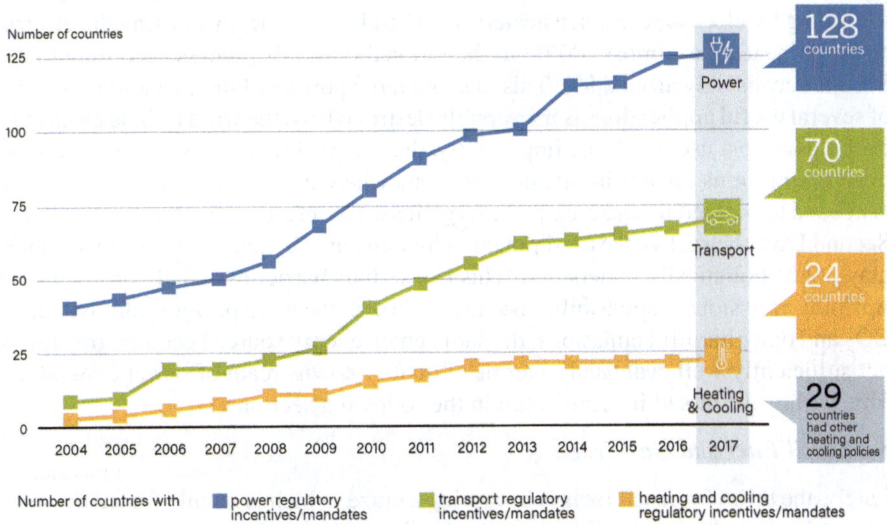

Fig. 17 The number of countries with renewable energy regulatory policies by sector (REN 2019)

Fig. 18 Sectoral breakdown of energy consumption-based global CO_2 emissions (IEA 2018)

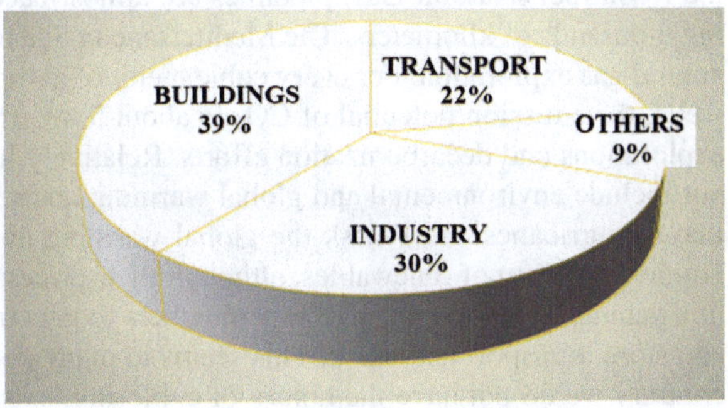

biggest share in the direct emissions, buildings become the most responsible sector, which further increases with the world population and modern urbanization.

4. ***Lack of Complete Interpretation of Renewable Power and Heat Thermodynamics***

All ratings and economic feasibility of renewable power (electricity) are based on the *quantity* of energy utilized from renewables in the built environment. This is the First Law of Thermodynamics. This law does not question downstream of the power generation about where this renewable energy is used and how it is used in terms of *quality* of energy both on supply and demand sides. This is the subject matter of the Second Law. Electric power has indeed the highest quality of energy (Exergy) and must be allocated to quality-balanced demands deserving such high quality. For example, if the electric power generated from a wind turbine is directly used to heat a building by electric resistance heaters, the First Law efficiency of using the electric power in heating is almost 100% but the rationality in this process according to the Second Law is only around 5%. This means a large portion of the execution potential of several useful applications is irreversibly destroyed and the wind turbine electricity becomes almost useless. More importantly, the exergy destructions directly translate to CO_2 emissions responsibility, because somewhere else some other fuel spending is needed to be used to make-up this exergy loss. Therefore, without reference to the Second Law, the real virtues and potential hidden environmental risks of renewables may not be holistically understood. This is a serious barrier especially on the global warming emissions responsibility because although the wind power is almost free of CO_2 and other harmful emissions, the same green energy source becomes grey if it is not sufficiently well evaluated in terms of quality downstream of power generation. This will be discussed in more detail in the following sections.

5. ***Fossil Fuel and Gas Craze***

Lately, the thirst for fossil fuels is becoming a craze, which is mainly fueled by international political interests. The same also holds true for nuclear energy. The situation is so critical that many countries are facing serious conflicts and sanctions. Some countries like Turkey have become a transitway on both land and sea, so much that the fossil fuel (Natural Gas) pipelines are almost becoming a gas highway extending thousands of kilometers. The Mediterranean Sea has become a focal conflict for natural gas exploration. For every cubic meter of natural gas surfaced from deep-sea wells, the emission potential of CO_2 is about 5 kg. This is a great conflict between explorations and decarbonization efforts. Relatively low fossil fuel prices (They do not include environmental and global warming costs with all its consequences like floods, hurricanes, etc.) mask the global warming emergency and reduce the economic feasibility of renewables, although their prices are dramatically dropping. In this manner, for example, no one remembers to generate hydrogen with renewables and store, transport and use it. This seems to many people a myth or luxury. On the contrary we do not have the luxury of exploring, extracting, and using fossil fuels. This argument also holds true for nuclear energy.

6. ***Deliberate Ignorance of the Second Law***

The Second Law, which deals with the quality of energy (Exergy) is considered to be a game-changer and treated in most cases as a threat to the economy and industry. The industry is not alone in this kind of resistance to the rationality of energy use. For example, World Energy Statistics Report does not refer at all to exergy (IEA 2018). Energy Efficiency Indicators released so far deal with only the quantity of energy (IEA 2018). In the EU there is only one concerned scientist group that proclaims the importance of exergy: The only European initiative about the inclusion of exergy in EU energy analyses is published by Science Europe (Science Europe 2016). In quote:

> Educators, researchers, policymakers, stakeholders, and citizens are urged to consider energy and natural resources on the basis of exergy, and in doing so understand that:

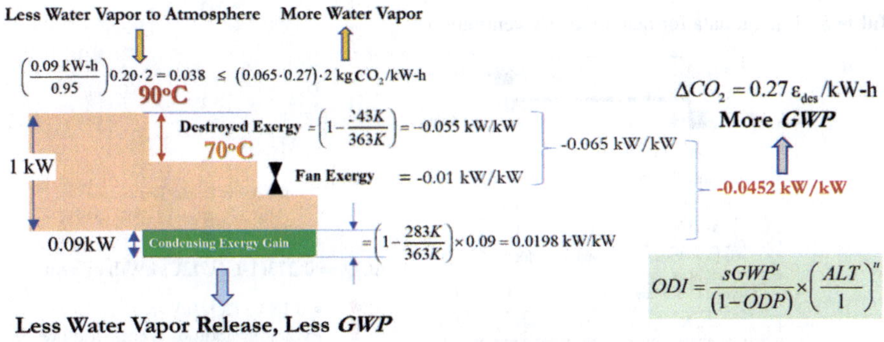

Fig. 19 Net exergy gain is negative in a condensing boiler

- exergy measures energy and resource quality;
- exergy-destruction foot-printing improves industrial efficiency;
- exergy offers a common international energy-efficiency metric;
- Optimal use of limited mineral resources by the application of exergy rarity;
- and exergy should be integrated into policy, law and everyday practice.

This proclamation shows how important it is to factor in exergy for 100% renewable transition. Many HVAC systems and equipment known and sold as *energy-efficient* and *green* are not energy rational at all. Two examples are given below:

a. Condensing boiler: Condensing boilers are unfair alternatives to renewables due to their very high First Law efficiency up to 97% peak. However, Figure 19 shows that actually, they consume natural gas by 23% more CO_2 responsibility compared to a conventional boiler. Therefore, an exergy rationality metric is essential like ψ_R for comparison purposes.

First, they reduce the combustion temperature in order for condensation. Then, they need a heat exchanger and a small additional (or oversized) fan.

b. Waste Heat Recovery: It is claimed that waste heat recovery from the exhaust ventilation air in buildings is highly efficient and cost-effective by referring to the First Law. Without considering the unit exergy difference between the electricity consumed by extra fan power in order to overcome the waste-heat exchanger pressure drop, and the thermal exergy recovered the *COP* indeed seems to be quite preferable, but *COPEX* value is much less than 1. A typical data is given in Table 3. The solution could be a better optimization of the heat exchanger and to utilize more efficient fans and electric motors with variable frequency drive in order to optimize the operation from an exergy point of view (Kilkis 2020).

7. *The Economy is not Everything*

We are paying for the quantity of energy but we are using only the quality of energy

Prof. Peter Novak, ASHRAE

Table 3 Typical data for heat recovery ventilation

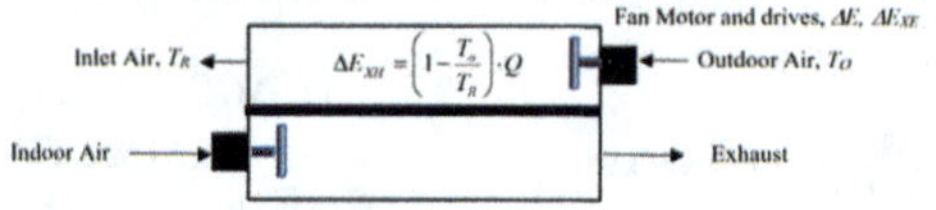

T_o: 283K, T_{ref}: 283K, T_R: 295 K, $E_{XH} = (1 - T_o/T_R) \cdot Q$ = 0.319 kW

Derived Data is Based on Air Properties at Room Temperature. Unit exergy of electricity is taken to be one.

Model	1	2	3
Air Flow Rate, m^3/hour	3000	4000	5000
Thermal Efficiency	0.65	0.51	0.46
T_R, K	290.8	289.1	288.5
Q, kW	7.86	8.20	9.24
E_{XH}, kW	0.319	0.173	0.176
ΔE, kW	2 x 0.45	2 x 0.55	2 x 0.55
First Law-*COP* $Q/\Delta E$	8.73	7.45	8.4
Second-Law *COPEX* = $E_{XH}/\Delta E$	0.35	0.30	0.34
Approximate Weight, *W* kg	112	135	145

Just due to Exergy Destruction

$\Delta CO_2 = 0.273(1\text{-}COPEX)$/kW-h of heat

= 0.18 kg CO_2/kW-h
Avoidably additional emission due to exergy destruction

$$CO_2 = \left(\frac{c}{\eta_T}\right)\frac{\psi_R \varepsilon_{sup}}{COPEX} = 0.57\frac{\psi_R \varepsilon_{sup}}{COPEX}$$

Total Emission

Solar panels are simply sold with €/m^2 or €/kW$_{peak}$ price based on the 1st Law, which does not make much sense since the type and the function of solar systems are much different. How much quality of energy they deliver is not asked. A cheap FPC only generates hot water with low exergy, a more expensive PV panel generates electric power with high exergy but wastes the remaining heat collected, a PVT system, generates both heat and power with higher exergy than the others, at a higher cost. Total efficiency is not a simple addition of power and heat, because of their different unit exergy. 1 kW of heat and 1 kW of cold also have different exergy. In order to avoid this confusion, Exergy-Levelized Cost, *ELC* has been developed, combining exergy rationality, price, *PC*, embodied costs, *EM*, panel area, *A*, and weight, *W* (Kilkis 2019).

$$ELC = \frac{\left(\frac{PC + EM\,(W_p)}{I_{test}A_p\left[\varepsilon_{\sup E}\eta_E + \varepsilon_{\sup H}\eta_H\right]}\right)}{\psi_R A_p} \quad \{\text{Euro/kW}_{\text{EXpeak}}/\text{m}^2\} \tag{19}$$

W_p is panel weight (kg), *PC* is in Euros. *EM* is in Euros/kg. Subscripts *E* and *H* stand for electric and thermal outputs respectively. *ELC* corresponds to design (test) conditions.

8. *Irrational Exergy Allocation*

US allocation of electricity for residential and commercial sectors in 2008 is shown in Fig. 20. In the residential sector, space heating, cooling, and DHW are responsible for 42.5% of the total consumption. This is a large waste of exergy, because most of the demand, including some of the cooling, may be achieved by non-electric systems with waste or solar thermal. At least a solar trigeneration system could be used for

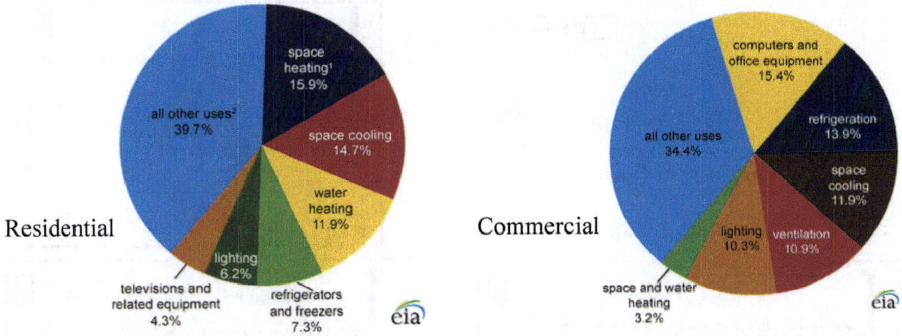

Fig. 20 US electricity consumption in residential and commercial buildings (OECD/IEA 2018)

minimum exergy destructions. Even if renewable electricity is generated first, on-site or near-site, the exergy question pends whether it is the best way to use it for heating. In commercial buildings, HVAC demand is about 40%, of which many of them may be satisfied by non-electric systems. This is a problem of misallocation of energy sources at the wrong applications in the wrong order.

6 Is Renewable Electricity Good for Heat? Not so: Electricity Was not Created for Heating

The use of high-exergy electricity for low-exergy applications is a chronic habit, which causes large exergy destructions. For example, China has recently developed a strategy about heating with wind electricity in her Northern provinces. Indeed, in cold climates, heating demand high but wind potential is abundant. This seems a good idea but the use of wind energy in space heating is a serious exergy misallocation. The problem is compounded by the low urbanization density and it is rather difficult to implement district energy systems and to rely on waste heat of near-by factories. On the contrary, in hot climates, electricity demand for cooling is increasing due to global warming and heat island effect, especially in dense urban settlements. In either case, extreme climates mean longer heating seasons with dispersed demand or longer cooling seasons with intense demand. The more the heating or cooling energy demand is, the more are the CO_2 emissions from conventional HVAC systems in buildings. Therefore, for cleaner air and less global warming, transformation into renewable energy systems gains prime urgency. These targets must be achieved by a holistic approach with the most exergy-rational objective. For example, almost 95% of the electrical energy may be transformed into many useful applications. However, if wind energy is first transformed into heat and used only for heating, a large loss of opportunities about remaining potential uses are lost (Irreversible loss of quality). This causes additional but avoidable CO_2 emissions. Figure 21 shows five methods of wind-to-heat strategy and hydrogen economy on a city level (Kilkis 2019). Method

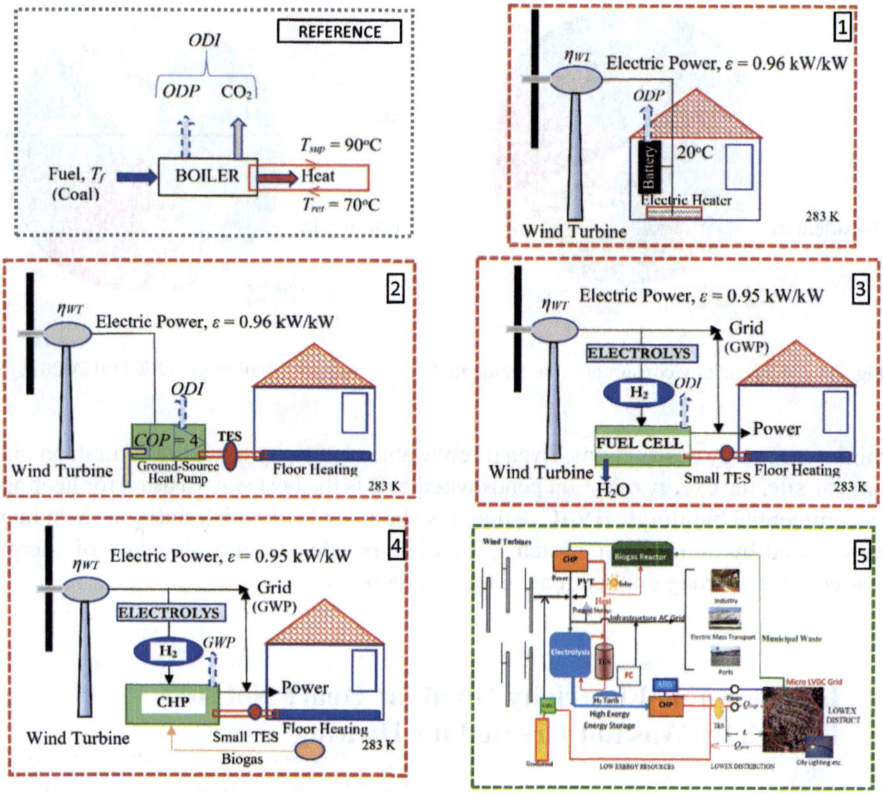

Fig. 21 Five wind-to-heat methods. The reference case is a coal-fired boiler or stove (Kilkis 2019)

1 is direct resistance heating with wind electricity. Method 2 is heating (or cooling) with a ground-source heat pump, driven by wind electricity. Method 3 is hydrogen production by electrolyzing water and then storing and using in a fuel cell for both heat and power again. The system may be interconnected with grid power. In this case, *GWP* exists. Method 4 is hydrogen- biogas fueled combined heat and power (CHP). Cooling is possible by adding an absorption system. Ultimate utilization of wind power is a hydrogen city, which is Method 5. Table 4 subjects their sustainability to 12 metrics.

$$\mathrm{PEF} = 1/PER \tag{20}$$

$$PER = COP = \eta_I \eta_{WT} \tag{21}$$

$$TI = \eta_i \psi_R (1 - ODI) \tag{22}$$

Table 4 Results of the performance metric data for five alternatives (Kilkis 2019)

Metric		Fossil fuel		Wind				
		Coal		Electric		Hydrogen		
		Reference		Renewable methods				
		Boiler	Stove[a]	1	2	3	4	5
1	η_I (or η_{WT})	0.60	0.35	(0.45)[b]	(0.45)	0.38	0.40	0.55
2	η_{II}	0.0378	0.0122	0.016	0.56	0.27	0.29	0.47
3	*PEF*	–	–	1	1	1	1	0.90[c]
4	*PER*	0.60	0.35	0.45	1.12	0.38	0.40	0.55
5	*COP*	0.60	0.35	0.45	4	0.38	0.40	0.55
6	*COPEX*	0.0378	0.022	0.016	0.56	0.27	0.29	0.47
7	Ψ_R	0.063	0.063	0.035[d]	0.14	0.72[e]	0.72[f]	0.85
8	CO_2	0.83	1.43	0[g]	0	0	0	0
9	ΔCO_2	0.78	1.34	0.80	0.71	0.232	0.24	0.12
10	$\sum CO_2$	1.61	2.77	0.80	0.71	0.232	0.24	0.12
11	*ODI*	0.115	0.115	0.14	0.135	0	0.023	0.015[h]
12	*TI*[i]	0.0336	0.0196	0.0135	0.121	0.2736	0.280	**0.460**

NOTES for Table 4
[a]Assuming that the same type of coal is used
[b]Typical electro-mechanical wind turbine efficiency
[c]Due to hydrogen and DC power district network losses
[d]This is a value, which is lower than a boiler case
[e]Excluding PV system contribution
[f]Without the ORC type of bottoming cycle
[g]Excluding embodied CO_2 emissions
[h]With industrial CO_2 capture and use for refrigerating cycles
[i]*TI* is the transformation index, which combines the First and Second Laws and represents harmful emissions by *ODI*. See Eq. 22

$$\eta_{II} = \psi_R \eta_I = COPEX \tag{23}$$

7 Sustainable Solutions for 100% Renewables

This section identifies ten steps for a sustainable future with 100% renewables, based mainly on the Second Law. These ten steps are explained below.

Step 1. Holistic Approach for Sustainable Build Environment

Today, we should not only speak about sustainable buildings but sustainable and smart cities for 100% renewables. A smart city is not just a city of smart automation controls, smart transport, driverless buses, and electric cars, etc. A smart city is a complex combination of individual buildings that generate, share, and utilize the

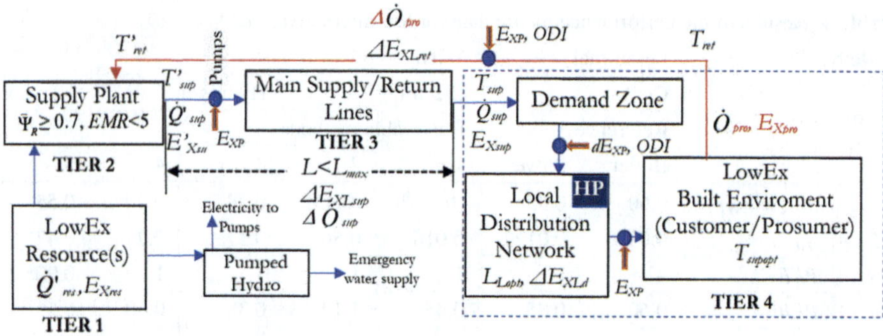

Fig. 22 Four tiers of the holistic district energy model with LowEX renewables (Kilkis 2019b)

energy with the maximum use of the quality of energy as well as the minimum use of the quantity of energy, while the environment is sustained. For example, the city of Delft is planning to rent the suitable roof areas of townhouses to install photovoltaic (PV) and photo-voltaic-thermal (PVT) panels to generate, share and utilize solar energy collectively at city-wide and grid-wide, not only in terms of power but also the heat of different forms. Therefore, we cannot speak any more about an individual building but a collection of buildings. The term *collective* is a keyword for future cities and buildings. Therefore, one needs to analyze buildings with a holistic approach integrated with the built environment at large with all its upstream and downstream. In such an approach, energy/exergy and water nexus, by using the Rational Exergy Management Model will be successful. Figure 22 shows such a holistic view of a district energy system in four tiers, starting from the renewable energy sources, which is the first ring of the exergy flow and ending at individual buildings, which are also prosumers, coupled with heat pumps of optimum capacity achieved by optimally insulating the building and oversizing the heating or cooling equipment to improve the *COP* of heat pumps (Kilkis 2019b). *EMR* is the payback period for the embodiments of renewable energy systems and equipment by their energy and CO_2 savings during operation. If a heat pump is necessary for upgrading the supply temperature, T_f to be compatible with the conventional equipment, an optimum T_f may be determined, which comprises partial oversizing of both the heat pump and the equipment.

By ignoring the incremental changes in the power demands of pumps (or fans), Eq. 24 is used, where C_{eq} and C_{hp} are the life-cycle cost-design operating temperature factors. T_a is the dry-bulb indoor air temperature (Kilkis 2000). The negative term is for cooling.

$$T_{opt} = \pm\sqrt{\frac{C_{eq}}{C_{hp}}} + T_a \tag{24}$$

Step 2. Decide Whether Net-Zero Energy or Net-Zero Exergy Comes First

In the quest of reaching the goals of the recent Paris Agreement for reducing CO_2 emissions, net-zero energy buildings (NZEB) and net positive-energy (NPEB) buildings are becoming common (IEA 2018). Although universally accepted several definitions for NZEB have been published, there are other issues to be resolved (US DOE 2015). A major issue that has not been addressed in the building and energy sector is the fact that with the increasing share of renewable energy resources and systems in the built environment at different exergy levels, their exergy differences and the need for exergy balance between the supply (resource) and the demand points (built environment, such as buildings) need to be identified and their importance in optimum and net-positive solutions have to be acknowledged (Kilkis 2011). In this respect, the shortcomings of the First- Law are demonstrated in the following two examples.

Example. The Net Zero Exergy Building (NZEXB) definition becomes more important than the NZEB definition while renewable energy systems become more diversified and connected to the district. The temperatures of the heat received from and supplied to a district system by a presuming building or the cold service received from the district and supplied to the district must be taken into account, exergy-wise. For example, in Fig. 23, an exergy exchange deficit occurs if the presuming building delivers 30 °C water from its solar FPC system to the district but receives 40 °C water from the district in the same quantity over a given period of time. A similar deficit occurs for cooling because exchange temperatures of the chilled water supply and return between the district and the prosumer building are different. This building is a net-*zero energy building* while it exchanges both electrical and thermal energy in equal quantities.

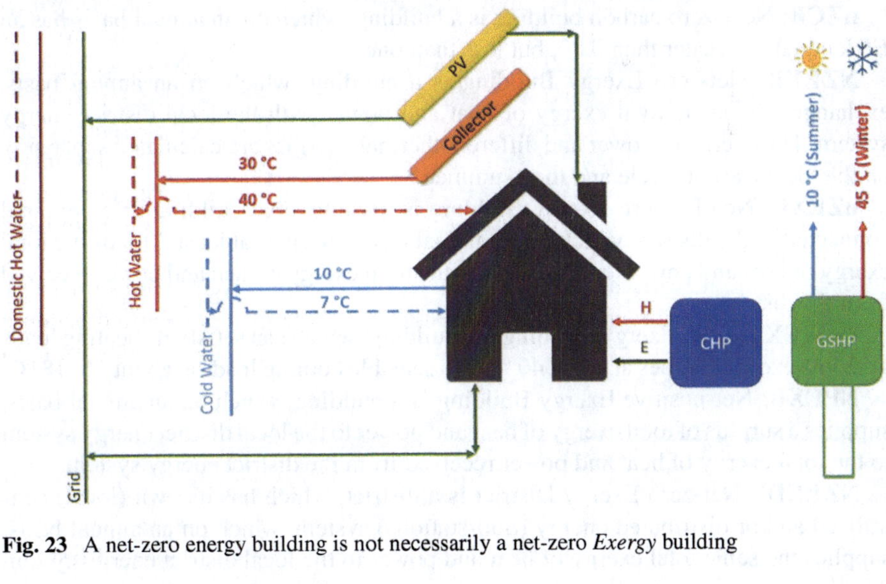

Fig. 23 A net-zero energy building is not necessarily a net-zero *Exergy* building

If for example, on an annual basis, the building receives 10,000 kWh of AC electrical energy with RMS of 5% and provides 10,000 kWh AC electrical energy with RMS of 10%. There is an electric power quality difference in terms of RMS. In the same period, the building receives 15,000 kWh of heat from the district (supply) at an average temperature of 353 K (80 °C) and provides heat of 15,000 kWh to the district (return) at an average temperature of 343 K (70 °C). In terms of the ideal Carnot Cycle, thermal exergy exchanges are:

$$E_{xsup} = (1 - 283/353) \cdot 15000 = \mathbf{2974.5kWh},$$
$$E_{xxet} = (1 - 283/343) \cdot 15000 = \mathbf{2623.9\,kWh}.$$

Obviously, although the net energy (Quantity) exchange is the same, there is net annual exergy (Quality) exchange deficit of 350.6 kWh. A similar condition holds true for the cooling season. Obviously, this building is not a net-zero-exergy building, because the supply and return exergies are not equal.

Consequently, the district should be equipped with exergy meters rather than calorimeters in order to establish a fair distribution of charges among the energy and power supplier (district), subscribers and prosumers in the district. This is especially important for-low exergy districts and buildings utilizing low-exergy sources.

Step 3. Implement New Metrics Leading to 100% Renewables

So far, exergy is mentioned and used only at a scale of a building, which defines only a low-exergy building. The following definitions were developed in order to expand the concept to district and city levels.

NZCB: Net-zero carbon building is a building, which on an annual basis has an *EDR* of 1.

nZCB: Near-zero carbon building is a building, which on an annual basis has an *EDR* equal or greater than 0.80, but less than one.

NZEXB: Net-zero Exergy Building is a building, which on an annual basis, exchanges the same total exergy of heat and power with the local district energy system. The exergy of power and different thermal energies are calculated separately on the ideal Carnot Cycle and then summed.

nZEXB: Nearly- zero Exergy Building is an individual building or compound connected to the district, which on an annual basis provides at least 80% of the total exergy of heat and power to the district as the total exergy of heat and power received from the district.

LoWEXB: Low-Exergy Building is a building, which can satisfy its heating loads with low exergy sources at about 40 °C and sensible cooling loads at about 15–18 °C.

NPEXB: Net-positive Exergy Building is a building, which on an annual basis, supplies a surplus of total exergy of heat and power to the local district energy system to the total exergy of heat and power received from the district energy system.

NZEXD[a]: Net-zero Exergy District is a district, which has its own (local) centralized and/or distributed energy (Sub-stations) system, which on an annual basis, supplies the same total exergy of heat and power to the local district energy system

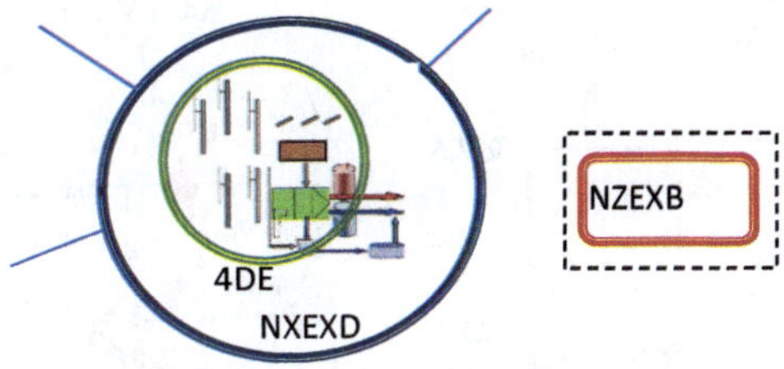

Fig. 24 NZEXD and NZEXB in 100%-renewable build environment

as the total exergy of heat and power received from them. This is generally known to be the 4th Generation District Energy System or beyond (> 4DE). Figure 24 shows the physical relation between NZEXD and NZEXB.

MtoEX: Another major universal flaw and energy-related literature is the use of Mtoe (Megaton of oil equivalent), which is a First-Law definition for a given quantity of energy. Crude oil has a standardized adiabatic flame temperature, about 2373 K and the unit exergy, ε_{sup} is 0.881 W/W in terms of the ideal Carnot cycle. Its exergy, E_X is 10.25 MW-h for an energy amount, Q of 11.63 MW-h for one ton of crude oil. Now consider 11.63 MW-h of hot water supplied by solar energy at 50 °C (ε_{sup}= 0.124 W/W). According to Mtoe definition, this amount is exactly 1 Mtoe. In contrary to this misleading equivalency, 11.63 MW-h of 50 °C water has only 1.44 MW-h exergy (0.124 W/W × 11.63 MW-h), which is equivalent to only 0.14 ton of oil (1.44/10.25). Evidently, the exergy equivalency of solar heat and crude oil does not exist.

Step 4. Decarbonization With the Second-Law

Referring to Fig. 11 and Eq. 11b, five major parameters may be identified for improving the overall efficiency and CO_2 emissions. These are:

1. Type of fuel or renewable energy source
2. Equipment and plant efficiency (First-Law)
3. *Exergy Rationality*, defined by ψ_R.
4. Plant and grid power transmission efficiency, transformer losses, etc.
5. Power loads.

Step 5. Avoid Mistakes in Renewable District Energy Systems

District energy systems can be of the best instruments towards 100% renewables because they allow optimally clustered and exergy-rational hybridization of renewable energy systems. Furthermore, central heating and cooling through a district network are generally more efficient. But this latter statement may not hold true from the exergy point of view. Therefore, a careful exergy rationality analysis is in order. Geothermal district heating is one of them.

Geothermal District Heating: In the City of Afyon geothermal energy at a wellhead temperature of 96 °C (369 K) is used for district heating (Şahin C and Gürler 2019). According to Fig. 25, the re-injection temperature is 48 °C (321 K). The wellhead flow rate is such that 6 MW of thermal power is available. Before utilization

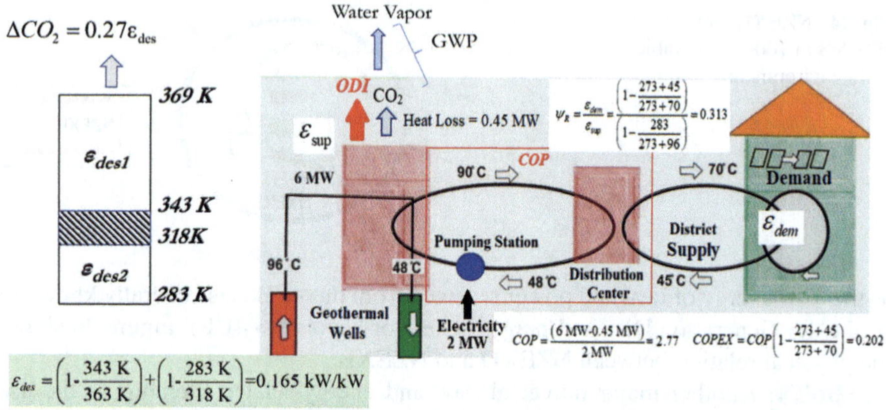

Fig. 25 Afyon geothermal energy district heating system layout (Şahin C and Gürler 2019)

of the heat, the geothermal brine is kept in an open-air pool for cooling down the supply heat to 90 °C (363 K), because the piping material used in the district network cannot withstand higher temperatures. This causes exergy destruction of at least 0.45 MW to the atmosphere with large amounts of CO_2, which has a global warming potential (*GWP*). *GWP* also translates to *ODP* (Eq. 3). Evaporation further reduces the available geothermal power and increases the *GWP*. District loop pumping station requires 2 MW of electric power. The result is a low *COPEX* value, which in this case is 0.202 if the system is treated to be a simple heat pump. On contrary, the *COP* value is 2.77, which is high enough to mislead that the system is efficient enough despite Ψ_R is only 0.313, which means that large amounts of exergy destructions take place in the system, which reflects avoidable but unaccounted CO_2 emissions by the First Law. How the system may be improved for minimum exergy destructions and better utilization of the geothermal energy towards 100% renewables, another study of the Author analyzed ORC and ground-source heat pump combination, which is shown in Fig. 26. In order to make the situation more exergy-critical, a lower well-head temperature (exergy) of 80 °C was chosen. Net exergy is the sum of electrical output and the thermal output minus the sum of exergy demand of circulation pumps and the cooling fan.

$$E_{XT} = E_{XE} + E_{XH} - E_{XP} - E_{XF} \tag{25}$$

$$COPEX = \frac{E_{XT}}{(E_{XP} + E_{XF})} \tag{26}$$

The power is split between the building and a ground-source heat pump, *X* providing comfort heat (or cold). If *X* is equal to zero, then this is the all-electric case, if *X* is equal to one, then this is an all-heat case. Calculations for ψ_R by using Figs. 27 and 28 show that at such a low source temperature it is slightly more rational to

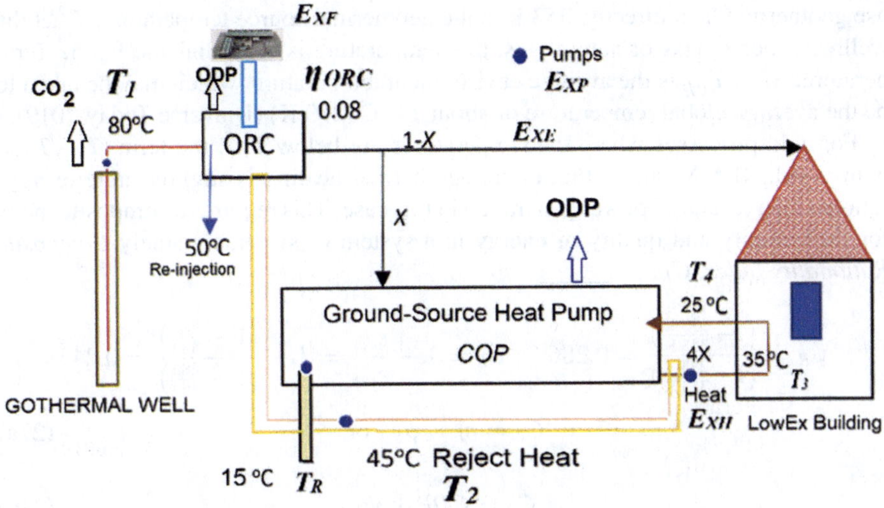

Fig. 26 Geothermal energy, ORC, and heat pump in a district (Kilkis 2019a)

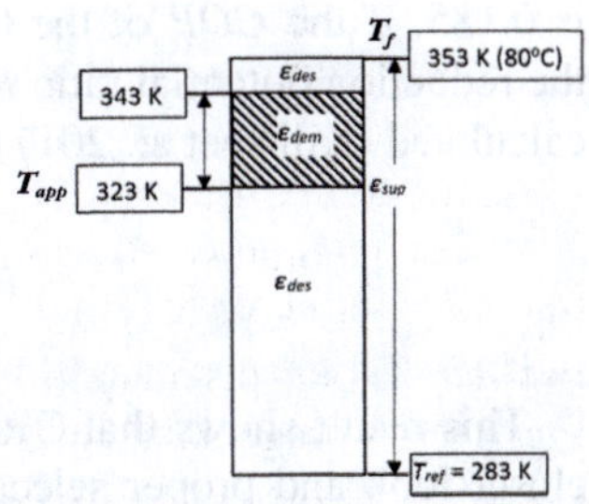

Fig. 27 Direct geothermal district heating

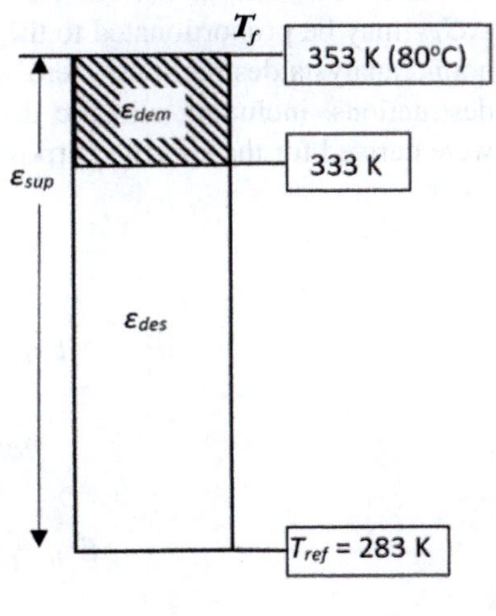

Fig. 28 Geothermal power with ORC

use geothermal heat directly. 353 K is the geothermal source temperature, T_f at the wellhead. For biogas or natural gas, this temperature is the Adiabatic Flame Temperature, *AFT*. T_{ref} is the average environment temperature, which may be taken to be the average global temperature of about 14 °C (290 K) (Universe Today 2019).

For cold processes where the temperatures are below T_{ref}, the term (T_{ref}/T_{app}) is inversed. REMM shows that direct geothermal heating is slightly more exergy-rational than just ORC power generation in this case. This requires a composite index for the quantity and quality of energy in a system of systems, namely *Composite Rationality Index*, C_R:

$$\psi_R = \frac{\left(1-\frac{333\text{ K}}{353\text{ K}}\right)}{\left(1-\frac{283\text{ K}}{353\text{ K}}\right)} = 0.286 \qquad \psi_R = 1 - \frac{\varepsilon_{des}}{\varepsilon_{\text{sup}}} = 1 - \frac{\left(1-\frac{283}{333}\right)}{\left(1-\frac{283}{353}\right)} = 0.243$$

$$C_R = \eta_I \times \psi_R \quad \text{or,} \tag{27a}$$

$$C_R = COP \times \psi_R \tag{27b}$$

For the geothermal district heating-only case, if the efficiency η_I is 0.65, then C_R is 0.185. If the *COP* of the ORC system is 0.10, then C_R is 0.024. Consequently, the reduction potential ratio *R* of avoidable CO_2 from the carbon stock may also be calculated (Kilkis et al. 2017):

$$R = \frac{(2 - C_R)_{ORC}}{(2 - C_R)_{District}} = 1.2 \tag{28}$$

This result shows that ORC and district cases have similar exergy rationality, a closer look and proper selection at low supply temperature save about 20% from avoidable CO_2 emissions. For the economic analysis, the exergy rationality cost, ΔC_{EX} may be proportionated to the number of exergy destructions is used for economic analysis destructions. Here, *c* is the average unit cost attributable to exergy destructions, including climatic disasters. The following exergy-based equations were derived for the specific purpose of the analysis.

$$\Delta C_{EX} = c\frac{\sum \varepsilon_{des}}{\varepsilon_s} \tag{29}$$

$$E_{XE} = \eta_{ORC}(1 - X) \tag{30}$$

$$\eta_{ORC} = a + bT_1 \tag{31}$$

$$E_{XH} = XCOP\left(1 - \frac{T_3}{T_4}\right) \tag{32}$$

$$COP = g + h(T_R - T_3) \tag{33}$$

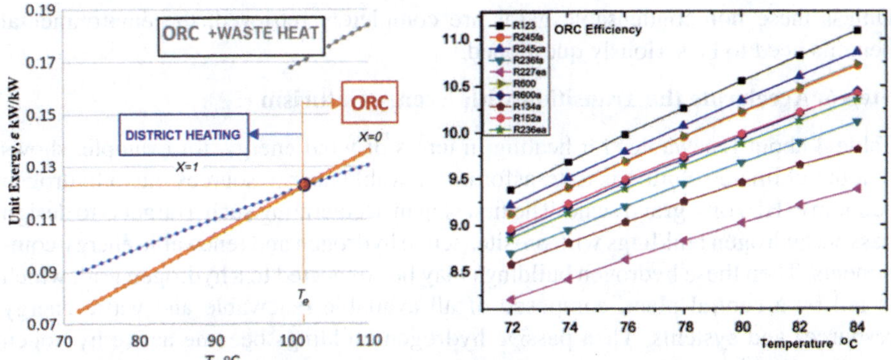

Fig. 29 Break-even temperature for heating or power (Eq. 37) (Kilkis 2019a)

$$E_{XP} = c + dX^2 \tag{34}$$

$$E_{XF} = \left(e\eta_{ORC}^{-f}\right) \tag{35}$$

$$X_{opt} = \frac{COP - \eta_{orc}}{2d} \quad \{0 \leq X \leq 1\} \tag{36}$$

If COP, η_{ORC}, and d are 4, 0.1, and 3 respectively, then X_{opt} is 0.65. Although the First-Law, in this case, gives the same X_{opt} with a positive energy quantity, the exergy is negative, meaning that the system is not rational yet towards 100% renewables. Figure 29 shows that there are two regions regarding the optimum solution in terms of the break-even temperature, T_e. If the geothermal source temperature is higher than T_e, then ORC system is preferable if waste heat is utilized. Otherwise, the system must be an all-heat system ($X = 1$). Equation 37 solves T_e.

$$T_e = \frac{COP\left(1 - \frac{T_3}{T_4}\right) - a}{b} \tag{37}$$

X_{opt} shows that COP, η_{ORC} and coefficient d of the power demand of pumps and fans are the major players in the success of the renewable energy system. Often the pump and fan exergy demand are ignored but sometimes included in energy analysis. According to Eq. 37, T_e gets more relaxed with an increase in a and b, which improves the ORC efficiency thus ORC becomes more favorable. If, COP increases then T_e increases in favor of the heat-only case.

It must also be noted that both the ORC unit and the heat pump emit ozone-depleting chemicals (ODC) and this must be kept in mind in questioning how ORC systems are environment-friendly when coupled with geothermal energy and heat pumps. Geothermal energy sources also emit chemicals and H_2S gas besides CO_2.

Unless these non-condensable gases are completely retrieved, the environmental benefits need to be seriously questioned.

Step 6. Accelerate the Transition with Exergy Futurism

Table 4 about renewables for heating in terms of wind energy, for example, shows that the ultimate solution is to transform renewable energy sources into a hydrogen economy (HE) on a grand scale. The first step in accelerating such a target is to design passive hydrogen buildings with on-site, active hydrogen and renewable energy components. Then these hydrogen buildings may be connected to a hydrogen grid, which is fed by a central plant, composed of all available renewable and waste energy resources and systems. Then passive hydrogen buildings become active hydrogen buildings, which exchange energy and water, completing an energy/exergy/water nexus (Prosumer). According to Fig. 30, a hydrogen house with its own active mini-hydrogen economy, its Ψ_R is 0.72. With the partial contribution of PV cells, Ψ_R value reaches 0.85. Then the corresponding CO_2 emission is only 0.166 kg CO_2/kWh of heating excluding PV system contribution. This is almost a ten-fold CO_2 emission reduction potential, compared to a coal-fired boiler case. Figure 31 shows the Hydrogen house.

$$\Sigma CO_2 = (0) + 0.83 \times (1 - 0.80) = 0.166\,\text{kg}\,CO_2/\text{kWh of heating.}\ \{\text{Without PV System}\}$$

$$\Sigma CO_2 = (0) + 0.83 \times (1 - 0.85) = 0.12\,\text{kg}\,CO_2/\text{kWh of heating.}\ \{\text{With PV system contribution}\}$$

$$\psi_R = 1 - \frac{\left(1 - \frac{313\,\text{K}}{333\,\text{K}}\right) + \left(1 - \frac{283\,\text{K}}{303\,\text{K}}\right)}{\left(1 - \frac{283\,\text{K}}{515\,\text{K}}\right)} = 0.72$$

According to Eq. 10, the virtual Carnot-Cycle equivalent supply temperature, T_f of the wind turbine is 515 K for an average First-Law efficiency, η_I of 0.45:

$$T_f = \frac{T_{ref}}{(1 - \eta_I)} = \frac{283\,\text{K}}{(1 - 0.45)} = 515\,\text{K}.$$

Step 7. Transform to Hydrogen Cities

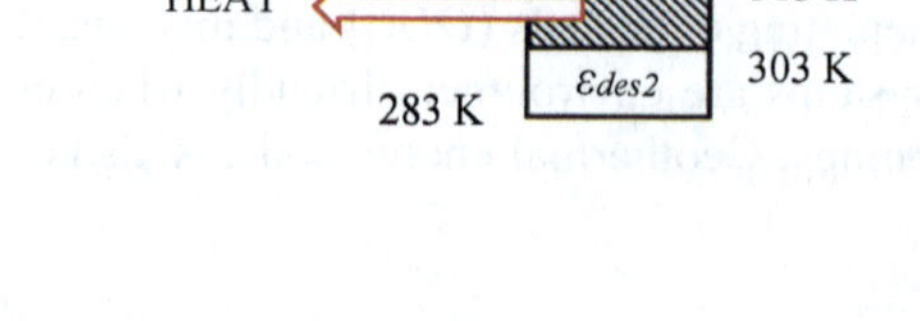

Fig. 30 Wind to hydrogen for space heating with fuel cell

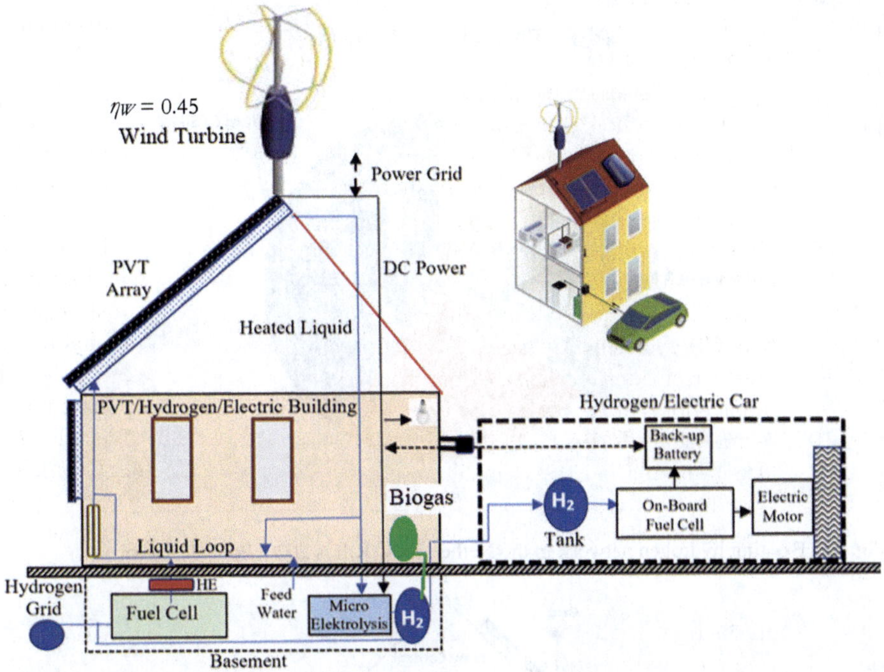

Fig. 31 Integration of wind and solar for heat and power in a hydrogen building (Kilkis 2019)

This step is the grand solution for the long-run strategies for renewable energy systems.

A system of hydrogen homes, which are prosumers be blended into hydrogen cities for maximum exergy rationality, in terms of ψ_R value, which may reach 0.95 with carefully optimized designs. Hydrogen is distributed in the district with pipes, which consume much less power than hydronic thermal distribution. Existing natural-gas pipelines may be upgraded for medium-pressure hydrogen transport. Figure 32 shows the already existing hydrogen network in the Netherlands. For heating purposes, fuel cells or micro CHP units (smaller than 50 kW power capacity) are installed on-site. The success of the hydrogen economy depends on the development of new and retrofit hydrogen energy buildings. Figure 33 shows the active hydrogen house concept, which exchanges hydrogen and power with the grid and water with the municipal water network.

Impact of Hydrogen on the Ozone Layer: There are claims that hydrogen fuel could widen the ozone layer due to leakages of the process chain and utilization (Jacobson 2008). In spite of these claims, there is a scientific consensus that the CO_2 equivalence of hydrogen is 4.3 megaton of CO_2 per 1 megaton emission of hydrogen over a 100-year time horizon. The plausible range between 0 and 9.8 express 95% confidence. On this basis, although the impact of hydrogen emissions on the global climate is very unlikely to be zero and it is very likely that it will be small. If a

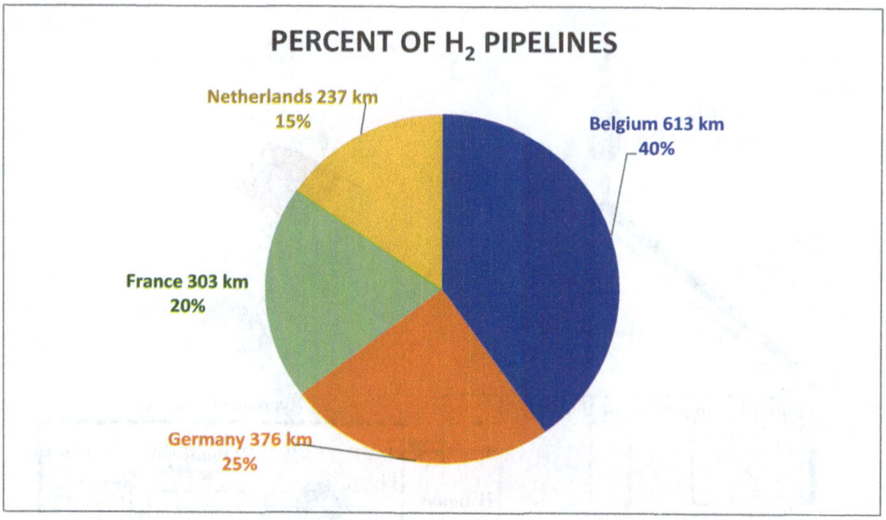

Fig. 32 Existing hydrogen network in the Netherlands (Kilkis 2020, Polar Teknoloji)

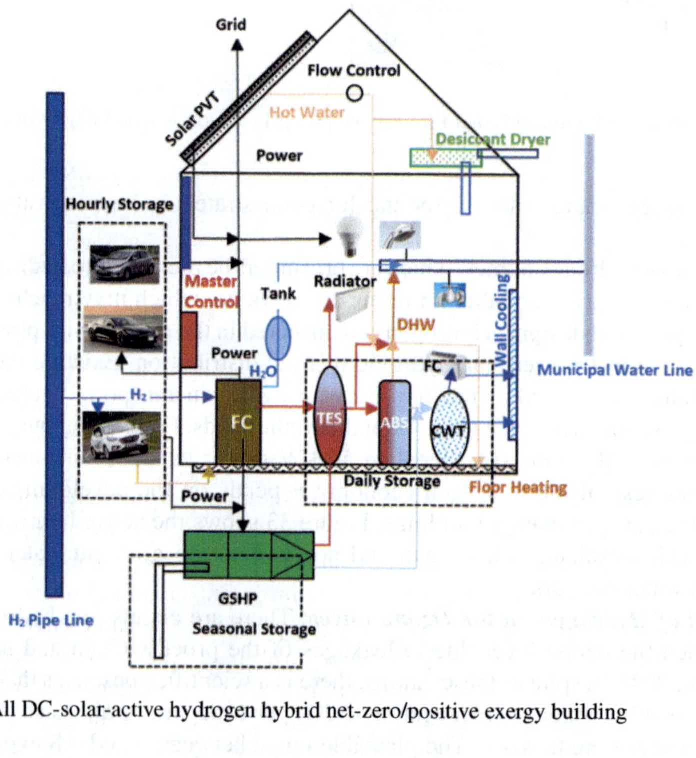

Fig. 33 All DC-solar-active hydrogen hybrid net-zero/positive exergy building

1% leakage for the entire hydrogen system is considered, then the global warming consequences are indeed very likely to be small. However, the *ODI* of hydrogen gas leakage is not zero and is roughly one-fifth of CO_2. $ODI_{H2} = ODI_{CO2} = 0.115/5 = 0.023$. Hydrogen cities with small *ODI* with an associated small *GWP* and 1% leakage seems to be the best alternative for storing and utilizing wind energy along with other renewables.

Active Hydrogen Home: Figure 33 shows an active hybrid house connected to the hydrogen grid. This building is an active hydrogen house because it generates electric power on-site with solar PVT and exchanges it with the grid. Its on-board fuel cell converts grid hydrogen to both heat and power. It stores hydrogen gas (not shown) and also water. An electrically operated GSHP heat or cools by radiant panels. ABS or ADS system may also be used for cooling.

Step 8. Make The Right Choice In Solar Energy- FPC, PV, or PVT?

There are several ways to harness solar energy, depending upon what the main demand is. The simplest is a solar flat plate collector, which only generates hot water in summer.

Solar Flat-Plate Collector (FPC): An FPC delivers only heat between supply and return temperatures, T_{sup} and T_{ret}, respectively. Figure 34 shows that exergy is destroyed between T_f-T_{out} (ε_{des1}) and T_{sup}-T_{ref} (ε_{des2}). Major exergy destruction (ε_{des1}) is upstream of the useful work because no electricity or any other useful work with higher-exergy is generated. In this case, the Rational Exergy Management Model (REMM), gives the rationality metric by, ψ_R.

$$\psi_R = \frac{\varepsilon_{HW}}{\varepsilon_s} = \frac{\left(1 - \frac{T_{ret}}{T_{sup}}\right)}{0.557} = 0.24$$

Solar PV: PV cells generate only power at a higher exergy level. This is shown by the Exergy Flow Bar in Fig. 35. Starting from the frame temperature of the PV panel, T_E, major exergy destruction takes place downstream of the useful application. If for the same solar insolation level, the PV panel frame temperature, T_E is 353 K, then

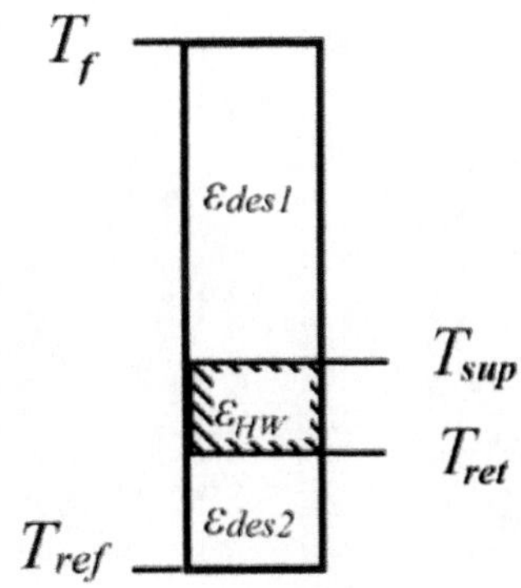

Fig. 34 FPC with T_f = 638.8 K, T_{ret} = 288 K, T_{sup} = 333 K, T_{ref} = 283 K

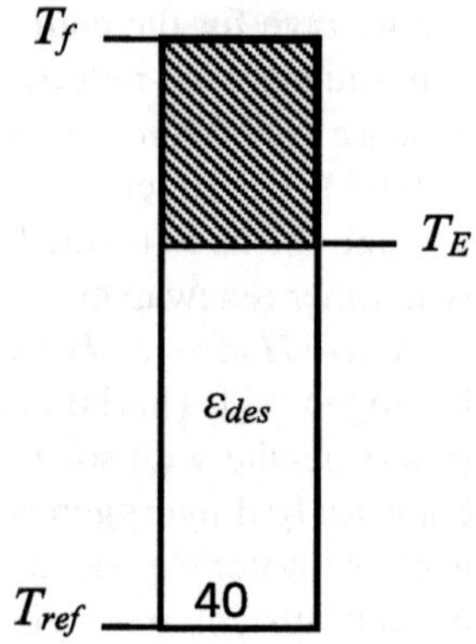

Fig. 35 Simple solar PV system

ψ_R will be 0.64.

$$\psi_R = 1 - \frac{\varepsilon_{des}}{\varepsilon_{solar}} = 1 - \frac{\left(1 - \frac{283}{353}\right)}{0.56} = 0.64$$

Conventional Solar PVT: Whether exergy is destroyed upstream or downstream of the useful work, a substantial portion of the available solar exergy is destroyed. However, when PV and FPC systems are coupled, exergy destruction decreases with two smaller exergy destruction points left. The effect of active cooling of the PV cells is revealed in Fig. 36 by a lower T_E (330 K). However, T_{sup} needs to be low for effective PV cooling (313 K). For T_{ret}= 303 K:

$$\psi_R = 1 - \frac{\varepsilon_{des1} + \varepsilon_{des2}}{\varepsilon_{solar}} = 1 - \frac{\left(1 - \frac{313\,\text{K}}{330\,\text{K}}\right) + \left(1 - \frac{283\,\text{K}}{303\,\text{K}}\right)}{\left(1 - \frac{283\,\text{K}}{638.8\,\text{K}}\right)} = 0.79$$

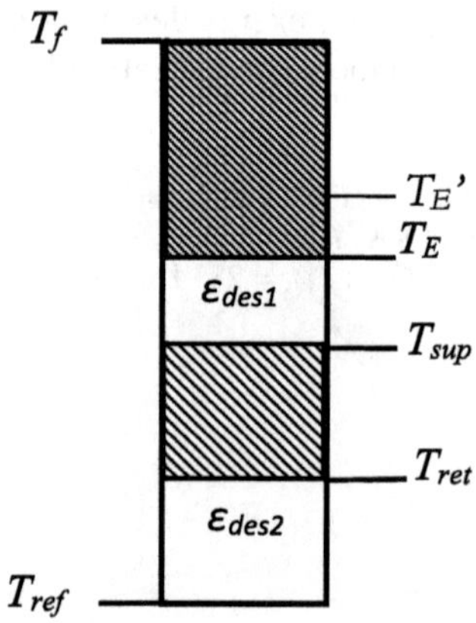

Fig. 36 Conventional solar PVT system without circulating pump losses

For actively cooling a PVT, a coolant fluid is circulated in the panel and its hydraulic ancillaries by a pump or fan. This parasitic loss is mapped to the exergy flow bar by T_E'.

$$\left(1 - \frac{T_E}{T_f}\right) \cdot (1 - c) = \left(1 - \frac{T'_E}{T_f}\right) \tag{38}$$

The term c is the ratio of the parasitic power exergy demand to the PV output. If c is 0.05, then, T_E' is 336 K, meaning a decrease in the ψ_R from 0.87 to 0.79. Such a decrease like 10% requires a major reduction or elimination of pumping need. Due to the interrupted nature of solar energy, an external thermal storage system (TES) is necessary.

In order to understand the mechanism better, a holistic model was developed, which is shown in Fig. 37. This figure shows how much the overall PVT system is complicated with five interconnected tiers, namely the PVT panel itself (Tier 1) the fluid circulation system (Tier 2), a thermal storage system with its own ancillaries (Tier 3), temperature peaking system (Tier 4) and load demand points in the building (Tier 5). The net total exergy output, E_{XT} of such a composition is given in Eq. 39. E_{XE} is the power exergy output, E_{XH} is the thermal exergy output, E_{XP1} is the power exergy demand of the primary fluid circulation system, and E_{XTES} is the exergy destruction due to heat loss of the thermal storage system to its environment. The insulation level may be another subject of optimization, which is excluded in this model. The symbol E_{XTP} stands for the exergy required for temperature peaking by the unit TP in Tier 4, if necessary, on the demand side. All electrical systems in Tiers 1 to 5 operate on DC electricity without conversion to AC, thus eliminates inverter (IN) losses. All tiers depend upon the volumetric fluid (or gas) flow rate, $\dot{V}$, which must be dynamically controlled and optimized for maximum E_{XT} (EU 2016). The exergy demand for fluid circulation for PV cooling is crucial such that the exergy

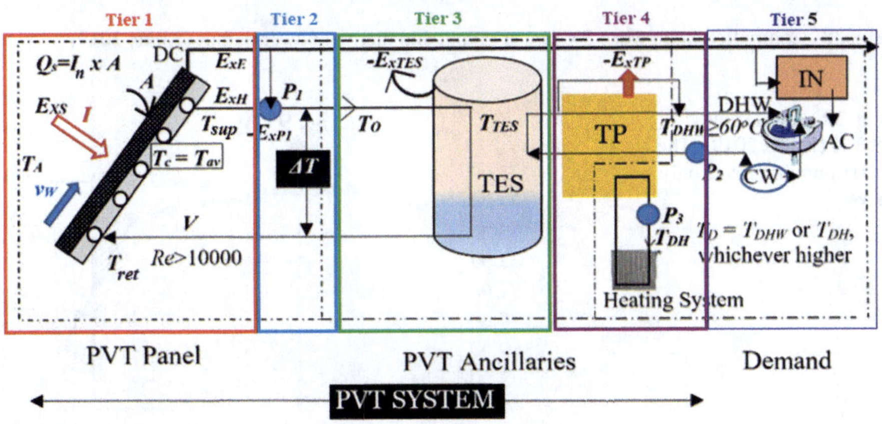

Fig. 37 Holistic model for a conventional PVT system (Kilkis 2019)

gain by preserving PV efficiency may sometimes be less than the exergy demand of the pump.

$$E_{XT} = E_{XE} + E_{XH} - E_{XP1} - E_{XTES} - E_{XTP} \quad \{\text{Maximize}\} \tag{39}$$

Tier 1 Tier 2 Tier 3 Tier 4

<u>For Tier 1:</u>

$$E_{XE} = \eta_{IPV} \times A \times I_n \times 0.95 \tag{40}$$

Figures 38 and 39 show that the maximum total exergy output E_{XT} (74.5 W) is less than the output E_{XPV} that a PV system alone could provide (97 W) (EU 2016). Performance metrics for different solar panels are summarized in Table 1,

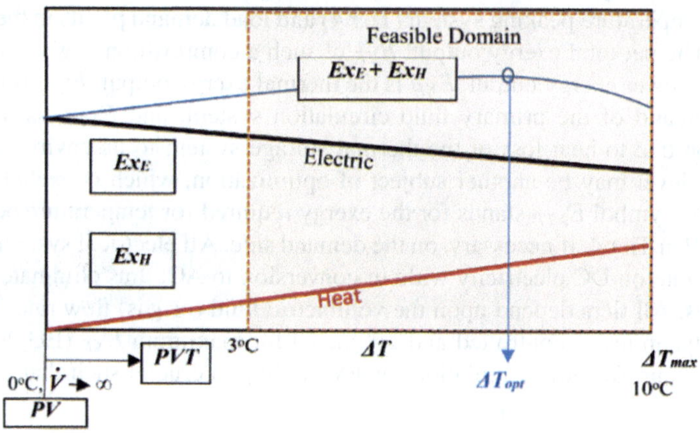

Fig. 38 Contradiction between PV power output and thermal power output (Kilkis 2019)

Fig. 39 A case study, T_{sup}= 293 K (Kilkis 2019). The maximum improves only 4 W

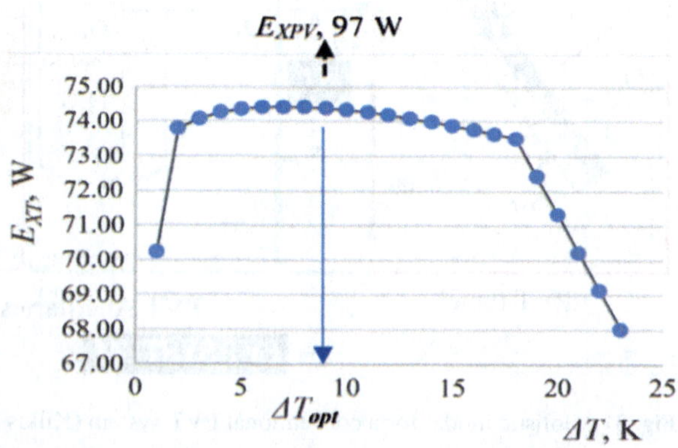

which shows that, if the primary goal is to decarbonize the heating sector, then the obvious choice will be advanced PVT for small applications but for large-scale district applications, the Hydrogen City option is preferable for 2050 targets (Figs. 40, 41, Table 5).

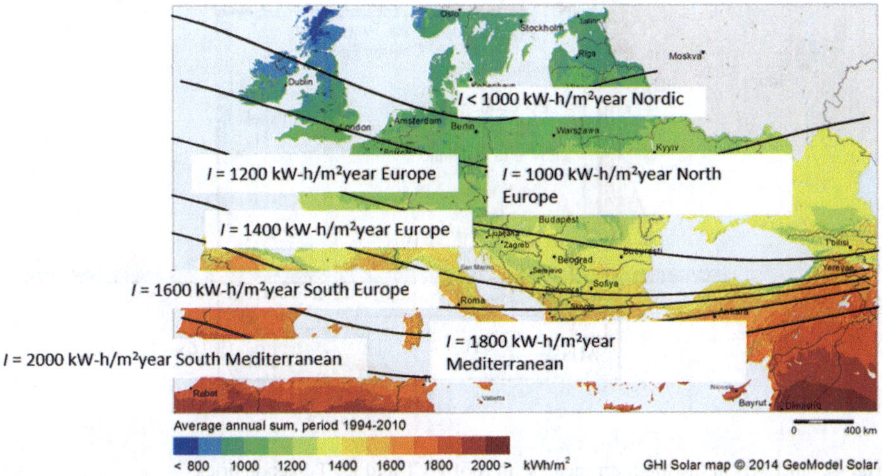

Fig. 40 Feasible regions for PVT technology

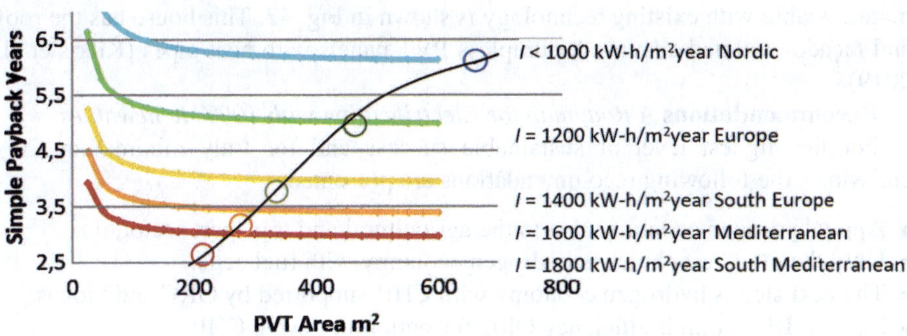

Fig. 41 Payback periods with project size and geography

Table 5 REMM efficiency and total exergy output of different solar panels

REMM efficiency	FPC	PV	PVT[a]	Advanced PVT
Ψ_R	0.24	0.64	0.79	0.88[b]
Total exergy per unit solar input, ε_T	0.09 W/W	0.20 W/W	0.26 W/W	0.35 W/W

[a]Excludes the power exergy demand of the PVT-dedicated circulation pump. Sometimes the net exergy output may be negative

[b]Operated with embedded heat pipes []

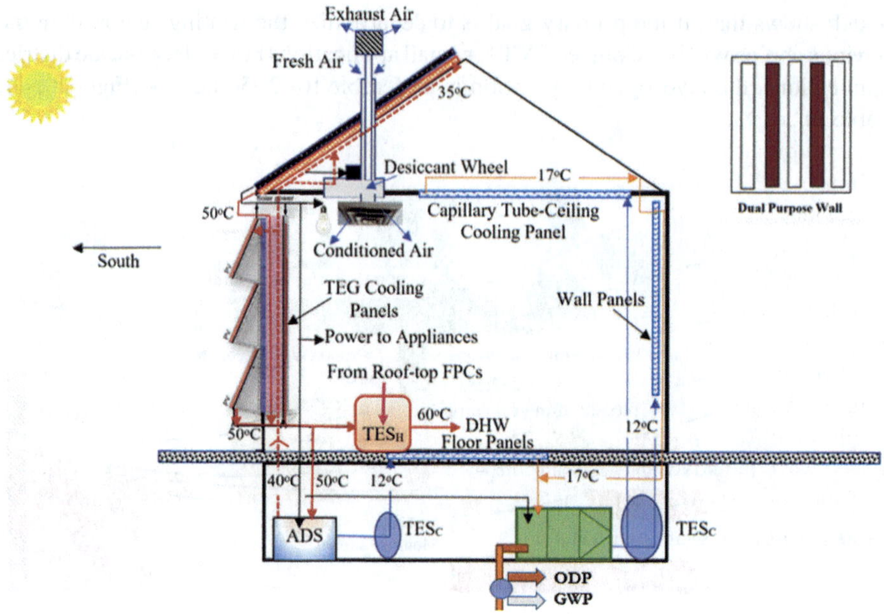

Fig. 42 nZEXB in the mediterranean with roof PVT and PVT Façade brick

Passive Hydrogen House: A net-zero exergy solar house for hot and humid climates, doable with existing technology is shown in Fig. 42. This house has the roof and façade-mounted advanced, pumpless PVT panels with heat pipes (Kilkis et al. 2019).

Recommendations *A Roadmap for Electrification with 100% Renewables*

For the highest level of sustainable success and for truly minimizing CO_2 emissions, the following recommendations are presented.

- Apply hydrogen economy also to the agricultural and transport sector,
- Until then the best choice is hydrogen economy with fuel cells,
- The next step is hydrogen economy with CHP, supported by ORC and biogas,
- Support RD for high-efficiency ORC for bottoming-cycle CHP,
- Never use electricity by direct resistance heating,
- Defer the use of heat pumps until *COP* increases and buildings are low-exergy type,
- If heat pumps need to be used, then support them by solar PV. In milder climates and warmer provinces, PVT systems may be used,
- If office buildings, schools, etc. need cooling and humidity control even in Northern climates at a certain period of the year, adsorption-cycle chillers may be used. This requires optimum sizing of the wind and other renewable energy systems,
- Use CO_2 instead of ozone-depleting refrigerants,

- CO_2 from coal-fired plants must be captured and used/marketed for refrigerant replacements. Indeed, CO_2 is the most viable and safe refrigerant, especially in heating. Therefore, this option becomes especially attractive for the Northern provinces,
- First upgrade buildings to low-exergy and energy-conserving type,
- Improve the efficiency of fuel cells and develop cheaper ones without rare-earth elements,
- Blend wind energy with solar, biogas, and geothermal,
- If feasible collaborate with industrial waste heat,
- Retrofit natural-gas systems with hydrogen,
- Develop optimization programs to optimize the wind power to heat options for given climatic conditions, building typology, load demands, available infrastructure, and size of the number of buildings. Even the building heights matter for minimum emissions.
- The waste heat of power plants must be used in district energy systems.
- If the cogeneration system is preferred over fuel cells, then the bottoming cycle for additional power generation must be used.
- In the transition process of electrification and heating into a hydrogen economy, existing coal and lignite power plants must capture H_2S gas and convert it to hydrogen for use.
- Later, convert district energy systems into hydrogen piping. In order to accomplish that conversion, the piping network must be designed so that they may be convertible to a large extent with less need for upgrading and infrastructure modifications.
- For better multi-variable optimization and system-of-a system-type of approach, new rating parameters must be developed and implemented, like the ones presented herein.

Step 9. Consider Other Sectors

100% Renewable Farm: Besides large-scale solar and wind energy installations, renewable energy systems should be integrated with smaller-scale applications in agriculture and animal farms. Figure 43 shows a 100% renewable system, involving,

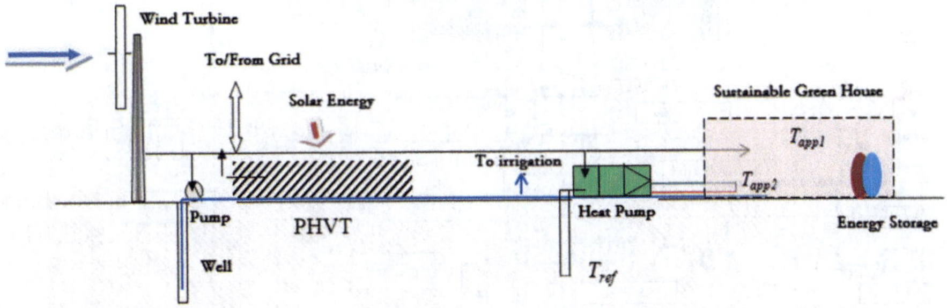

Fig. 43 100% renewable farm and greenhouse (Çolak et al. 2013)

solar, wind and ground heat, where PVT panels are coupled with wind turbines. Renewable electricity drives irrigation pumps. In this case, ψ_R is almost 1 (electricity-to-electricity). Another part of the electricity drives a GSHP for root-zone heating or for greenhouse cooling. Thermal storage both in heat and cold forms is available. PVT panels are cooled by the well water, but not so much that irrigation is not compromised. Computer-controlled, variable-speed pumps are used.

Step 10. Use Alternative Fossil Fuels Wisely-*If You Absolutely Need Them.*

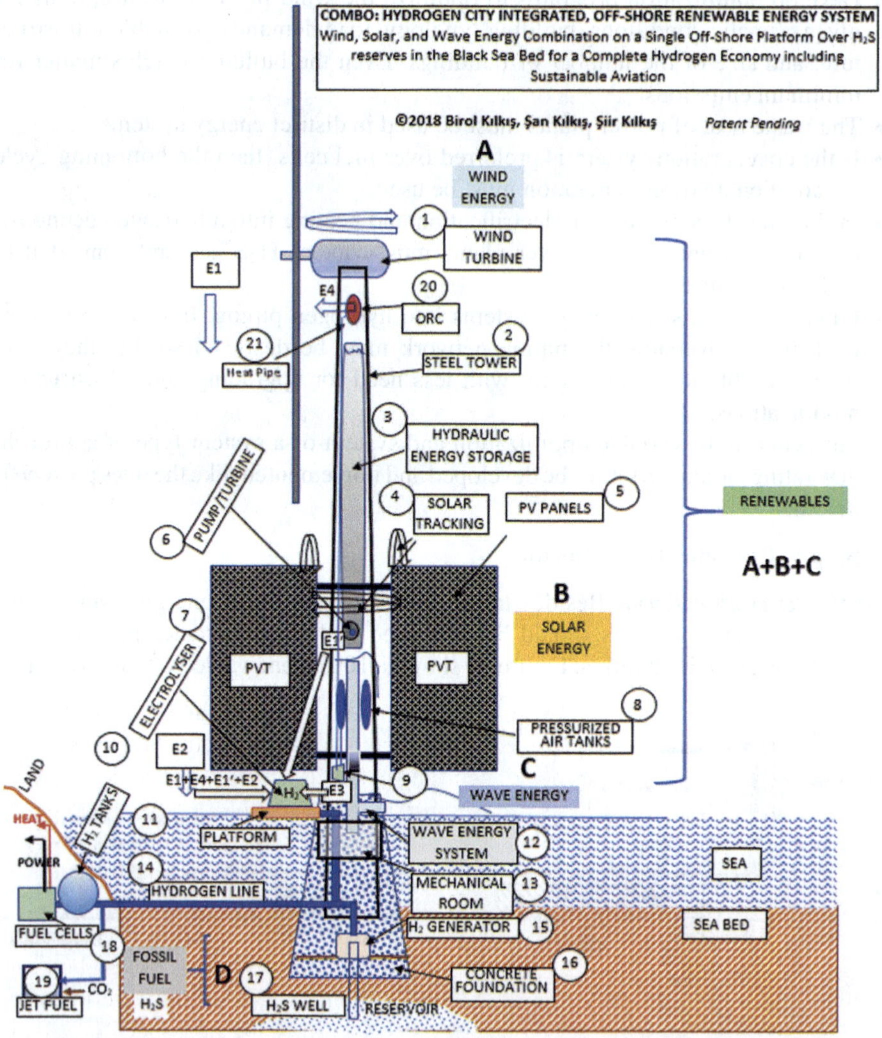

Fig. 44 Off-shore 100% renewable power plant with H_2S gas from the black sea bed

100% Black Sea Off-Shore Power: Fig. 44 shows an off-shore, 100% renewable energy system on a converted old oil drilling platform, which processes the H_2S gas to hydrogen (Kilkis 2019). This design encompasses wave, wind, and solar (PVT) energy systems with ORC bottoming cycle using PVT heat with compressed energy storage. The system converts H_2S to H_2. On the land, part of the hydrogen may be further processed to regular jet fuel with 100% renewables, if H_2 obtained from 100% renewables and wastes:

$$(2n+1)H_2 + nCO = C_nH_{(2n+2)} + nH_2O \tag{41}$$

The hydrogen gas input to this process is sourced from two different sources, namely by direct electrolysis of (sea) water using electricity obtained from renewables at a solar, wind, wave energy off-shore compound platform (SWWCP) in Fig. 44 and from the drilling system of the same SWWCP system tapping the hydrogen sulfide (H_2S) gas deposits, which is abundant in the Black Sea Bed. The following chemical process is used (WEF 2019):

$$2COS + O_2 = 2CO + SO_2 \tag{42}$$

Carbonyl sulfide (COS) is found in many industrial process streams, such as those associated with petroleum refineries and coal gasification plants. Carbon disulfide (CS_2) is a greenhouse gas. This chemical is used to form CO and dry sulfur dioxide (SO_2) (WEF 2019). Thus, an added value is generated while greenhouse gas emissions are sequestered from the industry (Fig. 45).

$$2CS_2 + 5O_2 = 2CO + 4SO_2 \tag{43}$$

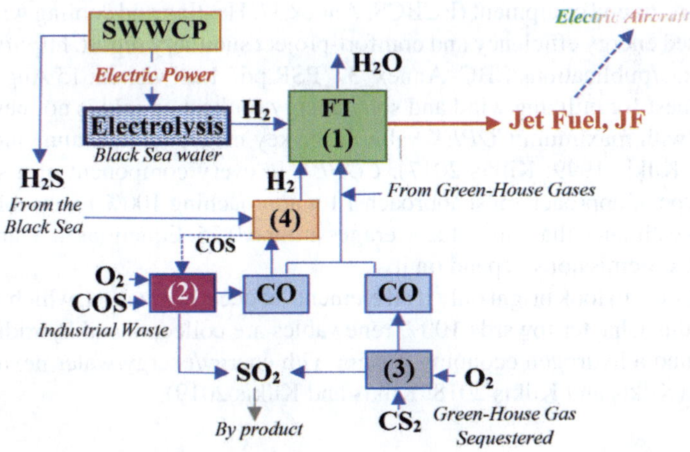

Fig. 45 Jet fuel but 100% renewable (Kılkış et al. 2019)

$$CO + H_2S = H_2 + COS \tag{44}$$

8 Conclusions

Exergy is both a game-changer and a game maker in the name of the 2nd Law. It first changes almost all games in the built environment, by pointing out that global warming, ozone depletion, climatic disasters are the prime responsibilities of human-made systems and equipment in the built environment. Many every-day applications have been shown hereby that they generate even more avoidable CO_2 emissions than direct emissions, which we can observe and calculate with the 1st Law. Maybe this unawareness makes exergy an unwanted term by many, despite the fact that it also shows ways and means to avoid exergy blunders, which are exemplified here. If a 100% renewables target needs to be achieved in a decoupled format of sustainable development and harmful emissions, exergy is the only rational way.

If we have any hope of keeping climate change within safe boundaries, global emissions need to fall to zero within the next three decades. That was the message of the Intergovernmental Panel on Climate Change in 2018. So just how far have we got to go? So far only seventeen nations and 34 major companies are planning to or have already set targets to reach net-zero (WEF 2019). That is how far we can go with the 1st Law techniques, at the face value of 100% renewables target. This chapter has shown that with quantified discussions and examples, that why the 2nd law is a game maker. In order to put the 2nd Law in sustainable actions before 2050, district energy systems must be implemented beyond their 4th generation (> 4DE) (Kilkis 2016). This approach starts with LowEx buildings (Kılkış 2012) and low-exergy systems and equipment (ECBCS, Annex 37. Heating and cooling with a focus on increased energy efficiency and comfort-project summary report. http://www.iea-ebc.org/Data/publications/EBC_Annex_37_PSR.pdf. Last visited: 15 Aug 2019). In fact, the quest for utilizing wind and solar energy for heat or cold is not new, but the right way with maximum *COPEX* value is the key metric for optimum and rational solutions (Kilkis 1999; Kilkis 2017). *COPEX* in every component of a system of systems type of approach must approach 1 for approaching 100% renewables target. Today it is estimated that this value averages around 0.25. Equations 4, 5, and 6 show that total CO_2 emissions depend on it.

The future will look bright only if all elements of decarbonization, which are exemplified in this Chapter towards 100% renewables are collected and hybridized optimally around a hydrogen economy system with energy/exergy/water nexus, shown in Fig. 46 (Kılkış and Kılkış 2018; Kılkış and Kılkış 2019).

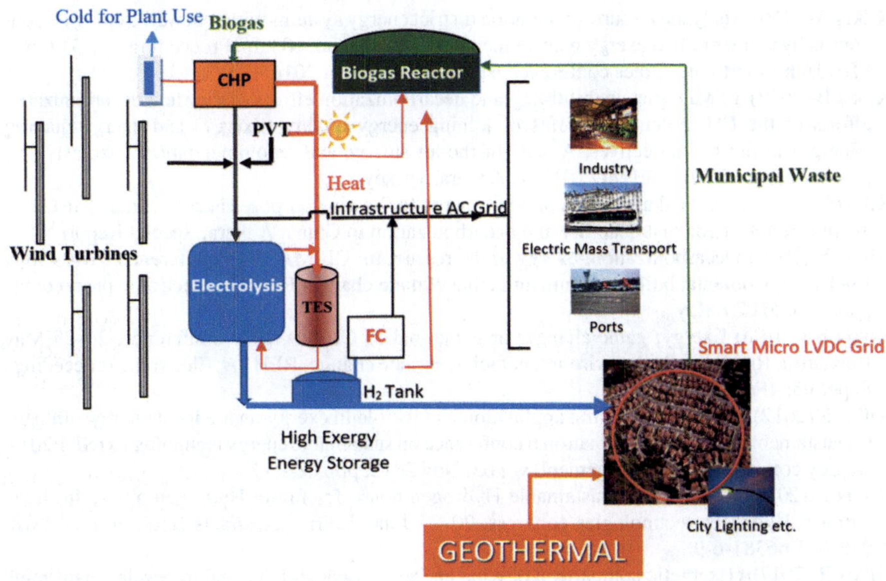

Fig. 46 A 100% renewables system of systems approach with hydrogen economy

References

Çolak L, Kılkış B, Yaltay CE, Dalarslan M (2013) Solar assisted pumping by a PHVT system in rural regions, 2013 ISES solar world congress. Elsevier, Energy Procedia

Cowan D, Gartshore J, Chaer I, Francis C, Maidment G (2009) IOR. Real zero- reducing refrigerant emissions and leakage-feedback from the IOR project, Proc Inst R 2009–10, pp 7–1 to 7–16

EU (2016) Communication from the commission to the European parliament, The council, The European economic and social committee and the committee of the regions. An EU strategy on heating and cooling, SWD (2016) 24 Final, Brussels, 16 Feb 2016 COM (2016) 51

ECBCS, Annex 37. Heating and cooling with a focus on increased energy efficiency and comfort-project summary report. http://www.iea-ebc.org/Data/publications/EBC_Annex_37_PSR.pdf. Last visited: 15 Aug 2019

EEA (2018) Air quality in Europe-2018, Report no: 12/2018

EIA (2018) US energy information administration, Monthly energy review, Table 2.1, 2018

EPA (2019) Basic ozone layer science. https://www.epa.gov/ozone-layer-protection/basic-ozone-layer-science. Last visited 14 Aug 2019

IEA (2018) Key world energy statistics 2018. https://webstore.iea.org/key-world-energy-statistics-2018. Last visited 14 Aug 2019

Jacobson MZ (2008) Effects of wind-powered hydrogen fuel cell vehicles on stratospheric ozone and global climate. Geophys Res Lett 35:L19803. https://doi.org/10.1029/2008GL035102

Kilkis İB (1999) Utilization of wind energy in space heating and cooling with hybrid HVAC systems and heat pumps. Energy Build 30:147–153

Kilkis BI (2000) Rationalization and optimization of heating systems coupled to ground-source heat pumps, ASHRAE Trans vol 106, Pt. 2, 817–822

Kilkis S (2011) A rational exergy management model to curb CO_2 emissions in the exergy-aware built environments of the future. In: Doctoral thesis, division of building technology school of architecture and the built environment. KTH Royal Institute of Technology, Stockholm, Sweden

Kilkis B (2016) Analysis of fourth-generation district energy systems with renewable energy cogeneration by using rational exergy management model, paper no. 103. In: Proceedings, pp 512–522, SBE 16th Smart metropoles conference Istanbul, 13–15 Oct 2016
Kilkis B (2017) TTMD, sustainabilitbility and decarbonization efforts of they and decarbonization efforts of the EU: potential benefits of joining energy quality (Exergy) and energy quantity (Energy) in the EU Directives, A state-of-the-art survey, and recommendations, exclusive EU position report ©2017, TTMD 2017–1, Ankara, Turkey
Kilkis B (2019c) Rational utilization of wind energy for heating purposes in cold climates in China, Technical note 1 for sustainability and decarbonization in China. Ankara, Special Report
Kilkis B (2019) Decarbonization: exergy to the rescue. In: CLIMA 2019 conference, 26–29 May, Bucharest, Romania, built environment facing climate change: REHVA, electronic proceedings, paper no: 5H/29 May
Kilkis B (2019a) Exergy: game changer or game maker, CLIMA 2019 conference, 26–29 May, Bucharest, Romania, Built environment facing climate change: REHVA, Electronic proceedings, Paper no: 4F/28 May
Kılkış Ş (2012) A net-zero building application and its role in exergy-aware local energy strategies for sustainability.In: 10th international conference on sustainable energy technologies (SET 2011) Energy conversion and management, vol 63, Nov 2012, pp 208–217
Kilkis B (2019g) Design of a sustainable Hydrogen house for future Hydrogen cities, 4th International Hydrogen technologies congress, 20–23 June. Edirne, Abstracts Book on line. ISBN 978-605-66381-6-9
Kilkis B (2017b) Exergetic comparison of wind energy storage with ice making cycle versus mini-hydrogen economy cycle in off-grid district cooling. IJHE 42(28):17571–17582
Kilkis B (2019) Development of a composite PVT panel with PCM embodiment, TEG modules, flat-plate solar collector, and thermally pulsing heat pipes. Sol Energy J, In print
Kilkis B (2020) Exergy-optimum coupling of heat recovery ventilation units, with heat pumps in sustainable buildings. JSDEWES, In print
Kilkis B (2019b) An exergy-based holistic urban development model for the distance limit between a renewable energy plant and its low-exergy district, 14th SDEWES 2014 conference, 1–6 Oct 2019. Dubrovnik, Croatia
Kılkış B, Kılkış Ş (2018) Hydrogen economy model for nearly net-zero cities with exergy rationale and energy-water nexus. Energies, 11(5)
Kılkış Ş, Kılkış B (2019) An urbanization algorithm for districts for minimized emissions based on urban planning and embodied energy towards net-zero exergy targets. Energy 179:392–406
Kilkis B, Kilkis Ş, Kilkis Ş (2017) Optimum Hybridization of wind turbines, heat pumps, and thermal energy storage systems for near zero-exergy buildings (nZEXB) using rational exergy management model, paper no. 2. In: 12th IEA heat pump conference, 15–18 May, Rotterdam 2017. Papers online, also, abstracted in print, pp 179–180
Kilkis B, Kilkis Ş, Kilkis Ş (2019) PV and TEG integrated PVT system with PCM layer for solar power and heat generation, PCT/TR2018/050382 18 07 2018. Application no: 2017/10622
Kılkış B, Kılkış Ş, Kılkış Ş (2019) A simplistic flight model for exergy embodiment of composite materials towards nearly-zero exergy aviation. Int J Sustain Aviat 5(1):19–41
NASA (2019) NASA ozone watch. https://ozonewatch.gsfc.nasa.gov/. Last visited: 14 Aug 2019
OECD/IEA (2018) Energy efficiency indicators-highlights, 191 p, Paris, France
REN (2019) REN21 renewables 2019 global status report. https://www.ren21.net/reports/global-status-report/. Last visited 14 Aug 2019
RHC (2019) RHC-ETIP position paper on horizon Europe work program, 22 July 2019. https://www.rhc-platform.org/rhc-etip-position-paper-on-horizon-europe-work-programme/. Last visited 14 Aug 2019
Science Europe (2016) In a resource-constrained world: Think Exergy, not Energy. https://www.scienceeurope.org/media/0vxhcyhu/se_exergy_brochure.pdf. Last visited: 14 Aug 2019
Universe Today (2019) What is the Earth's Average Temperature? <https://www.universetoday.com/55043/earths-temperature/> Last visited: 15. 08. 2019

Ulutürk Y. Şahin C, Gürler İ (2019) Geothermal energy district heating systems: afyon karahisar case (Main body of the text is in Turkish with English abstract). In: 14th National sanitary engineering congress, TESKON, 17–20 April 2019, Izmir, Geothermal Energy Seminar Proceedings, pp 229-237

US DOE EERE (2015) A common definition for zero energy buildings, Prepared for the U.S. Department of energy by the national institute of building sciences, 22 p, Sept 2015

WEF (2019) Zero by 2050: How the world's economy has planned to battle climate change. https://www.weforum.org/agenda/2019/07/zero-emissions-target-climate-change-impact/?fbclid=IwAR3bdOqyXZn0V8R9tRnrH2uYWeSICYRfwK8hjn5wxMmtJr4F4JkWL96fmFg. Last visited: 15 Aug 2019

Role of IRENA for Global Transition to 100% Renewable Energy

Elisa Asmelash, Gayathri Prakash, Ricardo Gorini and Dolf Gielen

Abstract The global energy transformation is more than a simple transformation of the energy sector—it is a multi-faceted transformation of our societies and economies. The transition towards a decarbonised global energy system can be realised much more cost-efficiently than previously thought due in part to the rapidly falling costs of renewable energy technologies. This chapter highlights the urgent need for an accelerated energy transition to 2050. Since the signing of the Paris Agreement in 2015 and despite the growth of renewable energy technologies, energy-related CO_2 emissions have risen by around 4%. In this context, the next years and decades are critical and the revisions of the NDCs in 2020 in combination with Long-term Strategies must yield a convincing outcome for an energy transition that puts the world on a global pathway to reduced emissions. Technologies for these systems are available today, are deployable and cost-competitive at a large scale and there are quite a lot of studies exploring 100% RE scenarios, indicating that it is clearly topic of growing interest. On this context, the analysis—part of International Renewable Energy Agency's (IRENA) latest global energy transformation roadmap details an energy transition pathway for the global energy-system to meet the Paris Agreement of "well-below 2 °C". By 2050, renewable energy in the power sector could reach 86%, while representing two-third of total primary energy supply mix. The pace of energy transition can be ramped up by several inter-related factors, ranging from technologies to socio-economics, to institutional drivers and different forms of finance.

Keywords Energy transition · Renewables · Decarbonisation · Sustainable development and energy policy

---

E. Asmelash (✉) · G. Prakash (✉) · R. Gorini · D. Gielen
International Renewable Energy Agency (IRENA), Innovation and Technology Centre (IITC), Willy-Brandt-Allee 20, 53113 Bonn, Germany
e-mail: easmelash@irena.org

G. Prakash
e-mail: gprakash@irena.org

T. S. Uyar (ed.), *Accelerating the Transition to a 100% Renewable Energy Era*, Lecture Notes in Energy 74, https://doi.org/10.1007/978-3-030-40738-4_2

1 Why 100% RE Systems

1.1 Rationale and Drivers

The Intergovernmental Panel on Climate Change (IPCC) *Special Report on Global Warming of* 1.5 °C (SR1.5), estimates that human activities have already caused approximately 1.0 °C of global warming above pre-industrial levels. Unless significant counter measures are taken, global warming would not be limited to/stabilise at 1.5 °C between 2030 and 2052 (IPCC Special Report on Global Warming of 1.5 °C 2018).

Pathways limiting global warming to 1.5 °C require rapid and extensive transitions in all sectors (i.e., energy, agriculture, urban infrastructure and buildings, transportation, and industrial systems). These system transitions are unique in terms of scale and more pronounced in terms of speed and they require cross-sector emissions reductions, a wide portfolio of mitigation options and a significant increase in investments. In addition, efforts to limit warming to 1.5 °C are closely linked to sustainable development, which balances social well-being, economic prosperity and environmental protection.

Reducing energy-related CO_2 emissions is the heart of the energy transition. Rapidly shifting the world away from the consumption of fossil fuels causing climate change toward cleaner, renewable forms of energy is key if the world is to reach the agreed-upon climate goals. There are many drivers behind this transformation.

Firstly, **falling costs of renewable energy (RE)**, which have continued to decline rapidly. As an example, electricity costs from utility scale solar photovoltaic (PV) projects since 2010 has been remarkable—between 2010 and 2018 the global weighted-average levelized cost of energy (LCOE) of solar PV declined to 77%. With the right regulatory and institutional frameworks in place, recent record low auction prices for solar PV in Dubai, Mexico, Peru, Chile, Abu Dhabi and Saudi Arabia have shown that an LCOE of USD 0.03/kWh is possible in a wide variety of national contexts (IRENA Renewable Power Generation Costs in 2018 2019). Similarly, in Europe, offshore wind can now compete at market prices, while in the US, non-hydroelectric renewable energy resources such as solar PV and wind are expected to be the fastest growing source of electricity generation in the next two years.

Secondly, **air quality improvements**. Air pollution is a major public health crisis, mainly caused by unregulated, inefficient and polluting energy sources, namely fossil fuels. The switch to clean renewable energy sources would bring a greater prosperity, improving the air quality in cities, preserving and protecting the environment.

Thirdly, **reduction of carbon emissions**. The transformation of the global energy system needs to accelerate substantially to meet the objectives of the Paris Agreement, which aim to keep the rise in average global temperatures "well below" 2 °C and ideally to limit warming to 1.5 °C in the present century, compared to pre-industrial levels.

Finally, transforming the global energy system will also **improve energy security and enhance affordable and universal energy access**. For countries heavily dependent on imported fossil fuels energy security is a significant issue, and renewables can provide an alternative by increasing the diversity of energy sources through local generation and thus contribute to the flexibility of the system and resistance to shocks. Similarly, energy access is an area of great inequality and renewable energy technologies can be adopted and applied in the rural areas where the national grid has not yet been extended through rural electrification, energy community projects, as well as through distributed renewable energy resources (DER).

Also, The United Nations Sustainable Development Goals (SDGs), adopted in 2015, provide a framework for assessing links between global warming of 1.5 °C or 2 °C and development goals including poverty eradication and reducing inequalities (IPCC, 2018). SDG 7, which calls for ensuring "access to affordable, reliable, sustainable and modern energy for all" by 2030, has a strong connection with the majority of SDGs, illustrating how energy is central to fostering the pathways necessary to keep the world well below 2 °C of warming and meet a wide range of SDG targets.

The decarbonisation of the energy sector and the reduction of carbon emissions to limit climate change is at the heart of IRENA's energy transition series, which examines and provides accelerated and feasible low carbon technology deployment pathway towards a sustainable and clean energy future (IRENA Transforming the energy system and holding the line on the rise of global temperatures 2019) (Fig. 1).

1.2 Mixed Progress in the Energy Transition

Despite clear evidence of human-caused climate change, support for the Paris Agreement on climate change, and the prevalence of clean, economical and sustainable energy options, the world is still not on track and efforts and progress are still well below the levels needed. Indeed, the world is starting from a baseline that it is still far away from what is needed for the decarbonisation of the energy sector. Recent trends are also not encouraging, as they show slow progress and slow improvements towards the final objective.

The Fig. 2 summarises the need for acceleration by looking at five key indicators, namely: (i) renewable energy share in power generation; (ii) total final energy consumption per capita; (iii) share of electricity in final energy consumption; (iv) emissions per capita; and (v) energy intensity improvement rate.

Some indicators do not show positive trends. In the period from 2010 and 2016, the share of renewable energy in final energy consumption stayed at roughly the same levels, electrification of final uses of energy has mostly stagnated, global CO_2 emissions from the energy sector increased by almost 13% and estimates indicate that emissions continued to rise and may have reached a new record high of 34.3 Gt CO_2 in 2018 (Carbon Brief Analysis: Fossil-Fuel Emissions in 2018 Increasing at Fastest

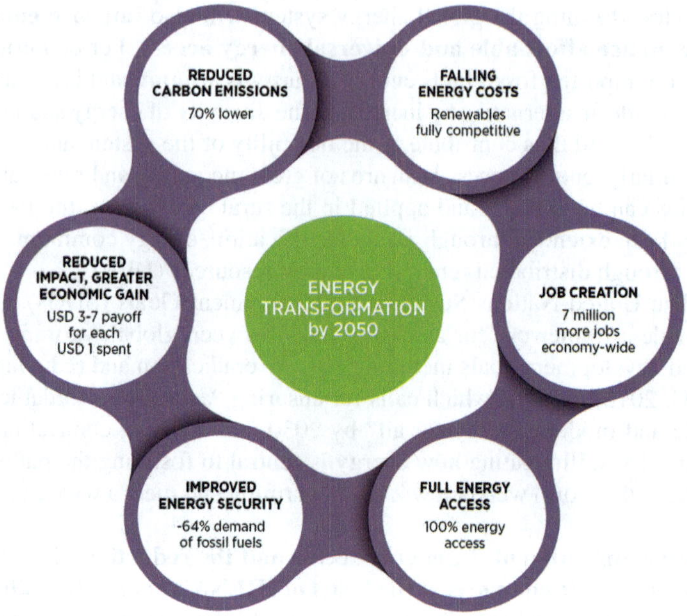

Fig. 1 Needs and opportunities

Rate for Seven Years 2018). In addition, investment in renewable energy declined in 2017 after several years of growth (IEA World Energy Investment 2018). Despite an increase in investment in energy efficiency, the combined investment in renewable energy and energy efficiency showed a slight reduction of 3% in 2017, compared to the previous year. That is unfortunate in a world where a strong acceleration in investments in energy efficiency and renewable energy is needed. Partly because of that decline and partly due to a modest increase in fossil fuel investment, the share of investment in fossil fuels in the energy supply increased in 2017 (IEA World Energy Investment 2018).

However, **despite the very slow progress, there are two positive trends. First, in the power sector, the share of renewable energy in electricity generation has been increasing steadily**. Renewable electricity generation share increased from around 20% to nearly 24% from 2010 to 2016 (or 3.1% per year on average) (IRENA Hydrogen from Renewable Power: Technology Outlook for the Energy Transition 2018). An estimate for 2018 indicates a further increase to 26%. **The second positive sign is the consistent improvement in the energy intensity of GDP**.

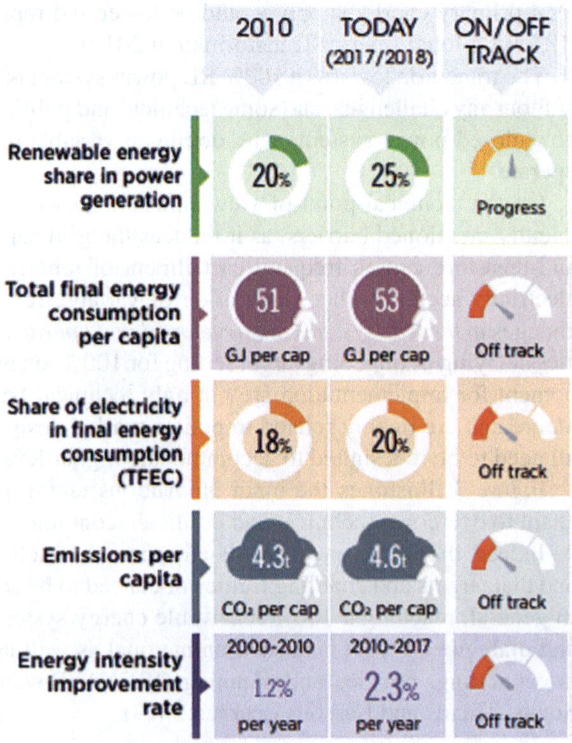

Fig. 2 Summary of recent trends and required levels of selected indicators

1.3 What Exactly is 100% Renewable Energy: Implications for Supply and Demand Sectors

There are quite a lot of studies exploring 100% RE scenarios, indicating that it is clearly a topic of growing interest. However, there seems to be some confusion between 100% RE in energy supply mix and 100% RE in power generation mix, which are indeed very different concepts. Having 100% RE in energy supply mix implies the complete phase out of fossil fuels in the complete energy sector (including power, transport, buildings, industry) and the creation of an energy system that runs entirely on renewable energy sources. On the other hand, 100% RE in power means that the entire generation of electricity will be covered by renewable energy sources, while fossil fuels could only be used as back-up in extreme circumstances or even not used with emerging low-carbon technologies such as hydrogen and other flexibility measures like energy storage.

Given the above, IRENA sees possibilities for many more countries with 100% RE power in the coming decades but for many countries 100% RE energy is unlikely for economic and technical reasons. In fact, according to IRENA's latest global REmap analysis, renewable energy in the power sector could reach levels towards 100%, representing 86% of the total power generation by 2050, while renewable energy in total primary energy supply would be lower and represent 67% of the total supply (IRENA Global Energy Transformation 2019).

The transition towards a 100% RE power system is not smooth and does not come without any challenges, and some technical and political uncertainties also remain on how closed power systems with dominant variable renewable energy shares would operate.

On the technical point of view, *inflexible power system* is among the most frequently mentioned barriers, as it reduces the grid capacity available for renewables and therefore causes frequent curtailment of renewable energy. Moreover, for the electricity sector, higher penetration of variable renewables requires a number of changes in *the way systems are developed and operated*. However, barriers and obstacles delaying or impeding target setting for 100% renewable energy and policy development for implementation are not only technical but also of a political/economic nature and are mainly related to policy, market design and business models, which all need to be redesigned to accommodate higher levels of VRE.

Figure 3 illustrates the main innovations taking place in the electricity supply chain to overcome technical and political/economic challenges.

Indeed, there is no one-size-fits-all approach to achieving 100% renewable energy, and that targets and enabling frameworks need to be adjusted to local circumstances. In general, reaching a 100% renewable energy system will require further analysis and dialogue on what is needed on national as well as sub-national level regarding target setting, policies and planning (IRENA Towards 100% Renewable Energy: Status, Trends and Lessons Learned 2019).

2 How to Achieve 100% Renewable Energy System: IRENA's Pathway for Global Transition to Renewable Powered Future

2.1 IRENA's Energy Transition Study

The global energy transformation is more than a simple transformation of the energy sector—it is a multi-faceted transformation of our societies and economies. As such, the direction and future shape of a Paris Agreement compatible energy system will be determined by several inter-related factors, ranging from technologies to socio-economics, to institutional drivers and different financial instruments.

The findings in this section are based on latest analysis conducted by the International Renewable Energy Agency (IRENA)'s "Global Energy Transformation—A Roadmap to 2050", which details an energy transition pathway for the global energy-system to meet the Paris Agreement aim of "well-below 2 °C".

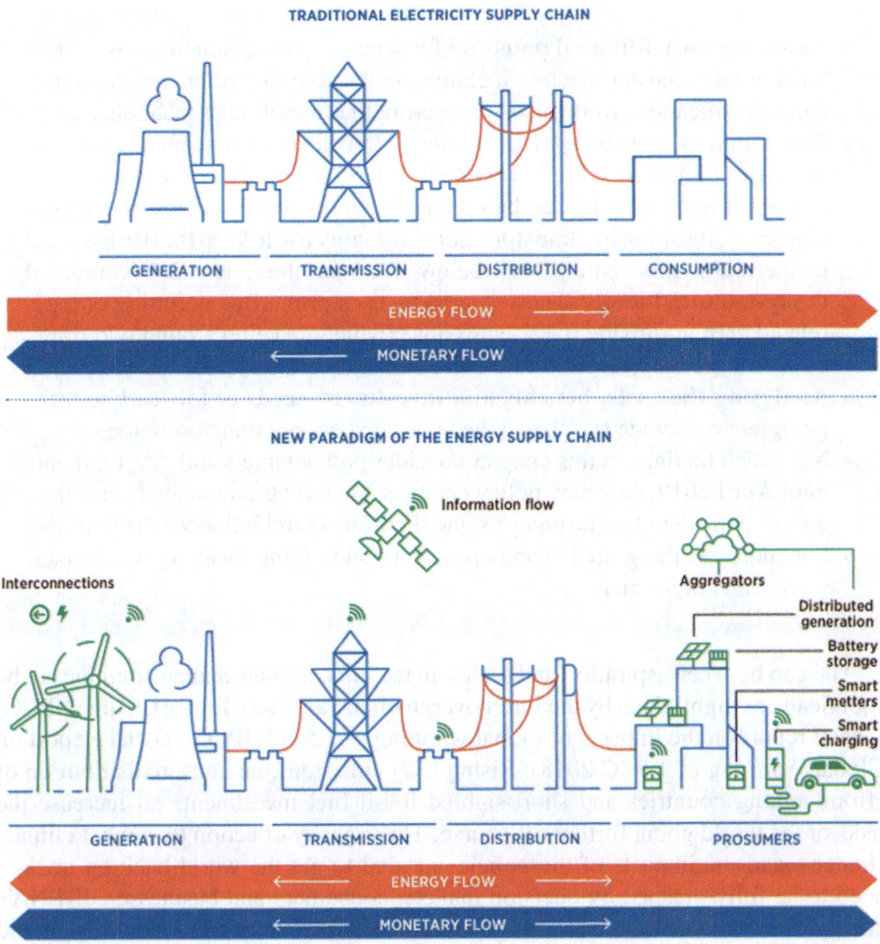

Fig. 3 Innovations taking place in the electricity supply chain

Box 1: Practical Options for Global Energy Decarbonisation

IRENA's renewable energy roadmap, or REmap approach [1] and analysis, includes several key steps:

- Identifying the current plans for global energy development as a baseline scenario (or Reference Case) as far as 2050. This presents a scenario based on governments' current energy plans and other planned targets and policies, including climate commitments made since 2015 in Nationally Determined Contributions under the Paris Agreement.

- **Assessing the additional potential** for scaling up or optimising low-carbon technologies and approaches, including renewable energy, energy efficiency and electrification, while also considering the role of other technologies.
- Developing a realistic, practical energy transformation scenario, referred to as **the "REmap Case"**. This calls for considerably faster deployment of low-carbon technologies, based largely on renewable energy and energy efficiency, resulting in a transformation in energy use to keep the rise in global temperatures this century well below 2°C and closer to 1.5°C compared to pre-industrial levels. The scenario focuses primarily on cutting energy-related carbon-dioxide (CO_2) emissions, which make up around two-thirds of global greenhouse gas emissions.
- **Analysing the costs, benefits and investment needs** of low-carbon technologies worldwide to achieve the envisaged energy transformation.
- Note: The findings in this chapter consider policy targets and developments until April 2019. Any new policy changes and targets announced since then are not considered in the analysis and therefore could influence the findings.
- For more on the global roadmap and its underlying analysis, see https://www.irena.org/remap.

The gap between aspiration and reality in tackling climate change continues to be significant, as highlighted by the Intergovernmental Panel on Climate Change (IPCC) special report on the impacts of global warming of 1.5 °C (IPCC Special Report on Global Warming of 1.5 °C 2018). Rising CO_2 emissions, an uneven distribution of efforts among countries and short-sighted fossil fuel investments all increase the risks of the world going further off course. The urgency of action to combat climate change—and the impacts of the policies needed to get the world back on track—need to be fully grasped by decision makers, consumers and businesses. IRENA's energy transition study shows that global fossil fuel production under current and planned polices of the Reference Case will peak between 2030 and 2035; whereas for a pathway aligned with the Paris Agreement goals, the peak would need to occur in 2020.

In 2017 and 2018 energy-related CO_2 emissions rose, driven largely by increased use of fossil fuels; on average, energy-related CO_2 emissions have risen around 1.3% annually over the last five years (Carbon Brief Analysis: Fossil-Fuel Emissions in 2018 Increasing at Fastest Rate for Seven Years 2018). **If governments' long-term plans, including their Nationally Determined Contributions (NDCs), were followed, annual energy-related CO_2 emissions will decline only slightly by 2050, and will put the world on track for at least 2.6 °C of warming by mid-century, and higher warming after**. Based on a carbon budget from the latest IPCC special report on the impacts of global warming of 1.5 °C (IPCC Special Report on Global Warming of 1.5 °C 2018), the Reference Case indicates that, under current and

[1] https://irena.org/remap.

planned policies, the world will exhaust its energy related CO_2 emissions budget in 10–18 years

To set the world on a pathway towards meeting the aims of the Paris Agreement, energy-related CO_2 emissions would need to be scaled back by at least an additional 400 gigatonnes (Gt) by 2050 compared to the Reference Case; in other words, **annual emissions would need to be reduced by around 3.5% per year from now until 2050 and continue afterwards**. Energy-related emissions would need to peak in 2020 and decline thereafter. By 2050 energy-related emissions would need to decline by 70% compared to today's levels. While the REmap analysis is focused only on energy-related CO_2 emissions, additional efforts are needed to reduce emissions in non-energy use (such as using bioenergy and hydrogen feedstocks); industrial process emissions; and efforts outside of the energy sector to reduce CO_2 emissions in agriculture and forestry (Fig. 4).

IRENA's REmap Case presented in this chapter outlines an aggressive, yet technically feasible and economically beneficial, route for accelerated climate action. It shows that the accelerated deployment of renewables, combined with deep electrification and increased energy efficiency, can achieve over 90% of the energy-related CO_2 emissions reductions needed by 2050 to reach the well-below 2 °C

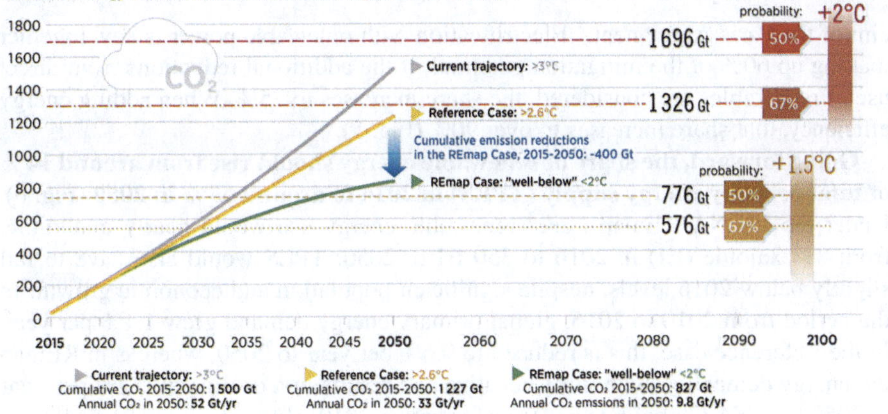

Fig. 4 REmap offers a pathway for a well-below 2 °C climate target, towards 1.5 °C. *Notes* (1) Taking into account 2015–2017 emissions on top of the budget provided in IPCC (2018) (Table 2.2—with no uncertainties and excluding additional Earth system feedbacks); (2) Budgets exclude industrial process emissions of 90 Gt; for this study, the assumption is that CO_2 emissions from land use, land-use change and forestry (LULUCF) fall from 3.3 Gt in 2015 to zero by mid-century. LULUCF subsequently becomes a net absorber of CO_2 over the remainder of the 21st century, and, as a result, cumulative CO_2 emissions from LULUCF between 2015 and 2100 are close to zero; (3) Current trajectory shows the recent historical trend line, assuming the continuation of the annual average growth in energy-related CO_2 emissions from the last five years (2013–2018) of 1.3% compound annual growth up to 2050; (4) Emissions budgets represent the total emissions that can be added into the atmosphere for the period 2015–2100 to stay below 2 or 1.5 °C at different confidence levels (50 or 67%) according to the IPCC (2018) report

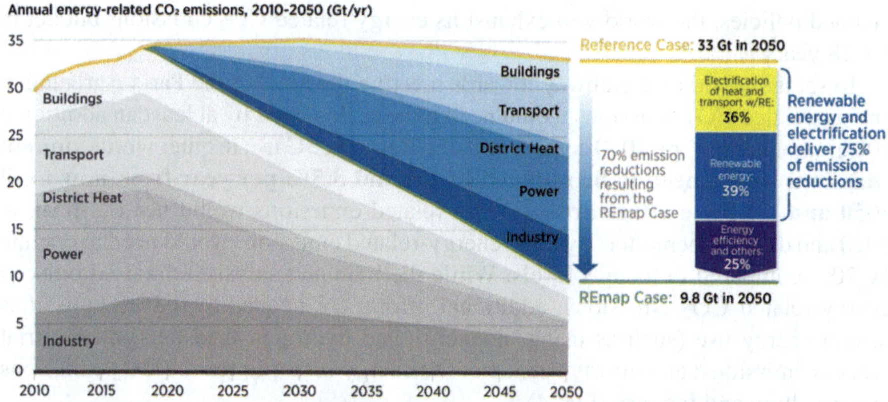

Fig. 5 Renewables and energy efficiency, boosted by substantial electrification, can provide over 90% of the necessary reductions in energy-related carbon emissions. *Note* "Renewables" implies deployment of renewable technologies in the power sector (wind, solar PV, etc.,) and end-use direct applications (solar thermal, geothermal, biomass). "Energy efficiency" contains efficiency measures deployed in end-use applications in industry, buildings and transport sectors (e.g., improving insulation of buildings or installing more efficient appliances and equipment). "Electrification" denotes electrification of heat and transport applications, such as deploying heat pumps and EVs

aim of the Paris Agreement.[2] Electrification with renewable power is key, together making up 60% of the mitigation potential; if the additional reductions from direct use of renewables are considered, the share increases to 75%. When adding energy efficiency, that share increases to over 90% (Fig. 5).

Going forward, the share of renewable energy should rise from around 14% of total primary energy supply (TPES) in 2016 to around 65% in 2050 (Fig. 6). Under the IRENA REmap Case[3] renewable energy use would nearly quadruple, from 81 exajoule (EJ) in 2016 to 350 EJ in 2050. TPES would also have to fall slightly below 2016 levels, despite significant population and economic growth. In the period from 2010 to 2016, global primary energy demand grew 1.1% per year. In the Reference case, this is reduced to 0.6% per year to 2050, whereas in REmap the energy demand growth turns negative and results in a decline of 0.2% per year to 2050 (IRENA Global Energy Transformation 2019) (Fig. 6).

[2]According to the IPCC, 67% 2 °C up to 1.326 Gt; the REmap case, with 827 Gt by 2050 is well below the 2 °C pathway, and towards the 50% 1.5 °C. More information about the carbon budget, and assumptions for non-energy greenhouse gas emissions, is included in the full report available online at www.irena.org.

[3]This analyses the deployment of low-carbon technologies, largely based on renewable energy and energy efficiency, to generate a transformation of the global energy system which for the purpose of the REmap analysis has the goal of limiting the rise in global temperature to below 2 °C above pre-industrial levels by the end of the century (with a 66% probability). For more information about the REmap approach and methodology, please visit www.irena.org/remap/methodology.

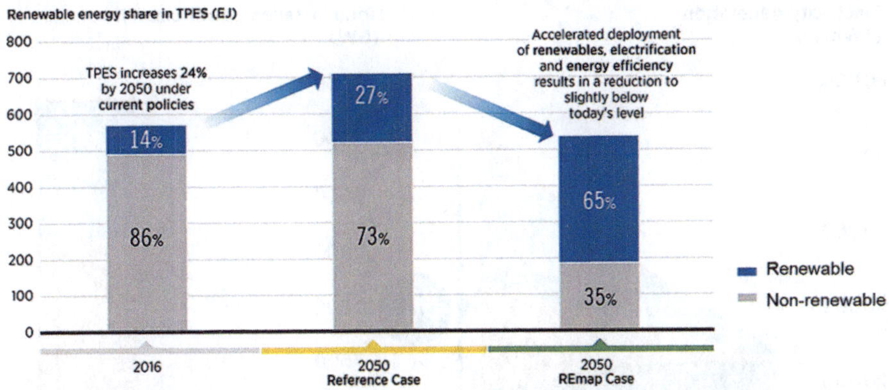

Fig. 6 The global energy supply must become more efficient and more renewables

2.2 *Pathway for the Electricity Sector: Towards 100% Renewables Power*

Delivering the energy transition at the pace and scale needed would require the almost complete decarbonisation of the electricity sector by 2050. This can largely be achieved by using renewables, increasing energy efficiency and making power systems more flexible. Under the REmap Case, electricity consumption in end-use sectors would increase 130% by 2050, to over 55 000 TWh, compared to 2016. By 2050, the share of renewable energy in generation would be 86%, up from an estimated 26% in 2018. Meanwhile, the carbon intensity of electricity generation would decline by 90%.

By 2050, variable renewable energy, mainly wind and solar PV, would account for three-fifth of total global electricity generation rising from their current shares of 7 % and 3 %, to 35 % and 25 % respectively (Fig. 7). These sources would lead the way for the transformation of the electricity sector, rising from around 564 GW of wind capacity and 480 GW of solar PV in 2018 to over 6 000 GW and 8 500 GW by 2050, respectively. In addition, strong growth in geothermal, bioenergy and hydropower would be seen as well (Fig. 7).

Investment in new renewable power capacity would increase to over USD 650 billion per year over the period to 2050. Transforming the power system to produce around an 86% share for renewable power would require investments in infrastructure and energy flexibility of another USD 350 billion per year (a total of USD 12 trillion for the period 2016–2050). In all, investment in decarbonisation of the power system will need to reach an average of nearly USD 1 trillion per year to 2050. Over the period between 2016 and 2050, investments in renewable power generation capacity would total USD 23 trillion in the REmap Case, more than double the investment requirements in the Reference Case of USD 11 trillion. Three-fourths of the additional investments are required to deploy variable renewables, mainly wind and solar PV (Fig. 8).

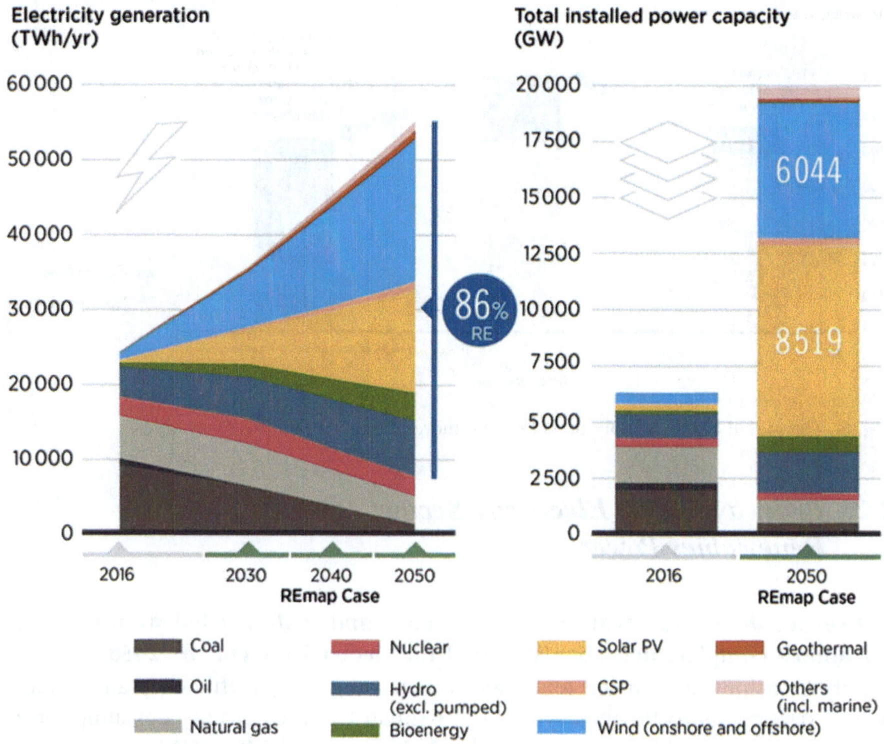

Fig. 7 Wind and solar power dominate growth in renewable-based generation. *Note* In electricity consumption, 24% in 2016 and 86% in 2050 is sourced from renewable sources. CSP refers to concentrated solar power

2.3 Electricity: The Central Energy Carrier

The most important synergy of the global energy transformation comes from the combination of increasing low-cost renewable power technologies and the wider adoption of electric technologies for end-use applications in transport and heat. Electrification of end-use sectors utilising renewable power would lead the transition. The renewable energy and electrification synergy alone can provide two-thirds of the emissions reductions needed to set the world on a pathway to meeting the goals of the Paris Agreement.

Overall, the share of electricity in final energy would need to increase from just 20% today to almost 50% by 2050. On sectorial level, the share of electricity consumed in industry and buildings would double. In transport it would need to increase from just 1% today to over 40% by 2050 (Fig. 9).

The transport sector sees the largest transformation. As performance improves and battery costs fall, sales of electric vehicles, electric buses and electric two- and three-wheelers are growing. By the end of 2018, over 5 million light electric cars

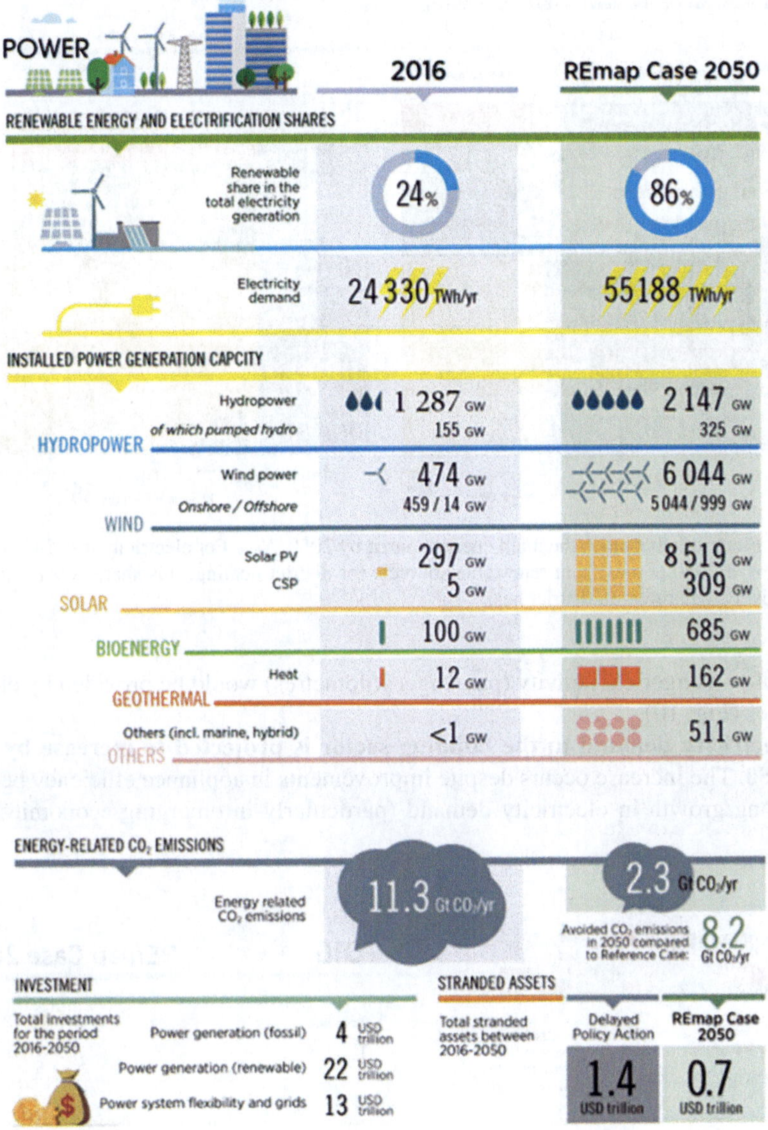

Fig. 8 Power sector key indicators

were on the road (IEA Global EV Outlook 2018). Under the REmap Case, the number would increase to over 1 billion by 2050. (That number could double if it includes all types of electric two- and three-wheelers). To achieve this, most of the passenger vehicles sold from about 2040 on would need to be electric. Under the REmap Case, while over half the stock of passenger vehicles would be electric by 2050, closer to

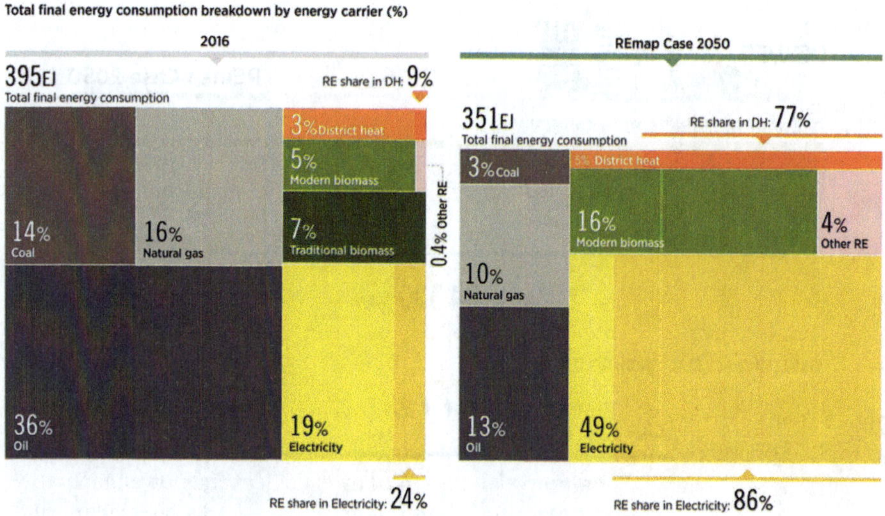

Fig. 9 Electricity becomes the main energy source by 2050. *Note* For electricity use, 24% in 2016 and 86% in 2050 comes from renewable sources; for district heating, this share is 9% and 77%, respectively. DH refers to district heat

75% of passenger car activity (passenger-kilometres) would be provided by electric vehicles (Fig. 10).

Electricity demand in the building sector is projected to increase by 80% by 2050. The increase occurs despite improvements in appliance efficiency because of strong growth in electricity demand (particularly in emerging economies) and

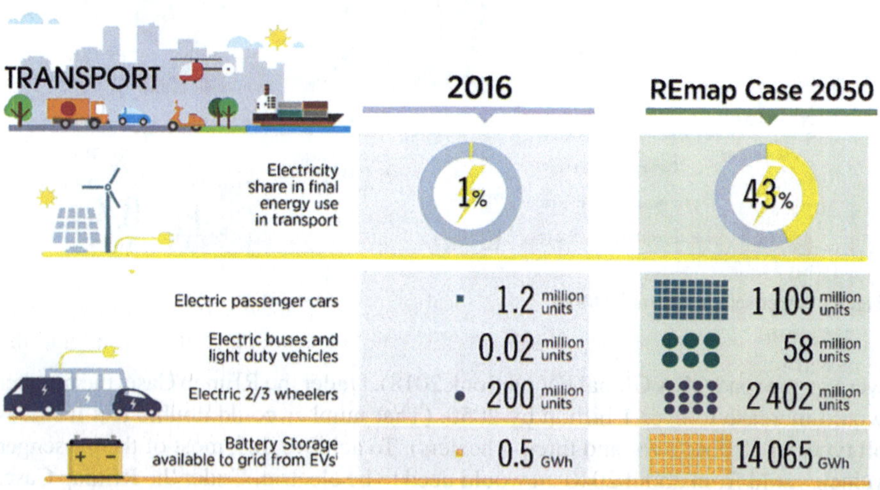

Fig. 10 Transport sector key indicators

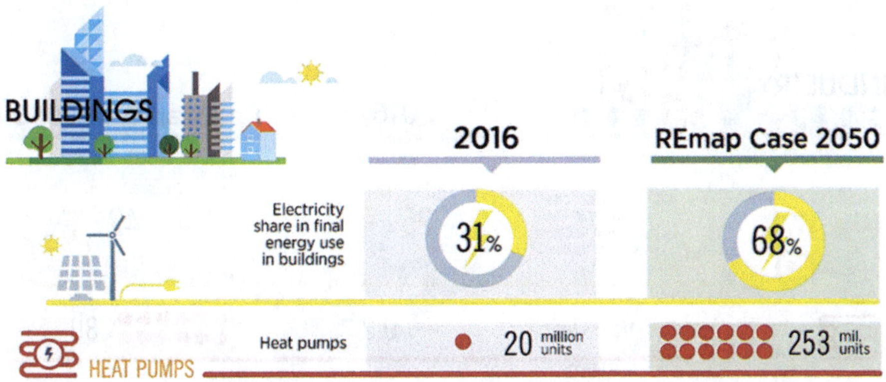

Fig. 11 Buildings sector key indicators

increases in the electrification of heating and cooling. The REmap Case considers deployment of highly efficient appliances, including smart home systems with advanced controls for lighting and air conditioning, improved heating systems and air conditioners, better insulation, replacement of gas boilers by heat pumps and other efficient boilers, and retrofitting of old and new buildings to make them more energy efficient.

Developing and deploying renewable heating and cooling solutions for buildings, urban development projects and industries is also key. Heat pumps achieve energy efficiencies three to five times higher than fossil-fuelled boilers and can be powered by renewable electricity. Under the REmap Case, the number of heat pump units in operation would increase from around 20 million in 2016 to around 253 million units in 2050. They would supply 27% of the heat demand in the buildings sector (Fig. 11).

Under the energy transition, electricity would meet more than 40% of industry's energy needs by 2050. By 2050, 80 million heat pumps would also be installed to meet similar low-temperature heat needs, more than 80 times the number in use today (Fig. 12).

Direct electrification in these sectors can be challenging for certain uses, however, unless renewable-based power can be further converted and stored via other energy carriers. **One such promising energy carrier is hydrogen** (IRENA Hydrogen from Renewable Power: Technology Outlook for the Energy Transition 2018). The production of hydrogen by splitting water into hydrogen and oxygen using electricity could be significantly increased. **As an energy carrier, hydrogen made from renewables could be seen as complementary to electricity since it offers a way to transport renewable energy over long distances**. It has the technical potential to channel renewable electricity to subsectors in which decarbonisation is otherwise difficult. The REmap Case shows that by 2050 hydrogen has the potential to supply nearly 29 EJ of global energy demand, two-thirds of which would come from renewable sources.

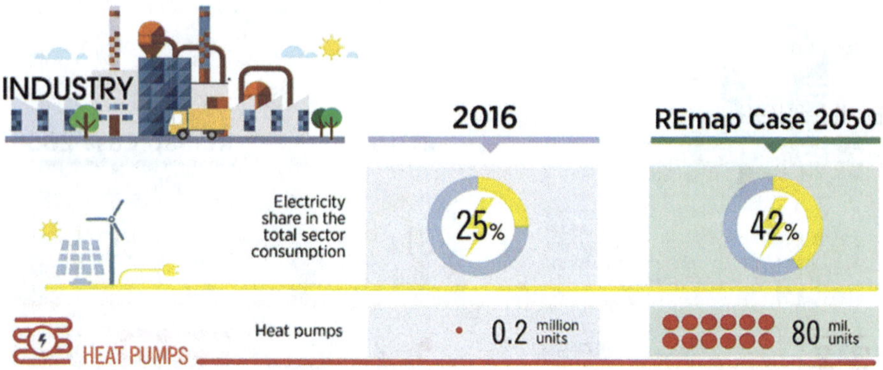

Fig. 12 Industry sector key indicators

Box 2: Best Practice Example—Sweden: 100% Renewable Electricity by 2040

IRENA is establishing a dialogue of best practice and is advising on innovative solutions to aid the development of national 100% renewable electricity strategies for Sweden (IRENA Innovative solutions for 100% renewable power in Sweden 2020).

In 2016, the Swedish government concluded an agreement on Sweden's long-term energy policy. The agreement consists of a roadmap for a transformation of the energy system including a target to reach 100% renewable electricity production by 2040.

In order to achieve such an ambitious target, **a new Climate Act** entered into force in 2018, as part of a climate policy framework imposing on current and future governments the obligation to pursue a climate policy in line with its climate goals, present a climate report every year and develop a climate policy action plan every four years to monitor progress. There are several instruments driving the overall renewable energy transformation. Firstly, **the carbon tax**. Initially introduced at USD 30/ton CO_2, the tax on fossil fuels based on carbon content has increased to USD 140/ton CO_2 at current levels. Secondly, **mandatory renewable energy quota system**. This aims at further increase the share of renewables in the electricity system, as it requires consumers to cancel renewable electricity certificates in proportion to their consumption. The certificates are generated by producers of renewable electricity from new plants and sold in an open market.

Since the carbon tax was implemented, the economy has grown by 75% while the country's emissions declined by 26%. Renewable energy technologies already contribute more than half (54%) of Swedish energy use, and hydropower is the largest renewable electricity source in Sweden, followed by wind and biomass. Hydropower generation in recent years has varied between

62 and 78 TWh, and renewable energy has contributed 60–75% of electricity consumption.

As the country moves towards 100% RE, no increase in hydropower generation is expected, and owners have decided to close two nuclear reactors by 2020. Since 2000, wind power has increased from 240 MW to 6 520 MW in 2017, passed 10% of generation and is projected to double by 2021. To promote implementation, the government in 2015 appointed a co-ordinator of Fossil Free Sweden. The initiative is open to all relevant stakeholders in Sweden, and more than 350 actors have signed up for the initiative. So far, a set of roadmaps for the development of different carbon-intensive industries has been produced. Although other challenges remain, such as decarbonising the transport sector, the success of carbon pricing provides an example for other countries of how to develop a competitive market for bioenergy and for other renewable energy technologies.

3 Actions Needed Now to Transform to 100% Renewable Energy System

This chapter makes clear that an energy transition is urgently required, and that developing 100% renewable energy systems is a key cornerstone in this process. Technologies for these systems are available today, are deployable at a large scale quickly and are cost-competitive.

The Paris Agreement was signed in 2015. Since then energy-related CO_2 emissions have risen by around 4%. The coming years are critical: there is a need for a leap in national collective ambition levels. The revisions of the NDCs in 2020 in combination with Long-term Strategies must yield a convincing outcome for an energy transition that puts the world on a global pathway to reduced emissions, despite differing views on the mitigation measures needed and the rapid evolution of renewable technologies

- The power sector needs to be transformed to accommodate growing shares of variable renewables.
- Digitalisation is a key enabler to amplify the energy transformation.
- Accelerating the electrification of the transport and heating sectors is crucial for the next stage of energy transformation.
- Hydrogen produced from renewable electricity could help to reduce fossil-fuel reliance.
- Supply chains are key to meet growing demand for sustainable bioenergy.
- Decarbonising the global energy system requires swift and decisive policy action.

Table 1 summarises some of the key decisive actions that are needed now for fostering the transition to 100% renewable powered future.

Table 1 Actions now

Key actions needed now towards 100% renewable energy future			
Policy shift	**Energy Planning shift**	**Investment shift**	**Cognitive shift**
Institution/ R&D/Innovation	**Planning/regulation/operation**	**Finance/markets/business models**	**Society**
Promote holistic policies design approach involving coordination among various energy and environmental ministries. Co-operate with and strengthen international programmes (such as IRENA, the IEA, and its Technology Collaboration Programmes and Mission Innovation) to define a joint agenda for setting long term energy strategies and renewable technology innovation.	Establish long-term energy planning strategies that aligns both climate and energy needs, considering action plans in the power sector and in each end-use sectors (couple with the SDGs and the NDCs).	Promote financing schemes for accelerating deployment of renewables and energy efficiency measures for energy demand and supply projects.	Consumers can evolve from passive energy users, to active stakeholders engaged in the development of new social practices around energy use.
All stakeholders will need new management and governance skills to enhance transparency, accountability and enforcement of clean energy policies.	Identify and map renewable energy resources and develop a portfolio of financeable projects over the medium and long terms. Set up a platform providing comprehensive, easily accessible and practical information, tools and guidance to assist in the development of bankable renewable energy projects.	Align the financial system needs with broader sustainability and energy transition requirements.	Promote" Do it Ourselves" approaches with adequate traning and education programmes targeting both consumers and citizens involved in energy communities to practively involve in energy transition.
Promote actions towards circular economy (material recycling, waste management, improvements in materials efficiency and structural changes such as reusing and recycling).	Build no new coal power plants and accelerate the decommissioning of existing coal capacity.	Promote community-based finance that can help lower the overall cost of borrowing for the transition, while simultaneously facilitating the involvement of society in the transition.	Use information communication technology and digitalisation, along with demand side management, to reduce peak electricity demand, lower the need to invest in power capacity, and reduce operational costs.

(continued)

Table 1 (continued)

Combine energy efficiency and renewable energy measures (for example, public sector policies that integrate renewable technologies in the renovation of public buildings).	Promote systemic innovation by creating a regulatory environment that enables smarter energy systems through digitalisation (eg., artificial intelligence, the Internet of Things, blockchain), to promote the coupling of sectors through greater electrification and to embrace decentralisation trends. This innovation needs to be expanded beyond technology and into markets, regulations, new operational practices in the power sector and new business models.	Set up specific funds for innovation and innovative financing and investments mechanisms for the implementation of DER technical solutions. These include, inter alia: combo loans, crowdfunding platforms, financing solutions in partnership with multilateral agencies, and blended finance schemes.	Actively engage communities in the design, construction, operation and maintenance phase of the projects in order to increase community buy-in and enhance sustainability.
Create a central goal of a just transition policy must be to create structures that enable individuals, communities and regions that have been trapped in a fossil fuel energy system to participate in the benefits of the transition.	Revise tariff structures and price regulations. Support regulatory and pricing policies including the right to generate and sell electricity, tariff regulation and grid-arrival policies. Adjust regulations to increase space and time granularity of system operation and pricing.	Potential stranded assets should be internalised in overall risks assessment.	Raise awareness and understanding of the potential of DER, not only for direct users, but also for the society as a whole to foster wiser behavioural changes.
Align public policies with private sector initiatives.	Adapt regulations and aid in the development of an active market to allow energy consumers to participate in ancillary service markets.	Facilitate competitive environments in which reduction in the cost of energy is both rewarded through the right to deliver new projects and supported through the provision of targeted public research and development (R&D) funding.	Promote initiatives that provide additional income (i.e. lease to land owners, creating jobs during project installation and operation) and clean energy for its members, also redirecting funds into educational visits to renewable plants (i.e. wind and solar) for social understanding and acceptance.
Create and promote education and training policies, including an assessment of the occupational patterns and skill profiles in rising and declining industries, and how workers might most successfully be retrained.	Deploy cost-reflective tariff structures by properly readjusting the balance between volumetric charges (USD/kWh), fixed charges (e.g., USD/meter-month) and, where applicable, demand charges (USD/kW).	Redesign power markets to enable the optimal investments for systems with high levels of VRE and enable sector coupling.	Promote equitable distribution of economic benefits and costs.
Establish carbon pricing, together with the elimination of fossil fuel subsidies, not only provide important signals to the market in favor of decarbonisation of the economy, but can also generate significant additional revenue flows.	Consider establishing community energy authorities with the sole purpose of supporting community energy projects through providing advisory services and funding opportunities. By facilitating stakeholder engagement and increasing public awareness, authorities could significantly accelerate their development.	Create stable and predictable market conditions for investment in clean energy to facilitate the reallocation of capital toward low-carbon solutions and to minimise the spectre of stranded assets and avoid long-term lock-in into a carbon intensive energy system.	Create a dedicated platforms and networks of experts as well as groups for exchanging knowledge and expertise for the discussion and exchange of renewable energy and DER best practices and benchmarks at a global level.

(continued)

Table 1 (continued)

Encourage the development of internationally harmonised technical standards and quality control standards to facilitate cross-border trade and exchange of innovative technologies.	Planners, regulators and operators must develop new skills and competences.	Promote institutional investors (pension funds, insurance companies, endowments and sovereign wealth funds) and growth of new capital market instruments, such as green bonds, through which investors can more easily invest in the energy transition.	Accelerate modal shift from passenger cars to public transport (electric railways or trams or electric buses).
Set up a stable and supportive policy framework for emerging technologies such as hydrogen.	Prioritize to improve flexiblity of power system (with flexible supply, storage, demand response, power-to-X, electric vehicles, digital and information and communciation technologies technologies, etc.,). Update grid codes.	Promote innovative business models that enhance the system's flexibility and incentivise deployment of renewable technologies. Examples include virtual power plants, innovative forms of power purchase agreements, platform business models such as peer-to-peer trading, and business models that enhance demand side response.	Cooperate with grid operatirs to schedule charging of EVs and batteries during off-peak hours
Concentrate RD&D efforts to assist sectors that lack commercially available decarbonisation solutions. Relevant sectors include energy intensive industries (iron, steel and cement production) and transport (freight, aviation and shipping).	Deploy microgrids to improve resilience of the grid and energy access rate with renewable sources. Deploy super grids to strengthen the interconnections among countries within a region.	Create business models that focus on mobility services rather than car ownership – such as car sharing, Uber and autonomous vehicles – could transform the way private and public mobility operates.	Mainstreaming gender perspectives through audits and awareness training, gender targets and quotas, better work-life balance, creating networks and supporting mentorship and transparent workplace practices.

Sources IRENA Global Energy Transformation: A Roadmap to 2050 (2019), IRENA Innovation Landscape for a Renewable-Powered Future: Solutions to Integrate Variable Renewables (2019), IRENA, IEA and REN21 Renewable Energy Policies in a Time of Transition (2018), IRENA Climate Change and Renewable Energy: National Policies and the Role of Communities, Cities and Regions (2019), IRENA Renewable Energy: A Gender Perspective (2019)

References

Carbon Brief Analysis: Fossil-Fuel Emissions in 2018 Increasing at Fastest Rate for Seven Years (2018)

IEA Global EV Outlook 2018 (2018) International Energy Agency (IEA), Paris

IEA World Energy Investment 2018 (2018) International Energy Agency (IEA), Paris

IPCC Special Report on Global Warming of 1.5 °C (2018) IPCC, Geneva

IRENA Climate Change and Renewable Energy: National Policies and the Role of Communities, Cities and Regions (2019) Report to the G20 Climate Sustainability Working Group (CSWG) International Renewable Energy Agency, Abu Dhabi

IRENA Global Energy Transformation: A Roadmap to 2050 (2019 edition) (2019) International Renewable Energy Agency, Abu Dhabi

IRENA Hydrogen from Renewable Power: Technology Outlook for the Energy Transition (2018) International Renewable Energy Agency (IRENA), Abu Dhabi

IRENA, IEA and REN21 Renewable Energy Policies in a Time of Transition (2018) IRENA, OECD/IEA and REN21

IRENA Innovation Landscape for a Renewable-Powered Future: Solutions to Integrate Variable Renewables (2019) International Renewable Energy Agency, Abu Dhabi

IRENA Renewable Energy: A Gender Perspective (2019) IRENA, Abu Dhabi

IRENA Renewable Power Generation Costs in 2018 (2019) International Renewable Energy Agency, Abu Dhabi

IRENA Towards 100% Renewable Energy: Status, Trends and Lessons Learned (2019) International Renewable Energy Agency, Abu Dhabi

IRENA Innovative solutions for 100% renewable power in Sweden (2020) International Renewable Energy Agency, Abu Dhabi

IRENA Transforming the energy system and holding the line on the rise of global temperatures (2019) International Renewable Energy Agency, Abu Dhabi

The Renewable City: The Future of Low-Carbon Living

The Istanbul Protocol

Peter Droege

Abstract This Protocol is dedicated to the city of Istanbul and its civic achievements, and the liberation of all cities and humankind from fossil and nuclear energy, in a world that is harmonious, peaceful, connected, democratic, bound- and boundaryless, ubiquitously demilitarised and united in the quest to pursue the United Nations Sustainable Development Goals, in order to jointly confront and avert the imminent self-immolation of humankind by anthropogenic global heating.

This paper was originally formulated for and commissioned by the Cooperative Research Centre for Low-Carbon Living, Australia, and borrows from my edited book, Urban Energy Transition—Renewable Strategies for Cities and Regions. It is an homage to best and next practice in renewable city building, and an expression of *hope for a future, sic*.

There is a great potential contribution to be made to transforming how we think about our immediate living environment and general place in the world—and the opportunities posed by lowering carbon emissions ('carbon' here always used as short for *carbon dioxide equivalent greenhouse gas emissions*: not all GHGs actually contain carbon) embodied in the production of and generated in powering, heating and cooling our residential environments, work spaces and the built environment in general. Commercial energy is to a large extent applied in the building and transport sectors, hence the focus on urban living in shifting the energy paradigm is both astute and profound. Energy renewability, embodiment, efficiency and sufficiency continue to form a magic quadrangle from which to draw instruction for action. Embodiment in particular presents an important growth perspective: as low-carbon living (LCL) gives way to what I would like to call *ultralow-carbon life* (ULCL) it is essential to lower the quantities of 'carbon'—greenhouse gases—in the atmosphere to keep the well-tempered greenhouse from sliding into a hothouse state.

This paper is also an urgent call to heed the need for rapid proliferation of LCL principles and projects, and their mobilization across the built environment production system. It is a call to build an open market for this by creating the required regulatory and policy frameworks, and to remove all the overt and hidden ways in which fossil content is subsidized. This is no longer just urgent but has now become

---

P. Droege (✉)
Liechtenstein Institute for Strategic Development, Vaduz, Liechtenstein
e-mail: droege@eurisd.org

T. S. Uyar (ed.), *Accelerating the Transition to a 100% Renewable Energy Era*, Lecture Notes in Energy 74, https://doi.org/10.1007/978-3-030-40738-4_3

manifestly overdue, as a result of political delays and incumbent industry inertia. And given the importance, even primacy of cities and urban areas in global human settlements, the Renewable City—urban environments, economies, movements and systems entirely relying of renewable energy resources—is now an essential precondition to any hope to stabilize the global climate. The future of low carbon living lies in ultra-low carbon cycle balance, and, consequently, highly carbon retentive cities and regions. Or better yet: a truly carbon negative built and cultural environment, one that removes, sequesters, stores and binds greenhouse gas already in the atmosphere. This cannot be enough: a massive regenerative action agenda needs to ensue to attempt at 'global gardening', the un-development and re-nurturing of Earth's biosphere.

1 Introduction

Humans can be brilliant in short-term, even medium-term planning by individuals or small groups, but modern humankind has proven spectacularly ill-equipped in devising and exerting conscious and constructive, collective, long-term agendas. The delays in critical action over the past generation—in full view of the risks—meant that today, seemingly paradoxically '100% renewable is no longer enough' and 'the zero emissions target is too high'. In recognising and meeting this challenge lies the very meaning and future of low-carbon living. We now know what some of us have long suspected: the UNFCCC targets and frameworks were always far too loose and narrow, and the IPCC projections scientifically naive—influenced by both political pressure and wishful thinking. They implied that there was a 'carbon budget' to work with, foolishly disregarding even the possibility that it had already been long blown by the time this very image of a 'carbon budget' was implanted. Many in today's climate-aware community understand this now. Another long and stubbornly held myth concerned the mechanical systems thinking that led to linear projections and simple-minded graphs showing how lowered emissions would directly correspond to lowered temperatures by the global climate and carbon systems.

Everyone bought into and promoted these comforting illusions, built into popular 'inconvenient truth' presentations and reports as convenient emotional escape hatches by major and minor climate celebrities: from Al Gore to Prince Charles, and from Sir Nicholas Stern to the last carbon compensation and offset scheme operator riding the airline passenger guilt wave. The unnerving notion of trading inside a great carbon bubble was made safe by the sober notion of 'the budget'. The shocking truth is: that budget was already blown back on June 10 1986 when James E. Hanson, another reluctant purveyor of the magic carbon budget myth, correctly testified before a US Senate committee that we will find ourselves in the very hot water that we are in now, although neither the long-blown budget nor the feedback problem were not then and are not now quite acknowledged.

2 100% Is not Enough—and 0 Is Too High

The most positive future to aspire to is a massive propagation of LCL developments, innovations and findings, in buildings, neighbourhoods and communities—and perhaps most importantly and challengingly the existing building stock. Elevating the retrofitting and refurbishment of the energy wasting building and plant stock to a national priority and making it the very foundation of building, construction and planning regulations is fundamental. It pales only in comparison to the equal second priority of decarbonising its energy source, and to dramatically shift away from coal, oil and gas, and retire the dying nuclear industry.

The carbon budget myth has given a wide berth to the widespread complacency that is the most tragic hallmark of modern society. As a result the equal third and equally urgent challenge lies in endowing the built and (agri-)cultural environment with the ability to also withdraw copious amounts of excess GHGs from the atmosphere and bind them in soils and materials, support biodiversity and sustainably manage increasingly scarce water resources.

Sustainability principles, once quaint aberrations in a landcape of business-as-usual have emerged as urgent survivability measures, without all really noticing it yet: relying on renewable energy, ending the combustion of fossil resources, transforming carnivorous food culture and industrial agriculture, lowering atmospheric GHGs and binding them in soils and materials, shrinking lifestyle footprints, revolutionising water management, and shoring up biodiversity are essential elements in ULCl actions and demands for the built environment. Ultra low carbon life means that target horizons shrink to 2020 and aims emerge as 150%—rather than 80–100%—emission reductions. The built, agri/cultural and natural environments together must achieve the net absorption of atmospheric CO_2 and other greenhouse gases.

3 From Low Carbon Comfort to Ultra Low Carbon—Historical Context—Paths Out of a Very Predictable Predicament

Earth is not a spaceship, contrary to the seminal sustainable development expression and literary image Spaceship Earth launched in the 19th century and popularised in the 1960s (George 1879; Ward 1966; Boulding 1966; Fuller 1968). It is a fairly ancient planet of 4.5 billion years on a fixed orbit—and yet it presently is also on a rapid journey in its climatic behaviour. Its degree of habitability is transforming in front of our very eyes: it is becoming a different kind of planet. For anyone who thought that she or he might not be so special: the perhaps mockingly named Anthropocene is an unusual moment in this planet's biospheric and human evolution. It would be a very unique moment if Earth were to morph from Goldilocks habitability into what would be a very difficult planet to colonise with enormous technological and financial resources, let alone with the actually quite limited means at our disposal.

Interested in interplanetary travel? Instead of racing for eight months through space to get to Mars, we need to do nothing, continue business as usual and wait for only a few more years to arrive at a wholly new and ferociously alien world: Terra 2 would feel a bit like Venus. This transformation does not come easy, and is being exerted by a century of concerted geo-engineering effort in the form of fossil energy injection, fossil carbon release and other means of transforming land, water, inherited populations, systematically eliminating biodiversity and resilience. We don't know what this planet will look like, or how long its final transformation would take. We hope we can halt it, that it still depends on how effective current and immediately future measures are and will be to halt the slide into the unknown. Halting the current mutation will take a massive change of exactly the kind presented in this forum as pioneering and lab like projects, but as a basic requirement for going on, not just an aspirational goal. Some astrophysicists like Carl Sagan or Stephen Hawking have long mused about the worst-case possibility of a finally stable terrestrial surface temperature of 250 °C, rather than the present 15, when the stunning temperature-maintaining phenomenon of a living Earth would have been boiled off.

4 Low Carbon—Existing Context—Actions for a Sustainable Predicament

Halting this slide is the new meaning of Sustainability, which has always been about Survivability. The rising number of innovations provide the methods and projects for this platform—part of a sustainable development trajectory that is fast becoming a global paradigm. It is the core, the seed of overdue emergency action agenda. The initiatives presented in this paper all propose to highlight the many extraordinary successes and advances in the proliferation of renewable energy have been made: the powerful feed-in tariffs; the rise of 100% renewable buildings, communities and regions; the broad march of solar and wind into many countries' power mixes; the revolution of national policies to embrace energy transitions; and the rise of renewable energy investment—which has long become the dominant mode in annual capital expenditures in new power generation capacity world-wide.

A renewable city supports, thrives on closer cooperation between the city, its hinterland, the state and beyond. It relies on intelligent renewable energy networks that monitor a constellation of decentralised renewable energy plants and generators at varying scales. It will need improvements and extensions to existing energy supply infrastructures to improve integration, connection and most importantly, increase accessibility to different types of renewable energy. Favourable and compatible spatial planning policies and guidelines at the urban, regional, federal, and even EU levels will be sought to achieve equitable, safe and reliable flows and access to such energy sources.

The following ten groups of important initiatives offer views into a sizeable and growing pool of constructive and tested practice that is to be transformed into mainstream. Presented are cases, themes and topics that epitomise the transformation from the Fossil to the Renewable City. What used to be an aspiration goal now has become an existential necessity.

5 Current and Emerging Approaches to Low Carbon Living

5.1 Renewable Nuclei: Active Homes

Low Carbon Sustainable Building: Josh's House, Perth, Australia

Carbon Positive Building: B10 Active House, Stuttgart, Germany

Model buildings are powerful ways to develop, study, demonstrate and inspire about the deployment of buildings providing their own energy from local renewable sources and to share excess power with neighbours or the grid. Two buildings from opposite ends of the world highlight two very different approaches. Josh's House in Fremantle is a large and lived-in demonstration family home with garden, and 'Werner's House' in Stuttgart is a model research exhibition of the future: a modes sized, modular, prefabricated unit for bungalow style or stacked apartment deployment. They perhaps also highlight differences in our attitudes about life and culture, and the production of the built environment (Figs. 1, 2).

Josh's House is a residential 'living lab' near Fremantle demonstrating that high performance, energy positive housing is accessible now as a mainstream market offering. Completed mid-2013, the 10 Star NatHERS rated home was originally fitted with a 3 kW solar PV system, a gas boosted solar hot water system and conventional energy efficient appliances. Monitoring of the operational energy requirements of the home demonstrated that energy usage was less than half of the local

Fig. 1 Josh's House, Perth. © Josh Byrne, Perth 2014

Fig. 2 B10 Active House, Stuttgart by Werner Sobek. © Zooey Braun, Stuttgart 2014

average and power generation was double to that consumed. The subsequent inclusion of a grid connected battery system enabled 81% self-reliance in solar power whilst continuing to export surplus to the grid. The house also includes a range of sustainable water management features such as rainwater harvesting, greywater reuse and water sensitive landscaping, resulting in a 90% reduction in mains water usage. The latest upgrade is the inclusion of an electric vehicle (EV) for family transport, along with an electric heat pump hot water system and an induction stove, enabling disconnection from the gas supply. The solar system has been upgraded to 6.4 kW of PV with a 5 kW inverter to meet the additional energy loads and remain energy positive. All aspect of this project have been extensively documented on video and factsheets, with these resources made openly available through the project website (www.joshshouse.com.au) along with live performance data. The principles of how to build this house and its power and water systems have been used in the scaled-up projects below that enable the research team to examine the opportunities provided by shared infrastructure (Josh Byrne).

B10 is 'Werner (Sobek)'s House' built by a prefabricated home manufacturer in Stuttgart's Weissenhof Siedlung. The techniques and the innovations developed and tested range from the building's design, production and assembly to the harvesting, storage and targeted local distribution of energy from renewable sources. B10 was completely prefabricated, with all building elements modularised for waste-free disassembly and reuse, based on an rigorous approach to industrial production combined with a high degree of customization. The building's timber structure is lightweight and entirely recyclable, just as the façade and other features. Surplus energy harvested by PV panels and solar heat is either stored in B10 or passed on to a neighboring historical monument. Energy distribution is managed by an intelligent automation system specifically developed for B10. B10 generates 200% of its energy consumption: the premise is that the process of rebuilding and upgrading existing neighbourhoods is far too slow and therefore new infill buildings should be designed and built as generators and batteries to serve conventional neighbourhoods. This is a prototype supported by the German government, designed to boost existing neighbourhoods' renewable energy supply: B10 provides power to the neighbouring Weissenh of Museum, a Le Corbusier designed building (adapted from Sobek 2018).

6 Renewable Cities and Quarters

City Energy Transitions: 100% Renewable Plans for Frankfurt and Munich, Germany

Active City-House for the Renewable Metropolis: Active City House, Frankfurt, Germany

The 100% renewable energy region movement is a growing force in Germany. Cities, too increasingly wish to control their energy destiny. Several German metropolitan cities adopt the regenerative resolve and initiative shown by the smaller 100% renewable villages and districts. The efforts by large metropolitan cities like Frankfurt am Main and Munich in combining bi-partisan political support, strategic planning and partnerships across sectors and administrative boundaries, raise the share of renewable energy, boost energy savings, and enable the testing of innovative energy infrastructure technologies. Critical were the urban planning and sustainable building frameworks these cities have put in place to guide their shaping of a 100% renewable city (Figs. 3, 4).

The urban planning centred approaches by the metropolitan cities of Frankfurt and Munich in Germany to achieve 100% renewable energy reveal the dynamics of the broader 100% renewable movement, its impact on villages, towns, cities and regions, and its relation to advancing the urban sustainability agenda. A well-developed system of governance for the deployment of renewable energy systems (RES) has evolved at several spatial planning levels in Germany, giving rise to a rich energy relevant domain of spatial considerations for city planners, urban designers and architects in the design of RES infrastructure for metropolitan areas that are based

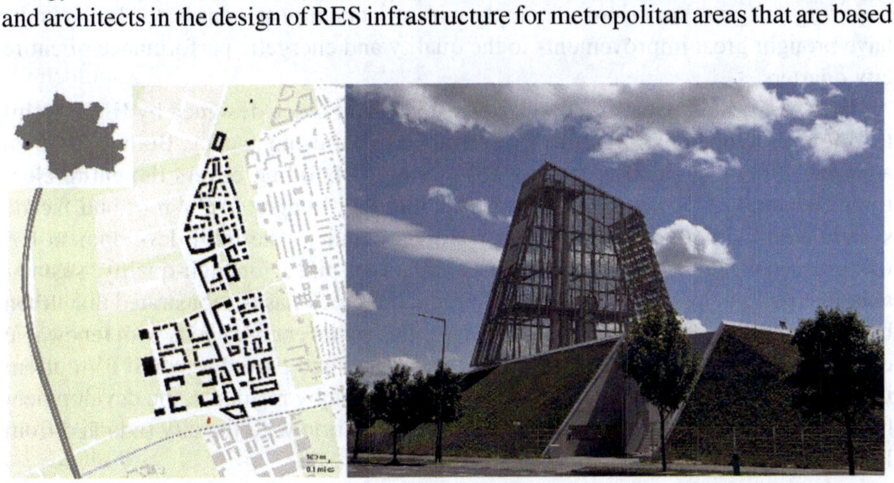

Fig. 3 Freiham-Nord Masterplan, as part of Munich's 100% renewable energy city concept. © Anis Radzi (2018)

Fig. 4 Active-city House, Frankfurt by HHS Architects. © Anis Radzi (2018)

entirely on local and regional sources. Successful examples in both cities illustrate how respective energy concepts are achieved, overcoming a range of specific regulatory, technical and methodological challenges. Frankfurt, to date, has the largest number of energy-efficient high-rise buildings in Germany with >1500 apartments, and >300,000 m^2 of surface area built to a passive house standard, including schools, day nurseries, and sports halls (Stadt Frankfurt am Main, 2013). Growth in energy efficient building has been propelled by municipal obligations that require passive houses to be built on city land or land purchased from the city. It has also been boosted by the city's "Green Building Frankfurt" prize, which is awarded to architects, planners, and construction firms for building innovative sustainable green structures. By requiring design submissions to comply with strict energy standards, the built results have brought great improvements to the quality and energetic performance of entire city quarters.

The "Aktiv-Stadthaus" is a cooperative apartment block designed by HHS Architects and developed by Frankfurt's own municipal housing agency. Built in 2016 on a former inner-city parking lot, the "plus energy" building covers the entire electricity demand of 74 apartments through a building-integrated PV roof and facade system coupled with an on-site energy storage facility. The infill development has transformed the image of the inner-city area as a place for high-quality, sustainable living. As a "power-and-storage" station, the block has demonstrated that urban buildings can help ease the ability of the city to cover its energy needs with renewable energy sources. And by using the locally-generated and partially-stored PV current in the block to power electric vehicles for shared use by residents, the development has contributed to raising their mobility and interaction with the city (adapted from Radzi 2018).

7 Replicable Renewable Energy Districts

Net-Zero Energy Districts: An Integrative Business Model for US cities

Renewable Malls: Transforming Shopping Centres Into Flexible, Decarbonized Urban Energy Assets

The Rocky Mountain Institute has developed an integrative business model for developing net-zero energy (NZE) or ultra-low energy districts in a way that is attractive to the district developer, parcel developer, and tenants, as well as beneficial to the local electric grid and neighboring community. While many elements are broadly replicable, this business case was first modeled specifically for the developer of a proposed 180-acre, 6 million square feet mixed-use NZE development located on a former industrial site in a midsize US city. The RMI uses the term net-zero energy to describe the general concept where the energy consumption of a building or multiple buildings is offset by renewable energy on an annual basis and should not be taken as implying alignment to any one specific, more granular, definition (Fig. 5).

Urban systems de-carbonization is achievable if supported by measures for energy efficiency and integration of renewable energy sources (RES). In this context, a key role can be played by shopping malls. They are usually identified as "icons of consumer society," but they also have a huge energy retrofitting potential. Moreover, they can have an active role in the future smart grid, connecting buildings and energy

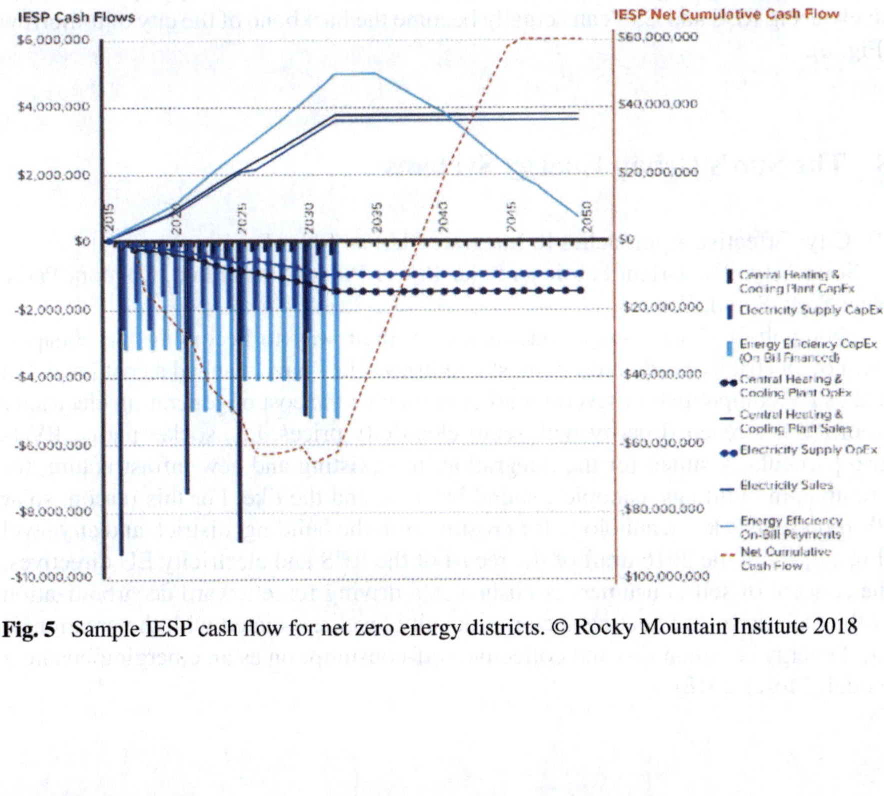

Fig. 5 Sample IESP cash flow for net zero energy districts. © Rocky Mountain Institute 2018

Fig. 6 Robinsons place Dumaguete, Manila. © ABS-CBN News 2018

infrastructures. Photovoltaic (PV) and energy storage systems (ESS) play a fundamental role in exploiting such potential, and can very quickly become a cost effective solution contributing to emissions reduction, as demonstrated in the presented case study. Considering the short economic perspective of investors, the capital expenditure associated with retrofitting could be a barrier, and the evaluation of suitable economic indicator of primary importance to choose among several retrofitting strategies. Despite the fact that overall legislative frameworks and regulations do not promote shopping centers as key energy and social infrastructures to achieve ambitious targets in the ongoing urban transformation, energy-efficient shopping malls massively using RES and ESS can actually become the backbone of the city of tomorrow (Fig. 6).

8 The Sun's Urban Energy Systems

PV City: Effective Approaches to Integrated Urban Solar Power

Solar City: The Urban Fabric as Solar Power Plant: Amsterdam, London, Paris, New York, Seoul, Tokyo

Photovoltaic (PV) energy systems are on their way to becoming the cheapest form of electricity production in most countries. They have reached a cost level that makes PV competitive in several market segments: the cost of generating electricity from PV has reached parity with retail electricity prices, i.e., socket parity. PV is also particularly suited for the integration into existing and new infrastructure, for example, in buildings, canopies, sound barriers, and the like. For this reason, solar PV represents a key technology for prosumers at the building, district, and city level (Figs. 7, 8). In the 2016 draft of the recast of the RES and electricity EU directives, the concept of self-consumers is pushed as a driving force toward decarbonization of the electricity and heating sector at the city level associated with the creation of local energy communities and collective self-consumption as an emerging business model (Moser 2018).

Fig. 7 Veteran Freiburg Solar Settlement, Vauban, Germany. © plusenergiehaus.de 2012

Fig. 8 Renovation of Halle Pajol by Jourda Architectes, Paris. © apur.org 2017

The degree of urbanisation has made the decarbonisation of cities and urban areas a paramount importance. The implementation challenge is particularly poignant in the urban energy services sector, where conventional infrastructure-scale solutions deepen climate injustice. Conventional infrastructure-scale energy production components (i.e., fossil fuel powered power plants) must be contrasted with the "solar city," where the urban fabric is reconstituted as a platform for decentralized production of solar electricity at a scale that advances the twin aims of sustainability and justice, yielding the economic, social, and environmental implications of urban decarbonization. The issuing of solar bonds and other financing means has been modeled to show that a city of the size and nature of New York, studied along with Amsterdam, London, Munich, Seoul and Tokyo, could be financed with a 10–12 year investment maturity horizon. The city has a potential solar energy to power conversion capacity—calculated on reasonably available roof areas only—of nine GWp or 11 TWh per year, providing about a quarter of its annual power consumption on average: Manhattan: 6%, the Bronx 31%, Brookly, 35%, Queens, 42% and Staten Island 48% (Byrne 2018).

9 Large-Scale Urban Regeneration Programs and Systems

Renewable Wilhelmsburg: recruiting the International Building Exhibition to fight climate change

Implementing the Heating Sector Transition

Integrated Urban Infrastructure: Energy Storage and Sector Coupling

Germany has a long history of building exhibitions, beginning in 1901. A building exhibition is always more than a showcase for architecture: building exhibitions drive urban development (IBA Hamburg, 2009). Building exhibitions concentrate and coordinate private and public spending on construction in an area or region with specific problems, as well as specific opportunities. So, they represent a treasure trove of more than a hundred years' experience when it comes to finding innovative solutions for the most pressing problems of urban community life. Many ideas still live on today. From 2007 through 2013 Hamburg hosted an international building exhibition (IBA) on Europe's largest river island, Wilhelmsburg (Figs. 9, 10). It initiated 70 building and 14 social and cultural projects (IBA Hamburg, 2014) in order to demonstrate what is possible when an entire city district is remodeled according to social and environmental considerations. These projects were designed to show what the future of modern environment friendly town planning might be, and how cities could be remodeled in a climate-friendly or even climate-neutral way. Another advantage was that a neglected district of Hamburg with a negative image could be reinvented as a pioneer of energy-efficiency and social inclusion. To realize the "Future Concept Renewable Wilhelmsburg," IBA specified four operational fields. First of these was the refurbishment of the existing building stock, second, 'new buildings of energetic excellence', third, 'local district heating', and fourth, 'local renewable energies' (Hellweg 2018).

The heating sector represents one of the biggest challenges to achieving climate neutrality in regions with cold winters. Local municipalities will have an important

Fig. 9 IBA Wilhelmsburg. © IBA-Hamburg 2014

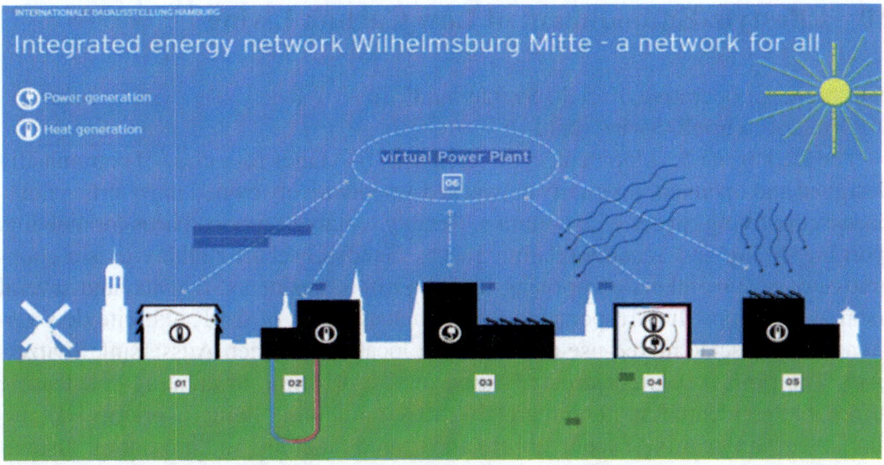

Fig. 10 Renewable Wilhelmsburg. © IBA-Hamburg 2014

role to play in this process. In Germany challenges arise with the transformation of the heating sector and the possibilities for action facing local municipalities. There are two main areas of action: the energy-efficiency retrofitting of existing buildings, and the transformation of district heating systems. With respect to the energy-efficiency refurbishment of building stock, it is important that cities and municipalities look at private, as well as public properties, and address concerns about historical preservation and social issues. District heating grids allow for the cost-efficient integration of renewable energy and waste heat sources; however, operating conditions, and in some cases network structures, must be modified accordingly (Sparber et al. 2018; Weiß et al. 2018).

The generation and transmission processes of renewable energy still takes mainly part in rural and peri-urban areas—not yet in urban centers at sufficiently significant scale. Many technologies easily implemented in lower density districts such as wind, water or biogas production are not yet adopted for application in urban areas—one exception is solar power and thermal energy. But even solar electricity is more widely applied in lower density and rural areas although its application in urban areas matches the existing grid infrastructure well. In contrast to rural areas with grid integration problems energy infrastructure in urban centers is already well prepared for renewables integration. There are almost unlimited energy storage possibilities with enormous capabilities but also large differences. There is a fundamental necessity to combine and couple the different energy sectors for electricity, heat, cold, gas and transport. Only in coupling the energy sectors and using cheap and efficient energy storage options from one energy sector to solve challenges within another the energy transition process can be managed in an efficient way (Stadler and Sauer 2018).

10 Lifestyle-Embodiment of Low Carbon Living

Kalkbreite Cooperative, Zurich, Switzerland

Ortoloco, Zurich, Switzerland

The Kalkbreite development, completed in 2014, stands for a class of communally designed and operated, internally networked, cooperatively owned inner-urban flexspaces: core living spaces with common, flexibly rentable support spaces for meeting, working, living and accommodating guests. Genossenschaft Kalkbreite is a cooperative of 850 members comprising neighbourhood residents, tenants, and several local associations. It is responsible for the management of the Kalkbreite development, a large-scale mixed-use urban project located in Zurich-Aussersihl, Switzerland. The project was constructed based on the principle of creating and leasing affordable residential and commercial spaces, combining living, working and culture, integrating a mix of social groups and shared facilities, and promoting sustainable development. All of which had to be in line with the objectives of the so-called 2000-W Society. The principles sustain vibrant urban life within one neighbourhood block (Figs. 11, 12).

The building is organised around internal network spaces. The cooperative works well with Zurich wide agricultural cooperatives like Ortoloco. This is one of several urban and regional agricultural cooperatives serving the metro-city of Zurich. These own and lease agricultural space and operate it on biological and agroforestry principles. Reducing food miles and global consumption Ortoloco and a series of other similar cooperatives are good match and expression of emergeing lifestyles that serve city dwellers that become vested in and support through shared cultivation, harvesting, storage, distribution and consumption largely based on city dwelling communities within bicycling distance.

Ortoloco is a self-managed vegetable cooperative based in Dietikon, Zurich. As a joint initiative of farmers and consumers, Ortoloco leases 1.4 hectares of arable land from organic farm Fondlihof in the Limmat Valley of Zurich to create their own

Fig. 11 Kalkbreite housing cooperative by Müller Sigrist Architekten. © Martin Stollenwerk 2018

Fig. 12 Ortoloco community events. © Ortoloco 2018

community vegetable farm. Over 60 types of vegetables and a variety of edible wild herbs are harvested weekly, distributed and consumed by members of the cooperative.

The cooperative operates on the principle of producing quality local food, ensuring fair working conditions, and maintaining sustainable production methods. Members of the cooperative participate in making important decisions, and are involved in the operation of the farm. This means that costs, risks, and profits are shared. The cultivation of vegetables takes place according to the directives of Bio Suisse, a Swiss certification for organic produce. The range of produce available to consumers change according to the seasons, and is distributed through established food "depots" and consumed weekly though vegetable bags subscribed for at least one year by members from Zurich, Dietikon and surrounding areas.

Ortoloco was inspired by the book Neustart Schweiz, which described a vision for a local economic model that ensured a community a wide range of services such as a bakery, dairy, and kindergarten amongst others, all of which were managed by local citizens. Individuals involved are expected to lend their specific skills to the neighbourhood. Whole neighborhoods instead of individual households would jointly organize projects, especially in the development of local food supplies.

11 Fremantle as Renewable Laboratory

From White Gum Valley (WGV) to Smart City Renew Nexus

Ten House Living Lab Study

The objective of the 10 House Living Labs project was to understand which factors influence house performance, as previous research has shown that houses designed to be energy and water efficient often do not perform as intended. While the design is important to minimise resource consumption, the way houses are used can have an equal effect on performance (Figs. 13, 14).

Fig. 13 WGV living lab. © CRC Low-Carbon living laboratory 2018

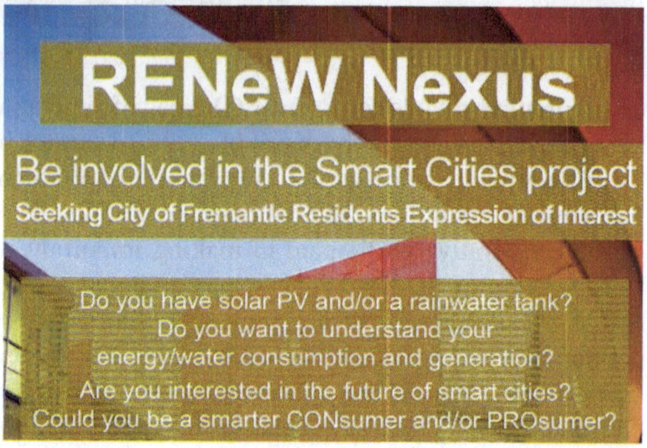

Fig. 14 RENeW Nexus. © Curtin University 2018

The project consisted of a longitudinal study of ten detached suburban family homes located in the City of Fremantle. While these houses had a mix of demographics and building designs, they all presented energy or water efficient features, such as solar panels, solar hot water and rainwater tanks. The houses also presented elements of climate sensible design. The ten houses had their electricity, water, gas, solar energy generation, rainwater use and internal temperature monitored for two years. The first year of data collection was used to establish a baseline in terms of occupant behaviours and practices and to evaluate the homes from a design perspective. During the second year, a behaviour change program was implemented, providing each household with a series of tailored tools designed to increase their awareness and facilitate a reduction of water and energy while enabling occupants to maintain a high-quality lifestyle.

This project demonstrated that while the energy efficient houses perform better than standard Australian dwellings, they do not operate to their full potential. Overall house performance is attributed not only to construction quality, maintenance and technology but also to everyday house operation. The latter is driven by occupant practices, which are reproduced sequentially as part of an established routine. The results of the 10 House Living Labs project were used to develop the concept of the

Home System of Practice, which explains occupant dynamics within a home and enables the creation of innovative technologies to improve resource efficiency while fulfilling occupants' needs.

Mainstreaming Zero Energy Homes

This national project is working with major land developers and builders to deliver a series of Net Zero Energy Homes (ZEH) in residential display villages around Australia with the aim of better understanding cost and market perception barriers to this offering whilst increasing engagement amongst industry players.

A key stage of the project includes collaborative design charrettes with builders to ascertain the steps required to get their 'base design' to ZEH status whilst tracking cost and trade capacity implications. The second stage involves surveying visitors to the display homes to understand the level of interest in the design and technologies underpinning the ZEH performance, versus interest in other house features. These activities are being captured in a series of videos and industry reports for wide dissemination.

To date, ZEH display homes have been built in Townsville (Queensland), and Melbourne (Victoria), and others are underway in Perth (WA) and Canberra (ACT). Findings from the design and building of the ZEH homes are that major energy efficiency gains were obtained mainly from additional insulation, glazing upgrades and energy efficient appliances (hot water systems and air conditioners in particular). In addition, only a relatively small sized PV system (4 kW) is required to cover the net needs of a typical Australian household provided that the building envelope is designed appropriately for the climate and the appliances are energy efficient.

WGV Living Lab

WGV is a 2.2 ha medium density, 100 dwelling residential infill development located in the City of Fremantle. Led by the WA State Government's land development agency LandCorp, WGV demonstrates design excellence on many levels by incorporating diverse building typologies (detached houses and apartments), climate sensitive considerations, solar energy generation and storage, innovative water management and creative urban greening strategies. The project has received international certification as a 'One Planet Living' community.

Researchers are following the WGV development process from construction through to occupancy, using a 'learn-by-doing' approach to research where innovations are tested in real-life settings with the aim of informing policy and industry outcomes. Research activities include the monitoring of different dwelling types to assess design performance, as well as the impact of technology choice and occupant behaviour on energy use and carbon emissions. The project is also exploring the inter-relationships between developers, local government, builders and purchasers, low carbon aspirations and outcomes, and how these can be better aligned.

As of late 2018, WGV is approximately 60% completed and occupied. Monitoring is underway across the development and data is being utilised by residents, researchers and industry. Learnings from the WGV project are being shared with industry and government through tours, speaking events and technical publications.

A 10 part web-based video series called 'Density by Design' (www.densitybydesign.com.au) documents the WGV Living Lab story in detail as a means of sharing the project journey and research outcomes to a broad audience.

WGV is demonstrating that low carbon residential developments are technically feasible and commercially viable in today's market with 'As Built' modelling and early building performance data indicating that WGV will meet its design goal of being a Net Zero Energy precinct. What's more, the collaboration between industry and researchers, which has been enabled by the CRC for Low Carbon Living, has led to a ground-breaking trial for shared solar power and battery storage technology on strata-titled developments.

Beyond WGV

Project is examining how to extend WGV innovations in scale across a brownfield redevelopment land in Fremantle. The area is the focus of research by both the CRCs for Water Sensitive Cities and Low Carbon Living. There is an opportunity to prototype a precinct scale sustainable neighbourhood across the whole precinct. The vision is to incorporate a range of innovations such as community batteries, water sensitive urban design, and a potential Trackless Tram transit system linking the precinct to the city centre in Fremantle.

'Beyon WGV' will seek to remove major barriers to new energy and water solutions for future redevelopments in the Knutsford precinct. This presents the real world challenge for decisions about establishing a business case and alterative governance model for these alternative systems. Further, investigations to date suggest that there are also no major technical barriers to the installation of renewable energy and water systems at a precinct scale, the major challenges sit in the areas of (a) developing alternative/integrated water and energy infrastructure that can align with uncertainties of multiple land owners and incremental development patterns; (b) engaging the stakeholders including business and citizens over the long term; (c) keeping the vision of what is being created, focused for 20–50 years; (d) developing appropriate governance and business models to support the rollout; and (e) in relation to the above marrying with the Local and State Government decision making frameworks and planning systems. This researchers are also in discussion with City of Fremantle and LandCorp about how to manage the urban renewal process over the long term to realize the vision of Knutsford as a leading sustainable development precinct.

RENeW Nexus

The RENeW Nexus project is supported and funded by the Australian Government through the Smart Cities and Suburbs Program. The program aims to help local governments and communities use smart technology and increase the accessibility and use of public data so that cities, suburbs and towns become more liveable, productive and sustainable and urban service delivery becomes more efficient and effective.

The project sponsor is Curtin University, and project partners include Murdoch University, City of Fremantle, Landcorp, Power Ledger, Western Power, Synergy, energyOS, CISCO and Data 61/CSIRO. The project has begun to explore the theory that the value of solar energy can benefit the grid and community through peer-to-peer trading across the grid. Further, the value of water can be captured if potable water can be supplemented by utilising rainwater and greywater through the subsurface aquifer to other household and landscape systems.

For both solar energy and water systems the focus is on peer-to-peer trading across the power grid or within a micro grid environment, and through a water balance across the aquifer in a development, or storage facilities in an established urban development. This is a distributed system where the transaction layer is blockchain based, and can be deployed with smart metering capability to create a dynamically connected smart city environment. The first trial of this system is being carried out in the City of Fremantle and includes the energy grid provider and retailer, a water provider and retailer, the City, a state developer and most importantly citizens as prosumers and consumers of energy and water.

Data analytics will determine trading logic and suitable conditions that are required, stakeholder satisfaction, utilisation of assets and efficiency gains achieved through peer to peer trading. The project trial will also allow for demonstration of the proposed distributed energy and water ledger that will be implemented at East Village at Knutsford. This project is enabling research and data analysis to support and inform the transition to a new energy and water network.

12 Climate Design for Social Justice

Solar for Gaza—Low Conflict Living As An Existential Human Right: Gaza Sustainable Neighbourhood Design in Developing Countries

Peaceful cooperation is a precondition for being able to effectively counter the causes of climate change, and successfully adapt to its effects. Solar for Gaza (S4G) was developed in the spirit of the ClimateforPeace.org campaign, a global call for the cessation of armed conflict in order to address one of the world's greatest common enemies: manmade global warming and growing fossil and nuclear energy risks. Solar for Gaza emerged in response to Israel's Gaza Strip bombings between December 27, 2008 and January 18, 2009—officially known as Operation Cast Lead—when a group of architectural students in Liechtenstein took up a long tradition of "engagement design," seeking to support academic and civic efforts in Israel and Gaza to address—from afar—the suffering of ordinary people caught in a seemingly permanent and tightening cycle of violence. Sketched out a Gaza and its wider region entirely based and prospering on renewable energy. It specified geographical, social, economic, technical, organizational and political factors supporting the incorporation of renewable energy into various phases of relief, recovery and regeneration (Figs. 15, 16). It was also advanced as embedded in an inspired regional initiative—Solar for Gaza and Sderot—in a collaboration between the Arava Institute for Environmental Studies, the Institute for Global Leadership at Tufts University and the Chair for Sustainable Spatial Development at the University of Liechtenstein (adapted from Droege et al. 2018b).

Fig. 15 Solar installation in Gaza. © Jersualem Post

Fig. 16 Village solar installation. © Greenpeace

Considering that almost 90% of global urbanization is projected to occur in countries of the developing world, many with substantial social justice, equity and poverty challenges, cities' growth in these countries will both have a significant impact on global GHG emissions, seriously threatening any effort to reduce them—and deepen their dependence on debt inducing fossil and/or nuclear energy dependency. Energy consumption is mainly determined by the building and transport sectors, but is also influenced by other issues that urban growth has to face: rising demand for food and greater demand for potable water, combined with changing rainfall patterns and depletion of aquifers. To cope with all these challenges, a paradigm shift is required, that is, a different urban design approach that affects the urban form, texture, and land use, and the way the basic urban services, such as energy, water, food, and waste treatment, are designed and provided, with a holistic view. A sustainable neighborhood design process in tropical climates is discussed, outlining the importance of adopting a systems perspective and considering infrastructure interconnections (adapted from Butera 2018).

13 Modelling Renewable Resource and Value Flows

STAR—Mapping regional negative carbon performance—Renewable energy self-sufficiency and positive income streams: Lake Constance Alpine-Rhine Region, Switzerland-Liechtentein-Austria-Germany

Value-Added and Employment Effects of Renewable Energies and the Energy-Efficiency Refurbishment of Existing Housing: Berlin, Germany

The Space Time and Renewables (STAR) model has the ability to model and map on GIS platforms such time variable scenarios for regions, cities, towns, and neighborhoods. It was developed for the Lake Constance Alpine-Rhine Energy (LACE) Region to not only test when and how it can become entirely independent energetically, or whether it can sequester more atmospheric carbon than it emits, and how much financial value creation can be achieved—but also to demonstrate the model's ability to test time sensitive scenarios for urban planning purposes. The LACE region not only has the potential to become carbon neutral, but even to become a CO2 sink that binds more carbon than it emits. The investments into the transformation process were assessed and juxtaposed with the fossil/nuclear energy savings (Figs. 17, 18). A shift of the systematic differential costs in time can be observed: initially the energy costs will rise, but in the longer term—from 2030 onwards—they will begin to fall dramatically. The savings gained from avoiding fossil and nuclear energy can be invested within the region to generate jobs and prosperity (adapted from Droege et al. 2018a).

Fig. 17 The potential for regions to become energy self-sufficient is illustrated in the Lake Constance Alpine-Rhine Region in Central Europe. © Droege/oekom

Fig. 18 Energetic refurbishment of apartment buildings in Germany. © Jens Wolf/DPA

STAR used a methodological 'plug-in' for ascertaining the communal value capture coefficient. The Berlin based Institute for Ecological Economy Research developed a model to quantify the value added and employment effects generated by renewable energies at the local level, to fill the gap concerning the identification of local and regional economic effects of the energy transition. Discussions of climate protection and the energy sector transition are still dominated by cost considerations; however, decentralized efforts such as the expanded use of renewable energies at a communal level yielded new 'value-added' financial flows—from salaries, tax income and coporate revenue in the renewable energy planning, manufacture and servicing sectors. But energy-efficiency improvements to existing buildings, too, can have a positive economic impact with respect to local and regional value added and employment. For this the 'renewables value added' approach has been extended to account for the economic effects of energy-efficiency refurbishment of existing dwellings (from Heinbach et al. 2018).

14 Urban Carbon Sequestration

The 4M Project: Increasing carbon sequestration and storage in city greenspaces, Leicester UK

Low-Carbon Building Materials: The Soft House, Hamburg

The 4M project used a spatial modelling approach to identify areas and landcover types in the city of Leicester, United Kingdom with the greatest potential for increased carbon storage. Davies et al. (2011) found that trees accounted for around 97% of the carbon stored in aboveground vegetation, and based on the spatial models, discovered treat potential for tree planting in residential and non-residential settings to increase aboveground carbon storage in the city, as well as to produce biomass that could be used to substitute for fossil fuels (Figs. 19, 20). The tree planting modelling assumed

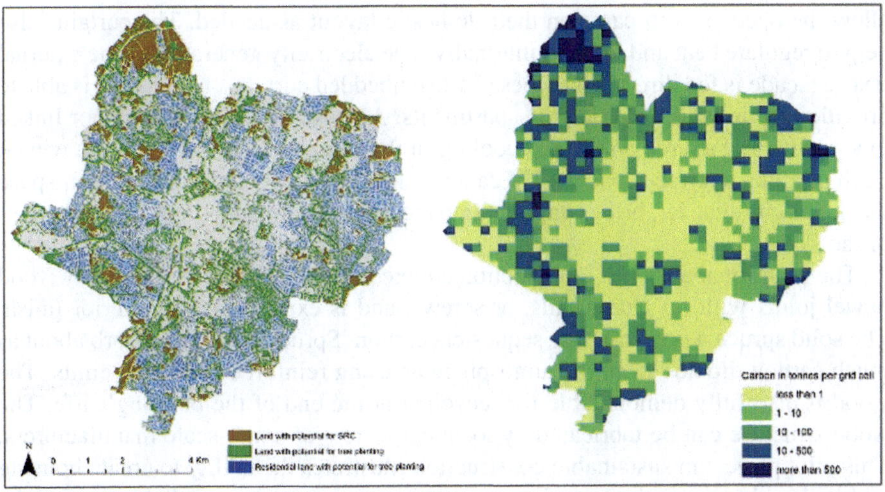

Fig. 19 Land potential for tree planting and additional carbon stroage potential in Leicester, UK. © Davies et al. (2011)

Fig. 20 The Soft House, Hamburg. © Sheila Kennedy, MIT

planting of mixture of species already existing in the city, and the potential carbon sequestration over 25 years. This is compared to an alternate scenario where short rotation coppice is used to generate woody biofuel (Davies et al. 2011).

The Soft House, designed by Kennedy and Violich Architecture, is an innovative work/live row housing project which demonstrates novel concepts in sustainable construction and domestic renewable energy generation. Located in Hamburg, Germany, it consists of four apartments units which uses a dynamic textile façade to harness sunlight, alongside solid wood construction. The energy-harvesting textile façade is responsive to movement of the sun, while inside the apartments transparent curtains allow the occupants to partition their domestic layout as needed. The curtains also help to regulate heat and warmth internally. The electricity generated by the external textile façade is fed directly into these LED embedded curtains, which thus is able to provide additional light inside the apartments. A dense wooden radiant floor linked to a geothermal source distributes cooling in the summer and heating in the winter. Each unit has a terrace space, a PV canopy, and a vertical convection atrium space that helps circulate air, brings daylight into the ground floor, and offers vertical views of the sky.

The traditional all-wood construction adopted in The Soft House uses only wood dowel joints with no glues, nails, or screws, and is exposed as the interior finish. The solid spruce wood structure sequesters carbon. Spruce is said to absorb about as much carbon dioxide from the atmosphere as using reinforced concrete emits. The wood is also fully demountable for recycling at the end of the building's life. The wood structure can be fabricated by local carpenters or small-scale manufacturers. This pilot project in sustainable construction illustrates the ability to create housing with reduced embodied material energy, that retains its soft natural character of the building material (Stauffer 2013).

15 Conclusion: Launching the Istanbul Protocol

In conclusion: local, national and global proliferation of the principles underlying the documented initiatives has long become critical—without current policy, regulatory and market frameworks having been yet fully adjusted. To articulate the call for action, we propose the *Istanbul Protocol*: paradigms and principles to shift to cities and regions regenerated through renewable energy, individual and collective innovation. It is a call to the nation—and nations elsewhere—for states, cities and regions to rise and support fundamental transformation in its economy, institutions and governance to enable the systematic replacement of inherited energy systems with distributed renewable energy infrastructures fully founded on new technologies and community benefits. This is also of meaning to finding ways of regenerating and retrofitting existing neighbourhoods and their building stock.

Acknowledgements This research analysis and compilation was funded by the CRC for Low Carbon Living Ltd supported by the Cooperative Research Centres program, an Australian Government initiative.

Peer Review Statement The CRCLCL recognises the value of knowledge exchange and the importance of objective peer review. It is committed to encouraging and supporting its research teams in this regard.

The author(s) confirm(s) that this document has been reviewed and approved by the project's steering committee and by its program leader. These reviewers evaluated its originality, methodology, rigour, compliance with ethical guidelines, conclusions against results conformity with the principles of the Australian Code for the Responsible Conduct of Research (NHMRC 2007), and provided constructive feedback which was considered and addressed by the authors.

Contributors Primary other contributors: Anis Radzi, Peter Newman, Josh Byrne, Mike Mouritz, Greg Morrison, Kayla Fox-Reynolds - and the authors of the second edition of Droege, P. 2018 Urban energy transition: renewable strategies for cities and regions. Elsevier – listed below and referenced throughout this paper. © 2018 Elsevier.

References

Boulding KE (1966) The economics of the coming spaceship earth. In: Jarrett E (ed) Environmental quality in a growing economy. Resources for the Future/Johns Hopkins University Press, Baltimore, MD, pp 3–14

Butera FM (2018) Sustainable neighborhood design in tropical climates. In: Droege P (ed) Urban energy transition—renewable strategies for cities and regions. Elsevier

Byrne J, Job T (2018) Utilizing the Urban fabric as the solar power plant of the future. In: Droege P (ed) Urban energy transition—renewable strategies for cities and regions. Elsevier

Davies et al (2011) Mapping an urban ecosystem service: quantifying above-ground carbon storage at a city-wide scale. J Appl Ecol 48:1125–1134

Droege P, Dieter DG, Ariane R, Matthias S (2018) Building regenerative regions rapidly: the STAR energy model as regional planning tool. In: Droege P (ed) Urban energy transition—renewable strategies for cities and regions. Elsevier

Droege PS, Teichman, Cody Valdes (2018) Solar for gaza: an energetic framework for renewable peace and prosperity for gaza and its greater region. In: Droege P (ed) Urban energy transition—renewable strategies for cities and regions. Elsevier

Fuller B (1968) Operation manual for spaceship earth. Lars Muller

George H (1879) Progress and poverty: an inquiry into the cause of industrial depressions and of increase of want with increase of wealth: the remedy. Robert Schalkenbach 1942 Edition

Heinbach K, Bernd H, Steven S (2018) Value-added and employment effects of renewable energies and the energy-efficiency refurbishment of existing housing—case study: Berlin, Germany. In: Droege P (ed) Urban energy transition—renewable strategies for cities and regions. Elsevier

Hellweg U (2018) Renewable Wilhelmsburg, Hamburg, Germany: using the international building exhibition to fight climate change. In: Droege P (ed) Urban energy transition—renewable strategies for cities and regions. Elsevier

Moser D, Marco L, Laura M (2018) Photovoltaic city: effective approaches to integrated urban solar power. In: Droege P (ed) Urban energy transition—renewable strategies for cities and regions. Elsevier

Radzi A (2018) The 100% renewable energy metropolis: governing the design of cities for renewable energy infrastructures. In: Droege P (ed) Urban energy transition—renewable strategies for cities and regions. Elsevier

Sobek W (2018) Buildings as renewable power plants: active houses for the electric city. In: Droege P (ed) 2018 Urban energy transition—renewable strategies for cities and regions. Elsevier

Sparber W, Roberto F, Chiara D (2018) Thermal city: comprehensive guide to the heating and cooling of Urban areas. In: Droege P (ed) Urban energy transition—renewable strategies for cities and regions. Elsevier

Stadler I, Sauer M (2018) Urban energy storage and sector coupling. In: Droege P (ed) Urban energy transition—renewable strategies for cities and regions. Elsevier

Stauffer NW (2013) Building façades that move, textiles that illuminate: A pathway to flexible, resilient architecture. http://energy.mit.edu/news/building-facades-that-move-textiles-that-illuminate/. Accessed 20 Oct 2018

Ward B (1966) Spaceship earth. Columbia University Press

Weiß J, Elisa D, Bernd H (2018) Implementing the heating sector transition in our cities—challenges and problem-solving approaches based on the example of municipalities in Germany. In: Droege P (ed) Urban energy transition—renewable strategies for cities and regions. Elsevier

Assessment of Prerequisites and Impacts of a Renewable-Based Electricity Supply in Austria by 2030

Gustav Resch, Gerhard Totschnig, Demet Suna, Franziska Schöniger, Jasper Geipel and Lukas Liebmann

Abstract In the final version of the Climate and Energy Strategy, the Austrian Federal Government postulated an ambitious target for the domestic expansion of renewable energy sources (RES) in June 2018: The goal is to generate electricity by 2030 to the extent that the national total electricity consumption is covered 100% (at a yearly balance) from renewable energy sources. This chapter provides information on how the transformation to an Austrian electricity system based almost exclusively on renewable energy generation can function and look like from a technical and economic point of view. Apart from that, we shed light on some of the requirements to and impacts of achieving this transition: On the one hand, a comprehensive economic reassessment of the expansion of renewable electricity supply in Austria by 2030 and the corresponding investment and support expenditures is presented. Apart from economic impacts, we also shed light on the impacts concerning supply security that come along with the strong uptake of renewables, specifically due to the massive expansion of volatile electricity generation from variable renewables like wind, solar and run-of-river hydropower. In conclusion, it should be noted that although the #mission2030 goal for the expansion of renewable energies appears

G. Resch (✉) · F. Schöniger · J. Geipel · L. Liebmann
Energy Economics Group, Institute of Energy Systems and Electrical Drives, Technische Universität Wien (TU Wien), Gusshausstrasse 25/370-3, A-1040, Vienna, Austria
e-mail: resch@eeg.tuwien.ac.at

F. Schöniger
e-mail: schoeniger@eeg.tuwien.ac.at

J. Geipel
e-mail: geipel@eeg.tuwien.ac.at

L. Liebmann
e-mail: liebmann@eeg.tuwien.ac.at

G. Totschnig · D. Suna
Austrian Institute of Technology (AIT), Vienna, Austria
e-mail: Gerhard.Totschnig@ait.ac.at

D. Suna
e-mail: Demet.Suna@ait.ac.at

T. S. Uyar (ed.), *Accelerating the Transition to a 100% Renewable Energy Era*, Lecture Notes in Energy 74, https://doi.org/10.1007/978-3-030-40738-4_4

extremely ambitious, it can nevertheless be classifiable as feasible. Massive investments in renewable technologies would occur, causing only a moderate increase in electricity prices since support expenditures would remain at a politically and socially acceptable level. Concerning supply security, it can be stated that expected increase in consumption due to the development of new consumption poses major challenges for the entire energy system. The model analysis shows that these are however manageable, but it requires considerable effort and proactive action. Rapid political action is consequently needed in order to be able to follow the ambitious path in a timely manner so that the planned increases in the generation stock as well as in necessary flexibility provision are available both in the early years after 2020 and later.

Keywords Renewable electricity generation · Austria · Investments · Support expenditures · Supply security · Flexibility needs

1 Introduction

This chapter provides information on how the transformation to an Austrian electricity system based almost exclusively on renewable energy generation can function and look like from a technical and economic point of view. Apart from that, we shed light on some of the requirements to and impacts of achieving this transition.

1.1 Background and Policy Context

In the final version of the Climate and Energy Strategy (BMNT 2018), the Austrian Federal Government postulated an ambitious target for the domestic expansion of renewable energy sources (RES) in June 2018: The goal is to generate electricity by 2030 to the extent that the national total electricity consumption is covered 100% (at a yearly balance) from renewable energy sources.

In the previous years, a broad set of assessments have been undertaken in this thematic context:

- In November 2015, the Association of Austrian Electricity Companies, namely Oesterreichs Energie (OE), published the electricity strategy "Empowering Austria". It goes beyond the year 2020 and plans to increase electricity generation from renewable sources by 20 TWh by 2030 compared to the current renewable electricity generation (2015). According to this, production from hydropower, wind power and photovoltaics should increase by 6–8 TWh in 2030 and the stock of electricity produced from biomass CHP (2 TWh) would be secured. In 2030, the share of renewable energies in electricity generation should amount to 85%.
- In April 2016, OE commissioned Ecofys and TU Wien to analyse which support systems for renewables would be the most cost-effective way to achieve the goals

outlined in the electricity strategy. In doing so, it should take into account the feasible expansion potential in Austria, comply with European requirements and refer to trends in other member states of the European Union. From that point of view, an extensive evaluation and analysis of the expansion of renewable energies and the required investments and support expenditures was carried out (Tiedemann et al. 2017).

In view of the current political circumstances, according to which an even stronger uptake of renewables is envisaged, a comprehensive economic reassessment of the expansion of renewable electricity supply in Austria by 2030 and the corresponding investment and support expenditures appears useful for the political debate. Apart from economic impacts, it appears of key relevance to shed light on the impacts concerning supply security that comes along with the strong uptake of renewables, specifically due to the massive expansion of volatile electricity generation from variable renewables like wind, solar and run-of-river hydropower.

All this was done in the context of two studies commissioned by OE,[1] aiming to support the implementation of the national integrated climate and energy strategy of the Federal Government (#mission2030) and the preparation of the energy security action plan. On the one hand, the study "Mission#Impact - Ökonomische Neubewertung des Ausbaus und des resultierenden Investitions - und Förderbedarfs erneuerbarer Energien in Österreich" (Resch et al. 2019) undertaken by TU Wien focusses on economic impacts and political needs. Complementary to that, on the other hand, the study "Versorgungssicherheit und Flexibilität bei 100% erneuerbarem Strom in Österreich im Jahr 2030 mit Hinblick auf 2050" (Suna et al. 2019) derived by AIT/TU Wien provides a first, comprehensive overview of the need for flexibility of the Austrian electricity system. Both studies focus on 2030 but provide an outlook beyond that (up to 2050).

2 Methodology and Key Assumptions

Comprehensive model-based assessments of the electricity sector in Austria and its neighbors have been undertaken in the course of both studies. The techno-economic analyses are thereby based on a transparent presentation of the results as well as the underlying assumptions. The sources of the model assumptions are thematically relevant preliminary studies on the part of OE as well as existing databases at TU Wien and AIT—especially with regard to costs and the dynamically available deployment potentials of renewable energies in Austria. Individual elements, such as the detailed design of the support policies for renewable energies, have been adapted specifically for this study.

[1]The authors gratefully acknowledge the financial and intellectual support provided by Oesterreichs Energie. For details on both studies we refer to https://oesterreichsenergie.at/the-world-of-electricity.html.

2.1 The Applied Modelling System

The quantitative analyses carried out are based on the use of two complementary energy system models developed by TU Wien:

- HiREPs: HiREPS is a power system simulation and optimisation model that allows for detailed assessments of the physical integration constraints of the power system. Thanks to a suitable grid representation also related limitations and/or requirements can be assessed. The model is used for the detailed modeling of the use of conventional energy sources and the effects of the coupling of electricity and heat generation by means of power-to-heat (P2H), the charging of electric vehicles and the necessary (pump) storage. The supply of renewable energy sources and the demand for electricity are included in the model as hourly (feed-in) profiles in accordance with the scenario specification.
- Green-X: The Green-X simulation model identifies the required market incentives, funding needs and funding design, and provides a detailed analysis of how different energy policy instruments work. The model aims at indicating consequences of RES policy choices in a real-world energy policy context. In principle, it allows for conducting in-depth analyses of future RES deployment and corresponding costs, expenditures and benefits arising from the preconditioned policy choices on country, sector and technology level on a yearly basis, in the time span up to 2050.

To maintain consistency, the models are coupled by an interface between detailed power system modeling on an hourly basis (HiREPs) and market research from today until 2030 and beyond (Green-X). Figure 1 gives an overview on the interplay of both models. Both models are operated with the same set of general input parameters, however in different spatial and temporal resolution. Green-X delivers a first picture of renewables deployment and related costs, expenditures and benefits on a yearly basis (2020–2050). The output of Green-X in terms of country- and technology-specific RES capacities and generation in the electricity sector for selected years (2020, 2030, 2040, 2050) serves as input for the power-system analysis done with HiREPS. Subsequently, the HiREPS model analyses the interplay between supply, demand and storage in the electricity sector on an hourly basis for the given years. The output of HiREPS is then fed back into the RES investment model Green-X. In particular the feedback comprises the amount of RES that can be integrated into the grids, the electricity prices and corresponding market revenues (i.e. market values of the produced electricity of variable and dispatchable RES-E) of all assessed RES-E technologies for each assessed country.

3 Key Results and Findings

This section is dedicated to shed light on key results and findings. We start with a brief recap of the modelled uptake of renewables in Austria's electricity sector.

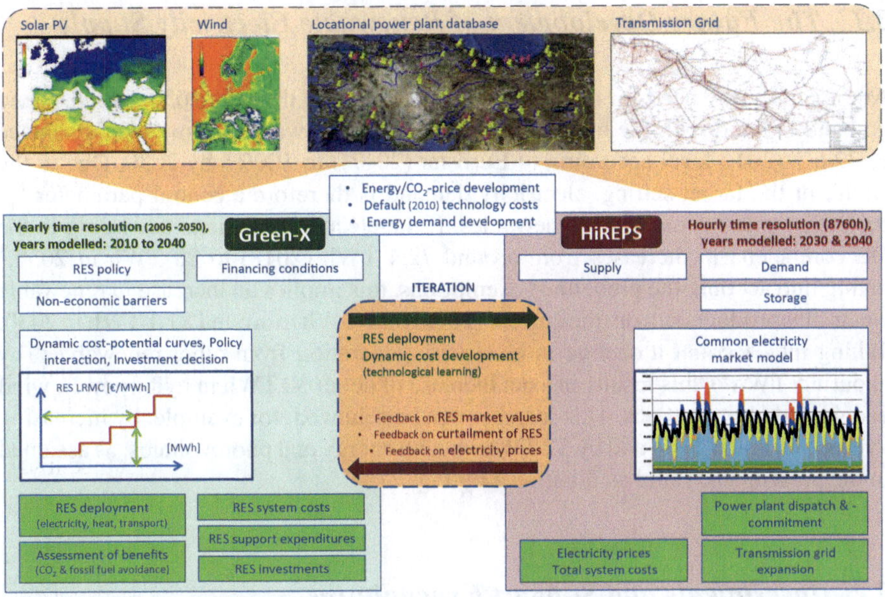

Fig. 1 Model coupling between Green-X (left) and HiREPS (right)

Next, economic impacts are illustrated, indicating the necessary investments and the support expenditures required to refinance the RES uptake. Finally, a closer look is taken at supply security, in particular at the required system flexibility that comes along with the strong expansion of volatile renewables like wind and solar (Fig. 2).

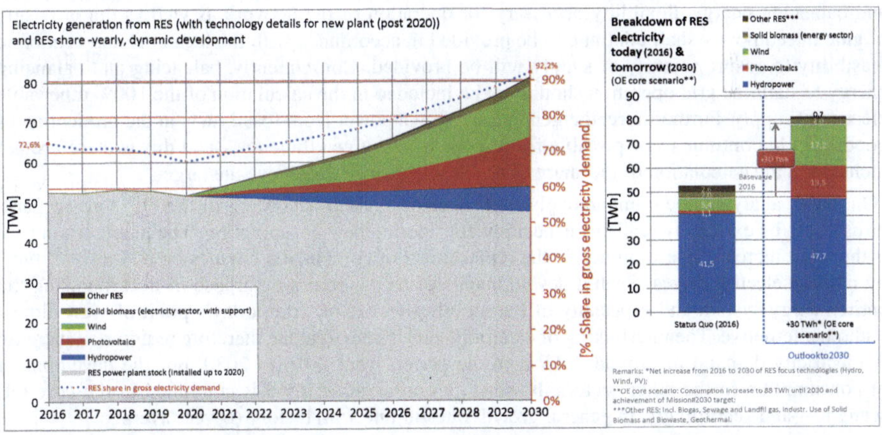

Fig. 2 Breakdown of the future development of renewable energy generation by construction period (left) and technology (right) according to the developed OE core scenario (*source* Green-X)

3.1 The Future Development of Renewable Electricity Supply

At the outset, the need for renewables expansion up to the year 2030, which would go hand in hand with the increase in the share of renewables from today's balance of 72.6% (2016 level according to Eurostat (2017)) to 100%[2] by 2030. Due to the nature of the target setting, electricity demand is therefore a central parameter for determining the demand for renewables. If the electricity demand according to the OE core scenario increases from around 72.4 TWh (2018) to 88 TWh in 2030,[3] taking into account the predefined exemptions, this implies an increase in renewable electricity production from the current (2016) 52.6 TWh to around 81.1 TWh in 2030. Taking into account a decline in electricity generation from other renewables (by about 1.5 TWh[4]), this results in a net increase of about 30 TWh in hydropower, wind energy and photovoltaics. This increase can be achieved, for example, by increasing hydropower by 6 TWh and by 12 TWh in wind energy and photovoltaics, as assumed in the present study and as illustrated by Fig. 2.

3.2 Investments and Support Expenditures

In the underlying study (cf. Resch et al. 2019), an economic reassessment of the future expansion and the resulting investment and promotion needs of renewable energies in Austria was made. In concrete terms, relevant aspects of the implementation of

---

[2]According to the Austrian Climate and Energy Strategy #mission2030 of June 2018 balancing and balancing energy, flexibility necessary for operation of the network as well as the provision of guaranteed power shall continue to be provided in accordance with the technical and economic feasibility to ensure security of supply will be provided. Consequently, balancing and balancing energy to stabilize grid operation should not be included in the calculation of the 100% renewable electricity supply. Furthermore, the generation of own power from fossil fuels in the production of goods should continue to be possible for reasons of resource efficiency—and this is therefore not included in the calculation of the share of renewable energy to achieve the target.

[3]On closer analysis, the comparatively high demand growth according to the OE core scenario turns out to be extremely realistic, especially considering new consumption. The main reason here is the rapid increase in e-mobility, ie the replacement of combustion engines based on fossil fuels by modern electric drives. Such a development seems to serve the purpose of decarbonising the entire energy system and especially of the mobility sector, or from today's point of view almost without alternative. The interlinking of electricity and transport alone therefore requires an increase in electricity demand of around 7 TWh in the period from 2016 to 2030. For the development of consumption in the other sectors, however, a conservative trend is continued in the core OE scenario—and accordingly, in general growth rates in line with history (6-year average).

[4]Particularly for biogas and according to the model analysis also for industrial biomass utilization, a significant reduction in production compared to today is to be expected due partly to expiring subsidies and due to the assumed persistently low energy price level. This sum totals 1.5 TWh comparing electricity generation from other renewables (i.e., biogas, sewage gas, landfill gas, geothermal, biogenic fraction of household waste and industrial biomass use) in 2016 and 2030.

the #mission2030 target for renewable energies in Austria's electricity sector are examined.

The investment needs induced by the expansion of renewables are significant (€ 2.6 billion per year according to the core scenario)

The massive expansion also requires significant investment in renewable energy technologies. In line with the net increase in hydropower, wind and PV power generation by around 30 TWh, on average over the coming demand yearly investments of around € 2.6 billion can be expected. The investment requirement correlates with the renewables ambition—if, for example, it increases from 30 to 35 TWh due to increased demand growth, this will increase the investment requirement to around € 3 billion and vice versa.

According to the core scenario, the support requirement is within a range of € 0.4–1.3 billion per year—depending on the general development of electricity prices

The analyzed need for support shows the demand for renewables, which is relevant from the consumer's point of view. In the case of floating market premiums (net) support represents the difference between the (competitively determined) price for electricity from a given RES technology and the market value of the injected electricity into the grid.

Since the market value of renewable electricity reflects the revenue situation on the wholesale electricity market, the clear dependence on the general development of electricity prices is given here. As illustrated above in Fig. 3 (left), three trend scenarios were considered in the modeling. According to the middle trend scenario, where a moderate rise in electricity prices is postulated, to around € 50/MWh by 2030, the annual average support requirement for the coming decade is around € 929 million (see Fig. 3, right). (Significantly) lower electricity prices, as postulated in the low-price scenario, would require a substantial increase (about 36%) in support expenditures. The same applies to the high-price scenario—if electricity markets follow this trend, this would result in a reduction of support costs by a considerable 53% compared to the core scenario of medium prices.

The need for support is sensitive to the level of ambition of the renewable expansion

On the other hand, the level of ambition regarding the expansion of renewable energies proves to be decisive. If, for example, renewables need to be increased by only 25 instead of 30 TWh net by 2030 due to a lower increase in electricity consumption (compared to the OE core scenario), this would reduce the average support expenditure by around 11%. An analogous statement (with opposite sign) naturally also applies in the case of a stronger renewal expansion in comparison to the OE core scenario serving as a reference.

Further sensitivities provide information about the influence of the design of the funding instruments

Within the framework of the underlying study, other relevant aspects influencing the resulting funding requirements were also examined in detail:

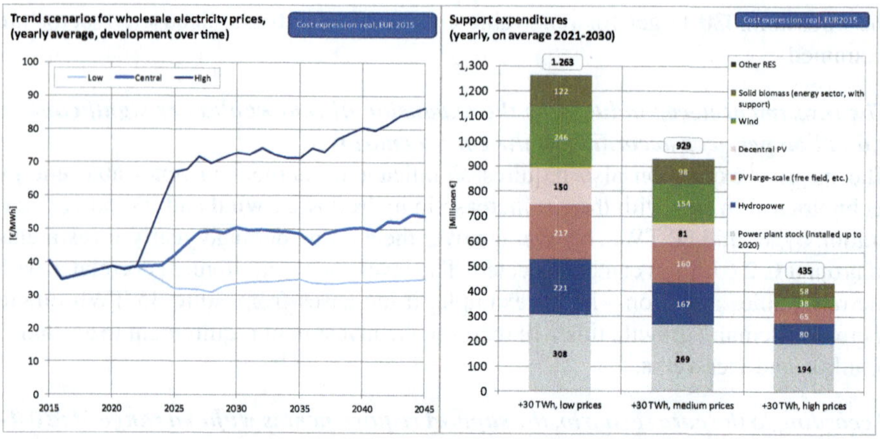

Fig. 3 Electricity price trend scenarios (left) and breakdown of the resulting demand for electricity from renewable electricity plants by construction period and technology according to the developed OE core scenarios (low, medium and high electricity prices) for the years 2021–2030 (right) (*Sources* OE Expert Advisory Council (2018) (left) and Green-X (right))

- Strategic auction bidding as an expected consequence of inadequate competition, according to the model calculations, resulted in an increase in support requirements of 9% on average compared to the respective comparison scenario, where perfect competition was assumed.
- A shorter guaranteed duration of support in accordance with current practice in Austria, i.e. 13 years instead of the across Europe common practice (20 years), leads to a significant increase of the support expenditures in the coming decade. The average annual funding requirement in the period from 2021 to 2030 would, according to the model calculations, increase by around 10% compared to the OE core scenario. This is in part a bringing forward of funding amounts that would otherwise have incurred later. In view of the expected increase in electricity prices in later years, but also to ensure the operation of RES plants over a longer period, even in the case of low electricity prices, it seems highly advisable to extend the duration of support.

3.3 *Supply Security and Flexibility Provision*

The high proportion of wind energy and photovoltaics leads to weather-related massive fluctuations in power supply. Flexibility is needed to compensate in the short term and to compensate for longer-term (seasonal) differences between production and consumption. Suitable options, which differ in their application due to various technical and economic parameters, are examined in the context of this study.

On the one hand, the flexibility options considered are the flexible power plant technologies consisting of gas power plants and combined heat and power generation, which can adapt their generation to changing situations in the grid as needed. Furthermore, the storage technologies (pump) storage power plants, compressed air storage, lithium-ion batteries and power-to-gas are considered. Additionally, load management is also taken into account, covering power-to-heat, electromobility and industrial load management. Further flexibility options include cross-regional and cross-border exchange via the transmission grid and the option of reducing non-controllable renewable generation (wind and PV).

Within the underlying study TU Wien's HiREPS power and district heating model is used for calculating a broad set of scenarios spanning the 2030, 2040 and 2050 years as well as various aspects of flexibility needs and security of supply.

The following presentation focuses on the most relevant case for security of supply: the scenario "Extrema-LimitHydro-2030". In this scenario, extreme weather conditions, including the occurrence of a "dark calm", as observable in the underlying meteorological data of 2006, in combination with low water levels in hydro reservoirs are assumed. The scenario provides information on whether the flexibility options available to Austria (in the future) can help to ensure a cost-efficient and secure electricity supply. In this scenario, network expansion in line with the APG master plan or the TYNDP by ENTSO-E, as well as a substantial expansion of storage power in Austria, remain valid as flexibility options.

Determining the demand for flexibility

In order to analyze the consequences of increasing volatility in electricity supply and possibly also in electricity demand, the combined analysis of the load and the volatile, i.e. the non-controllable, renewable electricity supply by means of the residual load factor is used. The residual load is defined as the electricity demand of the end customer minus the generation of electricity from volatile renewable energies (wind, PV and run-of-river power). Although the total annual electricity demand (in public power supply, excl. for example industrial auto-producers) of 81.5 TWh in 2030 is essentially covered by the generation of these three renewable sources in 2030, the simultaneous feed-in occurring in each hour in relation to the load leads to temporary generation gaps (positive residual load) or to temporary generation surpluses (negative residual load). As applicable from Fig. 4, indicating the development of the residual load over time for the given scenario, only in the rarest cases is the hourly balance produced a priori. The residual load thus shows that part of the load that has to be covered by the controllable power plants or other forms that provide flexibility.

The need for flexibility is derived from the fluctuations in residual load. These fluctuations occur both in the short term and in the long term and generally require different solutions. In order to better distinguish the short and long-term need for flexibility, the need for flexibility is broken down into five periods and evaluated. The shortest period is one day, followed by one week, one month, one season, and the longest period is one year. However, the requirements of flexibility (day, week, month, season and year) that occur recurrently within certain time periods are not summable

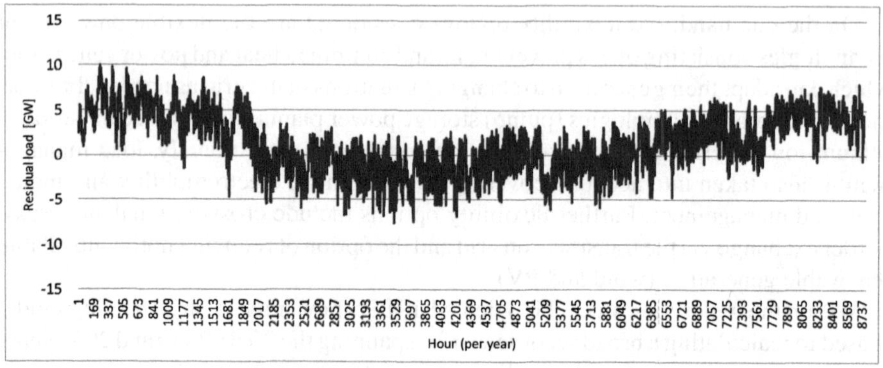

Fig. 4 Development over time of the residual load according to scenario "Extrema-LimitHydro-2030" (+shows deficit, −shows surplus)

but instead measure the variability of the residual load for the corresponding period. For example, the daily need for flexibility measures the hourly fluctuations in residual load within one day. It is calculated as the sum of the positive hourly deviations of the residual load from the respective daily mean value of the residual load. The result is then quantified as amount of energy per day. The summary of these positive daily differences for all 365 days of the year shows the total daily need for flexibility within one year.

In contrast, this study determines the annual need for flexibility as the cumulative sum (over all hours of a year) of the positive deviations of the seasonal mean values of the residual load from the annual average of the residual load. The annual need for flexibility thus characterizes the seasonal differences in residual load, for example in the case of an oversupply in the summer half-year and a shortfall in winter.

The calculations show that the Austrian power system of the future will be subject to strong dynamics in residual load and generation. It will be necessary to be responsive to the changes during system expansion and operation in order not to reach the limits of safe operation and thus security of supply. Remarkable is the strong increase of the negative residual load and the negative gradient (GW/h) of the residual load as a result of increasing temporary power surpluses. This is mainly due to the future high shares of volatile sources wind and PV in combination with a possibly dynamic load (where demand is responsive to price signals for e-car charge and power-to-heat by using electric boilers and heat pumps). Furthermore, the natural seasonal fluctuations of these plants must be managed. Depending on the scenario chosen, the need for flexibility is around 4.8 TWh per year, for example in the case of daily flexibility, and the annual flexibility requirement is around 10 TWh (cf. Fig. 5). The coverage of the need for flexibility as well as the use of power plants depend on price developments at the European electricity markets and the availability of the various flexibility options. It should also be borne in mind that merely meeting the need for flexibility, i.e. balancing variability, is not enough to guarantee a secure electricity supply. Here it is important to maintain the overall perspective and thus to balance

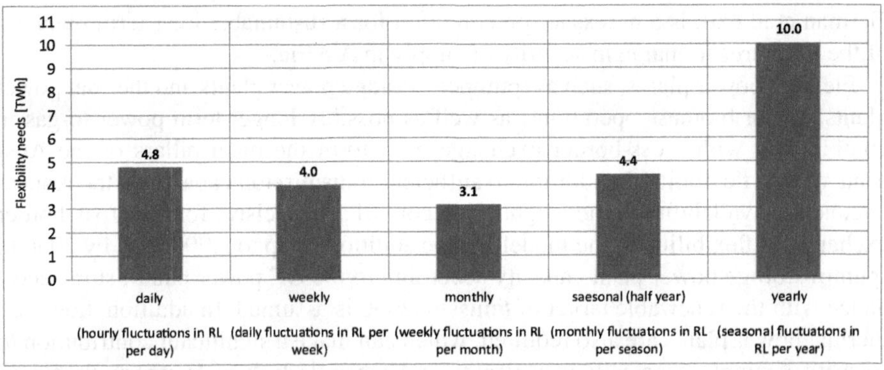

Fig. 5 Flexibility needs at distinct time periods according to the scenario "Extrema-LimitHydro-2030"

the residual load, both in a positive and in a negative direction. For this purpose, further measures should be provided in a suitable form.

Furthermore, the complete and on-schedule expansion of the (pump) storage power plants (construction of about 2,900 MW in the considered scenario) according to current OE power plant list until 2030 will not be sufficient to fully meet the need for flexibility in implementing the expansion targets for renewable electricity generation from their own resources.

4 Conclusions

The politically desired and promoted massive expansion of renewable energy as well as the expected increase in consumption due to the development of new consumption poses major challenges for the entire energy system. The model analyses show that these are however manageable, but it requires considerable effort and proactive action.

Active political action seems to be crucial in order to enable the planned expansion of renewable power station capacities by means of adequate framework conditions. This includes, for example, speeding up the implementation of power plant and pipeline projects in due time, as well as raising public awareness on the part of politicians regarding the necessary and useful measures for implementing the energy transition. To maintain system stability, the required network infrastructure must also be available. This implies the implementation of the grid expansion plans at the national level according to the APG master plan or at the European level according to TYNDP by ENTSO E. Since the international electricity exchange is and will be an important part of a cost-efficient electricity system, the timely recognition of energy policy developments in the neighboring countries, such as the solutions to the

German coal exit, is a necessary precondition for a sustainable long-term planning of the measures to maintain security of supply in Austria.

Flexible power plants, such as (pumped) storage power plants and thermal power plants, gas or biomass operations, as well as possibly longer-term power-to-gas in combination with cross-border exchange, will form the main pillars of the Austrian system flexibility. In addition to sufficient transmission capacities, the respective, actual availability in the neighboring countries is decisive for the cross-border exchange of flexibility. In the modeling, the addition of up to 2,900 MW by 2030 to (pump) storage power plant capacity according to the OE power plant list in accordance with the renewable target of #mission2030 is assumed. In addition, however, thermal power plants are also required, which can make a significant contribution to security of supply, especially in critical times such as dark skies. However, the future level of CO_2 prices probably plays a key role for their profitability. In the future, if electricity market revenues are not sufficient to ensure the continued existence of the required power plants or to allow new investments, alternative arrangements might be necessary to safeguard the availability of capacity reserves in critical hours and/or days.

In conclusion, it should be noted that although the #mission2030 goal for the expansion of renewable energies appears extremely ambitious, it can nevertheless be classified as feasible. Massive investments in renewable technologies would occur, causing only a moderate increase in electricity prices since support expenditures would remain at a politically and socially acceptable level. Rapid political action is consequently needed in order to be able to follow the ambitious path in a timely manner so that the planned increases in the generation stock as well as in necessary flexibility provision are available both in the early years after 2020 and later.

In the underlying studies, model analyses of the effects of the ambitious goals of the #mission2030 is carried out for the first time. They should thus serve as a starting point for further analyses that are necessary for a fact-based specification of the necessary conditions for the economic expansion and use of flexibility technologies in the energy system of the future.

References

Bundesministerium für Nachhaltigkeit und Tourismus (BMNT) (2018) #Mission2030 – Endfassung der österreichischen Energie- und Klimastrategie 2030. Bundesministerium für Nachhaltigkeit und Tourismus, Juni 2018. http://www.mission2030.bmnt.gv.at#Mission2020

Eurostat (2017). Online database from EUROSTAT on energy statistics, accessed in July 2018.

OE Expert Council (2018): Expert judgement (status 6 September 2018) on future electricity price trends, derived by OE's Advisory Board for the study Mission #Impact. Vienna, Austria, 2018.

Resch G, Liebmann L, Schöniger F (2019) Mission #Impact - Ökonomische Neubewertung des Ausbaus und des resultierenden Investitions- und Förderbedarfs erneuerbarer Energien in Österreich (in German). A study by TU Wien, commissioned by Oesterreichs Energie. Vienna, Austria

Suna Demet G, Totschnig C, Messner H, Aghaie J, Kathan W, Friedl G, Resch F Schöniger (2019) #MissionFlex – Versorgungssicherheit und Flexibilität 2030 (in German). A study by AIT and TU Wien, commissioned by Oesterreichs Energie. Vienna, Austria

Tiedemann Silvana, Klessmann C, El-Laboudy T, Resch G, Totschnig G, Welisch M, Liebmann L, Hiesl A (2017) Fördersysteme für erneuerbare Energien in Österreich (Endbericht – in German). A study commissioned by Oesterreichs Energie, done by Ecofys and TU Wien. Berlin, Vienna

History and Recent State of TIMES Optimization Energy Models and Their Applications for a Transition Towards Clean Energies

Kathleen Vaillancourt, Olivier Bahn and Nadia El Maghraoui

Abstract Mathematical models of energy-economy-environmental systems (E3) provide a rational framework for exploring the effects of energy and climate policies and support adequate decision-making. Numerous models have been developed over the years with different solution approaches, features, geographical scope and time resolution. There is no complete or ideal models but different models that answer different questions or similar questions with different perspectives. Developed since the early 1980s, the TIMES (The Integrated MARKAL-EFOM System) optimization models have contributed to support decision-making at various geographical scales from global to city levels. In this Chapter, we distinguish a set of national studies that performed TIMES model developments to study the energy transition and address the impacts of integrating high levels of renewable energies on the system. Each study follows a different approach with the sole purpose to optimize the energy used in order to reduce greenhouse gas (GHG) emissions. Examples of applications are provided to illustrate the rich potential of optimization models for assisting decision makers with climate change mitigation. In particular, a special attention is given to the electricity sector as electrification of end-uses and decarbonization of the electricity sector are consistent priorities of actions across studies.

Keywords Optimisation · Energy system models · Energy transition · Climate mitigation · Renewable electricity

K. Vaillancourt (✉)
ESMIA Consultants, Blainville, QC, Canada
e-mail: kathleen@esmia.ca

O. Bahn
GERAD and Department of Decision Sciences, HEC Montréal, Montreal, QC, Canada

N. E. Maghraoui
GERAD and Polytechnique Montréal, Montreal, QC, Canada

T. S. Uyar (ed.), *Accelerating the Transition to a 100% Renewable Energy Era*, Lecture Notes in Energy 74, https://doi.org/10.1007/978-3-030-40738-4_5

1 Modelling the Energy Transition

As part of the Paris Agreement, most countries committed to reduce their greenhouse gas (GHG) emissions with the ultimate goal of keeping the global temperature rise below 2 °C. Moreover, it is increasingly argued that the carbon neutrality of the global energy sector should be achieved in 2050 to maintain the temperature below 1.5 °C by 2100. Climate change mitigation is a very complex problem with many different combinations of possible solutions, which vary among sectors, among regions, and over time. In all cases, however, achieving these commitments requires important transformation of energy sectors.

The complexity of the links between the different energy sectors, as well as between the energy sector and the rest of the economy suggests the use of a systemic approach. In this context, mathematical models of energy-economy-environmental systems (E3) provide a rational framework for exploring the effects of energy and climate policies and support adequate decision-making. Numerous models have been developed over the years with different solution approaches, structures, intrinsic features, levels of details, geographical scope and time resolution. There is no complete or ideal models but different models that answer different questions and/or similar questions but different perspectives.

Optimization E3 models provide a rigorous analytical basis for defining decarbonization pathways that meet growing demands with progressive reductions in GHG emissions at a minimum cost. Among the optimization category of models is the TIMES (The Integrated MARKAL-EFOM System) model generator developed within a Technology Collaboration Programme of the International Energy Agency and used by a large number of organizations worldwide.

This chapter focuses on recent state of TIMES optimization energy models and their applications for a transition towards clean energies. Section 2 provides a brief overview of the main classes of E3 models with their main characteristics and roles for studying the energy transition. Section 3 takes a closer look at the optimization TIMES approach. Most recent developments in various TIMES models used worldwide are summarized in Sect. 4. Following these model improvements, examples of applications to a transition toward clean energies are provided in Sect. 5, before concluding in Sect. 6.

2 A Simplified Classification of E3 Models

For many years, several models covering energy, economy, and environment (E3 models) have been developed. However, each model may follow a different approach. In order to aid potential users of these models to find out which one is most suited for a certain purpose or situation, several studies have attempted to classify them using different criteria (Bahn et al. 2005; Beeck 1999; Boulanger and

Bréchet 2003; Morrison et al. 2015). The most common criteria used to differentiate between these models concerns their analytical approach, namely bottom-up, top-down and hybrid.

Bottom-up models use a techno-economic approach representing the technologies of the energy sector in detail while top-down models use a macroeconomic approach based on an entire economic description (Bahn et al. 2005). However, these two approaches become decreasingly relevant to the emergence of hybrid models in which the two approaches have been merged (Vaillancourt 2010). In addition, due to the recent advances in both model categories (bottom-up and top-down), the distinction between these categories tends to be somewhat confused (Loulou et al. 2004).

This review does not include integrated assessment models capturing the feedback between energy and climate systems.

2.1 Techno-Economic Models

Bottom-up models represent in detail the so-called Reference Energy System (RES) for one or more regions, including energy exchanges between regions. Resources, current and emerging technologies (extraction, production, consumption) and forms of energy are explicitly characterized by their technical and economic attributes (efficiency, emission factors, etc.) (Vaillancourt 2010). Bottom-up models calculate the production of primary and secondary energy and also the consumption of total final energy in order to meet energy demands as well as eventual emission reduction constraints (Bahn 2018). Demands are based on socio-economic (rather than energy) assumptions and are expressed exogenously from actual needs in physical units (number of houses, industrial production, vehicle-kilometers, etc.). We distinguish two main types of techno-economic models namely optimization models and simulation models (Vaillancourt 2010):

- Optimization models minimize the updated total cost of the system in order to meet final demands while respecting environmental constraints. They, therefore, calculate a partial equilibrium between energy supply and demand in a perfectly competitive market and determine the least costly technological combination to satisfy final energy demand, as well as the price of each form of energy. Policies are modelled through constraints on technologies, forms of energy or air pollutants.
- Simulation models focus on the representation of consumer behaviour (individuals, industries). They contain information on the competitiveness of rival technologies. The technological choices are determined by investments. Thus, the market shares of different technologies are not always based on optimal choices. Policies are modelled through constraints on market share and technology diffusion processes.

The principal limitation noted in bottom-up models lies in the fact that they do not represent the complete interactions that link the energy sector with the rest of the economy. Indeed, the optimal solution, in this case, corresponds to a partial economic equilibrium (Bahn 2018).

The first optimization models developed were MARKAL (MARKet Allocation) (Loulou et al. 2004), EFOM (The Energy Flow Optimization Model) and MESSAGE (Model for Energy Supply Strategy Alternatives and their General Environmental Impact) (IIASA 2019). They were developed during the 70s and 80s in order to optimize energy systems in the medium term. Improved versions were then combined in more recent models, in particular the TIMES model (Loulou et al. 2016). TIMES are reference analysis tools of the Energy Technology Systems Analysis Program (ETSAP 2019) of the International Energy Agency.

POLES (Prospective Outlook on Long-term Energy Systems) is a simulation model used by the European Commission (2019a). Similarly, the US Department of Energy uses NEMS (National Energy Modelling System) for annual energy outlook production (EIA 2019). MAED (Model for Analysis of Energy Demand) is yet another simulation model used to assess future energy demands based on assumptions on medium to long-term scenarios of socio-economic and technological development of a country or region (IAEA 2006).

2.2 *Macroeconomic Models*

Top-down models represent the whole economy; however, they describe energy technologies in an aggregated way. These models may represent more interactions that link the energy sector with the rest of the economy. The principal limitation noted in top-down models is that they do not represent precisely energy technologies (Bahn 2018). Two main types of macroeconomic models can be distinguished (Vaillancourt 2010):

- General equilibrium models that have a neoclassical view of the economic system describing the global economy through the behaviour of economic agents. They consider the feedback between quantities and prices and calculate an equilibrium price in each market assumed to be in perfect competition. They are useful for analyzing major structural effects in terms of supply and demand as well as long-term effects. They are used to analyze impacts of international climate policies on national economies and to proceed to a simulation of international cooperation strategies.
- Macro-econometric models are neo-Keynesian whose economy is demand-driven. They capture the medium-term dynamics of the national economic aggregates and its components. Prices vary with supply and partial market imbalances. They allow analyzing the global impact of climate policy on economic variables such as gross domestic product, employment and balance of trade.

The EPPA (Emissions Prediction and Policy Analysis) model of the Massachusetts Institute of Technology (MIT 2019) and the GEM-E3 (General Equilibrium Model for Energy-Economy-Environment Interactions) model of the European Commission (2019b) are examples of general equilibrium models used to analyze impacts of climate change policies. On the other hand, the Oxford model also called GMM

(Global Macro-econometric Model) is an example of a macro-econometric model that has been developed to perform more disaggregated analyses of the energy sector (Oxford Economics 2019). There also exist input-output models based on national data tables that are used to study the intersectoral effects of short-term climate policy at the national level, such as the effects of carbon taxes. The META-Net economic modelling system is applied to find the impacts of carbon taxes on national energy systems (Nakata 2004).

2.3 *Hybrid Models*

Hybrid models have been developed to better integrate macroeconomic factors into techno-economic models. For example, TIMES-MACRO (Loulou et al. 2016) is a technological model linked to macroeconomic modules. In another example, recent versions of GEM-E3 incorporate in their general equilibrium framework a techno-economic description of electricity supply (Capros et al. 2013). Finally, the Long-Range Energy Alternatives Planning System (LEAP) model combines different modelling approaches: techno-economic, simulation, macroeconomic and accounting techniques (Heaps 2008).

2.4 *Comparison of Approaches*

It should be recalled that there is no ideal or complete model, but rather many models whose choice will depend on the type of issues to be analyzed or the type of decisions to be made:

- Bottom-up models are used to identify an optimal configuration of energy systems. These models can assess the impact from a techno-economic point of view on the energy sector of energy and climate policies. They are also called energy system models.
- Top-down models are used to address the impact of energy and climate policies on macroeconomic variables such as employment and gross national product.

The main differences in characteristics, as well as the main role and limitations associated with these two categories of models, reside in their traditional forms. These differences are summarized in Table 1. We should note, however, that these distinctions are decreasingly noticeable with the development of hybrid models (Vaillancourt 2010).

In addition, we should note that this is not a comprehensive list of distinct features, as models differ in many dimensions, not only following the bottom-up and top-down classification but within each class of models as well. Bottom-up models for instance are characterized by alternative structures, geographical scopes, time resolution and assumptions about the future. Some models have a specific sector detailed (e.g. the

Table 1 Characteristics and roles of the two main families of E3 models

Category	Bottom-up models	Top-down models
Database	Explicit and detailed representation of energy forms and technologies in energy sectors Energy flows	More general representation of energy supply and demand in the main economic sectors Monetary flows
Approach	Techno-economic approach where energy balance conditions (supply = demand) are maintained throughout the system in physical units	A macroeconomic approach where economic feedback between the energy sector and other economic sectors are taken into account
Balance	Partial balance: energy sector	General balance: entire economy
Demand	Driven by the demand for energy services, exogenously specified by the user	The use of energy is defined as the result of economic equilibrium
Type	Optimization models Simulation models	General equilibrium models Macro-econometric models Input-output models
Examples	EFOM, MAED, MARKAL, MESSAGE, NEMS, POLES, TIMES	EPPA, GEM-E3, GMM, META-Net
Hybrid approaches (examples)	Bottom-up with link to macroeconomic: MARKAL-MACRO, CIMS	TB with a bottom-up description of some energy sectors: recent versions of GEM-E3
	LEAP: mix of bottom-up and top-down modelling techniques	
Role	Identify a configuration of energy systems and measure the technical and economic impacts of implementing policies	Evaluate the impact of policies on the global economy and macroeconomic variables such as employment and gross domestic product
Limit (traditional version)	Limited to the energy sector for policy analysis	Limited to a more general analysis of energy sector policies

Source Adapted from Vaillancourt (2010)

electric sector), while others include a full representation of the whole energy system. They are also characterized by different solution approaches which imply different assumptions about the decision process over time: some assume a perfect foresight context where actors can anticipate future events while making optimal decisions today and some assume they are myopic where decisions are made on the basis of current conditions.

3 A Closer Look at the Optimization TIMES Approach

We provide next more detail on the MARKAL/TIMES bottom-up approach, as the reminder of this paper will focus on developments and applications with TIMES models.

3.1 *MARKAL Model*

MARKAL model is one of the first bottom-up energy models that have been developed since the early 1980s after the first oil crisis. It has been widely used subsequently to study energy security issues after the first oil crisis, followed by environmental problems such as acid rain and now climate change mitigation.

It is cast as a dynamic mathematical model with perfect foresight. This implies in particular that investment decisions are made for each time period with perfect knowledge of future events. A period may, for example, be of a 5-year duration; the years of the same period are assumed to be identical.

Including a wide range of energy technologies, the model distinguishes each of them by describing its technical and economic parameters. The model calculates energy balances within the energy system to provide energy services at an overall minimum cost. MARKAL calculates then a (partial) balance of energy markets, meaning that energy producers deliver exactly how much energy consumers are willing to buy.

Contrary to earlier bottom-up energy models, demands for energy services are elastic to their own prices in MARKAL. This enables the model to maximize the total surplus of energy producers and consumers while facing certain constraints, and allows a more accurate computation of the balance between supply and demand (Boulanger and Bréchet 2003). However, assumptions of perfectly competitive markets may be relaxed by introducing specific assumptions, such as penetration limits for new energy technologies.

3.2 *TIMES Model*

The evolution of MARKAL is the TIMES model that combines the two MARKAL and EFOM approaches. TIMES is currently used by more than 80 institutions across 70 countries. It corresponds to a dynamic partial equilibrium model to analyze energy markets.

Similar to MARKAL, TIMES relies on linear programming to maximize the total surplus of energy producers and consumers, while respecting specific constraints. This is operationally done by minimizing the total discounted cost of the energy systems used to meet useful energy demands. The two models also share a multi-period, multi-regional, structure, to analyze a (potentially large) number of regions while capturing energy trade between them. But TIMES provides more flexibility to modellers, for instance through time periods of variable lengths, a flexible number of hierarchical time-slices, a more refine approach for representing vintage processes, or a flexible definition of energy processes (Loulou et al. 2016).

Moreover, the timing of investment payments is much more detailed in TIMES and allows distinct (and more realistic) capital flows whether it relates to large infrastructure (e.g. a new hydro dam) or smaller technologies (e.g. a car). TIMES acknowledges as well that global and technology-specific discount rates are time-dependent and not constant over time.

TIMES deals with the entire energy sector but can also apply to a single energy sub-sector (e.g. electricity or transport). To describe energy systems, the model uses inputs that represent inventories of equipment related to existing systems, but also the characteristics of (potential) future technologies. Similarly, TIMES represents current as well as future sources of primary energy supply. In addition to these inputs, the model also supports energy-environment policy analysis (Loulou et al. 2016). Furthermore, in order to explore future energy systems, in the long run, TIMES adopts a scenario approach. A scenario is based on a set of consistent assumptions. The model also recognizes that the demand for energy services is elastic in relation to its own prices. This makes possible the endogenous variation of demands in the policy scenarios relative to the baseline scenario, capturing behavioural changes and their impacts on the energy sector.

4 Recent Developments in TIMES Models Used Worldwide

We distinguish a set of studies conducted in different countries that performed TIMES model developments to study the energy transition and address the impacts of integrating renewable energies on the system. Each study follows a different approach and methodology with the sole purpose to optimize the energy used in order to reduce greenhouse gas (GHG) emissions.

Given the popularity of TIMES models for studying the energy transition, it was not possible to review all relevant methodological developments. For instance, the latest review report of the ETSAP (Vaillancourt 2018) includes over 200 references published during the 2014–2016 period only with developments and applications of more than 80 TIMES models.

We have selected few examples of recent papers to illustrate the numerous possibilities of TIMES in terms of analysis of the energy transition and renewable penetration. Moreover, we have restricted the selection of papers with analysis of deep decarbonization scenarios and very high renewable penetration rates. Consequently, other interesting developments of TIMES models are not necessarily covered in this chapter, including better representation of consumer behaviour and of other global challenges such as access to water and food.

4.1 Refine the Representation of Specific Sub-systems

Systems analysis is a central concept to the developments and applications of these models. System analysis is a problem-solving method that provides a better understanding of the behaviour of complex systems such as energy systems and serves as a basis for improved decision-making with access to better information. The core of the approach is to separate an overall complex system into various components or sub-systems with the most appropriate level of detail for adequately supporting the decision-making process that it is supposed to serve. A sub-system can be considered to be a specific sector in a specific region in a specific time period and its interactions with other sectors and the same sector in other regions and time periods. Policies and targets can apply the entire system, as well as to some or all of the sub-systems.

Given the complexity of energy systems, the large amount of data required to represent them in many details, and the numerous assumptions required to assign values to highly uncertain technical or economic parameters over time, developing specific sub-systems in existing large scale optimization models to study the energy transition is a methodological contribution in itself.

For example, Sgobbi et al. (2016) have developed a comprehensive hydrogen supply chain module in a multi-regional TIMES model for European countries to assess its role in climate mitigation scenarios with reduction targets up to 80% by 2050 compared with 1990 levels. They distinguished four types of hydrogen production technologies from various input fuels in centralized and distributed versions: gasification and pyrolysis, reforming, electrolysis and nuclear reactors. Hydrogen is also a by-product of advanced ammonia and chlorine production technologies. The capture hydrogen transport in liquid or gaseous forms from centralized facilities.

As another example, Vaillancourt et al. (2019) conducted a study to explore the role of bioenergy in Quebec's rapid transition to a low-carbon energy system through a TIMES model for Canada with a detailed representation of multiple bioenergy pathways. Model developments allow a comprehensive representation of supply chains with: (i) a large variety of feedstocks (crops, fatty residues, forest residues, agricultural residues, pulp and paper residues, dedicated crops of fast-growing trees, organic municipal waste, manure, sewage sludge, and landfill biogas), (ii) many conversion processes (fermentation, transesterification, combustion, gasification, hydrolysis, pyrolysis, anaerobic digestion, etc.), and (iii) numerous options for final usages of bioenergies.

4.2 Refine the Time Resolution

Traditionally, TIMES model are solved for a limited number of time periods over the 2050 or 2100 horizons with a limited number of annual time slice (e.g. 2–4 seasons and 2–4 intraday periods). Given the flexibility of TIMES for defining the time resolution as well as improvements in computing capabilities, few authors have tested the impacts of having a greater disaggregation of the annual time dimension for studying the decarbonization of the electricity sector.

In Krakowski et al. (2016), a TIMES model was used to evaluate the penetration of renewable energy in the French electricity system ranging from 40% to 100%. The purpose of this study was to broaden the debate on whether such high renewable energy penetration rates were feasible. TIMES provides a realistic representation of electrical systems and plausible options for their long-term development. The model was completed with a thermodynamic description of electrical systems to assess their reliability. In addition, the 2012–2050 horizon has been divided into 13 annual periods. Each period was further divided into seven seasonal periods (six monthly periods plus one period representing a potential winter week), each seasonal period was divided into two typical days (working days and weekends) and each typical day has been subdivided into six periods including two periods for the night, two for the morning, one for the afternoon period and the sixth period for the maximum demand).

Kannan and Turton (2011) developed the Swiss TIMES electrical system model (STEM-E) to generate insights on long-term development of the electricity sector under a cost-minimization framework. The main objectives were to analyze electricity generation at the hourly level taking into account the availability and operational constraints of the interconnected system elements, and elucidate the problems associated with the integration of intermittent renewable energy technologies. To achieve these objectives, STEM-E was calibrated on historical data from 2000 to 2009. Key inputs included past and future electricity demand, existing technology stocks, national and imported energy resources, technical and economic characteristics of future electricity and heat generation technologies.

Another study done by Drouineau et al. (2015) used a TIMES model to analyze the capacity of the Reunion Island to reach its autonomy in electricity by 2030 with a fine resolution of time periods: each year was divided into two seasons, namely; the summer season and the sugar season and each day has been subdivided into eight time slices. The approach adopted focused on conducting a prospective study, which provides future production mixes under different scenarios. This approach has been associated with a quantitative assessment of the reliability of the power supply using two reliability indicators indicating that intermittent sources can strongly develop and thus worsen their reliability.

4.3 Linking TIMES with Other Models or Tools

Another approach consists to link TIMES models to other models or analytical tools in order to capture additional dimensions of the problem to be solved and provide more complete solutions.

A soft-link approach with a simulation model has been used by few authors as it provides a more realistic picture of technology stocks turnover than optimization models and sometimes include more details and have a finer time resolution (one-year steps). Thellufsen et al. (2019) have analyzed cleaner solutions for district heating in Ireland in a future low-carbon system using the Irish TIMES model. To this end, EnergyPLAN was used to study in addition to the operation of the heating system, the feasibility of district heating compared individual heating solutions. EnergyPLAN allows time simulation of the heating system and therefore, it includes the operation of combined heat and power plants, boilers and storage facilities in relation to the electrical system, on a chronological basis on the year.

Similarly, Vaillancourt et al. (2017) have used a multiregional TIMES model to explore deep decarbonization pathways for Canada in a soft-link framework with a simulation model that is calibrated with historical data from 1978 and enables projections to 2050 and beyond in one-year steps. The soft-links work iteratively in both directions where the simulation model provides projections of key macroeconomic drivers and service demands and decision variables from the optimization model are integrated back in the simulation model for further refinements to input variables.

Another area where a soft-link approach with another model has proved to be particularly useful is the analysis of the optimal electricity generation mix with high penetration of renewables over time. Indeed, some electricity sector models provide a more precise representation of electric systems than TIMES namely regarding the optimal dispatch. Tigas et al. (2015) used a TIMES model by linking it to a probabilistic production simulation model (ProPSim) to study the decarbonization of the Greek electricity and transport systems by 2050. The ProPSim model calculates the residual charge duration curves, which are used to calculate the optimal extension of the dispatchable generation plants. This combination makes it possible to better manage the stochastic aspects related to the penetration of renewable energies.

Welsch et al. (2014) have assessed the effects of soft-linking the Irish TIMES model with a well-known unit-commitment and dispatch model: PLEXOS. It allows simulating the electricity market with a more detailed temporal resolution, thus enabling to minimize the expected costs of the electricity dispatch. PLEXOS accounts for additional operational details such as minimum stable generation levels and operating reserve requirements.

Finally, several attempts have been made for linking TIMES models to life-cycle analysis (LCA) tools. McDowall et al. (2018) aimed to flexibly link a LCA-based tool to a European TIMES model in order to determine how the inclusion of indirect emissions can change the optimal technological pathways for the decarbonization of the European energy system. The indirect CO_2 emission factors associated with the construction of power sector technologies have been calculated by means of a hybrid LCA approach. The method consists of a disaggregation of an input-output table and its environmental extension based on data from life-cycle inventories, and the use of Environmentally-extended Input-Output (EEIO) analysis to calculate carbon emissions. It allows overcoming the main limitations of each approach, i.e. the high aggregation of EEIO models and the incomplete system boundaries in LCA.

5 Applications to a Transition Toward Clean Energies

Once more, it was not possible to review the numerous publications with applications of TIMES models to study the transition toward clean energies. Examples are provided below to illustrate the rich potential of optimization models for assisting decision makers with such a complex problem that is climate change mitigation.

5.1 Decarbonisation of the Energy System

Multiple TIMES optimization models are used today to support the analysis of energy and climate policies all over the world. Geographical coverage varies from the global to the city level as does the spatial resolution within each model region (Vaillancourt 2018). Different models with different spatial resolution provides complimentary information for decision makers taking into account both internal regional differences and the global factors.

Due to the large diversity of energy systems within a specific region or even a specific country, detailed national or multiregional approaches are necessary to study the energy transition while capturing the needs for investments in infrastructure for clean energy transport and distribution within the country or between neighbouring regions (Vaillancourt et al. 2017). These investments are not captured in global models with continents or multiple countries represented as single aggregated regions. However, limiting assumptions are required in such national models regarding the potential evolution of the international demand for energy resources under different levels of commitments for climate change mitigation in the various countries.

Global models allow making more consistent and comprehensive assumptions regarding the evolution of international trade movements for energy commodities in such a mitigation context. As a global challenge, countries cannot be looked at in isolation when it comes to climate policies. Global models can also be used to study the international cooperation aspects, an important component of the global climate change agenda. This is a significant challenge for emerging countries as significant capital investments are required in order to transform the energy system without jeopardizing socio-economic growth.

As a technology-rich model with a flexible definition of the time dimension, TIMES is also well suited to investigate the techno-economic trade-offs for meeting medium-term targets at the least cost while taking into account the need for meeting even more stringer targets for the long term. Many studies have proved to be especially useful to define optimal pathways for achieving ambitious GHG reductions, while contributing to the growth of the economy and minimizing the risk of technological locked-in (Solano-Rodríguez et al. 2018).

Regardless of the geographical coverage or the time resolution, there are consistent observations across studies regarding priority actions for an optimal transition toward a clean economy: energy efficiency improvements in all sectors, electrification of end-uses (buildings, transports, industries, etc.), and decarbonisation of the electricity sector.

Following these transformations, various additional mitigation options are considered to achieve deep decarbonisation levels when electrification is not possible. Their role in the energy system depend on the geographical context and the characteristics of regional energy systems. For example, the role for biomass and biofuels is largely depending on the availability of biomass resources. Vaillancourt et al. (2019) conducted a study to explore the role of bioenergy in a Canadian province in rapid transition to a low-carbon economy using a TIMES model for Canada. The model calculates the most optimal energy sector configuration that would reduce GHG emissions to meet the official target by 2030 compared with 1990 levels (37.5%) and beyond (40%). They found out that in order to achieve the desired levels of GHG reduction, the energy transition required should include a larger role for bioenergy in 2030 (from 6% in a reference case to 18% in the most stringent scenario. This requires the access to a large diversity of biomass feedstocks and many improvements in efficiencies and costs of conversion processes.

Although very expensive, hydrogen could also play a role where others are limited and following technological development. For instance, Sgobbi et al. (2016) indicated that low-carbon hydrogen production technologies could become viable options for the transport and industry sectors as early as 2030 in a carbon mitigation context. Electrolysis technologies in particular provide flexibility to the system by absorbing electricity at times of high availability of intermittent sources.

However, it is increasingly argued that the carbon neutrality of the global energy sector should be achieved in 2050 to maintain the temperature below 1.5 °C by 2100. Reaching the carbon neutrality goal will require the integration of the most advanced technology innovations for the most energy intensive industrial sectors (aluminum, iron & steel, cement, chemicals, etc.) as well as a closer look to net negative GHG options such as the use of bio-energy with carbon capture and storage (Selosse and Ricci 2014) with all the social perception challenges it raises.

5.2 Integration of Renewable Electricity and Heat

Previous studies exploring deep decarbonization pathways showed that electrification of end-uses and decarbonization of the electricity sector were consistent priorities of actions for many countries. Consequently, any climate change mitigation plan requires a special attention to the electricity sector, in a systemic and dynamic view of the whole energy system given its complexity and regional diversity.

Some studies have proved to be especially useful to describe optimal pathways for decarbonizing the electricity supply mix while taking into account the capacity of the grid to integrate a high share of intermittent renewables.

For example, Amorim et al. (2014) have used a TIMES approach to analyze possibilities to fully decarbonize electricity generation in Portugal by 2050, in order to contribute to a deep decarbonization of the Portuguese energy sector. To this end, two cases were considered: (1) the Portuguese generation sector is optimized as an isolated system, and (2) the latter is part of an integrated Iberian system (Portugal and Spain). The study illustrates the benefits of optimizing the integrated system, which yields reduced energy costs. These cost reductions are achieved through a larger exploitation of renewable energy sources (such as solar and wind energy) in the Portuguese generation sector, together with new interconnections between Portugal and Spain (at the end of the model horizon, 2050) to accommodate higher electricity exports to Spain.

In a soft-link framework with a probabilistic production simulation model, Tigas et al. (2015) have analyzed the possibilities, for Greece, to reduce energy-related GHG emissions by 60–70% in 2050, from the 2005 level, in line with the European Union objective to abate its GHG emissions by 85% (in 2050, from the 2005 level). In particular, using the linked TIMES model, they have studied a scenario that envisions an almost 100% electricity generation from renewable energies. Besides, achieving an (almost) complete decarbonization of electricity generation would facilitate an intensive electrification of the transport sector that is currently responsible for a large amount of GHG emissions in Greece. However, this scenario implies higher overall energy costs compared to the situation where the TIMES model is free to choose least costly strategies to achieve the GHG reduction targets. In that case, besides decarbonizing to a large extent electricity generation, the optimal strategy also relies on the implementation of a number of targeted energy efficiency measures, and on a large-scale renewable energy system penetration in all end-use sectors.

A detailed analysis of the electricity sector often requires a finer resolution than for other sectors. In this regard, Krakowski et al. (2016) have analyzed different levels of renewable energy penetration, ranging from 40 to 100% by 2050, in the French electricity generation sector considering seven seasons, a distinction between working days and weekends and six intraday periods. They have relied for their analysis on a TIMES model, together with indicators of the power system reliability. The latter is likely to deteriorate due to the penetration of renewable energy sources, even at moderate levels (40%). However, the use of flexibility options, such as demand-response, shall help reduce the negative impacts of intermittent renewable energy systems. The authors conclude by highlighting the interest for decision-makers of such a study that would help them anticipate the aforementioned negative impacts.

At present, the (French) Reunion Island's electricity generation relies mainly on imported fuels, while it has significant potential for renewable energies. Drouineau et al. (2015) have used a TIMES approach, again together with indicators of the power system reliability, to assess from a techno-economic perspective whether Reunion Island could achieve electricity self-sufficiency by 2030. The authors conclude that self-sufficiency could indeed be achieved, by using in particular biomass (sugarcane, cane, and wood) but also intermittent energy sources. The latter would negatively impact the reliability of the power system. This effect could, however, be mitigated by imposing some legal limits on intermittent sources for instantaneous electricity production.

By linking a TIMES model with the unit-commitment and dispatch model PLEXOS, Welsch et al. (2014) could also study the evolution of the Irish electricity system in more details. They showed that long-term energy models can clearly underestimate the importance of flexibility in the electrical system if short-term operational requirements are not taken into account. The study highlighted some of the limitations of long-term energy systems models if they do not adequately take into account operational aspects. Energy strategies and policies may, otherwise, underestimate the costs of meeting climate change or energy security targets.

Other aspects not included in traditional analysis using TIMES models can change optional solutions if added to the analytical framework. For instance, McDowall et al. (2018) have conducted a study, using a European TIMES model, to determine the extent to which the inclusion of indirect effects on GHG emissions (from a life-cycle assessment perspective) could change the optimal technological pathway for the decarbonization of the European energy systems (EU28 member states plus Norway, Iceland and Switzerland). Although indirect emissions account for only a small part of the total emissions (less than 10%), their inclusion would lead to changes in the optimal configuration of energy sectors. In particular, some renewable energy technologies (notably solar photovoltaic) become relatively less attractive. But these changes are more pronounced than the reduction in the attractiveness of renewable energy as a whole. Besides, McDowall et al. (2018) also concluded that renewable energy sources remain an essential element of the European decarbonization strategy.

Finally, Thellufsen et al. (2019) have studied the impact of cleaner heating solutions, based on district heating from industrial waste heat and combined heat and power (CHP) plants, on achieving deep decarbonization (up to 80% reduction) in Ireland. They rely on an Irish TIMES model linked to EnergyPLAN, a model that enables to consider hourly operations of both heating and electricity systems. The study indicates that the district heating option is a more fuel-efficient solution than the individual heating one. In the Irish context, this increase in fuel efficiency yields more savings than the higher investment costs incurred by the district heating option. The authors conclude that district heating could play an important role in the transition towards clean energy systems.

6 A Need for Greater Transparency for Efficient Decision-Making

Mathematical E3 models are particularly useful for exploring the transition toward clean energies and by providing decision makers with rigorous insights on deep decarbonization pathway options and long-term mitigation strategies. The TIMES optimization models in particular, developed within the ETSAP program of the International Energy Agency, have contributed to support decision-making all over the world at various geographical scales from global to city levels.

As large scale energy system models, they describe the energy sector with a complete list of energy forms, as well as different existing, improved and emerging technologies. They allow for detailed accounting of all energy flows within the energy sector from primary and secondary energy production to final energy consumption and useful energy demands. These models are typically projected to 2050 or even 2100, which makes it possible to study the structural changes within the energy sector. In addition, they provide important additional features compared with other types of energy system models, such as simulation models. By following a linear programming formalism, these models make it possible to determine a optimal configurations of the energy sector which makes it possible to satisfy the total demand for energy services at lower cost, by respecting GHG emission limits or renewable penetration targets.

However, the use of such models to assist policy-makers with energy and climate policy design and implementation raises issues regarding the robustness of the solutions found by the optimization program, especially given the large number of assumptions are required to assign values to highly uncertain technical or economic parameters over time. The main critic addressed to optimization energy system models is indeed the "black box" aspect of the approach which affects the credibility of their outcomes and usefulness to support decision-making.

Different approaches have been used by the TIMES modeller community to address uncertainty issues such as sensitivity analysis, parametric scenario analysis, Monte Carlo simulations, and stochastic programming. These approaches have partially contributed to show policy-makers and other stakeholders that large-scale optimization energy system models can provide robust insights for policy making despite the large number of assumptions embedded in model databases. Nevertheless, enhancing the credibility further will follow a better understanding of the links between input parameters and output results. The only possible solution to overcome this limitation involves more transparency regarding model inputs, a better dissemination of model roles and possibilities as well as a reinforced communication links between science and policy.

References

Amorim F, Pina A, Gerbelova H, Pereira da Silva P, Vasconcelos J, Martins V (2014) Electricity decarbonization pathways for 2050 in Portugal: a TIMES (The Integrated MARKAL-EFOM System) based approach in closed versus open systems modelling. Energy 69:104–112

Bahn O (2018) The contribution of mathematical models to climate policy design: a researcher's perspective. Environ Model Assess 23:691–701

Bahn O, Haurie A, Zachary SD (2005) Mathematical modelling and simulation methods in energy systems. Encyclopedia of life support systems (EOLSS), p 15

Beeck NV (1999) Classification of energy models. Tilburg University & Eindhoven University of Technology, p 25

Boulanger PM, Bréchet T (2003) Une analyse comparative des classes de modèles. Modélisation et aide à la décision pour un développement durable, p 32

Capros P, Van Regemorter D, Paroussos L, Karkatsoulis P (2013) GEM-E3 model documentation. European Commission

Drouineau M, Assoumou E, Mazauric V, Maïzi N (2015) Increasing shares of intermittent sources in Reunion Island: impacts on the future reliability of power supply. Renew Sustain Energy Rev 46:120–128

EIA—US Energy Information Administration (2019) NEMS—national energy modelling system. https://www.eia.gov/outlooks/aeo/info_nems_archive.php

ETSAP—Energy Technology Systems Analysis Program (2019) Official website. https://iea-etsap.org

European Commission (2019a) POLES—prospective outlook on long-term energy systems. https://ec.europa.eu/jrc/en/poles

European Commission (2019b) GEM-E3—general equilibrium model for energy-economy-environment interactions. https://ec.europa.eu/jrc/en/gem-e3/model?search

Heaps C (2008) Long range energy alternatives planning system—an introduction to LEAP. http://www.energycommunity.org/documents/LEAPIntro.pdf

IIASA—International Institute for Applied Systems Analysis (2019) MESSAGE—model for energy supply strategy alternatives and their general environmental impact. http://webarchive.iiasa.ac.at/Research/ECS/docs/models.html#MES-SAGEMERGE

IAEA—International Atomic Energy Agency (2006) Model for analysis of energy demand (MAED-2). Computer manual series 18, p 189. https://www-pub.iaea.org/MTCD/publications/PDF/CMS-18_web.pdf

Kannan R, Turton H (2011). Documentation on the development of the Swiss TIMES electricity model (STEM-E). The Energy Departement PSI Bericht Nr. 11

Krakowski V, Assoumou E, Mazauric V, Maïzi N (2016) Reprint of feasible path toward 40–100% renewable energy shares for power supply in France by 2050: a prospective analysis. Appl Energy 184:1529–1550

Loulou R, Goldstein G, Noble K (2004) Documentation for the MARKAL family of models. Energy technology systems analysis program (ETSAP) of the international energy agency. https://iea-etsap.org/index.php/documentation

Loulou R, Goldstein G, Kanudia A, Lehtila A, Remme U (2016) Documentation for the TIMES model. Energy technology systems analysis program (ETSAP) of the international energy agency. https://iea-etsap.org/index.php/documentation

McDowall W, Solano Rodriguez B, Usubiaga A, Fernandez JA (2018) Is the optimal decarbonization pathway influenced by indirect emissions? Incorporating indirect life-cycle carbon dioxide emissions into a European TIMES model. J Clean Prod 170:260–268

MIT—Massachusetss Institute of Technology (2019) EPPA—emissions prediction and policy analysis. Joint program on the science and policy of global change. https://globalchange.mit.edu/research/research-tools/eppa

Morrison GM, Yeh S, Eggert AR, Yang C et al (2015) Comparison of low-carbon pathways for California. Clim Change 131:545–557

Nakata T (2004) Energy-economic models and the environment. Prog Energy Combust Sci 30:417–475

Oxford Economics (2019) GMM—global macro-economic model. https://www.oxfordeconomics.com/

Selosse S, Ricci O (2014) Achieving negative emissions with BECCS (bioenergy with carbon capture and storage) in the power sector: new insights from the TIAM-FR (TIMES Integrated Assessment Model France) model. Energy 76:967–975

Sgobbi A, Nijs W, De Miglio R, Chiodi A, Gargiulo M, Thiel C (2016) How far away is hydrogen? Its role in the medium and long-term decarbonisation of the European energy system. Int J Hydrogen Energy 41(1):19–35

Solano-Rodríguez B, Pizarro-Alonso A, Vaillancourt K, Martin-del-Campo C (2018) Mexico's transition to a net-zero emissions energy system: near term implications of long term stringent climate targets. In: Giannakidis G, Karlsson K, Labriet M, Ó Gallachóir B (eds) Springer lecture

notes in energy: limiting global warming to well below 2 °C: energy system modelling and policy development, pp 316–333
Thellufsen JZ, Nielsen S, Lund H (2019) Implementing cleaner heating solutions towards a future low-carbon scenario in Ireland. J Clean Prod 214:377–388
Tigas K, Giannakidis G, Mantzaris J, Lalas D, Sakellaridis N, Nakos C, Vougiouklakis Y, Theofidili M, Pyrgioti E, Alexandridis A (2015) Wide scale penetration of renewable electricity in the Greek energy system in view of the European decarbonization targets for 2050. Renew Sustain Energy Rev 42:158–169
Vaillancourt K (2010) Modèles énergie-économie-environnement. PRISME
Vaillancourt K (ed) (2018) Tools for analysis of a future energy revolution (TAFER): methodologies, tools and data bases. Final report for annex XIII of the energy technology systems analysis program (ETSAP) of the international energy agency. https://iea-etsap.org/index.php/documentation
Vaillancourt K, Bahn O, Levasseur A (2019) The role of bioenergy in low-carbon energy transition scenarios: a case study for Quebec (Canada). Renew Sustain Energy Rev 102:24–34
Vaillancourt K, Bahn O, Frenette E, Sigvaldason O (2017) Exploring deep decarbonization pathways to 2050 for Canada using an optimization energy model framework. Appl Energy 195:774–785
Welsch M, Deane P, Howells M, Ó Gallachóir B, Rogan F (2014) Incorporating flexibility requirements into long-term energy system models—a case study on high levels of renewable electricity penetration in Ireland. Appl Energy 135: 600–615

Electricity Grids for 100% Renewable Energy: Challenges and Solutions

Eberhard Waffenschmidt, Majid Nayeripour, Silvan Rummeny and Christian Brosig

Abstract On the way to a 100% renewable energy system electrical power grids face a number of new challenges: Big centralized power plants are being replaced by small distributed generators operating on renewable energy. Especially wind and photovoltaic power generation add new uncertainties to the generation of power. The distributed generation results in a different and even reverse power flow. On the other hand, the distributed generators also offer new possibilities like local power supply in case of emergencies. The local generation may even be beneficial for the reliability of the power supply. New types of electrical loads are emerging, like electric vehicles, heat pumps and battery storages. They may soon lead to local grid overloads. But fortunately, most of them are controllable loads. Thus, they can beneficially contribute to using the uncertainly generated renewable power. However, this requires a holistic view on the total energy system including a coupling of all involved energy sectors. Therefore, this chapter first gives an overview of existing publications about the impact on the electric power grid by the energy sectors and their inter-coupling. Then, a cellular power grid structure is proposed and described, which takes advantage of the distributed structure of renewable energy generation. Finally, some aspects on clustering and controlling such a cellular grid structure are presented.

1 Introduction

Eberhard Waffenschmidt

The climate change is propagating faster than experts have ever forecasted. The main contribution is the burning of fossil carbon for energy purpose. Therefore, the worldwide energy system has to be changed as fast as possible to a 100% renewable

E. Waffenschmidt (✉) · M. Nayeripour · S. Rummeny · C. Brosig
TH-Köln (Cologne University of Applied Science), Cologne, Germany
e-mail: eberhard.waffenschmidt@th-koeln.de

M. Nayeripour
Alexander von Humboldt-Stiftung, Bonn, Germany

T. S. Uyar (ed.), *Accelerating the Transition to a 100% Renewable Energy Era*, Lecture Notes in Energy 74, https://doi.org/10.1007/978-3-030-40738-4_6

energy system. This will be a nearly fully electrical power system, because the main contributors to renewable power are wind and solar power. On the way to such a system, especially electrical power grids face a number of new challenges:

Big centralized power plants are being replaced by many small distributed generators operating on renewable energy. Especially wind and photovoltaic power generation adds new uncertainties in the generation of power. Electrical and other energy storages are necessary as soon as possible to match generation and demand. Also controllable loads (see below) will help with this issue.

The distributed generation results in a different and even reverse power flow in the power lines and components. Especially distribution grids will behave totally different, because at sometimes they are used to collect power instead of distributing it. On the other hand, the distributed generators also offer new possibilities like local power supply in case of emergencies. The local generation may even be beneficial for the reliability of the power supply. These aspects require thinking of a different power grid structure, or at least a different organisation of the power grid.

New types of electrical loads are emerging, like electric vehicles, heat pumps and battery storages. They may soon lead to local grid overloads. In many areas this is much rather an issue for the local distribution grid than for the transmission grid connecting regions. But fortunately, most of the emerging loads can be considered as controllable loads. Thus, they can even contribute to beneficially use the uncertainly generated renewable power. However, this requires a holistic view not only on the electric, but on the total energy system including a coupling of all involved energy sectors. Such sectors are distributed electrical generation, heat demand, charging of electric vehicles and gas supply.

In addition, the control of power grids will change: The inertia of large generators will have to be replaced by virtual inertia. Primary control power is being taken over by fast controllable devices, especially batteries, and even completely new control structures without Primary and Secondary Control Power are proposed.

Therefore, this chapter first gives an overview of existing publications dealing with the impact of the different energy sectors and their inter-coupling on the electric power grid. Then, a cellular power grid structure is proposed and described to take advantage of the distributed structure of renewable energy generation. Finally, some aspects on clustering and controlling such a cellular grid structure are presented.

2 Effects on the Electrical Distribution Grids Due to Coupling of Energy Sectors

Christian Brosig

2.1 Introduction

The electrical energy system is only one part of the energy system to be decarbonized. Heating and mobility have a very high share of the overall energy consumption. In Germany today they account for 782 TWh and 728 TWh (Bundesministerium für Wirtschaft und Energie 2018), while the whole consumption of electrical energy is only 520 TWh (Arbeitsgemeinschaft Energiebilanzen 2018). For mobility it is clear, that decarbonisation will particularly be met with an electrification and there is also a trend in heating towards heat-pumps, as they are much more efficient than conventional heating devices. If we look at the proportional distribution of energy consumption in the three sectors, it is clear, that this will be a huge challenge for the electrical distribution grid. This part of the chapter gives an overview of upcoming technologies and how they will affect the electrical distribution grid. It focuses on:

1. Photovoltaics (PV)
2. Power-to-heat—heat-pumps (HP) and combined heat and power (CHP)
3. e-mobility.

Power-to-gas is another emerging technology, which will be only marginally addressed. All results have been elaborated and gathered within the project ES-FLEX-Infra, funded by the European fund for regional development (EFRE).

2.2 Photovoltaics

Georg Kerber describes the limit for the feed in of PV in low-voltage distribution grids in his PhD-thesis 2010 (Kerber 2010). Problems arise above all in suburban and rural areas. The 3% voltage change criterion of the VDEW Guideline (VDE-AR-N 4105 2017) is a strongly limiting factor. Otherwise, adaptation measures such as reactive power control of inverters make between 60 and 90% of the PV power potential in low-voltage networks accessible, without serious adjustments (Kerber 2010). In urban areas, the PV potential on roof-tops often is not sufficient to imply problems for the grid. This is also shown in (Birk et al. 2018).

2.3 Power-to-Heat

The discussion about power-to-heat is largely driven by the use of HP. Here, too, numerous papers on their effects on the low-voltage networks exist. Particularly to emphasize, through a very well-founded research, a clear structure and detailed models is described in Protopapadaki and Saelens (2017). The paper discusses the influence of HP on rural low-voltage grids in Belgium, using a thermal model of the houses, as well as a stochastic behaviour influence of the inhabitants. Without a buffer storage, the authors come to the conclusion that a degree of penetration of 20–30% can be achieved, whereby the cables are the limiting factor in their models. Voltage problems only occur at higher penetration rates (Protopapadaki and Saelens 2017).

Navarro-Espinosa and Mancarella (2014) investigates the effects on suburban networks in Great Britain and refers to a series of sensitivity analyses. They find a degree of penetration of about 40% for air heat pumps and 50% for ground-source heat pumps. Here, too, thermal overloading of the cables at the outgoing end of the string is more problematic than voltage band violations (Navarro-Espinosa and Mancarella 2014).

Problems with heat pumps are load peaks which occur and which depend on the ambient temperature. This stresses the network with a high simultaneity factor.

Heat storages can help to compensate for peak loads. This has been demonstrated for example in Baeten et al. (2017) for the complete Belgian electrical network. The authors come to the result that the load peaks in the overall system can be reduced by about 11%, depending on the use of a storage (Baeten et al. 2017). The authors of Arnold et al. (2013) follow a different approach, which is to investigate the interaction of heat pumps, PV and CHP plants in a suburban low-voltage network. They assume that CHP plants are operated in a heat-driven way and thus flatten the exact peak load times of the heat pumps. Key findings are, that at 20% penetration of the heat pumps, the same additional penetration can be achieved with cogeneration units, which increase the minimum voltage in the network from 0.85 pu to 0.9 pu. PV has a neglectable compensating effect.

For a suburban network in the Cologne area, (Kusch et al. 2015) comes to the conclusion that 45% of the degree of heat pump penetration is already possible without storage and with storage even 90.6%. However, it assumes an application of domestic demand side management (DSM) in all considerations, whose influence on the results is not explained in detail and thus difficult to assess.

Shao et al. (2013) comes to far higher possible degrees of penetration in an urban low-voltage grid in Denmark. However, the authors of this study assume significantly lower installed power for the heat pumps used and use standard load profiles. This way, they neglect high simultaneity factors, which may appear in smaller grids. The control of heat pumps and battery electric vehicles (BEV) enables them to conclude in a feasibility of a 100% penetration of both technologies (Shao et al. 2013).

Within the framework of smart-grids, heat pumps are said to have an enormous potential for flexibility. Fischer and Madani (2017) provide a well-founded overview of relevant publications on the topic and derives some recommendations. Applications range from the integration of renewable energy, over the reduction of energy costs to grid-compatible operating patterns in the virtual power plant.

2.4 *E-Mobility*

The effects of e-mobility or BEV on the networks have already been worked out in a dissertation (Van Roy 2015), although the focus is on the coordination of local charging and integration in a building. In general, the different charging capacities must be considered separately. In the low-voltage grid, outputs of 2.3 kW (10 A), 3.68 kW (16 A, single-phase), 11.04 kW (16 A, three-phase) and in extreme cases 22.08 kW (32 A, three-phase) are usual. In addition, rapid charging stations are being discussed and are already being built at filling stations, which will be equipped with much higher charging capacities from 43 kW up to 135 kW.

For a distribution network in Gothenburg, Sweden, (Babaei 2010) comes to the conclusion that in commercial and industrial applications up to 56% or 90% (with 2.3 or 3.68 kW charging power) of the BEV can be charged if a complete switch to e-mobility has been made. In residential areas the figure is 64 and 102% respectively. The authors assume a very high simultaneity. In the medium-voltage network the cables and in the low-voltage network the transformers are identified to be the weakest spots in the system (Babaei 2010).

Marwitz and Klobasa (2016) analyses the effects on the basis of a rural low-voltage network. Without controlled charging, at a penetration rate of 20% already, the first network overloads occur. They assume a higher charge power of 10.8 kVA and investigate three different charging controls, from the central, smart control, to a price signal control, right up to a decentralised autonomous charge control on the basis of the voltage measured on site. They identify the central control to be the most effective—it can handle the 20% penetration—and the price signal control—which leads to an unacceptable voltage drop at 20% penetration—as most stressing for the grid (Marwitz and Klobasa 2016).

2.5 *Power-to-Gas*

Power-to-gas includes both electrolysis, i.e. electrical hydrogen generation, as well as methanation, the further processing of hydrogen into methane in order to integrate it into the natural gas network.

These are expensive and elaborate processes which, for cost reasons are only worth to be built on a large scale. Thus, impacts on the lower network levels lapse. Decentralised electrolysis for decentralised hydrogen feed-in is more cost-efficient, but can also only be fed into the gas grid on the high-pressure level (except for pure hydrogen grids). In general, Power-to-gas fulfils the task of absorbing excess electricity in order to seasonally store it in the gas network. Therefore, in the case of central installations and under the conditions mentioned above, no negative effects on the electricity grid are expected.

2.6 Interactions Between Different Technologies

As the new technologies emerge in parallel, they cannot only be considered separately, but must be analysed on interactions. In general, we can distinguish between technology that feeds-in electricity (PV and CHP) and such that draws electricity from the grid (HP and BEV).

PV and CHP (if it is heat-led operated) add up their peaks only marginally, as PV has its peak in summer and the CHP in winter. This is shown in Fig. 1. Figure 2 shows a close-up of the intermediate season, where overlapping takes place. Peaks do not occur at the same time. Nevertheless, in that season the two technologies have a certain simultaneity.

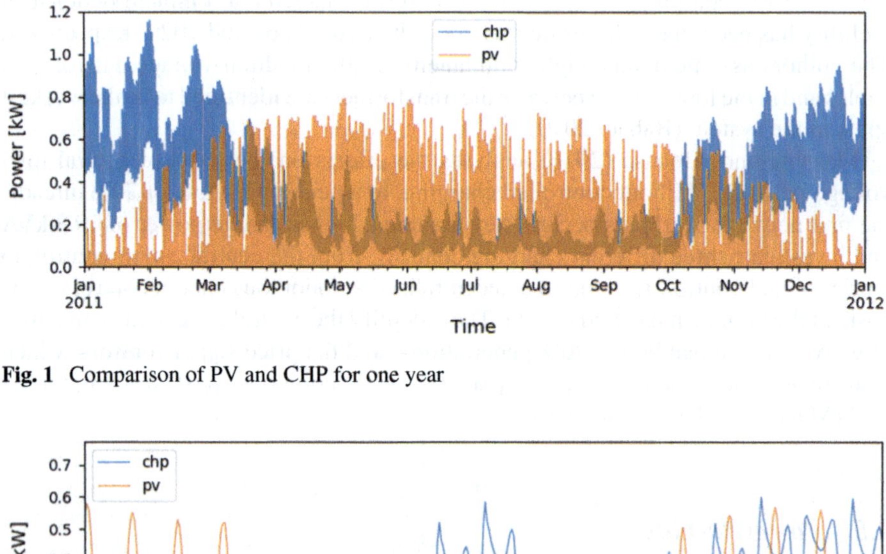

Fig. 1 Comparison of PV and CHP for one year

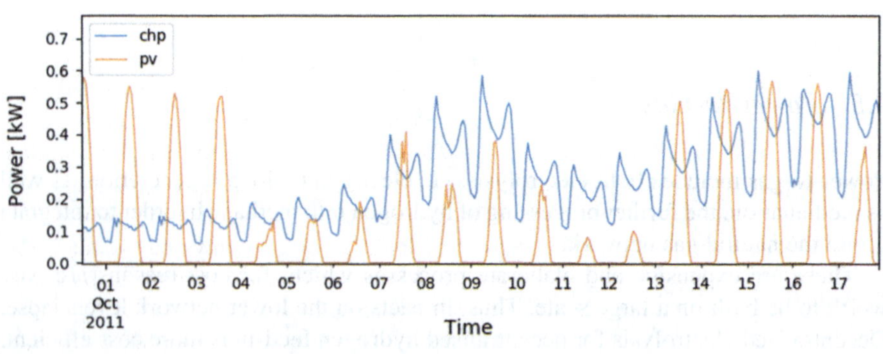

Fig. 2 Comparison of PV and CHP for beginning of October

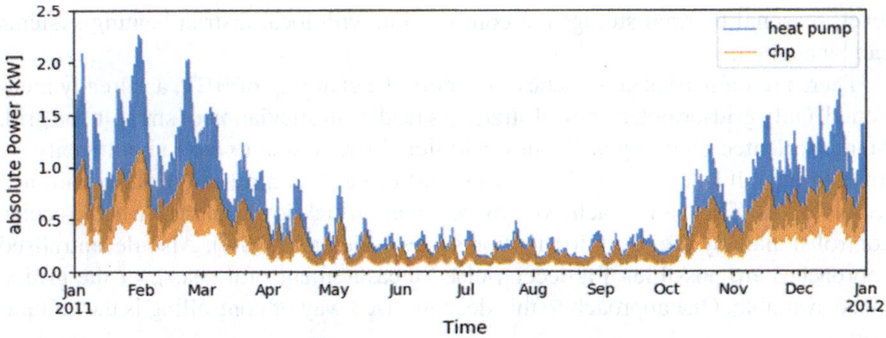

Fig. 3 Comparison of HP and CHP

HP and BEV do add up their influence, especially in winter. This has been shown in Birk et al. (2018), where only the combination of both technologies leads to an overloading of the transformer in an urban low-voltage grid.

As already mentioned, CHP can balance the peaks of heat pumps to a certain extent, if they are operated in heat-led mode. This is clarified, if we compare the load-profiles of an air-source HP and the feed-in profile of a CHP, as shown in Fig. 3.

Both profiles are based on a real temperature profile measured 2011 in Cologne by the LANUV (2019). The heat demand is set to 10000 kWh per year, with the same building as reference for both units and calculated based on the BDEW standardized daily demand for heating (BDEW et al. 2015). An hourly load-profile is applied, based on Bundesverband der deutschen Gas- und Wasserwirtschaft (BGW) (2006). Notable is, that this does not include the exact operation of the HP or CHP but assumes that they follow the standard curve in a partial load behaviour. Therefore, real measured load-profiles with a more detailed resolution would have higher peaks, also in summer. For the CHP, an electrical efficiency of 30% is assumed. The electrical load of the heat pump is calculated with a temperature dependent COP, derived from Staffell (2012). The COP is relatively low during wintertime, due to the low temperatures. Thus, the CHP is not able to buffer these peaks. The PV profile is taken from SMA. Das leistet Photovoltaik in Deutschland. (https://www.sma.de/unternehmen/pv-leistung-in-deutschland.html) for the same region as the temperature-profile and also for 2011.

2.7 *Solutions Towards Non-solved Challenges*

As mentioned, the evolving technologies are able to add flexibility to the system, provided they are equipped with the necessary managing unit and, in case of HP and CHP, thermal storages. For PV and CHP, battery-systems can help to reduce power peaks of the feed-in (as well as load peaks). CHP can balance HP to a certain extent, as shown above. To smoothen the seasonal heating demand on a regional level, seasonal thermal storages in combination with local district heating systems can be applied.

There are numerous approaches to control the charging of BEV, as already mentioned. Only grid-oriented control strategies lead to an alleviation of stress in the grid. Market oriented charging will cause additional stress, due to high simultaneity. In total, there will be a need for large-scale grid-oriented controlling of the mentioned technologies. This can be achieved by regional virtual power plants, which receive control signals by the local distribution system operator (DSO). Also decentralised approaches are possible, if a decentralised measurement of the state of the grid is made available. One approach to this decentralised way of controlling is the cellular grid.

Grid reinforcement is a last consequence, which in a lot of cases cannot be prevented if a steady overloading of utilities is present. This can be the reinforcement of single equipment, such as lines and transformers. In case of voltage range violations due to decentralised feed-in, already controllable local grid transformers can reduce the problem.

Not yet mentioned are voltage problems due to uneven loading of the phases in the grid. They can be avoided by connecting major loads (especially BEV) to all three phases. To offset the imbalance, linear regulators can be applied.

2.8 Conclusions

Emerging technologies will be challenging for the distribution grids, especially on the low-voltage level. Decentralised feed-in will be mostly a problem in rural areas, as their potential there is often higher, and grids have lower capacities. BEV and HP have very high potentials in cities, as the density of inhabitants is high. They will, especially in combination, be responsible for medium term load peaks. Regional virtual power plants or decentralised grid stabilizing operation of these technologies can reduce the impacts to a certain extent. CHP balance the peaks and cross-sectoral coupling to district heating systems with seasonal thermal storage could also reduce winter peaks from HP.

For high penetration rates also grid refurbishment measures have to be taken into account. A lot of operation strategies also are based on measurement. An increasing application of measurements and inside the grid is inevitable (sometimes referred to as “smart grid”).

3 Future Structure of the Power Grid: Cellular Grids

Silvan Rummeny

3.1 The Cellular Energy System

This part of the chapter is based on the activities of the working group "Energy Supply 4.0" of the VDE Association for Electrical, Electronic & Information Technologies in Germany. In addition to the author of this chapter, Josef Bayer should also be mentioned, who has made a substantial contribution to the content of this chapter.

3.2 Motivation for a Cellular Energy System Approach

In the transformation of our energy system from a central to a distributed energy production with low CO_2 emissions, there are fundamentally new demands on its design. Particularly in the electricity sector, the expansion of large supply lines or the installation of compensation systems (e.g. phase shifters) is one way to ensure load balancing, stability and security of supply in the Europe-wide electricity network. At the same time, more and more measures must be taken in the electrical distribution network (e.g. BDEW Cascade) to guarantee regional security of supply and at the same time contribute to the stabilization of the overall system. This is expressed even more clearly in the "Stadtwerkestudie 2017" (BDEW—Bundesverband der Energie- und Wasserwirtschaft and Stadtwerkestudie Summary 2017), in which distribution system operators are named "enablers and the backbone of energy system transformation". In a similar way, the CEER final report "Flexibility Use at Distribution Level" (CEER—Council of European Energy Regulators, Conclusion Paper 2018) underlines the growing requirements for distribution network operation. The VDE study "the cellular approach" (Studie 2015) shows how supply security and stability of the electrical grid can be guaranteed by conversion or storage into other forms of energy. A first overview of the idea of a cellular power grid is presented in Waffenschmidt (2015). From an economic point of view, cross-sector solutions that combine different forms of energy are also more efficient than individual solutions for certain forms of energy and their networks or subsystems. In the cellular energy system, the physical balance between energy supply and demand is as far as possible already established at regional, local level in accordance with the subsidiarity principle. In this way, the expansion of renewable energies can be pushed forward rapidly, and the electrical transmission grid can be relieved of measures for grid stabilization.

The aim is not only to optimize the expansion of the electrical grid, but above all to improve the efficient use and exploitation of energy at all levels—local, regional and supra-regional. In addition, a multi-level cellular system would increase the system resilience of the electricity, gas and heat infrastructure by allowing some cells to support emergency supplies. This chapter creates the basis for the planning and development of products, services (product/service roadmaps) and infrastructure measures for the design of a cellular energy supply.

3.3 *Definition of the Cellular Approach*

The cellular approach is a new organizational model for energy supply. In this interdisciplinary model, technical, economic, legal and political (and social) interests are taken into account. The central building block of this model is the energy cell. Figure 4 shows the schematic representation of an energy cell in the energy system.

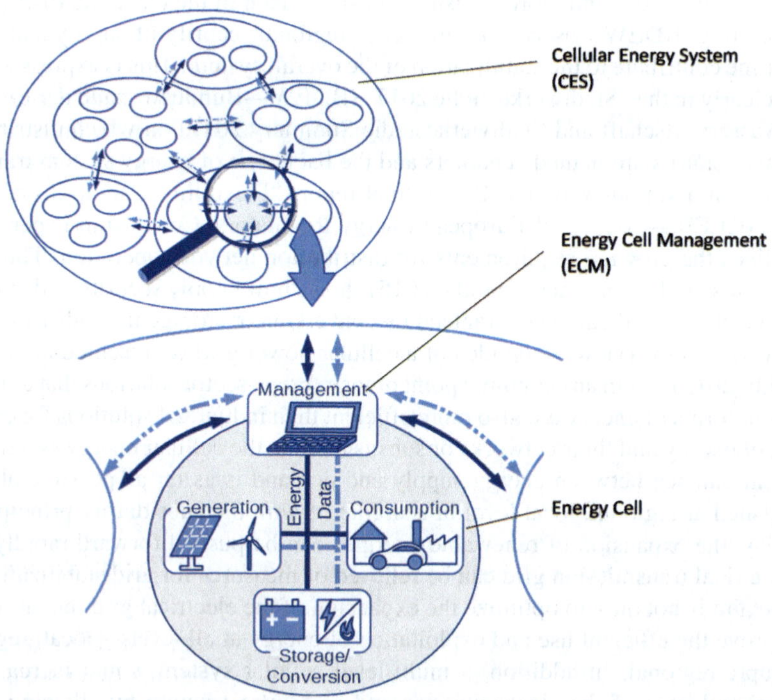

Fig. 4 General structure of an energy cell (below) in a Cellular Energy System (above)

Definition of a cell:
An energy cell consists of the infrastructure (1) for different forms of energy (2), in which the balancing (3) of generation and consumption is organized by an energy cell management (4) in possible coordination with neighbouring cells (5) over all existing forms of energy.

Explanations:

(1) The infrastructure consists of complete subdivisions of transmission or distribution grids and includes all resources used to convert, transport, distribute and store energy.
(2) The forms of energy considered include electricity, gas, heat and mobility. A cell may also contain only one form of energy.
(3) In terms of balancing, which can be carried out either seasonally or dynamically, the three states can result: balanced, oversupplied or undersupplied via all available forms of energy.
(4) The Energy Cell Management (ECM) includes all I&C equipment, including the necessary communication technology.
(5) Neighbouring cells can be arranged hierarchically. Thus there are cells on the same level as well as on superimposed and subordinate levels.

The essential characteristic of a cell is its ability to independently organize the energy supply within the cell. To do this, the cell physically needs the ability to balance energy supply and demand. As usual, this can be done by connecting the cell to the upstream grids—electricity, heat, gas—or by combining and using the cell's own generation, energy converter or storage units. The crucial factor here is that the so-called Energy Cell Management (ECM) is able to control the internal components (generation, consumer, storage, energy converter units) and at the same time negotiate offers and enquiries with neighbouring or upstream energy systems.

This requires that the relevant components within a cell are networked and that secure communication can be established at the same time. Within a cell, the data of the relevant components are aggregated. Only aggregated data is communicated to the outside world, which enables data to be handled economically and securely. This includes communication with control systems or the energy cell management of upstream network operators or with market participants from whom the cell wants to receive price information or with whom the cell wants to trade energy. The schematic structure of a cell is shown in Fig. 4.

These energy cells are now used to build up the Cellular Energy System (CES). The structure is repeated on all network levels and forms a fractal system.

3.4 Design of the Cellular Energy System (CES)

The cellular approach is ideal for organizing a large number of distributed plants, participants and tasks. The organization should be based on the principle of automation technology: a multi-level management system is proposed in which network parameters are regulated locally, regionally, supra-regionally, nationally and internationally depending on the network cell level. This is based on the principle of subsidiarity, which defines that control deviations or problems that occur are primarily managed directly at the source of the problem and are only secondarily managed in the next-neighbouring upstream or downstream grid areas.

The logical structure of the Cellular Energy System (CES) can mainly be derived from the physical structure of the networks. In the VDE study "Protection and automation technology in active distribution networks" (VDE-Studie 2016) it is recommended to define such local management systems within a galvanically coupled area of an electricity grid group. In such galvanically coupled areas of a grid group, there could also be managed several cells in parallel.

Figure 5 exemplarily describes in detail the interconnection of the cells, their plausible components and their integration into the infrastructure of electricity, gas and heat grids with all levels from transmission to distribution.

The following section suggests a distribution of responsibilities and mechanisms to safely operate such a cellular energy system. At the end, recommendations for actions are given on how to prepare for change and which challenges still need to be solved.

3.5 *Operation of Cellular Energy Systems*

3.5.1 Tasks and Responsibilities of the Energy Cell Management System (ECM)

An energy cell management system (ECM) is a new role in the cellular energy system. Such an "energy cell manager" has control over a limited, locally linked infrastructure that can include generators, consumers, storages and energy converters. However, an energy cell manager can also interact with subordinate energy cells in his own grid area. The tasks and responsibilities of the ECM is shown in Fig. 6.

These tasks and responsibilities always apply both to their respective grid access node and to their own infrastructure, which they control. A grid access node can, for example, be the house connection of a residential building cell, the local grid transformer station of a low-voltage energy cell, the transformer station of a 20-kV energy cell, etc.

As shown in Fig. 6, the ECM is responsible for the safe operation and suppression of interferences of the grid area. An energy cell can only be operated when a functioning grid protection concept is in place.

Voltage control includes the control of reactive power flow and the control of transformer tap (e.g. with adjustable local grid transformers). At the grid access node of the energy cell, a reactive power setpoint is controlled which can be demanded by the higher-level energy cell manager or network operator.

In normal operation, but also in particular to enable an islanding grid operation in the event of an interference, an ECM must also ensure the provision of frequency

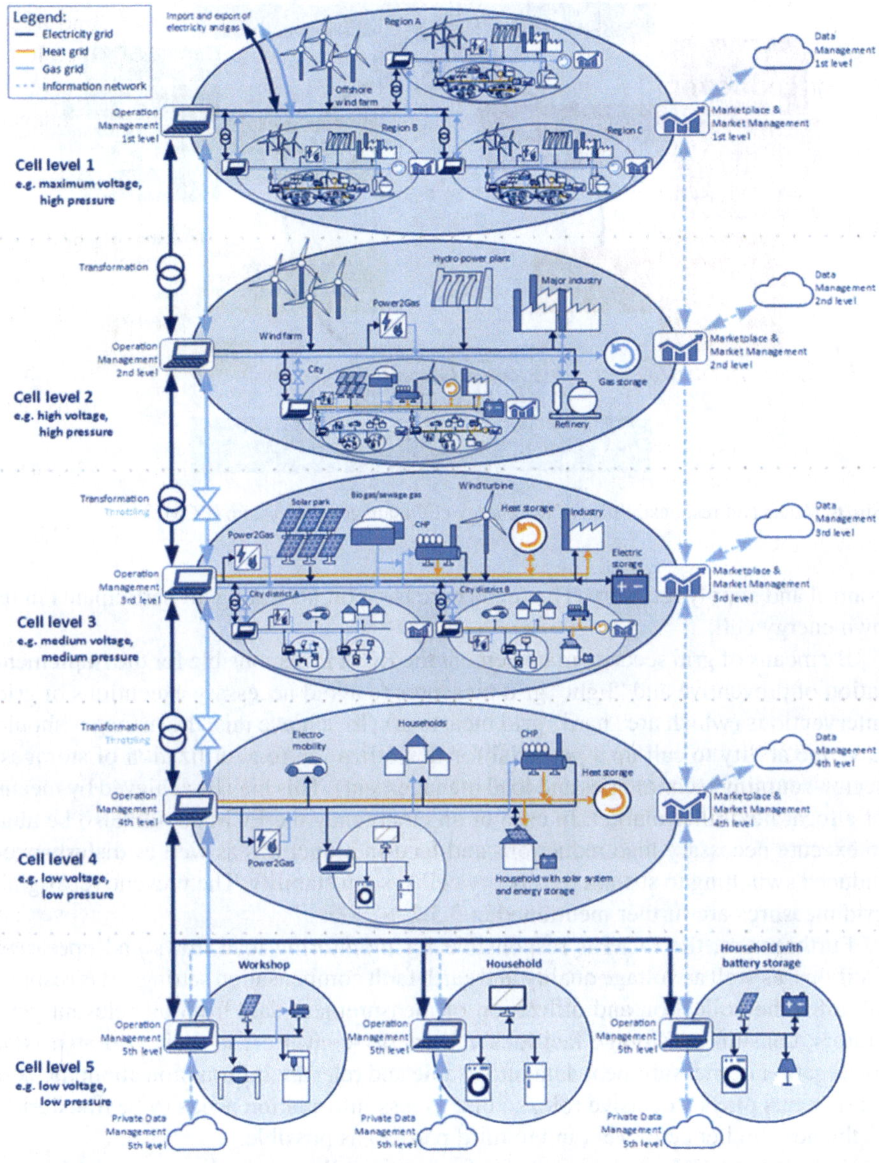

Fig. 5 Exemplary structure and interconnection of a cellular energy system (CES) including electricity, gas and heat grid from the transmission to the distribution level

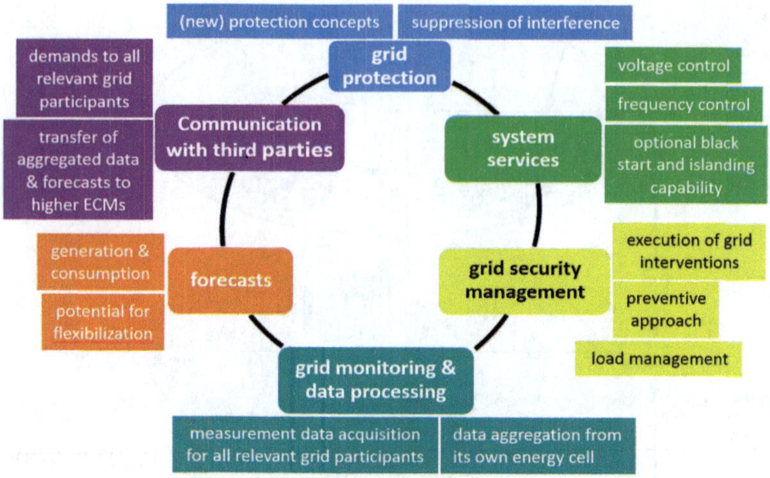

Fig. 6 Tasks and responsibilities of the energy cell management system (ECM)

control and supply recovery. The focus here is on orchestrating all participants in its own energy cell.

By means of grid security management the ECM is responsible for the implementation of preventive and "light" grid measures to avoid necessary executions of grid interventions (which are "hard" grid measures). To achieve this, the manager should have the ability to call up a potential for flexibilization (e.g. utilization of storages, sector-coupling technologies and load management). This may be achieved by means of a local flexibility market. In case of an emergency the ECM should also be able to execute necessary load reductions and feed-in reductions as well as disturbance-induced switching to sustain the energy cell system stability. The preventive, "light" grid measures are further mentioned in 3.5.2.

Furthermore, the ECM is responsible for monitoring load flows and operating facilities, as well as voltage quality and earth fault compensation settings. It is responsible for the collection and utilization of measurement data from all relevant generators, consumers, storage facilities and energy converters in their grid area. The aggregation of measurement data into usable and relevant information about the network status plays a decisive role, so that as less information needs to be transferred to the next higher cell level (and to third parties) as possible.

At least the ECM is responsible for forecasting all connected consumers and producers. In the future, the forecast of the (price-dependent) flexibility potential of all connectees (controllable loads, storage, energy converters and generators) will play a decisive role for the possible load flow control by market mechanisms or the ECM itself. Further, the ECM is responsible for passing on the aggregated information and forecasts about its grid area to higher level ECMs. In addition, ECM can pass on information and demands to the participants in their own energy cell as required. The ECM thus plays the most important role in the BDEW communication cascade

(BDEW Bundesverband der Energie- und Wasserwirtschaft 2014) and extends it to the end users.

In Conclusion, transmission system operators, distribution system operators, municipal utilities, but also industrial customers, neighbourhood and district system operators with their own grid operation could get such an energy cell manager within their own property boundaries.

3.5.2 System Stabilization During Normal Operation

In normal operation an energy cell should always operate grid tied. Since an electrical power grid has a much lower grid droop in grid-tied operation, the grid stabilization is significantly easier and therefore cheaper. Then it is proposed that the system stabilization is maintained by means of a three step mechanism in the respective energy cell:

- *Automated stabilization (instantaneously),* passively by means of (virtual) inertia and actively by means of droop-based control of fast adjusting flexible generators, loads, storages or sector-coupled technologies. At least every distributed power plant, especially inverter-based systems, should contribute to the automated stabilization
- *Preventive grid security management (within 1 min.)*, actively by means of the ECM control to call up further potential for flexibilization. This may already be controlled by means of a market dependent on the ECM to control supply bottlenecks.
- *Free market mechanisms (within 15 min.)*, organized independently from the ECM similar to today's electricity stock exchanges, but with consideration of general grid capacities.

This three-step mechanism should be applied for the case of unpredicted supply bottlenecks in the energy cell. In case of a predictable bottleneck, the mechanism should work vice versa (first try to solve by free market mechanisms, then preventive grid security management and automated stabilization).

3.5.3 System Stabilization During Backup-Operation for Emergencies

Unlike normal operation, interferences can occur in the cellular energy system. A long-lasting and large-scale power outage can interrupt the nationwide and demand-oriented supply of the population with essential goods and services. Such a situation poses a threat to public safety (Petermann et al. 2011).

Then and only then an energy cell shall continue to supply itself in a backup-operation for emergencies. An islanding back-up operation can be done either with or without a supply interruption:

- *Disconnecting with supply interruption* the energy cell has to get black started by a grid-forming power plant, which is sufficient to the supply of the energy cell.

- *Disconnecting without supply interruption* the energy cell has to contain a very fast reacting power source(s) or sink(s) which can cover the demand or supply instantaneously within the energy cell. For this option a voltage-guided source or sink is recommended, which can secondarily cover the demanded residual load of the energy cell.

There are higher requirements for the grid stabilization in islanding back-up operation, because of a higher grid droop. Therefore, the energy cell should not maintain in this back-up operation forever. As soon as the interconnected power grid is available again, the operation should be switched again to a grid-tied operation. Again, this can be done with or without a supply interruption:

- *Reconnecting with supply interruption* the energy cell has to execute a brown out (grid shut down) first. Afterwards it can be easily switched to the interconnected power grid again without any further requirements.
- *Reconnecting without supply interruption* the switch of the grid access node to the upstream grid has to be a synchronous clutch switch. Furthermore, a power plant has to be able to resynchronize the islanding frequency to the one of the upstream grids.

Not every energy cell will be able to back-up its supply by islanding operation. The first requirement is, that a certain demand can be covered by some generators, storages or sector-coupled plants, which can be controlled appropriately by the ECM. It is also possible that an energy cell in islanding operation no longer has a functioning grid protection concept. Then it either has to be adapted or operation is not possible. Further problems can occur with the star point treatment, when as often usual the star point isn't within the energy cell ownership border. In advance installed island grid detections of distributed generators can also affect islanding operation stability, when wrong set.

3.6 Conclusion

The cellular energy system is supported by decentralised management systems. So-called energy cell managers have control over a limited, locally linked infrastructure that can include generators, consumers, storage and converters. The logical cell structure can/must be derived largely from the physical structure of the networks. Each piece of information should be processed at the level where it is possible for the first time. Energy cell management systems thus have exact information about the network state and the flexibility potential of the energy cell.

Cells should be stabilized by automatic processes of decentralized plants alone. This includes the provision of passively stabilizing resources as well as the retrieval of actively stabilizing resources.

Energy flows are controlled by an international and perhaps cellular market with regional and local sub-markets. The local grid situation must be taken into account in these market mechanisms.

In general, it is possible for energy cells to continue to supply itself in an islanding grid. However, islanding grid operation in the cellular energy system should only be a back-up-option in an emergency. Normally, interconnected operation and solidarity are the standard.

4 Clustering and Control Structure of Power Grid Areas

Majid Nayeripour

4.1 Introduction

Expansion planning and sustainable growth in using renewable energy resources to reduce the greenhouse gas emissions are the main purposes of European Commission in Energy Road map Strategy 2050. In this regard, energy efficiency optimization in smart grids based on distributed control strategies and distributed optimization methods to facilitate the expansion planning in energy conversion with high reliability play a very important role to reach 100% electric power generation by renewable energy resources in this sector.

This chapter devotes to some problems and obstacles regarding to 100% electric energy power production via renewable resource in accordance to energy road map strategy and introduces the clustering and distributed strategy methods to handle problems in this area.

4.2 Motivation

Increasing penetration of high power non-dispatchable distributed generations (DGs) in distribution and transmission system, may change the amount of power flow in transmission lines substantially or reverse the direction of power flow (Khanbabapour et al. 2018).

In these systems a large amount intermittent active power of renewable DGs may transfer via High Voltage Direct Current (HVDC) and Alternative Current (AC) transmissions lines between microgrids (MGs), or exchange between microgrids. This characteristic should be compatible to IEEE Std 2030.8™–2018 (IEEE Standard for the Testing of Microgrid Controllers 2018). In such systems, intermittency in active power exchange between microgrids may also cause voltage and frequency deviations. In addition, it may be the cause of some other power quality problems, which may be extended to higher voltage level or violate the grid code requirements for some parts of system.

Moreover, there may be created some serious problems relating to power transfer capability (Lee et al. 2012; Mishra et al. 2017), voltage stability or even voltage collapse (Li et al. 2018; Robitzky et al. 2018; Pulcherio et al. 2016; Choi et al. 2018), which should be considered carefully for each special case.

In this high penetrated DGs environment, the operation of distribution systems and their assets will be important. Such assets may be controllers of DGs and load tap changers to coordinate with local controllers of transmission system in reactive support (Robitzky et al. 2018), supporting transmission system by DGs in lower voltage to increase the resiliency (Chowdhury et al. 2017), and supporting the other microgrids via active power support in contingency and continuing the minimum self-sufficiency of microgrids.

This means that all regional controllers of different voltage levels in an area can contribute to support and resolve the local or regional problems. This functionality may be handled more effectively by locally coordinated active and reactive power controllers. They should use a decentralized and distributed strategy that may get feedback from optimal active power flow (OPF) or economic dispatch (ED) controllers as well as power exchange between microgrids in their dynamic or static optimization processes. As a new recently applied method, the system may be partitioned to several areas based on some static and dynamic criteria (Nayeripour and Waffenschmidt 2019) by a centre of partitioning. In the next sections, static clustering and distributed control strategy as the main infrastructures and feature for actual and next generation of smart power system with high penetrated renewable DG will be introduced.

4.3 Static Clustering

A self-sufficient microgrid is a power system in distribution and sub-transmission voltage levels that balances the power inside the microgrid and has a certain degree of adequacy. To have more reliability constraints on adequacy and increase the self-sufficiency degree of microgrids as well as operation costs reduction, microgrids are connected to the main grid or other microgrids. In this state, although the microgrids may operate independently, the interconnected grid can be considered as a large-scale system with improved capability following a contingency or a shortage.

On the other hand, a large-scale power system can be considered as a number of self-sufficient sets which may be controlled independently. Partitioning is a general form of clustering and zoning, which may cover the operation of protection devices and control strategy for a particular purpose. The reason of partitioning a system to some areas is because of physical constraints and difficulty of this system to be controlled centrally. Moreover, it is also possible to have different degree of adequacy in each area via DG allocation. Imposing the bidirectional power constraints on boundary areas in the control strategy of each area adds to the allocation.

Partitioning of a system based on a similarity would be a solution to apply a common control strategy in each area. Then the controllers would be more effective and compatible with their areas with secure and reliable operation.

Clustering and zoning of a large-scale smart power system via a similarity of performance indices is a general objective for a multi objective optimization problem. Such performance indices may be e.g. loss, voltage deviation, reliability, self-sufficiency/adequacy, economic or energy efficiency. In addition, the placement of static breaker devices, location and size of DGs are further variables of optimization. Various methods such as game theory to increase the performance of system can be used for the optimization.

Although static clusters do not have any dynamic problems in normal operation, some proper dynamic objectives may also be considered in static clustering. This concept may play an important role in high penetrated DGs environment to have stable microgrids with self-sufficiency and dynamic stability following islanding process. In this challenging issue, intermittency of renewable DGs and load uncertainty in a proper period of variations based on available data are very important.

In this regard, the goal is to find a suitable similarity index in coordination to new static and dynamic similarity indices to determine the boundary areas. Therefore, a pareto algorithm should be used to compromise between results.

The results of the clustering depend on the selected similarity indices and the period of analyses. Two different solutions of two-level optimization and single-level (with dynamic indices as constraints) optimization can be considered to improve the performance of system regarding to transient and dynamic stability.

It should be noted that each static area may contain a part of transmission and distribution system models. Constant power or voltage source models for low and high voltage levels may no longer be valid.

The static partitioned areas prepare a suitable structure for distributed and decentralized control strategies. Dynamic consideration in their objectives guarantees the self-sufficiency and dynamic stability in each static area as well as microgrids creation following events in each area. In addition, it can be used for active and reactive coordination between two voltage levels and security assessment.

Considering the following aspect and assumptions may be useful in clustering of interconnected smart grids:

- It may be assumed that some microgrids exist physically before partitioning of the system and do not take part in partitioning, but they impose some bidirectional constraints on optimization. In this regard, the rest of system with considering to new constraints is partitioned.
- It is also assumed that the amount of power generation by renewable DGs in distribution and sub-transmission system is comparable to a controllable power plant in sometimes. Dynamic analysis and deriving new performance indexes relating to mutual effect on static and dynamic problems may be important in static partitioning with dynamic consideration.
- Battery energy storage systems (BESS) will not be investigated in the static partitioning method. However, optimal allocation of BESS in each dynamic cluster (microgrids) will be investigated in order to increase the self-sufficiency and dynamic stability in potential dynamic microgrids of each static area.

4.4 Control Structure

The operation control of the system has some restraints: If we could run the optimal active power flow (OPF), economic dispatch (ED) as well as optimal reactive power flow (ORPF) centrally at the same time and apply the result for real time power scheduling of controllable DGs and main loads, the optimal operation of system would be guaranteed. This plan is not practically implementable, and we must design a proper strategy which practically approaches the optimal operation in different voltage levels.

Continuous increasing penetration of high power DGs has moved the system structure from central power plants toward distributed units. Thus, it is more convenient to have a distributed and/or decentralized control strategies in each static partitioned area rather than a wide area management system to take the control actions on the overall system centrally. In this regard, the optimal operation of individual static areas (sub-systems) based on energy and power efficiency and using coordination methods between areas are preferred. A hierarchical strategy advocates this feature, which guarantees the near optimal operation of overall system.

In a comprehensive control system, the system is managed via a three-layer hierarchical strategy: In this strategy, a main OPF as the third level schedules and manages the main power of large-scale power plants based on load and generation estimation, economic and market analyses and constraints. Also, high power renewable DGs such as offshore wind farms are included. In this layer, OPF scheduling is run periodically (e.g. in every 15 min.) to determine the initial set point for the controllers of second layer in the static areas.

The main objective of the second control layer is to reach near optimal operation of static areas including dynamic microgrids following islanding operation between updating OPF signals. This layer may be considered as another OPF based controller in short time. It determines the set point of controllable DGs and loads. In this environment, it would be important to minimize the active and reactive power of controllable DGs via secondary controllers.

Considering grid code requirements as penalty factors in control strategy or at least as references for triggering the switches may play an important role in secondary control strategies to increase the degree of seamless transfer between grid-connected and islanding operation. The strategies may also consider some non-linear functions and constraints relating to operation management into control methods. These functions and constraints may be coordinated with modal analysis, small signal stability, security analyses and market signals to increase the performance and robustness feature generation of interconnected system high penetrated renewable DGs.

Unfortunately, the idea of using the secondary control level for optimization has some difficulty which have been prevented to be used yet.

Controllers of the primary layer try to balance the power inside each self-sufficient microgrids. Two following suggestions as primary controller can be applied in coordination with secondary controller: The first suggestion uses a modified adaptive droop method and the secondary controllers add the modulating signal to the primary controller. In the second suggestion, the secondary controllers work individually and determine the active and reactive power set point of DGs in each area. Then an energy storage system (ESS) or a synchronous generator (SG) balances the power mismatch dynamically. This ESS or SG acts as the primary controller and increases the security of microgrids.

If the communication link encounter to an error during the operation of the secondary controller, DGs will continue their operation with previous set point in power constant mode and the primary controller balances the power for short times.

Following consideration maybe helpful to be better familiar with analyses and problems relating to high penetration of renewable DGs and how they can be handled regarding to secondary controllers:

- Stability condition of microgrids about certain degree of self-sufficiency and dynamic condition should be investigated. One of the most challenging issue in distributed control strategy is to find the condition of converging the output variables relating to boundary buses. They should be calculated via two different distributed agents (Nayeripour and Waffenschmidt 2019). This gap has not been investigated yet and in all methods, it is assumed that they converge to each other. Initial simulations show that controllability from one side and observability from another side are one of the requirements of convergence.
- The average optimization run time of distributed secondary controllers should be less than a certain time (e.g. 1 s.). This is a practical consideration which is proper for on-line applications. In addition, it is also possible to insert the result of iterations during the performing of the algorithm because the methods converge.
- Each area should have its own cost function for secondary controllers.
- A structure of a bilateral data acquisition system for controllers inside the area should be created. This structure may also have access to some data from neighboured areas.
- Cyber security in the architecture of WAMS is important and the structure of the control system should be in such a way that the system is resilient and flexible.
- It is required to consider some technical problems regarding to the distributed control strategy of interconnected microgrids with comparable stiffness. These problems for the cases of microgrids relating to different DSOs are more important than virtual microgrids under supervision of one DSO which have determined via clustering. The first case is more vulnerable and may lead to some undesirable problems. For example, in connection of a stiff grid with high degree of adequacy to a weak grid, a certain band for adequacy of each grid based on an agreement should be imposed. Moreover, DSOs are not eager to share their data, which may be used in other microgrids or even in their coordination centre.
- The static areas may contain a part of the transmission system while DGs are usually connected to the distribution grid and sub-transmission levels of areas. To increase the security and self-sufficiency operation of interconnected power system and to reduce the interaction between different voltage levels, it is required to isolate the operation of microgrids from the transmission grid as much as possible via reinforcement of the coupling point.

- Non-controllable DGs track their Maximum Power Point (MPP) and it is assumed that they can take part in reactive power support based on their available capacities.

5 Summary

This chapter gives an overview of some solutions electrical power grids are facing at the transition to a 100% renewable energy supply.

The first part listed further literature describing the impact of different energy sectors and their coupling into the electric power grid. Concluding, the coupling of energy sectors is an important measure to mitigate the uncertainty between power generation and demand. However, power control is necessary on a local level in order not to overload the power grid on the last mile. Nevertheless, some problems still require grid refurbishments.

The second part describes a grid structure adopted to the distributed generation of power. It includes a cell-like structure. Each cell manages its power flow on its own task and only aggregated data is submitted to others. This minimizes the control effort and improves the security of the power control. Such a power cell should also be able to operate autarkic as microgrid in case of an emergency. However, in this case restrictions will apply.

Finally, considerations on how to partition such a cellular structure are presented. In addition, a three-level control for a future grid is proposed.

In summary, this chapter shows that there are new challenges for electrical power grids, but also possible solutions are presented. Therefore, the electrical power grid will be able to support a transition to a 100% renewable energy system.

References

Arbeitsgemeinschaft Energiebilanzen e.V (2018) Auswertungstabellen zur Energiebilanz Deutschland

Arnold M, Friede W, Myrzik J (2013) Investigations in low voltage distribution grids with a high penetration of distributed generation and heat pumps. In: 48th international universities power engineering conference (UPEC), pp 1–6. https://doi.org/10.1109/upec.2013.6714884

Babaei S, et al (2010) Effects of plug-in electric vehicles on distribution systems: a real case of Gothenburg. In: Innovative smart grid technologies conference Europe (ISGT Europe), pp 1–8. ISSN: 2165-4816. https://doi.org/10.1109/isgteurope.2010.5638947

Baeten B, Rogiers F, Helsen L (2017) Reduction of heat pump induced peak electricity use and required generation capacity through thermal energy storage and demand response. Appl Energy 195:184–195. ISSN: 0306-2619. https://doi.org/10.1016/j.apenergy.2017.03.055; http://www.sciencedirect.com/science/article/pii/S0306261917302854

BDEW Bundesverband der Energie- und Wasserwirtschaft e.V. (2014) Praxis-Leitfaden für unterstützende Maßnahmen von Stromnetzbetreibern. Springer, Berlin. https://www.bdew.de/media/documents/Awh_20141031_BDEW-VKU-Leitfaden-Massnahmen-Stromnetzbetreiber-3-0.pdf

BDEW, VKU, GEODE (2015) Abwicklung von Standardlastprofilen Gas. Springer, Berlin

BDEW—Bundesverband der Energie- und Wasserwirtschaft, Stadtwerkestudie Summary (2017) Der Verteilnetzbetreiber der Zukunft—Enabler der Energiewende, Ernst & Young Global Limited. Springer, Berlin https://www.bdew-kongress.de/files/bdew_kongress_verteilnetzbetreiber_ey_stadtwerke_studie_2017.pdf

Birk S, Brosig C, Waffenschmidt E (2018) Auswirkungen von Sektorkopplungsmaßnahmen im zukünftigen innerstädtischen Niederspannungsnetz, 15. Symposium Energieinnovation, Austria

Bundesministerium für Wirtschaft und Energie (2018) Energiedaten: Gesamtausgabe

Bundesverband der deutschen Gas- und Wasserwirtschaft (BGW) (2006) Anwendung von Standardlastprofilen zur Belieferung nichtleistungsgemessener Kunden. Springer, Berlin

CEER—Council of European Energy Regulators, Conclusion Paper (2018) Flexibility use at distribution level, reference DocID: C18-DS-42-04, Brussels

Choi J, Illindala M, Mondal A, Renjit AA, Pulcherio M (2018) Cascading collapse of a large-scale mixed source microgrid caused by fast-acting inverter-based distributed energy resources. IEEE Trans. Ind. Appl 54(6):5727–5735. https://doi.org/10.1109/tia.2018.2854748

Chowdhury T, Chanda CK, Chakrabarti A (2017) Resiliency improvement for a part of south Indian power transmission network. In: Australasian universities power engineering conference (AUPEC). IEEE, Christchurch, pp 1–5

Fischer David, Madani Hatef (2017) On heat pumps in smart grids: a review. Renew Sustain Energy Rev 70:342–357

IEEE Standard for the Testing of Microgrid Controllers (2018) IEEE Std 2030. IEEE Standards, pp 1–42

Kerber G (2010) Aufnahmefähigkeit von Niederspannungsverteilnetzen für die Einspeisung aus Photovoltaikkleinanlagen. PhD thesis, Technical University Munich, Munich

Khanbabapour S, Mohamad E, Hamedani G (2018) Synchronous DG planning to help high voltage systems. IEEE Trans Power Syst 33(3):2440–2451

Kusch W, Stadler I, Bhandari R (2015) Heat pumps in low voltage distribution grids by energy storage. In: 2015 international energy and sustainability conference (IESC), pp 1–6. https://doi.org/10.1109/iesc.2015.7384386

LANUV (2019) Validated Measurements VKCL Köln Clevischer Ring. https://www.opengeodata.nrw.de/produkte/umwelt_klima/luftqualitaet/luqs/konti_nach_station/OpenKontiLUQS_VKCL_vorjahre_validiert.csv. Accessed 01 July 2019

Lee S-S, Jang Y-S, Ahn S-H, Park G-P, Park J-K, Yoon YT (2012) Northeast Asia interconnection-based integration of DG, DR, HVDC, and nuclear load-following under smart grid in the South Korean power system. In: IEEE power and energy society general meeting, pp 1–7

Li Z, Guo Q, Sun H, Wang J, Xu Y, Fan M (2018) A distributed transmission-distribution-coupled static voltage stability assessment method considering distributed generation. IEEE Trans Power Syst 33(3): 2621–2632; Cited by: Papers (1) IEEE J Mag

Marwitz S, Klobasa M (2016) Auswirkungen von Ladestrategien für Elektrofahrzeuge auf den Investitionsbedarf in ein elektrisches Niederspannungsnetz. In: 14. symposium energie innovation, vol 10. Graz/Austria

Mishra M, Panda PC, Swain SC, Roy S (2017) Effect of length of transmission line on power system parameters during islanding for different load cases. In: 2nd international conference on communication and electronics systems (ICCES), pp 277–285

Navarro-Espinosa A, Mancarella P (2014) Probabilistic modeling and assessment of the impact of electric heat pumps on low voltage distribution networks. Appl Energy 127:249–266. ISSN: 0306-2619. https://doi.org/10.1016/j.apenergy.2014.04.026; https://www.sciencedirect.com/science/article/pii/S030626191400378X

Nayeripour M, Waffenschmidt E (2019) Distributed control design to improve the performance of modern distribution power systems toward smart sets of micro grids based on online optimization. In Research project, Alexander von Humboldt foundation

Petermann T, Bradke H, Lüllmann A, Poetzsch M, Riehm U (2011) Was bei einem Blackout geschieht—Folgen eines langandauernden und großflächigen Stromausfalls. Nomos Verlag, Berlin

Protopapadaki C, Saelens D (2017) Heat pump and PV impact on residential low-voltage distribution grids as a function of building and district properties. Appl Energy 192:268–281. ISSN: 0306-2619. https://doi.org/10.1016/j.apenergy.2016.11.103; http://www.sciencedirect.com/science/article/pii/S0306261916317329

Pulcherio M et al (2016) Evaluation of control methods to prevent collapse of a mixed source microgrid. IEEE Trans Ind Appl 52(6):4566–4576

Robitzky L, Häger U, Rehtanz C (2018) Modelling of active distribution networks and its impact on the performance of emergency controls. In: 2018 power systems computation conference (PSCC). IEEE, Dublin, pp 1–7

Shao N, You S, Segerberg H (2013) Integration of 100% heat pumps and electric vehicles in the low voltage distribution network: a Danish case story. In: 3rd international conference on microgeneration and related technologies. Bd. 3

SMA. Das leistet Photovoltaik in Deutschland. https://www.sma.de/unternehmen/pv-leistung-in-deutschland.html

Staffell I et al (2012) A review of domestic heat pumps. Energy Environ Sci 5(11):9291–9306. https://doi.org/10.1039/c2ee22653g

VDE Studie (2015) Der zellulare Ansatz, VDE/ ETG Publikation, Frankfurt. https://www.vde.com/de/etg/publikationen/studien/vdeetg-studiederzellulareansatz; https://shop.vde.com/de/copy-of-vde-studie-der-zellulare-ansatz

Van Roy J (2015) Electric vehicle charging integration in buildings —local charging coordination and DC grids. Dissertation, KU Leuven

VDE. VDE-AR-N 4105(2017–07): Eigenerzeugungsanlagen am Niederspannungsnetz (2017)

VDE-Studie (2016) Schutz und Automatisierung in aktiven Verteilnetzen, VDE Verband der Elektrotechnik Elektronik Informationstechnik e.V., Frankfurt a.M

Waffenschmidt E (2015) Cellular power grids for a 100% renewable energy supply. In: 5th international 100% renewable energy conference (IRENEC 2015). Istanbul, Turkey

The Sustainable Energy Transition Cities and Local Governments in Focus

Maryke van Staden

Abstract This paper presents an overview of key global developments relevant to local climate and sustainable energy action, and leading the transition to a sustainable energy future. In light of the urgency and necessity for such action, the role of local governments in cities and towns is explored, considering the need for strong leadership, direction and accelerated action. Local climate and energy action are intertwined with sustainable development, embedded in the Sustainable Development Goals (SDGs). It is unfolding through high impact global initiatives that support local action such as the Global 100% Renewable Energy Cities and Regions Network and the Global Covenant of Mayors for Climate & Energy (GCoM), as well as through city networks such as ICLEI that provide policy and technical assistance, deal with advocacy, and support integrated Measuring, Reporting and Verifying (MRV) of local action. People and leadership, responsibility, innovation, resilience, security and finance are a few key words illustrating how the sustainable energy transition is unfolding today. Cities of all sizes and their local governments are increasingly responding to the call for climate neutrality at the latest by 2050. They are using ICLEI's GreenClimateCities™ program, with its step-by-step methodology, tools and guidance, to shape their communities' future. By defining their sustainable energy transition roadmap and finding innovative approaches, local governments are making their cities and towns liveable, resilient and climate neutral. Here the underlying premise is access to affordable, reliable, clean, sustainable energy which is essential for sustainable development, quality of life and a just transition in a changing world where climate change and limited resources frame the context. The transition, when well-managed, offers multiple benefits, such as improved air quality and associated health impacts, reduced and avoided greenhouse gas emissions, and new (local) employment opportunities created, to mention but a few. In combination, these make this transition so relevant and interesting to different types of stakeholders at community level. Mayors, councillors and senior local government leaders have understood this, and are re-defining and implementing the sustainable

M. van Staden (✉)
ICLEI's Low Emission Development Pathway/ICLEI–Local Governments for Sustainability, World Secretariat, Bonn, Germany
e-mail: maryke.van.staden@iclei.org

T. S. Uyar (ed.), *Accelerating the Transition to a 100% Renewable Energy Era*, Lecture Notes in Energy 74, https://doi.org/10.1007/978-3-030-40738-4_7

energy transition in their governmental operations and communities. Yet, local governments acting in their territories—cities and towns of all sizes—do not work in isolation. They are embedded in a national context and are key action partners for other levels of government (subnational state or region, province, national), as closest to the local community and closest to where the action should be. Each level of government should have a clear, specific mandate and role to address climate and energy. However, practice often does not reflect this crucial component, with energy mostly remaining a national responsibility. Only by exploring all forms of renewable energy, including decentralised generation and use, can we achieve a climate neutral future. A constructive approach is needed to define what this could look like, exploring relevant scenarios and fully committing to move away from burning fossil fuels. The Talanoa Dialogues were launched in 2018 to address these issues, among others, through exploring improved multilevel governance and vertical integration. Here we addressed three key questions: "where are we", "where do we need to go", and "how do we get there". Moving from the unique Local Government Climate Roadmap—a major global climate advocacy process—that has helped to set the scene, the focus is now on exploring multilevel governance as a cooperation model between all levels of government, and further mobilising local governments to scale up local climate and energy action. This complex space and relevant developments are addressed in this paper.

1 Introduction

Climate change and declining natural resources are two important challenges facing the world today. These do not only pose environmental threats. They also have significant socio-economic, security and political impacts and present a range of challenges that will affect everyone. Yet, it will have a significant larger impact on the poorer section of the global population, who have fewer (or no) resources to prepare and respond to the unfolding effects thereof. These challenges require a global spotlight on the interplay between energy, climate change, as well as security and protection of people and the places they live, and the role of government to deal with these issues.

Alarming trends such as the global population growth rate, declining natural resources, increasing energy consumption, proliferation in the use of fossil and nuclear energy and the visible damaging impact of their use (on air quality, health, water), have drawn attention of policy-makers. It is necessary for governments at all levels to respond to the grave impact on energy-economy-society-security, and plan a roadmap that will enhance resilience, outline how to adapt to existing and expected future climate change impacts, and clearly define emissions reduction trajectories—with commitments to achieving short and long term goals to climate neutrality.

The local impact *on* climate change is clear. Energy use is a key contributor, as greenhouse gas emissions are emitted by burning fossil fuels, often to provide energy for urban areas but also for energy intensive sectors such as industry. This contributes to accelerating the natural phenomenon of climate change—and causes extensive problems for the global ecosystem. Climate scientists and experts in many disciplines are investigating this, building a valuable knowledge base which can help decision-makers today to guide us towards a sustainable future. Many local and national leaders are tackling climate change mitigation, though at this rate the tempo is still too slow to remain below the target of the Paris Agreement, well below the 2 degrees Celsius (°C) of global warming.

The local impact *of* climate change is already visible around the world: from changing rainfall patterns and rainfall intensity, with more and more intense storms and droughts, rising sea level, shifts in growing seasons and disease distribution, to mention but a few impacts. These combined developments are frightening, especially when considering their accelerated impact as global warming increases. The Intergovernmental Panel on Climate Change (IPCC) Summary for Policy Makers of 2018 (IPCC 2018) presents a clear message: "limiting global warming to 1.5 °C would require rapid, far-reaching and unprecedented changes in all aspects of society". All these aspects directly impact on humans and their environment, quite often in urban areas. There are many reports consistently showing alarming trends and drawing attention to the need for accelerated action.

2 Scaling up Local Action

There can be no solution to climate change without local climate and energy action. By approaching climate and sustainable energy action from the perspective of achieving a myriad of benefits, the case is compelling and clear from many different perspectives.

Why address urban areas and cities? They account for 37–49% of global greenhouse gas emissions and urban infrastructure accounts for over 70% of global energy use. Consider that the main climate impacts are on urban spaces, where today already more than 50% of the global population lives, with a substantial projected growth rate of urban areas. The Summary for Urban Policymakers (The Summary for Urban Policymakers 2018), written by several lead authors of the IPCC (Inter-governmental Panel on Climate Change) Special Report on Global Warming of 1.5 °C (SR1.5), translates the IPCC report's key scientific findings and policy observations for officials and policymakers of the world's cities and urban areas. From start to finish, the message is clear and succinctly captured in this sentence: "The tools are at hand. We possess the material basis and policy solutions for transformations and system changes in the direction of greater sustainability, inclusion, and resilience. Urban policymakers must seize the opportunity to meet the defining challenge of the planet, not in the distant future but, as the SR1.5 makes clear, within the next two decades."

The local community level, where the impact *on* and *of* climate change is addressed is typically governed by a local government (used as an overarching term as terminology may vary, e.g. municipality, district council, etc.). Local decision-making processes and implementation can be faster, and action can have immediate impact, compared to more policy-oriented action at other levels of government. This does not, however, mean other levels do not need to engage.

Effective local action requires a supportive and enabling national framework, to guide and actively aid local and other subnational governments in comprehensively tackling climate change mitigation and adaptation in their geographical territories, be these cities, districts or provinces. There is a need for improved communication, coordination and planning between all levels of government to deal with this local, national and global challenge. This is a new area of exploring intra-and inter-governmental coordination, moving away from the traditional top-down approach to an evolution of combining bottom-up and top-down that clearly defines the roles and responsibilities of each level of government, as well as processes and structures to ensure an optimal approach is enabled.

A key publication released in 2019, Climate Emergency, Urban Opportunity (2019) by the Coalition for Urban Transitions (CUT), outlines how national governments can secure economic prosperity and avert a climate catastrophe by transforming cities. The report suggests six key priorities on which national governments should act:

- Develop an overarching strategy to deliver shared prosperity while reaching net-zero emissions—and place cities at its heart.
- Align national policies behind compact, connected, clean cities.
- Fund and finance sustainable urban infrastructure.
- Coordinate and support local climate action in cities.
- Build a multilateral system that fosters inclusive, zero-carbon cities.
- Proactively plan for a just urban transition.

Citizens, business and industry look towards (all levels) of their government to show leadership, especially in such times of crisis, to guide and protect them, and to develop a clear roadmap for action that is inclusive and provides confidence to all, including the much-needed investors.

2.1 What Action Is Needed?

Sustainable energy is defined as combining the optimised use of (ideally local) renewable energy sources, energy efficient technologies and energy conservation. This is becoming the "new normal", the recognised way forward, also recognising the need to leapfrog dirty technologies and unsustainable approaches. Energy is critical for daily use, not only for citizens with their daily needs of electricity, space and water heating/cooling, but also in providing services and goods in an efficient and cleaner way. It is important to level the playing field as it relates to energy costs, removing subsidies for fossil fuels, and to make businesses and industries viable and successful over the long term by incorporating actual energy costs. This will support the transition to clean fuels and renewable energy resources, as the main element of sustainable development.

Energy conservation and improved energy efficiency (technologies and measures) are key action areas to reduce energy demand and save costs. Here immediate or fast financial benefits can be gained, e.g. where more efficient technologies are used the pay-back time on investment (amortisation) is short with substantial mid- to long-term savings, as energy demand goes down. This can easily be done for example by installing energy efficient lighting, but also with more expensive action areas such as complete building renovation to achieve modern energy standards in old buildings (low energy, passive house, zero energy or event energy plus). New buildings need to confirm to the highest possible standards, especially considering their potential long lifetime.

Switching to renewable energy sources is another key component in this sustainable energy transition. We need energy alternatives to fossil and nuclear energy, where fuels are safe and not so easily depleted (hence called "renewable"). The wide range of renewable energy sources available implies that nearly every location on Earth has some form of renewable energy that could be used. The ideal case is using an energy mix: from solar—and wind energy to waste-to-energy, tidal and wave power, geothermal energy, as well as sustainable bio-and hydro energy, and other. The renewable energy mix and using effective energy storage solutions are crucial to planning and rolling out a well-managed energy transition that is low-carbon and resilient—moving towards climate neutrality.

2.2 *Motivation for Action*

Safety, money, growth and environment—these are some of the issues drawing the attention of modern society. All these are inter-connected with the climate change emergency and our response to this global phenomenon. It is well known that global resources are limited. This requires us to reduce waste, optimise resource efficiency and generally be cleverer in applications that require energy and resources. The human impact on climate change is also now well understood. Many solutions are readily available. What is needed? Decisions and action. Our response requires a combination of applying technology and effective policy, as well as changing behaviour, with the latter the hardest angle to address in many cases.

Increasingly local communities—cities and towns of all sizes—are responding to these challenges by engaging in local climate and sustainable energy action. Why do local governments engage? Recognition of responsibility is typically a main driver for action. Local leaders and municipal staff assess the local climate risks and vulnerabilities, inventory the community-scale greenhouse gas emissions, and identify where energy needs are and what is imported (i.e. identifying dependency and finance flows away from the community). Further, they explore local climate and sustainable energy action options and associated (co)benefits, developing action plans with priority actions, and how these can be financed.

Local governments understand the local context and local needs. They are responsible for good governance and service delivery. The underlying motivation for action is typically how to retain or further develop "good quality of life" for people in the living and working space for citizens, businesses and industries, by tackling climate change and enhancing resilience of the community. For local governments a helpful starting point is to design the roadmap to become climate neutral in governmental operations (by 2050 at the very latest), not only to "practice what you preach" but also to optimize efficiency of services, reduce costs and be resilient to shocks and changes. Rising energy costs are also often a driver, as it is challenging to budget effectively over time with fluctuating prices.

Engaging in local climate and energy action seems to be a logical choice—but this choice clearly implies change. It requires change in the way energy is generated and distributed (moving away from centralised systems to decentralised, smaller inter-connected systems and safer fuels). It also requires change in the way energy used. Energy has become a valuable commodity! Yet, this is part of the transition challenge, as local governments often do not own energy utilities and typically do not have a mandate to address energy—but they understand the need for acting in this area. The leaders have started and use diverse reasons for this.

This local process is typically led by local governments, but also requires the involvement of many stakeholders: citizens, business and industry, non-profit organisations. The active support and engagement of many different groups is a proven key to success. In practise this means especially engaging with people in these stakeholder groups: municipal staff, councillors, Chief Executive Officers (CEOs), etc.... The value of individual "champions" in each of these stakeholder groups is a necessity to help drive the agenda, and ensure it remains "on the agenda".

Municipalities have different and multiple motivations to engage and reap multiple local benefits, which include, for example:

- exploring approaches for improving air quality and associated improved health;
- urban infrastructure improvement: anticipating the impact of climate change and adapting to this. 'local to global' responsibility: protecting people and the environment in a changing climate where resources are becoming scarcer, addressing the global common good;
- generating local profits and local taxes that can be re-invested in climate and energy action;
- green economy development: improvement by saving energy and reducing energy bills, making money for local sustainable energy generation, local job creation, stimulating the small and medium sized enterprise sector;
- keeping money in the local economy: local energy production, local energy use, and supporting circular development with involvement and benefits to the immediate neighbouring areas;
- social upliftment of poorer residents: by reducing their need for energy, e.g. through energy efficient housing renovation, and offering effective public transport options; and

- energy security: securing stable energy supply, ensuring that energy remains affordable for citizens and business, while supporting a good quality of life (making it interesting for people and businesses to stay).

2.3 Role of Local Governments

The role of local government (also referred to a local authority, municipality, council, administration, etc.)—as the level of government closest to citizens—is critical in the context of climate protection, climate change adaptation and resilience, and the transition to sustainable energy.

Local governments are usually responsible for defining *strategy*, implementing local *policy and regulations*, developing and maintaining *structures* that handle administration and provide services, providing a range of *services* to local inhabitants and businesses. This role differs from country to country but can include policing, health services, education, social services, transport-, water- and sanitation services, and sometimes also energy services (e.g. local sales from national grid). Further to this they often *own or manage infrastructures* such as buildings, roads, even electricity grids in their geographical territory.

In all of these cases local governments can thus shape and guide change among inhabitants, businesses, and industry as well as their own municipal operations. They can motivate and lead a change of direction in the whole community, to benefit the community. These are areas where local climate action is possible, with vast potential for reducing emissions, improving overall efficiency and quality of life.

Those cities and towns that are achieving success usually:

- have one or more local champions who can motivate people, draw attention to these issues, and constantly make sure climate/energy is on the political agenda. These include political representatives and senior municipal staff—highly recommended as a valuable driving force for local action;
- have a comprehensive and regularly updated (climate or energy) action plan which is being implemented and monitored;
- understand where challenges come from—i.e. where energy is being used and emissions released, and conduct regular greenhouse gas (GHG) inventories,
- review local renewable energy resources to assess where energy imports can be reduced, and how local resources can be optimised; and
- conduct a community SWOT analysis to identify Strengths, Weaknesses, Opportunities, and Threats (or similar exercise) in the relevant sectors and areas.

These actions and processes help to coherently address climate change mitigation and adaptation at the community level.

3 Global Developments

3.1 Expanding Partnerships

Many leading local governments have been engaging in climate action since the 1990s, when ICLEI started its Cities for Climate Protection™ (CCP) Campaign, the first global campaign supporting local governments in tackling climate change. Campaign participants numbered over 1000 cities and towns from around the world in the early 2000s. Since 2013, ICLEI transitioned the CCP Campaign to the GreenClimateCities™ (GCC) Program, helping to analyse, act and accelerate climate action, addressing local governments of cities of all shapes, sizes and at varying levels of development in the Global South and North with an updated and more holistic approach. It is linked to ICLEI's 2018 call for climate neutrality, connecting climate action to sustainable development, rolling out the 5 interconnected development pathways, each of which has a dimension relevant to climate change: low emission development, resilient development, nature-based development, circular development, and equitable and people-centered development.

The EU Covenant of Mayors for Climate and Energy brings together thousands of local governments voluntarily committed to implementing EU climate and energy objectives. The Covenant of Mayors was launched in 2008 in Europe with the ambition to gather local governments voluntarily committed to achieving and exceeding the EU climate and energy targets. Today this initiative gathers 9,000+ local and regional authorities across 57 countries drawing on the strengths of a worldwide multi-stakeholder movement and the technical and methodological support offered by dedicated offices.

The Compact of Mayors[1] was launched at the UN Climate Summit in September 2014 by C40, ICLEI and UCLG, in partnership with UN Secretary-General Ban Ki-moon, his Special Envoy for Cities and Climate Change, Michael R. Bloomberg, and UN-Habitat. As a global initiative, it aimed at recognizing new and existing local climate commitments, and make sure that these are recognized globally. This has helped to create a global political movement, with political commitments to local climate action combined with global accountability. The further evolution was the Global Covenant of Mayors for Climate and Energy,[2] which formally brings together the European Union's Covenant of Mayors and the Compact of Mayors—the world's two primary initiatives of cities and local governments—to advance city-level transition to a low emission and climate resilient economy, and to demonstrate the global impact of local action.

Another key development was the release of the Global Research and Action Agenda on Cities and Climate Change Science (CitiesIPCC Conference 2018), which outlines the first of its kind agenda, as a partnership effort of city networks, academia and researchers and many other actors.

---

[1]http://compactofmayors.org.

[2]https://www.globalcovenantofmayors.org/.

In parallel, thematic activities have also evolved, such as the Global 100% Renewable Energy Platform[3] (formerly called Campaign), established to initiate dialogue about 100% RE, build capacity and educate policymakers about the opportunities, also to release case studies and stories that are happening all over the world. The Global 100% RE Cities and Regions Network, managed by ICLEI, is part of this global platform, offering guidance and support to committed local and regional governments that are defining their 100% RE transition roadmap.

Access to finance is also a hot topic. The Cities Climate Finance Leadership Alliance (CCFLA),[4] announced in 2014 and revitalised in 2019, is a coalition of leading organisations and nations committed to deploying finance for city level climate action at scale by 2030. ICLEI's Transformative Action Program (TAP),[5] established in 2015, is connected to the CCFLA, offering a robust project pipeline of local projects. Many partners are jointly exploring how to scale up support for low carbon resilient infrastructure projects in accessing finance.

A host of new initiatives has since evolved, with exciting new developments seen that inspire hope. Below some highlights are reflected in a non-exhaustive list relevant to local climate and energy action.

3.2 *Unpacking Technical Assistance*

Local and regional governments need tailor-made guidance and assistance to address climate change and lead the transition to a sustainable energy future. Below is a snapshot showing how ICLEI deals with this, as one of the oldest and most comprehensive approaches available to cities, towns and regions.

ICLEI's GCC Program supports local and regional governments as key players in defining, guiding and monitoring integrated climate action and the transition process. Integrated climate action refers to reducing GHG emissions from high-to low-to zero (climate change mitigation) and addressing climate change adaptation and enhancing resilience. This means action both in government operations and at community level, in all sectors and engaging with all relevant local stakeholders—citizens, NGOs, business and local industry.

The Bonn Centre for Local Climate Action and Reporting, referred to as ICLEI's carbon*n*™ Center,[6] was established in 2009 as an international centre of excellence. This was a joint United Nations Environmental Programme (UNEP) and ICLEI initiative. Today the Center provides the overview of ICLEI's climate services and support, including

- GreenClimateCities (GCC) Program and associated tools and guidance,

[3]https://www.global100re.org/.

[4]https://www.citiesclimatefinance.org/.

[5]https://www.tap-potential.org.

[6]https://carbonn.org.

Guidance on MRV and integrated MRV (refer to reports such as Multilevel climate action: the path to 1.5° 2018) addressing:

- Encouraging the use of relevant standards and protocols to be able to assess developments in a systematic way, and compare progress against a proper baseline.
  These include for example the Global Protocol for Community-scale GHG Emissions Inventories (GPC) and the GCoM Data Standard. Standardised approaches, templates and guidance are provided, e.g. Climate Risk and Vulnerability Assessments, GHG Emission Inventories and Climate Action Plans.
- Measuring, Reporting and Verifying (MRV) of local climate action with a focus on local and subnational governments—mitigation and adaptation;
- Vertical integration of systems and processes addressing climate action planning and reporting among all levels of government and across different sectors;
- Facilitation of multilevel governance dialogues and discussions on how to more effectively cooperate across and between all levels of government (refer to the Cities and Regions Talanoa Dialogues of 2018, for example the Mexican Talanoa Dialogue 2018);
- Access to finance for subnational governments to support urban sustainable energy and climate-related project investments and calling for vertically integrated NDC implementation and investment plans.
  The work conducted by Project Preparation Facilities (PPFs) is one example of what is being studies, to identify good practice (e.g. Summary of good practice of successful project preparation facilities 2018).

• Reporting, to track progress, understand where action is effective but also to identify gaps and challenges, as well as the investment needs for local action.

  - Reporting is done through the CDP-ICLEI Unified Reporting System, where ICLEI validates data for its network, and provides feedback on data improvement or outlines further support offers. The CDP-ICLEI Unified Reporting System serves the Global Covenant of Mayors for Climate & Energy, the Under2 Coalition, and many other initiatives.

• Analysis services based on voluntary reporting by local and other subnational governments.

  - ICLEI uses reported data and consolidates advocacy messaging, in its capacity as focal point of the Local Governments and Municipal Authorities (LGMA) constituency at UNFCCC. Aggregated data is used for annual messaging towards national governments.

The Table 1 provides an overview of some key developments relevant to local climate action which impacts on local governments and how they engage.

Table 1 Chronology of relevant efforts in the global response to climate change (non-exhaustive list)

Year	Chronology of climate relevant efforts
1990	• First Assessment Report of IPCC (FAR) • Establishment of ICLEI and start of ICLEI's climate work through the 1st climate action project with committed cities
1992	• Adoption of UNFCCC (no specific definition of greenhouse gases at this stage)
1993	• Launch of ICLEI's Cities for Climate Protection (CCP) Campaign, following a 5-milestone process to tackle and track climate change mitigation (adaptation added in early 2000s)
1995	• Second Assessment Report of IPCC (SAR) published • IPCC 1995 Guidelines for GHG Inventories released • Local authorities accepted as observers at global climate negotiations
1996	• IPCC Guidelines and Good Practice Guidance for GHG Inventories released
1997	• Adoption of Kyoto Protocol (Annex-A lists specific GHGs and sectors) • First Global IEA Report on CO_2 Emissions from Fuel Combustion
1998	• Launch of the Greenhouse Gas Protocol
2001	• Marrakech Accords Presented • GHG Protocol Corporate Accounting And Reporting Standard (first edition) Third Assessment Report of IPCC (TAR) released
2004	• GHG Protocol Corporate Accounting And Reporting Standard (revised) released
2005	• Formation of the C20 Cities Climate Leadership Group under leadership of former London Mayor Ken Livingstone (now C40), to engage megacities around the globe • Launch of European Emissions Trading Scheme (ETS) • Launch of the US Conference Of Mayors Climate Protection Agreement
2006	• Release of ISO14064 Standard • IPCC Revised Guidelines Released • First Global Report of ICLEI's Cities for Climate Protection (CCP) Campaign • Partnership between C40 and Clinton Climate Initiative (CCI) announced
2007	• Fourth Assessment Report of IPCC published • Launch of Global City Indicators Facility and the Climate Registry in the US • COP13 in Bali: Launch of the Local Government Climate Roadmap,[a] an advocacy coordination approach to stress the urgency for action and role local governments can play in tackling climate change
2008	• Launch of European Covenant of Mayors,[b] funded by DG Energy and supported by European city networks • Release of ICLEI's US Local Government Operations Protocol (LGOP) • Launch of ICLEI-US/CDP Cities Pilot Project to explore reporting by local governments • COP14 in Poznan: Local Government Climate Sessions

(continued)

Table 1 (continued)

Year	Chronology of climate relevant efforts
2009	• Launch of the first global protocol for local governments on GHGs—the International Local Government GHG Emissions Analysis Protocol (IEAP) • Launch of Greenhouse Gas Regional Inventory Process (GRIP) • Launch of Bonn Center for Local Climate Action and Reporting (carbon*n* Center) to identify methodologies and guidance approaches to enable global reporting of local climate action • COP15 in Copenhagen: Local Government Climate Lounge, with the LGMA constituency bringing the largest subnational government delegation ever to a Climate COP
2010	• Kick-off for ISO/TR14069 • Launch of the Global Cities Covenant on Climate—the "Mexico City Pact", as a global initiative to collect climate commitments • Launch of carbon*n* Cities Climate Registry (cCCR), a global reporting platform (renamed carbonn Climate Registry in 2014) to serve the Mexico City Pact and other initiatives collecting commitments and targets • Kick-off for drafting of US community GHG protocol • Start of ICLEI's Resilient Cities Congress series in Bonn • COP16 in Cancun: Recognition of local governments as "governmental stakeholders"
2011	• Release of C40/CDP Cities Report • ICLEI-C40 MoU to develop the GPC, the Global Protocol for Community-scale GHG Inventories, to harmonize GHG accounting and reporting at city level • Release of GHG Protocol Corporate Value Chain (Scope 3) Accounting and Reporting Standard • Release of 2011 Annual Report of carbon*n* Cities Climate Registry • COP17 in Durban: Launch of the Durban Adaptation Charter
2012	Launch of the draft Global Protocol for Community-scale Greenhouse Gas Emissions Inventories (GPC), developed by WRI, C40 and ICLEI
2013	• Local Government Climate Roadmap Phase II started • COP19 in Warsaw: first "Cities Day organized at a COP"
2014	• UN Special Envoy for Cities and Climate Change appointed, namely Michael R. Bloomberg • UN SG Summit: Launch of the Compact of Mayors by the three global city networks: C40, ICLEI and UCLG, supported by UN-Habitat • UN SG summit: launch of the Cities Climate Finance Leadership Alliance (CCFLA) and several other important initiatives relevant to accelerate climate action • COP20 in Lima: Launch of the GPC
2015	• COP21 and Paris Agreement
2016	• Start of the merger of the Covenant of Mayors (EU partners) and Compact of Mayors (global partners)
2017	• Changing the Global 100% RE Campaign into the Global 100% RE Platform, stimulating new partnerships and action • Launch of the GCoM Data Standard at COP23 in Bonn • Announcement of CDP-ICLEI planned cooperation on reporting

(continued)

Table 1 (continued)

Year	Chronology of climate relevant efforts
2018	• Year of Talanoa Dialogues after call of Fiji, COP23 President, resulting in over 60 Cities and Regions Talanoa Dialogues around the globe • Cities and Climate Change International Conference (CitiesIPCC) in Edmonton, March 2018, with the Global Research and Action Agenda on Cities and Climate Change Science released • GCoM secretariat and partners calling for vertically integrated NDC implementation and investment plans
2019	• Launch of the GCoM Global Common Reporting Framework (CRF) • Launch of the CDP-ICLEI unified reporting system to support local and regional governments with climate action reporting

[a]http://old.iclei.org/climate-roadmap
[b]http://eumayors.eu

4 Conclusion

Increasingly local governments around the globe are starting and accelerating local climate and energy action. The reasons for this are manifold and include a growing recognition of responsibility to enhance local resilience, adapt to inevitable climate change and plan for a carbon neutral future while reaping many other benefits. This progress is being tracked through initiatives such as the GCoM, the Global 100% RE Platform, as well as other sectoral initiatives where commitments are made, assessments conducted and action plans developed—all with a view to define and implement a roadmap towards climate neutrality.

From an energy perspective, more leadership at community level is seen, with RE and EE at the heart of the sustainable energy transition. This means enabling access to sustainable clean energy for all, improving energy systems and grids, utilizing and constantly improving storage approaches, and generally ensuring stable and quality energy service provision to the community. The co-benefits of this include improved air quality, reduced GHG emissions, job creation opportunities, stimulating the local economy, circular development, the evolution of business models (community energy, exporting clean energy), and exploring an inclusive and just transition that ensures the vulnerable and poor are part of the sustainable transition roadmap. This outlines the unfolding of a much-needed win-win approach for the local/regional/national/global levels.

All tiers of government need to engage when dealing with climate and energy issues. These are not purely national issues, despite the strategic nature of energy. Synergy and cooperation between different levels of government is key to optimise processes and approaches when exploring climate neutrality. However, political leadership is only one element, albeit a key one. This should also be linked to engagement of other stakeholders to effectively co-design and implement comprehensive programs that deal with these major challenges in modern life.

Processes are in place that mobilise, support, offer guidance and even address advocacy with and for local governments around the globe. ICLEI's GreenClimateCities^{TM} Program is one of these, available to local and regional governments around the globe, developed to guide and support them in scaling up integrated climate action. A harmonized reporting system has been established in partnership between CDP and ICLEI to ease the reporting burden and provide services to self-reporting local and regional governments. These elements help track progress, celebrate successes and draw attention to local investment needs at the global level.

Every local government can and should combine energy conservation, energy efficiency and renewable energy—ideally moving towards 100% renewable energy communities that optimize policy, technology and people-power, to ensure zero emission, resilient communities are created around the globe. In this way their contribution can probably help scale up the level of ambition of their respective countries' NDCs, while addressing local concerns such as quality of life, economic livelihoods and protecting the environment.

References

Basil O, Katharina S-R, Charlotte B, van Staden Maryke (2018) CCFLA. Summary of good practice of successful project preparation facilities. http://urbanleds.iclei.org/fileadmin/user_upload/Resources/guidance_and_tools/Climate_Finance/Summary%20of%20good%20practice%20of%20successful%20project%20preparation%20facilities%20-%202018.pdf

Coalition for Urban Transitions (CUT) (2019) Climate emergency, urban opportunity: how national governments can secure economic prosperity and avert climate catastrophe by transforming cities

CitiesIPCC Conference (2018) Extended version: global research and action agenda on cities and climate change science. https://citiesipcc.org/beyond/global-research-and-action-agenda-on-cities-and-climate-change-science/

ICLEI—Local Governments for Sustainability (2018) Multilevel climate action: the path to 1.5 degrees. carbonn® Climate Registry 2017–2018 report. Bonn, Germany. Chang D-B, Dana V, van Staden Maryke, Ferreira AR. https://iclei.org/en/publication/multilevel-climate-action:-the-path-to-1.5-degrees

ICLEI's GreenClimateCities^{TM} (GCC) Program https://iclei.org/en/activities_database/GreenClimateCities-(GCC)-program

IPCC (2018) Summary for policymakers of IPCC special report on global warming of 1.5 °C approved by governments (SR1.5) (Global Warming of 1.5 °C, an IPCC special report on the impacts of global warming of 1.5 °C above pre-industrial levels and related global greenhouse gas emission pathways, in the context of strengthening the global response to the threat of climate change, sustainable development, and efforts to eradicate poverty). https://www.ipcc.ch/2018/10/08/summary-for-policymakers-of-ipcc-special-report-on-global-warming-of-1-5c-approved-by-governments/

National Autonomous University of Mexico (2018) Mexico Talanoa Dialogues. In: Gian Carlo Delgado Ramos, van Staden Maryke, Edgar VF (eds) Mexico City

The Summary for Urban Policymakers (2018) What the IPCC Special Report in Global Warming of 1.5 °C means for Cities

The Pathway to 100% Renewable Energy—A Vision

Rian van Staden, Filippo Boselli and Anna Leidreiter

1 Introduction

The Global 100% Renewable Energy Platform (Global 100% RE) is among the first of global initiatives that advocates for 100% renewable energy.

Global 100% RE connects the dispersed dots of renewable energy advocates and builds a global alliance to confirm that being powered by 100% sustainable renewable energy is both urgent and achievable. This unique campaign builds on initiatives already taking place on national, regional and local levels, and sets the global discourse on renewable energy towards 100% RE as the new normal. Since its establishment, the goal of Global 100% RE has been to initiate dialogue, build capacity, and educate policy makers on the benefits of 100% RE based on the ever-increasing number of case studies from around the world.

A substantially increasing number of municipalities, cities, regions and countries have committed to a 100% renewable energy future. As of late 2018, more than 350 cities, municipalities and regions including Frankfurt, Vancouver, Sydney, San Francisco, Copenhagen, Oslo, Scotland, Kasese in Uganda, Indonesia's Sumba island and the Spanish Island of El Hierro have demonstrated that transitioning to 100% RE is a viable political decision (go100re.net). Many of these municipalities and regions are setting the 100% RE target as they consider it not only a technically and economically feasible option but an ethical imperative in the face of global climate change. During COP 21 in Paris in December 2015, nearly 1000 Mayors and councillors pledged to reach the 100% Renewable Energy target within their municipalities (citiscope.org/habitatIII/news/2015/12/paris-city-hall-declaration-world-mayors-throw-down-gauntlet-climate). Given that most people are living in cities today, and urban areas are accountable for 70% of energy related

R. van Staden (✉) · F. Boselli · A. Leidreiter
Global 100% Renewable Energy Platform, Bonn, Germany
e-mail: rvanstaden@global100re.org

T. S. Uyar (ed.), *Accelerating the Transition to a 100% Renewable Energy Era*,
Lecture Notes in Energy 74, https://doi.org/10.1007/978-3-030-40738-4_8

CO_2 emissions (ghgprotocol.org/city-accounting), this commitment to a renewable energy transition is an encouraging trend.

But this movement goes far beyond the local level. Fifty-five countries with small-island states in the lead are planning to fully decarbonize their electricity system and achieve 100% renewable electricity within the next decades. At the COP22 in Marrakesh, 48 developing countries pledged to "strive to meet 100% domestic renewable energy production as rapidly as possible while working to end energy poverty, protect water and food security" (climatenetwork.org/press-release/historic-breakthrough-48-countries-commit-transition-100-renewable-energy).

This encouraging movement would not be as successful as it is without civil society, practitioners and businesses pioneering it. Indeed, all major non-state actor groups have joined forces to build the political will for 100% RE in the Global 100% Renewable Energy Campaign (go100re.net/about-us/supporters). Further, more than 80 corporations committed to 100% Renewable Electricity under the RE100 campaign (there100.org/companies).

And thousands of small and medium-sized companies, entrepreneurs and citizens are making the transition away from fossil fuels towards 100% RE, creating innovative business models to help accelerate the transition. For example, in Bangladesh, since 2003 more than 3.9 million rural households and shops have electricity from solar home systems, providing sustainable electricity for millions of people in rural areas (worldbank.org/en/news/feature/2016/10/10/solar-program-brings-electricity-off-grid-rural-areas). In Africa, M-Kopa's business innovation is using the mobile payment systems to deliver solar energy in off-grid regions (m-kopa.com/asante-sana-300000-east-african-homes-now-on-m-kopa). Also, in industrialised countries like in Germany, the energy system is undergoing fundamental changes with more than 800 community energy cooperatives forming the backbone of the pioneering Energiewende (worldfuturecouncil.org/parliamentarians-can-revive-energiewende).

While this is an encouraging development, crucial questions arise: What does 100% RE actually mean? What are the implementation steps to achieve the target? How do we measure success? And, how do we ensure that the transition to 100% RE is an instrument towards wealth redistribution, creation of social wellbeing and the protection of our ecosystems? In fact, urban energy planning and climate mitigation and adaption are not new fields and several methods and instruments exist (See for example ICLEI's GreenClimateCities MRV handbook or the local renewables process, climate cities benchmark by climate alliance, German 100% RE-regions criteria by deENet, or the energy transformation index by Fraunhofer ISE). However, a survey unveiled (go100re.net/wp-content/uploads/2015/11/DiscussionPaper_go100re_criteria.pdf) that they do not sufficiently address the transformation to 100% RE, the action's impact on local sustainable development or are not applicable in an international context.

Meanwhile, legislators, policy makers and community champions often find themselves in the position in which converting the energy system to 100% RE is about more than replacing fossil resources with renewable sources. Planning urban growth around 100% renewable energy targets calls for coordinated implementation strategies that can integrate with the myriad other priorities municipal staff are charged with addressing. This sharply contrasts stepwise implementation of one-off renewable energy projects that can be developed in relative isolation from other social, economic, environmental planning goals.

To navigate the complexity inherent in policy planning, frameworks and guidelines which are built on available practices and experiences can help decision-makers and community champions orient themselves to confront the challenges of achieving 100% RE.

While acknowledging that every jurisdiction faces unique challenges and is situated in a distinctive context, this chapter aspires to create a common and comprehensive set of recommendations to facilitate the building of bridges, overcoming political, cultural and social differences in places around the world.

2 The Vision: Steps on the Journey to 100% Renewable Energy

The following steps outline a vision of an achievable path to 100% renewable energy. They are based on the experience of progressive cities, nations and institutions. While not every entity will take all of these steps, they offer three important things—a place to start, ideas of steps to take, and a feeling for what the journey entails. The steps are based on and adapted from the Building Blocks for 100% Renewable Energy developed by the Global 100% Renewable Energy platform.

2.1 Assess the Local RE Potential

Assess the renewable energy potential of the particular region and ensure regional natural strengths are fully captured. The design of a renewable energy system strongly depends on the natural resources available for a neighbourhood's, community's or municipality's consumption needs; which renewable energy sources—solar (PV and solar thermal), wind, hydro, biomass or geothermal energy—should be used; and the extent to which the solution goes beyond being technological. It is important to remember that while a technology might be acceptable to one for additional benefit. Evaluate the resilience and flexibility of the current energy system and map the system's dependence on particular sources (for example from imported oil and gas). When fossil fuel resources are local, for example, this phase will enable communities to imagine their local economy thriving without these assets, thereby overcoming carbon lock-in and optimizing use of more sustainable local resources.

2.2 *Identify GHG Emissions by Sector*

Estimate total GHG emissions produced by sector and calculate relative shares of those sectors and sources. This is essential to understanding which sectors contribute the most to GHG emissions and how to strategically prioritize interventions in ways that will have the greatest impact on reduction and mitigation.

2.3 *Measure Current Energy Costs*

Estimate average annual per capita costs for electricity, heating/cooling and transport. In other words, determine how much each resident spends per year. It is important to compare the current costs with the costs projected for the 100% RE scenario to confirm whether costs are realistic and competitive.

2.4 *Measure and Quantify Externalities*

Assess externalities associated with the current energy system. These are the costs that are not accounted for, such as health and climate change costs, among others. A long-term cost analysis of proposed energy technologies should also be considered. Externalities are important yet difficult aspects account for when pricing energy. Yet incorporating them allows to appreciate how competitive renewable energy sources actually can be (iclei.org/fileadmin/PUBLICATIONS/Case_Studies/2_Chemnitz_-_ICLEI-IRENA_2012.pdf).

2.5 *Define the 100% RE Target*

Formulating a target that is time-bound and measurable, and whose scope and political obligations are well-defined, is essential to developing and implementing a comprehensive and coherent 100% RE strategy. As a general guideline, a 100% RE goal is fully achieved when the amount of renewable energy generated within or imported into the defined area equals or exceeds the annual energy consumed. Yet, this definition is left intentionally vague, as it does not specify which energy-use sector is included (electricity, transport, heating/cooling, etc.) nor does it set the scope of the target (whether community-wide or municipal). It also does not specify where the energy is to be produced or by whom, or still more ambitious, whether the technology will be sustainable in the broadest sense, for example: whether the solar panels are to be manufactured per strict environmental standards or whether the community accepts hydropower as environmentally or socially sustainable, etc.

Therefore, the specifics about a community's 100% RE target must be clearly and comprehensively defined each time a 100% RE target is set. The details of the particular 100% RE target depend on the specific case and they should be defined by local authorities in a transparent, open, accessible and comprehensible manner. The geographic area to which the target applies should also be clearly defined to avoid confusion.

Interim targets should also accompany the longer-term targets. Furthermore, other community targets should also be explored for how they could complement the 100% RE target. Among others, energy conservation and efficiency, GHG emissions and air pollution reduction targets would complement the 100% RE goal.

Lastly, it is important to highlight that setting the 100% RE target does not necessarily equate to energy independence or self-sufficiency. Regions should understand that a 100% RE Future can be a target best achieved through an interconnected system where energy is shared across territories and where nations and regions cooperate and integrate their local grids.

2.6 *Estimate the Potential Economic, Environmental and Social Benefits*

The results of the energy scenario can also help develop an estimation of the potential economic, environmental and social benefits that such an energy transition would entail. These should include specific estimates in terms of job creation, energy savings, local revenue production, opportunities for local industries, positive effect on human health, local air pollution, climate change mitigation potential and resilience. These are essential also to creating the necessary momentum and political will across different sectors, departments and jurisdictions.

In terms of economic impact, it is important to understand that the transition may strand assets. These include local resources, which cannot be used because of their carbon intensity (such as oil, coal or natural gas), as well as related infrastructure like coal or natural gas plants. The labour sector will also transform from traditional energy expertise and professions to renewable energy skills and careers. It is therefore warranted to have economic analyses and conversion strategies that also consider the impact the transition has on current social structures, including on employment.

2.7 *Fix Binding Targets*

After declarations that signal strong and widespread political commitment for the goal, it is essential that the 100% Renewable Energy target is legislated as binding and enforceable. Setting an ambitious, long-term renewable energy target demonstrates political commitment and provides investors, businesses, and residents with a clear long-term vision for the region, along with better understanding of how their roles within it. Providing an official mandate for action catalyses change and helps streamline efforts. Depending on the national context, local authorities might not have the legislative power or the capacity for this and implementation may fall to the jurisdiction of regional and national governments. Therefore, local communities should not work in isolation but collaborate as much as possible with other levels of government (regional, national and international) in order to be able to formalize declarations. Governments' representatives from municipal to national level should find pathways to work together and to set up the necessary policy frameworks and institutional bodies able to align purposes and streamline efforts across jurisdictions and constituencies.

Setting and communicating a 100% renewable energy target has a number of additional advantages: it can stimulate engagement with a wide range of stakeholders; it can ensure more efficient deployment of both technical and administrative resources and reduce duplication and the risk of developing competing policy goals; and it can give key stakeholders (such as utilities, or private investors) the confidence required to make large investments, such as upgrading transmission and distribution grids. By improving investment certainty, ambitious targets can also attract domestic and international investors, in turn making it easier to achieve the target. Experiences from around the world demonstrated that targets can also build awareness among external audiences as well as among the region's citizens. This awareness can be essential to building the public support needed to support and eventually reach the target.

2.8 Establish Relevant Institutionalized Bodies

Institutionalization also means establishing formal bodies or organizations to be responsible for designing, implementing and monitoring the transition towards achieving the target. These bodies should promote and facilitate multi-level governance, cross-sectoral collaboration and peer-to-peer cooperation between regions, cities and local governments. As an energy transition is a long-term endeavour, it needs to be well-rooted within a combination of institutional practices that are formal (e.g. regulations, laws and acts) and informal (e.g. round tables, task forces, energy days). A platform for permanent, efficient organization and allocation of the necessary resources can enable generational continuity of the transformation and coordination of local actors and projects. A local authority, an energy agency, or a third-party could provide such a platform, but to foster trust, it should be independent of corporate interests, influence of private investors and short-term political trends.

It is very important that these commissions and the policy framework that are established to support the 100% RE target can moderate potential swings in political trends related to changes in political leadership and mandates.

Instituting mechanisms to monitor the pace of progress is also needed. The appointed commissions or taskforces should be charged with evaluating progress and reporting on best practices and what has not performed to expectations.

2.9 Change Human Behaviour

A large decrease in energy consumption can come from non-technological measures that support a cultural focus on energy savings that lead to behaviour change. This can be done by promoting a culture of sustainability within the community, which is based not only on raising the level of awareness among citizens (e.g. through education and awareness campaigns) but also on increasing their level of engagement within their community (go100re.net/properties/byron-shire-council).

To promote sustainable behaviour, a much more systematic approach to the problem is needed (sustainablecitiesinstitute.org/topics/equity-and-engagement/community-engagement-sustainability-principles). Behavioural psychology has shown that a sense of belonging to one's own community and territory, fostered by engagement in decision-making processes, helps to nurture a feeling of empowerment and responsibility leading to more sustainable behaviours (environment.scotland.gov.uk//media/1653www.environment.scotland.gov.uk//media/16539/Understanding-Behaviour-Change.pdf). However, people can often feel disconnected from their community and environment, especially in cities (sustainablelifestyles.ac.uk/sites/default/files/motivating_sc_final.pdf). This can foster feelings of political apathy and increasingly individualistic behaviours that lead to indifference to the impact of personal action on the community and environment (ourworld.unu.edu/en/bridging-the-emotional-disconnect-between-people-and-cities). City and community councils behave as neighbourhood representatives closest to the community and therefore play a crucial role in enabling change.

For example, improved community engagement can be achieved by supporting or creating platforms for people to interact and undertake community activities (gardening, tree planting, energy saving initiatives, community energy projects). By providing a space for individual and collective participation and learning in a social context, community engagement platforms catalyse sustainable behaviour in three ways. Firstly, participatory decision-making processes help create a sense of identity, ownership and belonging to a community which impacts behaviour. Notably, if rules and regulations guiding behaviour are decided through a participatory process, people are more likely to act by these rules and model their behaviour to follow them. Secondly, community activities and active democratic participation create more conscientious and better-informed citizens who are aware of how dependent everyone is on one another. They should also be more inclined to consider their individual behaviours in the context of their community. Lastly, participation also means investing in public and shared spaces and promoting activities that fulfil our most human needs, diverting people from merely material or consumption-based actions.

Another way to promote sustainable behaviour can be achieved by inviting people to prioritize certain choices. This means that governments will need to consider how their particular laws and policies are affecting behaviour (smf.co.uk/wp-content/uploads/2008/06/Publication-Disconnected-Citizens-Is-Community-Empowerment-the-Solution.pdfenvironment.scotland.gov.uk//media/16539/Understanding-Behaviour-Change.pdf). For example, holistic and cross-sectoral considerations such as human-scale and integrated urban planning, dense mixed-use development, pedestrian and bicycle-friendly development are all solutions that can lead people to make energy-wise choices (researchgate.net/profile/Katie_Williams12/publication/242734000_Spatial_Planning_Urban_Form_and_Sustainable_Transport_An_Introduction/links/556d957b08aec2268305883a.pdf).

Similarly, providing attractive routes and planning for pedestrians and cyclists can lead to considerable savings in the amount of energy used for transportation. Creating a dense, multi-modal, affordable, accessible and well-functioning public transport network can also encourage car-free mobility and, overall, a more sustainable transport system. Similarly, mixed-use development can create denser urban areas where people work and live in the same area of the city, allowing them to walk or ride to work. All of these interventions impact behaviour and can significantly reduce the overall demand for energy.

2.10 Retrofit Existing Built-Environments

Buildings contribute a substantial amount of greenhouse gas emissions. For example, UK buildings contributed 37% of the total UK greenhouse gas emissions in 2012 (theccc.org.uk/wp-content/uploads/2013/06/CCC-Prog-Rep_Chap3_singles_web_1.pdf). Considerable amounts of energy and carbon emissions can be saved by aggressively retrofitting existing buildings (altenergymag.com/article/2015/04/retrofitting-buildings-to-improve-energy-efficiency/19349). Policies must establish strict standards for all new buildings and local governments should invest in retrofits of existing public building stock. The energy associated with the built environment include numerous factors including (nationalplatform.org.uk/filelibrary/RRCscopingstudy_final.pdf): embodied energy (energy required to extract, manufacture, transport, install and dispose of construction materials); operational energy required by the mechanical and electrical systems (amount and types of lighting, heating and cooling); passive energy conservation provided by the building envelope (the interface between the interiors and the outdoor environment); on-site energy generation (ideally by integrating renewable energy technologies into the building design); and other energy end-uses in buildings (these can decrease with state-of-the-art technologies substituting older and less efficient technologies).

Costs for retrofitting buildings can be substantial and lead to savings only in the long-term. The technologies and material involved are often more expensive than conventional solutions, and construction can be interruptive and result in lost earnings for the owner (greenbuildingadvisor.com/blogs/dept/musings/high-cost-deep-energy-retrofits). District energy fuelled by renewable energy sources can reduce the need for costly refurbishments.

Several studies demonstrate that significant improvements in energy efficiency coupled with renewable energy sources can actually reduce building emissions to zero (greenbuildingadvisor.com/blogs/dept/musings/high-cost-deep-energy-retrofits). Regardless of the amount of initial financial outlay, efficient buildings have been shown to yield energy and cost savings over the lifetime of the building, reducing utility bills by more than 50%, and also increasing property values (c2es.org/technology/overview/buildings).

Concern over "split incentives" commonly deters investment in energy efficiency retrofits. A split-incentive indicates the disconnect between the interests of the people who own, manage or operate a building and those who pay the utility bills, for example. Landlords may not have incentives to retrofit their properties because it is their tenants who may be responsible for paying the utility bills, they are often not motivated to retrofit their properties (c2es.org/technology/overview/buildings). Implementation of progressive sustainable building policies are essential to promoting climate-friendly buildings. They can include a range of strict standards and codes, financial incentives, information and education programmes, lead-by-example programs, and investment in further research and development (oecd.org/env/consumption-innovation/1936478.pdf).

2.11 Upgrade Infrastructures and Support Efficient Technologies

By upgrading infrastructure, energy conservation can be achieved (ice.org.uk/media-and-policy/the-infrastructure-blog/march-2016/the-smart-way-to-upgrade-britains-infrastructure). Technologies that enhance energy efficiency and save energy through improvements in infrastructure and efficient technologies include cogeneration systems (Combined Heat and Power), district heating and cooling systems (especially those designed to shift from using fossil fuels to biomass, geothermal energy or other renewable source as the primary fuel), decentralised electricity generation, smart grids and micro-grids, and recapturing industrial waste heat and other secondary heat sources.

For example, UNEP reports that district energy can result in a 30–50 per cent reduction in primary energy consumption (unep.org/energy/portals/50177/DES_District_Energy_Report_full_02_d.pdf). Denmark has seen a 20 per cent reduction in national CO_2 emissions since 1990 thanks to district heating. The district heating and cooling systems of Tokyo use 44 per cent less primary energy and emit 50 per cent less CO_2 compared to individual building heating and cooling systems (unep.org/energy/portals/50177/DES_District_Energy_Report_full_02_d.pdf).

2.12 Increase Renewable Electricity Generation

To develop a sustainable energy system that is meeting (or reducing) local energy demand, which itself varies hourly, daily, and seasonally, a smart, integrated use of RE, energy efficiency, demand management and energy storage technologies is necessary. Variable RE resources such as solar and wind play key roles for energy generation, as well as dispatchable renewables like bioenergy and hydropower. Coupling RE technologies with district heating and cooling can complement a 100% RE strategy and help regulate variable loads. New system technologies such as power-to-heat or power-to-gas are showing potential as useful means to convert renewable electricity generated into heat or gas for non-electric application uses, or to use excess electricity directly for heating or transport (EVs) (ida.dk/sites/default/files/klima_hovedrapport_uk_-_web_0.pdf).

2.13 Tackle the Built Environment Challenge (Heating/Cooling)

The built environment improvements are key to achieving energy efficiency and savings in heating and cooling and electricity, and participation by private homeowners and business owners and employees is essential. Working towards energy efficiency and saving energy in buildings is known as demand side management (DSM) and it can include smart metering, providing incentives for energy efficient appliances, lighting and energy systems (e.g. ventilation, heating, cooling, local district energy), as well as offering energy consultation and training.

New investment and financing models that de-risk initial capital outlay as well as innovative loan and other supporting programs also augment DSM activities. Local government staff together with policy makers from the regional and national government can mandate and enforce high energy efficiency standards, particularly for new buildings. It can also help increase the rate of retrofitting existing buildings using existing jurisdictional levers (citeseerx.ist.psu.edu/viewdoc/download?doi=10.1.1.471.5889&rep=rep1&type=pdf).

Lastly, considering relatively low natural gas prices, especially in certain regions of the world (e.g. North America), transitioning away from natural gas for heating purposes in building remains an economic challenge (europarl.europa.eu/RegData/bibliotheque/briefing/2014/140815/LDM_BRI(2014)140815_REV1_EN.pdf). Stronger financial incentives for cleaner technologies and adequate taxation of carbon emissions will be essential tools to favour renewable energy technologies over fossil fuel-based options (ren21.net/wp-content/uploads/2016/05/GSR_2016_Full_Report_lowres.pdf).

2.14 Tackle Mobility and Transport Challenges

Transitioning the transport sector towards 100% RE is one of the most challenging aspects of an energy transition, from policy, user behaviour and technological perspectives. To date, the share of renewable energy in transportation plays only a minor role in this sector, as many appropriate technologies to replace fossil resources still need to be tested and further developed (ren21.net/wp-content/uploads/2016/05/GSR_2016_Full_Report_lowres.pdf). Such technologies range from electric to hydrogen to biofuel-based solutions. With respect to biofuels, an unsolved challenge is to ensure regional and sustainable fuel sources (wikipedia.org/wiki/Issues_relating_to_biofuels). Additionally, very often efforts to make transport sustainable with biofuels challenges the need to promote a fundamental change in the underlying transport technology or drive train. Whether ethanol, biodiesel, or traditional gasoline, the fact remains that internal combustion engines are staggeringly inefficient. Biofuels have been a distraction for urban and other policy makers which should focus on driving the transition to truly sustainable transport and mobility technologies such as light rail powered by renewables, cycling and electric vehicles.

With respect to hydrogen cars, costs and efficient hydrogen production remain two key techno-economic challenges. With respect to biofuels, ensuring regional and sustainable fuel sources remains a significant challenge. The most accessible option as of 2017 remains electric cars (energypost.eu/hydrogen-fuel-cell-cars-competitive-hydrogen-fuel-cell-expert). Policies should focus on the electrification of private transport and on the phasing out of diesel and petrol cars by a certain date, for example as it is being explored in the Netherlands (theguardian.com/technology/2016/apr/18/netherlands-parliament-electric-car-petrol-diesel-ban-by-2025) and in Germany (arstechnica.com/cars/2016/10/germanys-bundesrat-votes-to-ban-the-internal-combustion-engine-by-2030), which are planning to ban sales of combustion engines cars by 2025 and 2030 respectively.

In parallel to the technological challenges, it is equally important to address infrastructure and cultural barriers. One way is to reduce the need for private cars all together by developing, improving, and spreading concepts and strategies that improve planning and reduce traffic (e.g. by inviting car sharing or by investing in a dense, smart public transport system that is integrated with dense cycling and pedestrian infrastructure). Other key challenges include developing feasible renewable energy-powered alternatives for interregional and international transport. This extends to the need to decarbonize shipping and air travel, which remains considerable global issue that must be addressed systematically by any 100% Renewable energy scenario.

While these issues fall beyond the sphere of influence of local governments, they should be debated at all levels and local plans should consider their roles within regional, national and international dynamics and how they can contribute in changing them (e.g. engaging in international networks, lobbying national and international negotiations, evaluating their community's dependency on international resources, improving the use of local resources as opposed to imported ones, etc.)

2.15 Increase Renewable Electricity Generation

Policies must support the integration of the technical and infrastructural changes needed to support an energy system fully powered by renewable energy sources. A policy framework will need to delineate clear actions to overcome the major technical hurdles specifically related to the flexible nature of RE and the necessary modernization of the energy grid.

In particular, to increase the penetration of renewable energy into the total energy supply, it is fundamental to increase the connectedness of the power, heating/cooling, and transport sector and to identify how these different sectors can be optimally integrated (fraunhofer.in/content/dam/indien/en/documents/FIT%20Platform%2020). Electric grids across borders should also be integrated to mitigate variability issues related to most renewable energy sources (irena.org/DocumentDownloads/Publications/IRENA-ETSAP_Tech_Brief_Power_Grid_Integration_2015.pdf).

There are five major infrastructural and technical challenges that will need to be supported by any 100% RE policy framework. Some of them may again require cooperation with upper level governments.

2.16 Enlarge and Improve the Network Infrastructure

A key technical challenge for achieving 100% RE is the intermittency of RE sources. One way to mitigate it consists of connecting a large number of variable sources across a wide geographic area smooth out the variability of output from the renewable sources (irena.org/DocumentDownloads/Publications/IRENA-ETSAP_Tech_Brief_Power_Grid_Integration_2015.pdf). In simple terms, an example may be, when the wind is not blowing on a wind farm in one region, it might be blowing very strongly in another part. If the grid is connected to both sources, electrical reliability is enhanced and requirements for storage and back up energy are minimized. A policy that facilitates the construction and management of a large electrical network is therefore recommended. It is important to point out that network infrastructure enlargement should be assessed against the possibility of creating off-grid, mini-grid and smaller scale smart grid solutions which are able to tackle the variability issues through autonomous storage systems. The decision of upgrading and enlarging the power grid as opposed to creating smaller scale off-grid systems should be based on a careful economic feasibility analysis and comparison between on-grid and off-grid options (researchgate.net/publication/257415166_Grid_vs_storage_in_a_100_renewable_Europe). For example, for areas with low on-grid connectivity such as in several developing countries, it is often more efficient and economically viable to establish off-grid solutions rather than to extend the grid to remote rural places (ensia.com/features/africa-energy).

2.17 Increase Generation Flexibility and Mix of Resources

Greater power reliability is provided by diversifying portfolio of RE technologies, such as wind, tidal, solar, geothermal and hydroelectric. For example, combining wind power with tidal power in certain regions can reduce by as much as 37% the cost of extra reserve to balance for variability as compared to a wind-only scenario (David 2010). Furthermore, small and flexible plants such as biomass thermal plants or combined heat and power plants, are ideal for the integration of variable renewable energy (VRE) and for reducing balancing costs. In contrast, much less flexible generating plants, such as coal or nuclear plants, are incompatible and unable to adapt rapidly to changes in output from renewable sources (renewablesinternational.net/nuclear-and-renewables-a-possible-combination/150/537/78260).

2.18 Demand-Side Management and Efficiency Improvements

Flexibility can come not only from diversifying the supply side mix, but also from improved interaction with the demand side and bi-directional power and information flow (producer to consumer and consumer to producer). Implementation of a smart grid and smart metering systems for demand-response management to help variable sources better match variable loads will be very important (ieadsm.org/wp/files/IEA-DSM-Task-17-Subtask-13-conclusions-recommendations-2016-09-27.pdf). The development of advanced communication technology with smart electricity meters linked to control centres will offer greater flexibility which will require consumer engagements in terms of changes in behavioural patterns, social acceptance and privacy/security issues.

Schemes and incentives will be needed to encourage consumer participation with power system operator schemes that require demand-supply smart interaction (Jacobson and Delucchi 2011). This may include distributed intelligent community grids or prosumer networks. Lastly, reaching the 100% RE target will also require major increase in energy efficiency and energy savings which will involve demand side interventions such as more energy efficiency lighting systems, more efficient building insulation, more efficient cooling and heating technologies and so forth.

2.19 Improved Operational, Market and Planning Methods

Current operational, planning and electricity market procedures, mostly designed around dispatchable and predictable energy load patterns, will need to change to facilitate integration of variable sources (ucsusa.org/clean_energy/smart-energy-solutions/increase-renewables/barriers-to-renewable-energy.html#.WInkVn3tVj8).

The nature of regulatory practices greatly affects balancing costs: If system balancing actions are determined close to real time, system balancing costs are minimised since variable output can be forecast with a high degree of accuracy within such timescales. In countries where balancing decisions are made a long period ahead, forecasts of VRE supply is much less accurate and therefore balancing costs are higher. Forecasting for improved predictability of variable output, such as improved wind forecasting and accurate prediction models, will also be essential.

2.20 Expand Storage

Storage is also core to the discussion around the transition towards 100% renewable energy. While it is true that energy storage will be necessary to achieve 100% renewable energy, it is also important to note that with a relatively well-interconnected grid (e.g. Germany), storage would only become necessary with penetration of variable RE greater than 60% or 70% (reneweconomy.com.au/german-grid-operator-sees-70-wind-solar-storage-needed-35731). Furthermore, with a well-integrated energy sector where electricity, heating/cooling and transport are increasingly interconnected (e.g. vehicle-to-grid options for grid balancing or smart district energy systems that integrated electricity and heating) the problems associated with RE variability and storage are reduced (ida.dk/sites/default/files/klima_hovedrapport_uk_-_web_0.pdf).

Storage can become extremely valuable assets when there are great fluctuations in demand (e.g. in high tourism areas where demands changes periodically (easthamptonstar.com/Lead-article/20151224/Soaring-Demand-Electricity-East-Hampton-Saps-Supplies) or in rapidly urbanizing areas) as storage can provide a prompt and stable supply of energy during peak demand while storing energy during low demand, which increases the overall efficiency of the energy supply system.

Many storage options available can be extremely useful in balancing variable loads with fluctuating sources (r-e-a.net/upload/rea_uk_energy_storage_report_november_2015_-_final.pdf). For example, pumped hydro storage allows water to be pumped up into a reservoir when demand is low but then can produce electricity when the water is released to drive generating turbines. While commercial options like small-scale electric batteries (e.g. Tesla domestic batteries Powerwall (tesla.com/en_GB/powerwall)) are becoming increasingly accessible and cost competitive, support from the government to research and develop improved and ever more efficient storage technologies is needed.

It is also extremely important to ensure that the benefits of storage options are carefully compared with those of other balancing technologies, such as increased interconnections, flexible spinning reserves, demand side management measures, among others (researchgate.net/publication/257415166_Grid_vs_storage_in_a_100_renewable_Europe). It is crucial that local communities understand the extent to which variable RE can be used (sometimes up to 70% (reneweconomy.com.au/german-grid-operator-sees-70-wind-solar-storage-needed-35731)) before storage becomes the most economic option for further integration.

2.21 Introduce Innovative and Alternative Financing Mechanisms

These include options such as public private partnerships (PPP), private financing initiatives (PFI), civic crowdfunding, cooperative funding models and local share-based cooperative models (worldenergy.org/wp-content/uploads/2016/10/Perspectives_Paper_World-Energy-Scenarios_Innovating-Urban-Energy.pdf, unepinquiry.org/wp-content/uploads/2016/06/Sustainable_Infrastructure_and_Finance.pdf). Open and accessible online tools to monitor public expenditures (e.g. participatory budgeting schemes) are also recommended (iclei.org/details/article/over-40-countries-commit-to-100-domestic-renewable-energy-production-and-iclei-supports-transformat.html). Local governments should work cohesively with national governments towards decentralized fiscal policies that ensure access to the tax revenues and financial instruments needed to make the necessary local investments (uclg.org/sites/default/files/_28fr29_uclgpolicypaperonlocalfinanceeng2.pdf).

2.22 Implement New Mechanisms to Internalize Externalities

Promote adoption of innovative, locally-based fee systems, such as a carbon tax, waste tax or pollution tax, and other financial mechanisms (iclei.org/details/article/over-40-countries-commit-to-100-domestic-renewable-energy-production-and-iclei-supports-transformat.html) to favour low-polluting in the marketplace alternatives over carbon and resource intensive processes—especially in certain sectors, such as in the building sector where natural gas heating, for example, remains cost competitive (worldfuturecouncil.org/how-to-achieve-100-renewable-energy)). This can help ensure that personal short-term interest (i.e. saving money) is closely aligned with society's longer-term self-interest (reducing waste, minimizing pollution, etc.).

2.23 *Establish Stable, Long-Term Support Schemes*

Create financial incentives that stimulate private investments and that encourage private individuals to opt for renewable energy options rather than conventional fossil fuels (worldfuturecouncil.org/record-breaking-year-for-renewables-and-community-energy). Feed-in Tariffs are an example of a tool that helps de-risk investments and provide price certainty. Consistent financial support from national, regional and local governments is essential to develop the renewable energy market and to stimulate the necessary participation of companies and private individuals in moving this transition forward. Financial support schemes should not only target renewable energy technologies but also energy efficiency and solution to reduce consumption.

2.24 *Ensure Accountability and Transparency*

Holding politicians responsible and ensuring an environment of trust among community members and political authorities are essential to raising and maintaining public engagement. Accountability and transparency are fundamental aspects of an effective, inclusive and "future just" transition that ensures citizens are motivated to take ownership of the 100% RE system.

Mechanisms to ensure accountability and transparency legitimize government commitments and, in turn harness the citizen support, trust and consensus needed to govern effectively. Greater accountability and transparency also ensure politicians legislate with the best interest of the community in mind and align their political commitments with a decentralised, participatory approach such that they promote energy democracy and equitable access to clean renewable energy. There are several resources available to support local governments in accountability and transparency efforts, such as carbonn® Climate Registry (cCR).

2.25 *Promote Inclusive Communication and Outreach*

Communication and outreach are also necessary—without a shift in awareness by the broader population, far-reaching energy transition processes cannot be launched. Citizens need to be involved in decision-making processes that lead to a shared 100% RE goal. Furthermore, information and consultation raise citizen awareness to motivate energy conservation.

A thorough communications network (e.g. via social media, newsletters, online networking, etc.) is necessary to inform citizens about how they can participate throughout planning and development processes. Local community groups and local governments should work to ensure citizens receive regular information about the objectives, strategies and interim achievements regarding the overall 100% RE strategy, as well as the actions and opportunities to participate. This can help initiate further projects and activities, as citizens and other local stakeholders might develop new methods, greater capacity and innovate ways to engage with the overall energy strategy.

Since not all citizens can be reached through electronic communications other materials need to complement the public engagement strategy including brochures, press articles and conferences. Friendly competitions between regions or groups and celebrations to award prizes are additional ways to motivate citizens to champion the 100% RE strategy. Targeted group activities, such as in education and training also can help promote participation by children, students, adults and practitioners (e.g. craftspeople, architects and engineers). Where significant portions of the community communicate in another language, it is important to provide outreach material in those languages.

2.26 Empower a Decentralized and Diversified Energy Transition

The shift from a centralized energy system based on fossil fuels to one that is decentralized and run entirely on renewable energy sources requires citizens and communities to evolve into "Prosumers"—not just consumers but also producers of energy. Citizens must gain access to the local electric grid, and ownership of renewable energy technologies at the household level needs to be simplified and rewarded financially. There is arguably no better way to ensure the long-term success of the energy transition than through broad-based ownership of the infrastructure and assets that underpin it.

Therefore, empowering new actors to enter the market and safeguard their right to produce energy and sell it to the grid operator becomes an extremely important aspect of the transition. In practice, this can be achieved by implementing specific open-access and inclusive policies such as feed-in tariffs, offering targeted incentives, and by creating long-term investment certainty for citizens, local businesses as well as for international investors. Governments should aim to create inclusive policy frameworks that allow new business models to emerge as well as new forms of citizen engagement. By providing market access to a wide range of stakeholders, policy makers can help build positive synergies across the region and help sustain the momentum required to achieve 100%.

By providing market access to new stakeholders that have not been part of the energy sector in the past, innovative business models emerge that help facilitate the transformation of the energy system. In other words, achieving a 100% RE target can enable policy makers to deliver simultaneously on a wide range of non-climate-related priorities. An inclusive energy system enables and strengthens cooperation, and a collective awareness of both the challenges, and the solutions available to overcome them. Moreover, as local opposition to energy infrastructure (in particular transmission projects) can be a major barrier to 100% RE, local and regional involvement of citizens and businesses help policy makers to overcome this hurdle and build public support.

Thus, the transition to 100% RE is not just a switch from the combustion of fossil fuels to renewables: it is also an opportunity to strengthen and diversify the energy market, stimulate new forms of socio-economic development, and enable a wider range of stakeholders and citizens to participate in the financing and ownership of energy infrastructure.

2.27 *Safeguard a Socially Just Transition*

The energy transition will not come without some level of socio-economic disruption. To guard against impacts of potential fall-off in traditional manufacturing and construction jobs, policies to enable families and businesses to alter their professions, business models and their consumption choices in responsive and effective ways must be in place.

Communities must be empowered to innovate and transform in a manner that ensures that all, including the most vulnerable members of the community, participate and benefit in an equitable way from this transition. For this reason, policy makers need to ensure that a transition to 100% RE really serves and benefits the greatest number of people and that is centred on community participation, engagement, accountability, and transparency of decision-making processes (jobscleanenergywa.com/wp-content/uploads/2016/04/Alliance-Policy-Four-Pager_final.pdf).

Therefore, safeguarding equity needs to gain a fundamental role within any 100% RE plan. Governments need to work together to ensure an equitable transition for businesses, workers, and communities, one where RE targets can be met without sending jobs and emissions out of state and where workers and communities are not disproportionately penalized by the transition off of fossil fuels. Policies should also protect vulnerable communities, such as low-income, which tend to be the most impacted by system changes as well as most directly affected by climate change (keepeek.com/Digital-Asset-Management/oecd/governance/cities-and-climate-change/multi-level-governance-a-).

2.28 *Improve Vertical Cooperation*

Cities, regions and sub-national governments cannot work in isolation and cannot achieve the 100% RE target without engaging the support across all levels of government. Building partnerships and intensifying coordination and collaboration throughout international, national, regional and local levels are critical actions to ensure policy coherence and integration (conceptual-framework_9789264091375-11-en#.WIn1VH3tVj-). Establishing multi-level governance approaches to policy-making, strengthening alliances and supporting a constructive dialogue among the different levels of government are therefore recommended for any successful 100% RE roadmap.

Vertical collaboration should be institutionalized in formal ways to support the coordination across levels of government and to establish a supervisory body that has a specific mandate and official executive and monitoring role. Examples of a supervisory body for multi-level governance include, a National Commission for Sustainable Development in Finland, which coordinates the sustainability agenda across various levels of government (ym.fi/en-us/the_environment/Sustainable_development/Societys_commitment_to_sustainability). The World Future Council has also proposed creating National Urban Policy Commissions (NUPCs) to coordinate governance up through all levels (worldfuturecouncil.org/wp-content/uploads/2016/06/WFC_2016_Towards-National-Urban-Policy-Commissions.pdf).

It is therefore important that local communities consider ways either to contribute to the creation of new platforms or that tap into existing local, subnational, national or international groups that can already support multi-level governance and vertical cooperation.

2.29 Cultivate Horizontal Cooperation

When initializing and developing a 100% RE roadmap it is important to make sure that the broadest possible coalition of actors is included in the process. Key actors within a local government's territory usually represent the administration, political parties, city managers, indigenous populations, business associations, citizen initiatives, research bodies, actors from the local economy, local energy suppliers, agriculture and forestry representatives, freight and trade agents, technical experts, banks, and so on. The more diverse the community participants in the 100% RE strategy are, the further reaching and reliable are the results. The actors can fulfil multiple functions ranging from supporting the strategy, organizing and steering the process, to contributing, spreading and implementing ideas. Each of these actors will add complexity but also strength to the process, bringing in their particular interests and needs, as well as their expertise and skills. The earlier the different actors are engaged in the process, the easier it will be to address and discuss potential conflicts and reservations.

Moreover, achieving 100% RE often requires the cooperation of different Ministries, or Departments that can have few opportunities to collaborate or might even have opposing or conflicting interests. Also, a coherent and robust policy framework must build on the integration of different sectors, departments and policy areas. These processes of cooperation should also be institutionalized (legislated and facilitated through supervisory bodies) to ensure the realisation of interdepartmental coordination is realized.

Horizontal collaboration is also about cooperation between regions, especially between urban and rural areas, which face different challenges and opportunities during an energy transition and in addressing climate change issues. That said, a closer look reveals that their challenges and opportunities are complementary in many ways, which leads to great potential in the relationship between cities and regions. Cooperation between a city and its surrounding area, for example, holds advantages for both sides in terms of sustainable energy development and climate protection. Large cities (characterized by dense infrastructure and a high population numbers) would not be able to fully meet their energy demands—even if reduced—by producing their own renewable energy. The available surface area, strictly within city limits (on roofs, for example) is relatively limited. Yet, cities are richly equipped with know-how, investment capital and pools of varied competencies (especially in the services sector), all of which are important in promoting energy efficiency, energy savings, climate-friendly mobility, and decentralized energy production.

By comparison, rural areas may have at their disposal relatively large areas to produce RE. Sustainable development of this resource offers investment opportunities for cities and revenue opportunities for regions to generate and sell the surplus energy they produce. Apart from this, there are many other ways in which cities and surrounding areas could cooperate to achieve a sustainable transition to 100% RE (e.g. city-hinterland mobility, and climate change mitigation and adaptation efforts). Cooperation is not only favourable for city-hinterland partnerships, but also for villages that are too small to produce sufficient or balanced mix of RE.

2.30 *Generate and Disseminate Specific Knowledge*

A number of activities can accelerate shared learning between actors from both research and practice. For example, hosting demonstration and pilot projects to test an idea, becoming part of a research alliance, building a training centre, hosting educational programs, or ensuring a continuous process of evaluation and monitoring will all provide increased opportunities to share knowledge.

2.31 *Make Knowledge and Data Accessible*

Policy makers and political leaders at the local level globally all stand to gain from a free and open exchange of lessons learned, best management practices and promotion of a further exchange of knowledge across jurisdictions, regions and countries around the world.

Private companies and utilities also stand to gain from locally produced energy transition projects. By providing open data and allowing local governmental access to pertinent data and information, a more circular knowledge loop will accelerate RE uptake. For these purposes, clear information sharing policies must be in place to ensure access to data, good communication and constructive exchange of information across sectors, especially between the private and public sectors.

The knowledge generated should also become easily accessible by the public and presented in a way that invites exploration and understanding of the transition process. Efforts to connect and promote sharing of policies and legislative efforts should also be developed between multiple municipalities and regions that share similar local characteristics. Tools to visualize, display, and exchange information in a user-friendly and accessible manner are crucial.

2.32 Promote Capacity Building and Training

The transition towards 100% RE involves major structural transformation from centralized, dispatchable, fossil fuel-based systems to decentralized, variable, renewable energy based-systems. This major restructuring cannot be achieved simply by swapping human capital and technologies, but by making sure that jurisdictions develop and reinforce their own local capacities and expertise that take advantage of their local human capital and fit their local contexts. Skills and training must be present or developed to support a structural transition of this scale and scope. To support this, capacity building projects should be developed to support both administrative and technical staff, allowing them to be well-prepared and able to create the necessary tools and mechanisms of their new energy system. Capacity building projects could include further technical training for local politicians or policy crafting workshops for local engineers, for example.

2.33 Form and Engage in Local and Regional Networks

At local or regional levels, interesting opportunities can often only be seized through common effort. Examples might include directing waste heat from a local industry to a local district heating system or developing a wind park where benefits are spread widely throughout the community: Local farmers earn income from leasing their lands, energy cooperatives grow their capital, local banks support investments and profit from their interest payments, and local authorities demonstrate successful interventions to their constituents. Beyond the local or regional level, other opportunities for networking and cooperation exist and must be expanded.

Exchanging experiences and know-how with other local governments and civil society groups can enable leapfrogging and can even manifest in a joint wind or solar farm, for example.

2.34 Participate in International Networks

Becoming part of an international networking platform not only supports constructive knowledge exchange and cooperation but can also enhance a local government's visibility and branding. Membership provides opportunities to promote a city's or community's efforts, emboldening political leaders to partake in a common 100% RE planning process.

International networks can provide support to even the smallest jurisdictions around the world by initiating cooperation with like-minded partners to achieve 100% RE targets. Local communities and governments should look to engage international networks for insights beyond what is available in a local community, thereby accelerating, legitimizing and substantiating their local efforts within a larger, worldwide movement.

3 Conclusion

As many more communities and governments across the world commit to moving towards a 100% RE future, they need to be equipped with a solid understanding of what 100% RE actually means and how to implement the necessary political targets within the framework of sustainable development. They also need a comprehensive methodology to structure a 100% RE plan that can tackle the diversity of issues that arise along the path effectively. The Building Blocks developed by the Global 100% Renewable Energy Platform, that this chapter is based on, offer one such a guiding framework. The intention is to provide a comprehensive yet adaptable tool to ensure effective and successful implementation of a 100% RE target.

The overall aim of the Building Blocks is not simply to support communities and governments to structure their 100% RE action plan but also to define the actual meaning of 100% RE. As the 100% RE movement grows continuously across the world, a multitude of definitions emerge which create confusion and uncertainty on its actual meaning. With these building blocks, the Global 100% RE Platform and its partners are finally providing a shared definition of 100% RE. In essence, setting a 100% RE target is not only about transitioning from one form of energy to another, but about a much broader and inclusive socio-economic transformation towards a cleaner, fairer, more equitable and sustainable future. By adopting these Building Blocks and by fully understanding the meaning of 100% RE, communities can guarantee that all benefits of such a transition are reaped by all, in an equitable and democratic manner.

References

go100re.net
citiscope.org/habitatIII/news/2015/12/paris-city-hall-declaration-world-mayors-throw-down-gauntlet-climate
ghgprotocol.org/city-accounting
go100re.net/map
climatenetwork.org/press-release/historic-breakthrough-48-countries-commit-transition-100-renewable-energy
go100re.net/about-us/supporters
there100.org/companies
worldbank.org/en/news/feature/2016/10/10/solar-program-brings-electricity-off-grid-rural-areas
m-kopa.com/asante-sana-300000-east-african-homes-now-on-m-kopa
worldfuturecouncil.org/parliamentarians-can-revive-energiewende
See for example ICLEI's GreenClimateCities MRV handbook or the local renewables process, climate cities benchmark by climate alliance, German 100% RE-regions criteria by deENet, or the energy transformation index by Fraunhofer ISE
go100re.net/wp-content/uploads/2015/11/DiscussionPaper_go100re_criteria.pdf
Please see all contributors in the Acknowledgements section of this report
go100re.net/wp-content/uploads/2014/01/KasselFinalreport_160303.pdf
renewablecities.ca/rc-wp/wp-content/uploads/Dialogue-Report-100-Renewable-Energy-in-Cities.pdf
https://www.google.de/url?sa=t&rct=j&q=&esrc=s&source=web&cd=2&cad=rja&uact=8&ved=0ahUKEwiP9ubA_OLQAhUFX
hQKHcNlC2oQFggjMAE&url= ttp%3A%2F%2F100kommunen.hessen-nachhaltig.de%2Fde%2FBerichte_aus_den_Foren.html%3Ffile%3Dfiles%2FKommunen%2Fdownloads%2FBerichte_Foren%2FForum%2520Nordhessen%2F2016-07-13_Bericht_Regionalforum_Nordhessen_Melsungen.pdf&usg =AFQjCNFETVbPznTk7R-kT8y8mOSN2AkdSA
oxfordcounty.ca/portals/15/Documents/News%20Room/5586_0_Jun_22_2016_Agenda_version02.pdf#page=51
oxfordcounty.ca/Portals/15/Documents/CAO/Zero%20Waste/Draft%20Zero%20Waste%20Plan%2020160928.pdf
iclei.org/fileadmin/user_upload/ICLEI_WS/Documents/Climate/100percentRE-CityProfiles-Issue1-web.pdf
vancouver.ca/green-vancouver/renewable-city.aspx. http://vancouver.ca/green-vancouver/greenest-city-action-plan-development.aspx
aspencore.org/programs/remp
iclei.org/fileadmin/PUBLICATIONS/Case_Studies/2_Chemnitz_-_ICLEI-IRENA_2012.pdf
irena.org/DocumentDownloads/Publications/IRENA_REmap_externality_brief_2016.pdf
carbonn.org
iclei.org/fileadmin/PUBLICATIONS/Case_Studies/7_Malmo_-_ICLEI-IRENA_2012.pdf
go100re.net/properties/frankfurt-am-main
iclei.org/fileadmin/user_upload/ICLEI_WS/Documents/Publications/REN21_Local_Renewables_Policies_2011.pdf
sciencedirect.com/science/article/pii/S0306261910000644
go100re.net/properties/byron-shire-council
sustainablecitiesinstitute.org/topics/equity-and-engagement/community-engagement-sustainability-principles
environment.scotland.gov.uk//media/16539/Understanding-Behaviour-Change.pdf
sustainablelifestyles.ac.uk/sites/default/files/motivating_sc_final.pdf
ourworld.unu.edu/en/bridging-the-emotional-disconnect-between-people-and-cities

smf.co.uk/wp-content/uploads/2008/06/Publication-Disconnected-Citizens-Is-Community-Empowerment-the-Solution.pdfenvironment.scotland.gov.uk//media/16539/Understanding-Behaviour-Change.pdf
researchgate.net/profile/Katie_Williams12/publication/242734000_Spatial_Planning_Urban_Form_and_Sustainable_Transport_An_Introduction/links/556d957b08aec2268305883a.pdf
theccc.org.uk/wp-content/uploads/2013/06/CCC-Prog-Rep_Chap3_singles_web_1.pdf
altenergymag.com/article/2015/04/retrofitting-buildings-to-improve-energy-efficiency/19349
nationalplatform.org.uk/filelibrary/RRCscopingstudy_final.pdf
greenbuildingadvisor.com/blogs/dept/musings/high-cost-deep-energy-retrofits
c2es.org/technology/overview/buildings
oecd.org/env/consumption-innovation/1936478.pdf
ice.org.uk/media-and-policy/the-infrastructure-blog/march-2016/the-smart-way-to-upgrade-britains-infrastructure
unep.org/energy/portals/50177/DES_District_Energy_Report_full_02_d.pdf
ida.dk/sites/default/files/klima_hovedrapport_uk_-_web_0.pdf
citeseerx.ist.psu.edu/viewdoc/download?doi=10.1.1.471.5889&rep=rep1&type=pdf
europarl.europa.eu/RegData/bibliotheque/briefing/2014/140815/LDM_BRI(2014)140815_REV1_EN.pdf
ren21.net/wp-content/uploads/2016/05/GSR_2016_Full_Report_lowres.pdf
wikipedia.org/wiki/Issues_relating_to_biofuels
energypost.eu/hydrogen-fuel-cell-cars-competitive-hydrogen-fuel-cell-expert
theguardian.com/technology/2016/apr/18/netherlands-parliament-electric-car-petrol-diesel-ban-by-2025
arstechnica.com/cars/2016/10/germanys-bundesrat-votes-to-ban-the-internal-combustion-engine-by-2030
fraunhofer.in/content/dam/indien/en/documents/FIT%20Platform%202014_%20Mr_Gerhard%20Stryi-Hipp_Fraunhofer%20ISE.pdf
irena.org/DocumentDownloads/Publications/IRENA-ETSAP_Tech_Brief_Power_Grid_Integration_2015.pdf
researchgate.net/publication/257415166_Grid_vs_storage_in_a_100_renewable_Europe
ensia.com/features/africa-energy
David E (2010) Sustainable energy: opportunities and limitations. Palgrave Macmillan, Basingstoke
renewablesinternational.net/nuclear-and-renewables-a-possible-combination/150/537/78260
ieadsm.org/wp/files/IEA-DSM-Task-17-Subtask-13-conclusions-recommendations-2016-09-27.pdf
Jacobson MZ, Delucchi MA (2011) Providing all global energy with wind, water, and solar power, part I: technologies, energy resources, quantities and areas of infrastructure, and materials. Energy Policy 39(3):1154–1169
ucsusa.org/clean_energy/smart-energy-solutions/increase-renewables/barriers-to-renewable-energy.html#.WInkVn3tVj8
reneweconomy.com.au/german-grid-operator-sees-70-wind-solar-storage-needed-35731
easthamptonstar.com/Lead-article/20151224/Soaring-Demand-Electricity-East-Hampton-Saps-Supplies
r-e-a.net/upload/rea_uk_energy_storage_report_november_2015_-_final.pdf
tesla.com/en_GB/powerwall
uclg.org/sites/default/files/_28fr29_uclgpolicypaperonlocalfinanceeng2.pdf
worldenergy.org/wp-content/uploads/2016/10/Perspectives_Paper_World-Energy-Scenarios_Innovating-Urban-Energy.pdf. unepinquiry.org/wp-content/uploads/2016/06/Sustainable_Infrastructure_and_Finance.pdf
iclei.org/details/article/over-40-countries-commit-to-100-domestic-renewable-energy-production-and-iclei-supports-transformat.html
oilprice.com/Energy/Energy-General/Cheap-Natural-Gas-To-Spark-Another-Wave-Of-Coal-Plant-Retirements.html

worldfuturecouncil.org/how-to-achieve-100-renewable-energy
worldfuturecouncil.org/record-breaking-year-for-renewables-and-community-energy
energydemocracyny.org/wp-content/uploads/2015/04/REEM_Document_FINAL.pdf
jobscleanenergywa.com/wp-content/uploads/2016/04/Alliance-Policy-Four-Pager_final.pdf
keepeek.com/Digital-Asset-Management/oecd/governance/cities-and-climate-change/multi-level-governance-a-
conceptual-framework_9789264091375-11-en#.WIn1VH3tVj-
ym.fi/en-us/the_environment/Sustainable_development/Societys_commitment_to_sustainability
worldfuturecouncil.org/wp-content/uploads/2016/06/WFC_2016_Towards-National-Urban-Policy-Commissions.pdf
worldfuturecouncil.org/file/2016/03/WFC_2015_Kassel_International_Dialogue_on_100_Renewable_Energy.pdf
go100re.net/cities-regions-network

Clean Energy Manufacturing: Renewable Energy Technology Benchmarks

Debra Sandor, David Keyser, Margaret Mann, Jill Engel-Cox, Samantha Reese, Kelsey Horowitz, Eric Lantz, Jon Weers, Billy Roberts, Stacy Buchanan, Doug Arent, Brian Walker and Robert Dixon

Abstract The manufacturing of clean energy technologies has become a global enterprise. An analysis of four clean energy technology manufacturing processes—wind turbine components, crystalline silicon solar photovoltaic modules, light-duty vehicle lithium-ion battery cells, and light-emitting diodes—was conducted to provide a reference point for comparison over time as this sector evolves. The manufacturing supply chain analysis included processing of raw materials, making required subcomponents, and assembling final products, using benchmarks that crossed 12 economies: Brazil, Canada, China, Germany, India, Japan, Malaysia, Mexico, South Korea, Republic of China (Taiwan), the United Kingdom, and the United States. Larger economies, with extensive manufacturing supply chains and high prevailing wages, tend to retain more value added from clean energy manufacturing than small economies. Manufacturing of clean energy technologies drives extensive trade among economies to support the widely distributed supply chain links. Production of wind turbine components and photovoltaic modules is more centralized than production of light-emitting diodes and lithium-ion battery cells. Across the four clean energy technologies evaluated, there was generally an excess of manufacturing capacity relative to global supply.

Keywords Clean energy manufacturing · Renewable energy · Energy efficiency · Global trade · Benchmarks

1 Introduction

Clean energy technologies are expanding rapidly and growing in significance with respect to contributing to the world's energy systems (Johnansson et al. 2012). The

---

D. Sandor · D. Keyser · M. Mann · J. Engel-Cox · S. Reese · K. Horowitz · E. Lantz · J. Weers · B. Roberts · S. Buchanan · D. Arent
National Renewable Energy Laboratory, 15013 Denver West Parkway, Golden, CO 80401, USA

B. Walker · R. Dixon (✉)
U.S. Department of Energy, 1000 Independence Avenue SW, Washington, DC 20585, USA
e-mail: robert.dixon@hq.doe.gov

T. S. Uyar (ed.), *Accelerating the Transition to a 100% Renewable Energy Era*, Lecture Notes in Energy 74, https://doi.org/10.1007/978-3-030-40738-4_9

manufacture of these technologies—including extracting and processing raw materials, producing required subcomponents, and assembling end products—has become a global enterprise (Yang et al. 2010). The *Benchmarks of Global Clean Energy Manufacturing* report analyzes the global state of clean energy manufacturing and establishes a 2014 baseline for tracking global changes in clean energy manufacturing (Benchmarks of Clean Energy Manufacturing 2016; Sandor et al. 2017). Four technologies were examined: wind turbine components (blade, tower, nacelle), crystalline silicon (c-Si) solar photovoltaic (PV) modules, light-duty vehicle (LDV) lithium-ion battery (LIB) cells, and light-emitting diode (LED) packages for lighting and other consumer products—across manufacturing supply chains that include processing raw materials, making required subcomponents, and assembling final products (Sandor et al. 2017).

The impacts of the manufacturing supply chain for these four technologies were assessed in terms of four common benchmarks—manufacturing value added, global trade flows, market size, and manufacturing capacity and production—across 12 economies, selected because they comprise the primary manufacturing hubs for the four technologies: Brazil, Canada, China, Germany, India, Japan, Malaysia, Mexico, South Korea, Republic of China (Taiwan), the United Kingdom, and the United States (Benchmarks of Clean Energy Manufacturing 2016; Sandor et al. 2017). The data and insights provided by these benchmarks can help guide research agendas, inform trade policies, and identify manufacturing opportunities by location and technology.

2 Framework

A piece of the larger clean energy economy, manufacturing is the linchpin between technology development and its deployment in the marketplace (Fig. 1). Clean energy manufacturing supply chain upstream links include innovation in the development

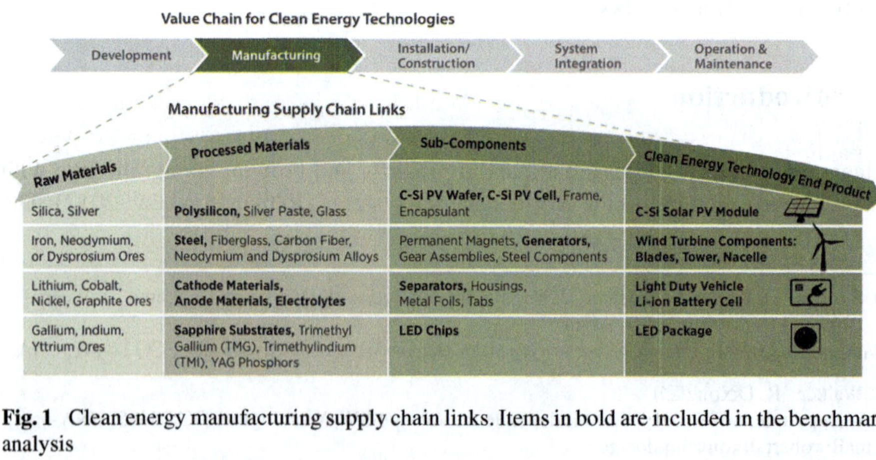

Fig. 1 Clean energy manufacturing supply chain links. Items in bold are included in the benchmark analysis

stage that has economic value in the form of intellectual property, research, and corporate management. Downstream, highly localized installation, systems integration, and operations bring economic value through employment, services, property taxes, and reduction of negative environmental impacts. While development and deployment of technologies make tremendous contributions to the economy, this analysis focuses exclusively on the manufacturing aspects of the larger clean energy value chain and examines each technology in terms of four manufacturing supply chain links: raw material, processed material, subcomponents, and end products (Benchmarks of Clean Energy Manufacturing 2016; Sandor et al. 2017).

3 Methodologies

Novel standardized methodologies were developed to generate the data sets for each benchmark, while accommodating the variations in clean energy technology manufacturing supply chains and data availability. Specifically, we established four common points of reference—benchmarks—to provide a standardized basis for (1) comparing key economic aspects of clean energy technology manufacturing on both a national and global basis, and (2) tracking changes as markets and manufacturing processes evolve. The methodologies are outlined here and detailed in the methodology report (Sandor et al. 2017; Goodrich et al. 2013; Fu et al. 2015; James and Goodrich 2013; Moné et al. 2015; Cotrell 2014; Chung et al. 2015, 2016).

3.1 Benchmark 1: Clean Energy Manufacturing Value Added

This benchmark provides insight into the contribution and importance of clean energy manufacturing to national economies. Value added is a key component of national gross domestic product (GDP). It has two components defined as:

- *Direct value added* is the amount that clean energy manufacturers themselves contribute to national GDP. This includes payments to manufacturing workers, property-type income such as profits earned by owners and investors, and taxes paid on production less government subsidies.
- *Indirect value added* is often referred to as the economic ripple effect. When clean energy manufacturers make products, they purchase inputs such as accounting services or raw materials. A generator manufacturer, for example, may purchase copper wiring from a domestic wire manufacturer. This wire manufacturer and its contribution to GDP would be included in the indirect effect.

We estimate manufacturing value added using a combination of cost analysis data, market data, and social accounting data from the Organization for Economic Cooperation and Development (OECD) Structural Analysis (STAN) Input-Output

(I-O) database (Sandor et al. 2017; Goodrich et al. 2013; Fu et al. 2015; James and Goodrich 2013; Moné et al. 2015; Cotrell 2014; Chung et al. 2015, 2016).

3.2 Benchmark 2: Clean Energy Trade

This benchmark provides insight into global clean energy trade activity and interconnectedness across the manufacturing supply chain. Trade connects the global community and can be a significant source of economic growth. Balance of trade (exports less imports) is another key component of national GDP. Trade flow data for the benchmark report are compiled from the United States International Trade Commission (USITC) and the International Trade Centre. Trade data are in U.S. dollars (USD) rather than local currencies. Fluctuation in trade that is measured in a standard currency, such as USD, can be caused by changes in the volume of trade or the value of the local currency relative to the USD. A relatively strong domestic currency makes exports more expensive in the international market, while a weaker currency makes them less expensive. While official trade data for the final products are often available, the upstream data are often intertwined with much larger industry sectors and difficult to extract for the specific technology of interest. Where not available, the balance of trade for upstream components was estimated using market data from secondary sources.

3.3 Benchmark 3: Clean Energy Market Size

This benchmark provides insight into the relative concentration of demand for clean energy technologies across the globe. Market size (or market demand) data were collected from existing secondary sources to estimate the market size for each technology across the manufacturing supply chain and in each economy. When available, actual production data for each subsequent downstream intermediate formed the basis of demand estimates for key supply chain intermediates. When data were not available, typically for smaller industries (LED packages and LDV Li-ion battery cells), the demand for intermediates was approximated by assuming that the production volume of the end product is equivalent to the demand for each upstream intermediate product. The monetary value of demand was determined by applying estimates of average global unit prices to allow comparison across technologies and economies.

3.4 Benchmark 4: Clean Energy Manufacturing Capacity and Production

This benchmark provides insight into the clean energy manufacturing capacity and production around the world and highlights opportunities for expansion to meet demand.

Manufacturing capacity and production were estimated to highlight the economies that make the largest contributions in each category and to understand where excess capacity is located around the world for each technology. Like market size data, data were collected from existing secondary sources, and monetary values were determined by applying estimates of average global unit prices to (1) allow comparison across technologies and economies, and (2) provide input for the value added benchmark based on the production value of each technology and intermediate.

4 Results

4.1 Larger economies, with more extensive manufacturing supply chains and higher prevailing wages, tend to retain more value added from clean energy manufacturing than smaller economies. Manufacturing value added generally tracks production revenues from manufacturing clean energy technologies, reflecting the size of an economy's clean energy manufacturing sector. Normalizing the direct clean energy manufacturing value added for each technology by production revenue (yielding an estimate of value added retained) provides insight on the extent that the manufacturing supply chain associated with these clean technologies is domestically sourced and shows how much workers, investors, and governments gain from each unit of production (Fig. 2).

Larger, more productive economies such as the United States, tend to retain higher percentages of clean energy manufacturing direct value added as a portion of production revenue over the entire supply chain than other economies. The larger economies tend to have more extensive domestic supply chains, and many of the economies that have higher values for this benchmark also have higher prevailing wages.

4.2 Manufacturing of clean energy technologies is a complex global enterprise, with extensive trade among economies to support the geographical distribution of production and demand across the links in the supply chain. Economies that are net importers of end products may be major exporters of upstream processed materials and subcomponents of those same technologies.

Trade is a significant component of GDP in many economies; balance of trade (exports less imports) is one element of GDP and is influenced by production capacity, capacity utilization, and domestic demand for manufactured products. Figure 3 shows the balance of trade for the four clean energy technologies. C-Si PV modules and LED packages are most heavily traded, yet trade of the end product is not the full story; for example, while major PV deployment markets such as the United States

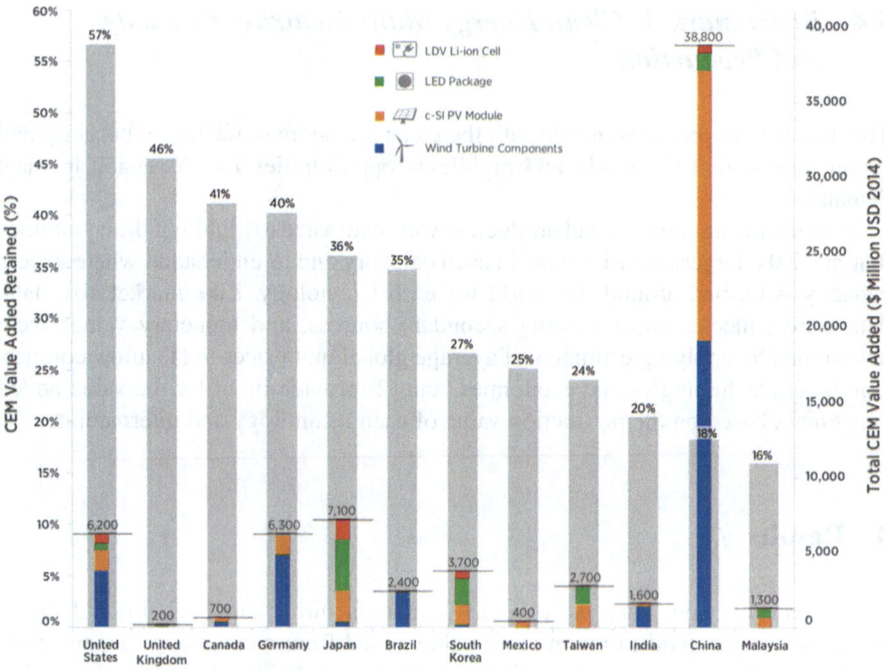

Fig. 2 Direct manufacturing value added retention for four clean energy technology supply chains, 2014. The colored bars (right axis) indicate the total clean energy manufacturing value added for each economy; the clean energy manufacturing value added retained within each economy (direct value added as a share of production revenue) is indicated by the gray bars (left axis). See methodology report for data quality discussion

and Germany are net importers for PV modules, they are also the largest exporters of polysilicon to make those modules, purchased largely by Japan and China (Goodrich et al. 2013; Fu et al. 2015) (Fig. 4).

The interactive trade charts highlight differences in the networks of trade among the benchmarked economies. PV modules and LED packages were most heavily traded, most likely due to the fact that they are more easily shipped than the other end products. Wind turbine components were the least heavily traded, due to their large size and the related practice of manufacturing them domestically or regionally (Goodrich et al. 2013; Fu et al. 2015).

4.3 Production of wind turbine components and c-Si PV modules is more concentrated than production of LED chips and LDV Li-ion battery cells. Wind components are typically made in the same economies that have high demand, but manufacturing and demand for c-Si PV modules, LED chips, and LDV Li-ion battery cells are less coincident.

Manufacturing activity and investment in new manufacturing facilities varies across economies based on demand in domestic markets, demand in export markets, and investment incentives. Figure 5 compares production and demand for the

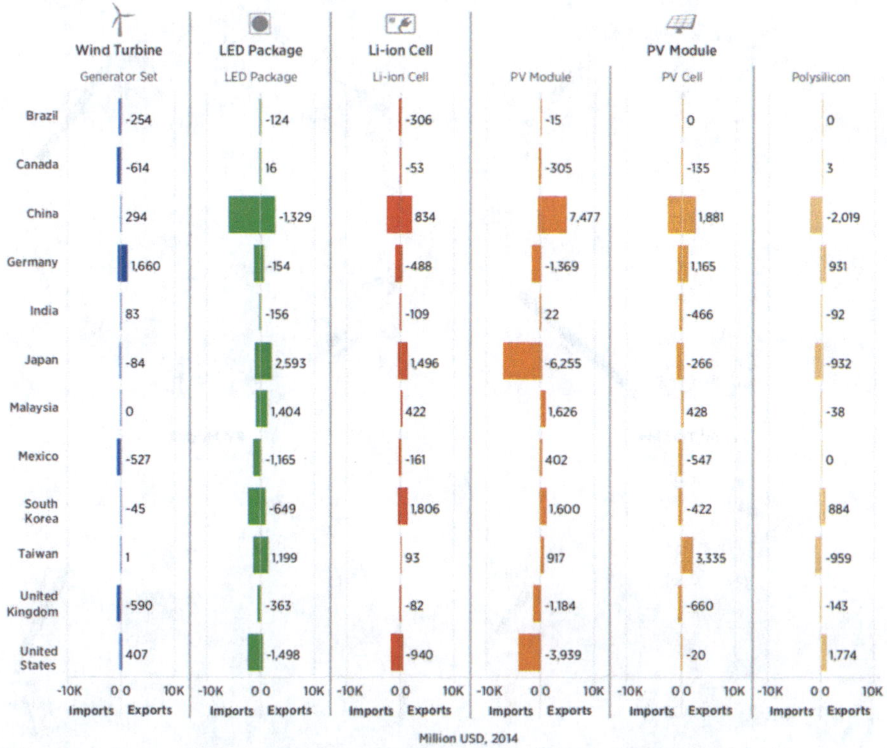

Fig. 3 Balance of trade in select clean energy technology end products and across c-Si PV module supply chain, 2014 (Millions USD 2014). The bars show the clean energy technology end product imports as negative values and the exports as positive values. The balance of trade is noted to the right of the bar. Interactive trade flow charts can be accessed at ManufacturingCleanEnergy.org/Benchmark. Tradedata are not dissaggregated for the specific clean energy technologies studied. For wind components, only data for wind generator sets (which consist of a nacelle packaged with blades) were available. Trade data for Li-ion battery cells were not dissaggregated by end use. See methodology report for data quality discussion

four clean energy technology end products. Wind turbine component production and demand are relatively balanced in each of the 12 economies considered. While China plays a dominant role in PV module manufacturing, production outside of China is generally dispersed across the economies included here (only Brazil and the United Kingdom have no PV module production). Production of LED packages and LDV Li-ion battery cells is more globally distributed than PV module production, although it is still concentrated in Asia.

Across the four clean energy technologies evaluated in 2014, there was generally an excess of manufacturing capacity relative to global demand.

Manufacturing production and capacity data suggest excess capacity existed across the 12 economies assessed in 2014 (Fig. 6). The average manufacturing capacity utilizations were estimated at 62% for wind turbine components, 55% for c-Si

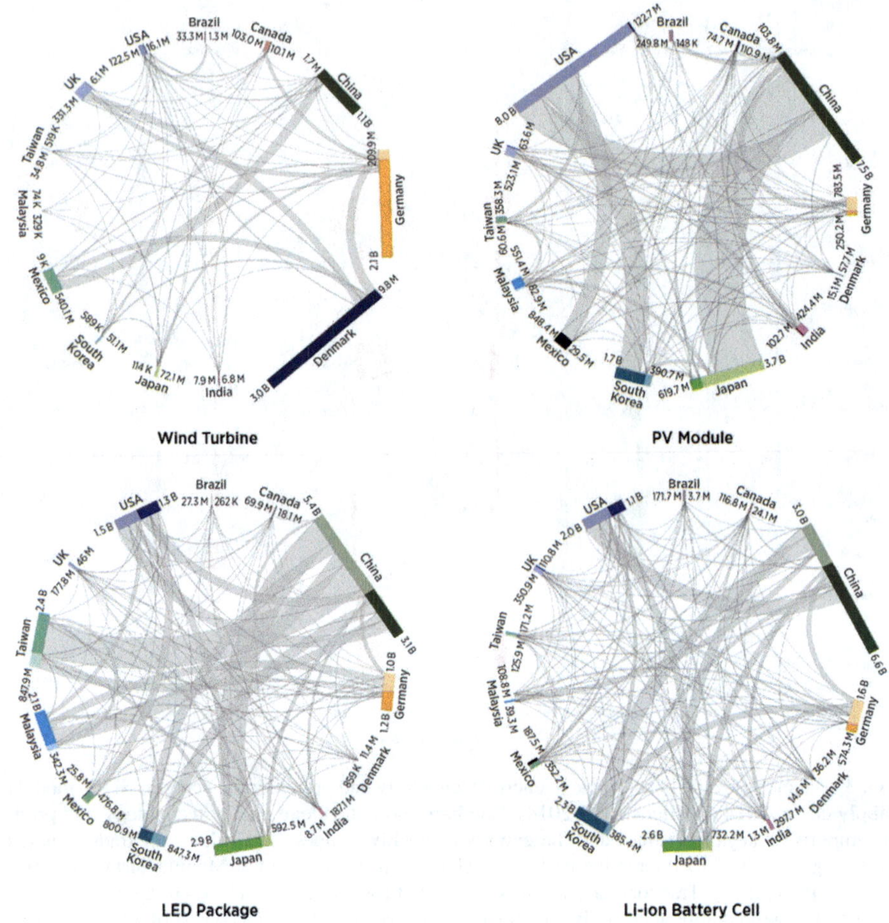

Fig. 4 Benchmarked clean energy technology end product trade flows. Charts shows 2016 trade flows among benchmarked economies for wind gensets (nacelle and blades), PV modules, LIB cells, and LED packages in US$(2014). Note that for the trade data, Denmark has been included due to its importance in wind component manufacturing. An interactive version of the chart is available at http://www.manufacturingcleanenergy.org/benchmark/wind.php

PV modules, 37% for LED chips, and 41% for LDV Li-ion cells. Excess capacity can be used to meet potential demand growth from increased technology adoption. However, without increased demand, persistent excess capacity can place downward pressure on pricing.

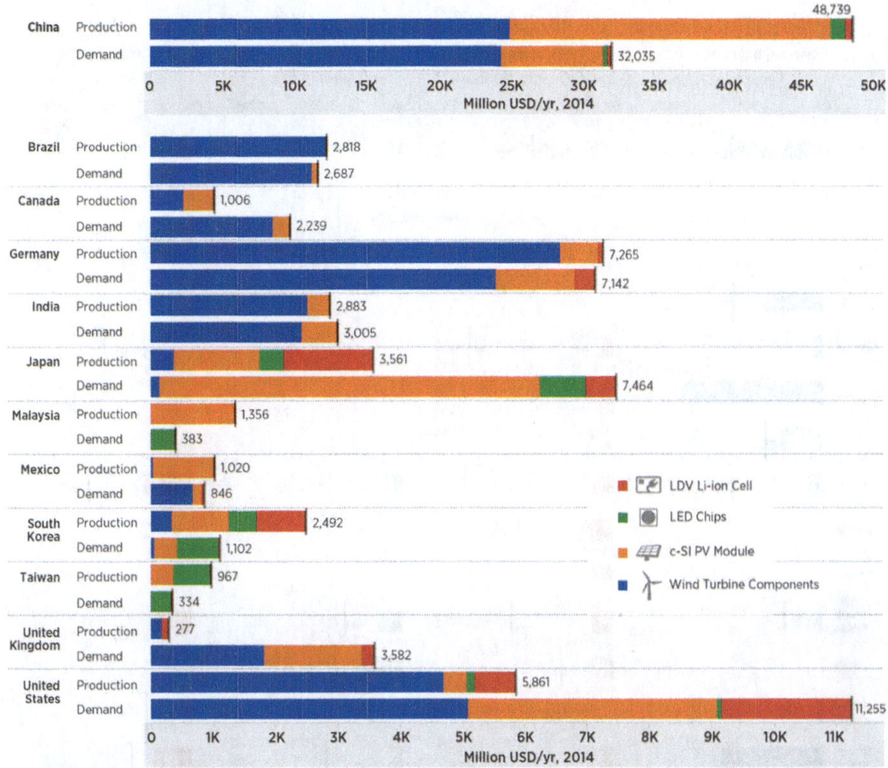

Fig. 5 Production and demand for four clean energy technology end products, 2014. *Note* LED chip (subcomponent), rather than LED package (end product) data reported, due to lack of economy-specific LED package production data, 2014. The bars indicate the production and market demand (consumption). See methodology report for data quality discussion

5 Next Steps

While the initial benchmark report focused on 2014 as a baseline year, an update to the benchmark report (forthcoming in 2019), will expand the analysis to summarize trends between 2014 and 2016. To address stakeholder feedback and incorporate additional years of data, the updated report will include new visualizations to present the key benchmark trends, with select visualizations published in the data book, and additional visualizations available online [*add link here*]. The report will also add Denmark to the group of manufacturing hubs examined, in recognition of its contribution to wind turbine component manufacturing and trade. The manufacturing value added analysis is expanded to look beyond the effect of domestic manufacturing and provide insight into the impact of global manufacturing as well. And finally, raw materials data will be integrated into the report, providing additional insight into potential supply chain risks and opportunities.

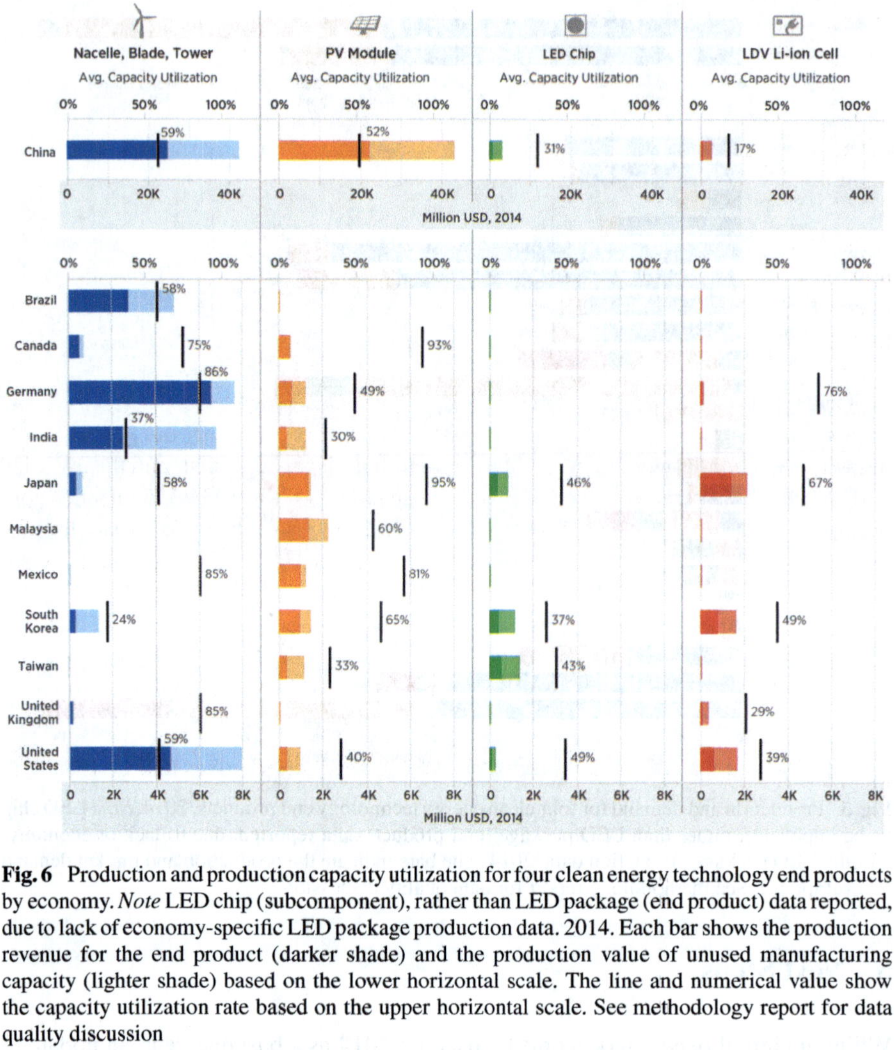

Fig. 6 Production and production capacity utilization for four clean energy technology end products by economy. *Note* LED chip (subcomponent), rather than LED package (end product) data reported, due to lack of economy-specific LED package production data. 2014. Each bar shows the production revenue for the end product (darker shade) and the production value of unused manufacturing capacity (lighter shade) based on the lower horizontal scale. The line and numerical value show the capacity utilization rate based on the upper horizontal scale. See methodology report for data quality discussion

6 Conclusions

The current state of clean energy trade reflects the cumulative dynamics of a high-growth decade in which both markets and manufacturing have grown significantly within an increasingly complex set of policy environments. Strong domestic markets have not necessarily been supplied by domestic manufacturing—particularly markets for those technologies that benefit from economies of scale and where incentives for manufacturing investment or output have been adopted, and markets for technologies where transportation was not a determining factor for manufacturing location, such as PV modules, Li-ion battery cells, and LED packages (Sandor et al. 2017; Goodrich et al. 2013; Fu et al. 2015; James and Goodrich 2013; Moné et al. 2015; Cotrell 2014; Chung et al. 2015, 2016).

The U.S. situation is notable, as clean energy markets have been particularly strong and are served by both domestic and imported end products. The United States is one of the top five manufacturing economies globally and retains the highest amount of manufacturing value added of the technologies evaluated. Even though the United States is a net importer to meet its large demand for the technologies evaluated, some U.S. clean energy technology manufacturers are net exporters of components upstream in the supply chains. China stands out as an example where policies have been implemented to support both domestic markets and the expansion of domestic manufacturing to serve both domestic and export markets. In Japan, both of these situations are apparent for specific technologies.

The country's strong domestic market for PV modules is served with significant imports, while its LED package manufacturing serves both its domestic and export markets.

For the clean energy technologies covered in this report and many others, technology innovation is anticipated to continue to drive relatively rapid turnover of technologies and associated manufacturing capacity. Such innovation creates significant opportunities to attract manufacturers that can serve domestic markets, compete effectively in other markets, and displace incumbent technologies.

Manufacturing activity and investment in new manufacturing facilities respond to a number of key drivers, including demand in domestic markets, demand in export markets, and investment incentives. Domestic markets can be an initial driver for domestic manufacturing, although as deployment increases and prices fall, there is no guarantee that manufacturing will be geographically aligned with demand, absent other policies or economic drivers. With the right combination of skilled labor and investment, manufacturing for export can become a second key driver, sometimes even without a local market. Irrespective of manufacturing, localized clean energy technology deployment, as well as multinational corporate headquarters and research facilities, both generate significant value in their own right. Increasing deployment of clean energy technologies provides manufacturers with a more stable demand and enables investment that drives down prices through economies of scale.

Our results also emphasize the importance of policymakers having a deep understanding of the entire supply chain of clean energy technologies, because even in cases where the end product manufacturing is concentrated, the upstream components and materials may come from many economies. Due to the complex influences across many sectors of national and global economies, considering the entire development, manufacturing, and deployment supply chain in investment and incentive decisions could be important.

Manufacturing of clean energy technologies is a global enterprise that changes in response to market forces and technology advances in new end products and also in advanced manufacturing equipment, processes, and materials used to generate these end products. Deeper knowledge of the product supply chains and market volumes can inform industry decisions related to the location of manufacturing facilities for extracting and processing raw materials, making the required subcomponents, and assembling the final product. This knowledge can also inform decisions around research and development and international trade. While the initial benchmark report focused on 2014 as a baseline year, an update to the benchmark report is forthcoming and will expand the analysis to summarize trends between 2014 and 2016. The report will also add Denmark to the group of manufacturing hubs examined, in recognition of its contribution to wind turbine component manufacturing and trade. The manufacturing value added analysis will be expanded to look beyond the effect of domestic manufacturing and provide insight into the impact of global manufacturing as well. Finally, raw materials data will be integrated into the report, providing additional insight into potential supply chain risks and opportunities.

Acknowledgements We thank Ed Balistreri, Rebecca Hill, Donald Chung, Chris Scarlata, Chris Mone, Tian Tian, and Scott Gossett for their contributions to the manuscript.

References

Benchmarks of Clean Energy Manufacturing (2016) Manufacturing analysis center (CEMAC). NREL/TP-6A20-66086. National Renewable Energy Laboratory, Golden, CO

Chung D, Elgqvist E, Santhanagopalan S (2015) Automotive lithium-ion battery supply chain and U.S. competitiveness considerations. Clean Energy Manufacturing Analysis Center. NREL/PR-6A50-63354. National Renewable Energy Laboratory, Golden, CO

Chung D, Elqqvist E, Santhanagopalan S (2016) Automotive lithium-ion cell manufacturing: regional cost structures and supply chain considerations. Clean Energy

Cotrell J (2014) Land-based wind turbine transportation and logistics barriers and their effects on U.S. wind markets. PR-5000-61780. Presented at the American Wind Energy Association (AWEA) WINDPOWER 2014 Conference, Las Vegas, Nevada

Fu R, James TL, Woodhouse M (2015) Economic measurements of polysilicon for the photovoltaic industry: market competition and manufacturing competitiveness. IEEE J Photovolt 5:2–10

Goodrich AC, Powell DM, James TL, Woodhouse M, Buonassisi T (2013) Assessing the divers of regional trends in solar photovoltaic manufacturing. Energy Environ Sci 10:85–95

James T, Goodrich A (2013) Supply chain and blade manufacturing considerations in the global wind industry. NREL/PR-6A20-60063. National Renewable Energy Laboratory, Golden, CO

Johnansson TB, Patwardhan A, Nebojsa N, Gomez-Echerverri L (eds) (2012) Global energy assessment: toward a sustainable future. IIASA, Laxenberg, Austria

Moné C, Stehly T, Maples B, Settle E (2015) 2014 cost of wind energy review. TP-5000-64281. National Renewable Energy Laboratory, Golden, CO

Sandor D, Chung D, Keyser D, Mann M, Engel-Cox J (2017) Clean energy manufacturing analysis center benchmark report: framework and methodologies. NREL/TP-6A50-67666. National Renewable Energy Laboratory, Golden, CO

Yang M, Dixon RK, Taylor PG (eds) (2010) Energy efficiency policies and strategies. Energy Policy 38(11):6391–6485

Development and Thermodynamic Analysis of a 100% Renewable Energy Driven Electrical Vehicle Charging Station with Sustainable Energy Storage

Abdulla Al Wahedi and Yusuf Bicer

Abstract Plug-in electric vehicles (PEVs) expansion is accelerating rapidly due to their massive contribution for reducing fossil fuel consumption and CO_2 emissions. However, to fulfil the charging requirements of millions of PEVs from the grid would overload the grid and introduce technical, environmental and economic burden on power sector. This study proposes and thermodynamically assesses a grid-independent and stand-alone multigeneration PEV charging station to fast charge 50 number of PEVs per day. The system consists of a hybrid solar and wind sub-systems with battery, hydrogen and ammonia storage units. Considering the site-specific conditions of State of Qatar, the system performance as well as energy and exergy efficiencies are investigated through parametric studies performed in Engineering Equation Solver (EES) software. The integrated energy system for PEV charging station produces almost 270 kW power from both wind turbine and PV power plant. In addition, the fuel cells, running on hydrogen and ammonia, generate almost 10 kW power for sustaining the operations. The energy and exergy efficiencies for the integrated system are found to be about 16.9% and 17.6%, respectively.

Keywords Plug-in electric vehicle · Charging station · Wind turbine · Solar system · Multigeneration system · Electrochemical storage

1 Introduction

According to International Energy Outlook studies, transport sector demand will increase by 54% until 2035, leading to substantial grow of world's demand for fossil fuels in case no green initiative is implemented in transport sector (Mozafar et al. 2017). The dominant share of oil is utilized by the transportation and this major consumption leads into severe consequences such as environmental deterioration,

---

A. Al Wahedi (✉) · Y. Bicer
Division of Sustainable Development (DSD), College of Science and Engineering (CSE), Hamad Bin Khalifa University (HBKU), Qatar Foundation (QF), Education City, Doha, Qatar
e-mail: amalwahedi@mail.hbku.edu.qa

Y. Bicer
e-mail: ybicer@hbku.edu.qa

T. S. Uyar (ed.), *Accelerating the Transition to a 100% Renewable Energy Era*, Lecture Notes in Energy 74, https://doi.org/10.1007/978-3-030-40738-4_10

economic vulnerability and climate change (Amjad et al. 2018). Using these fossil resources will not only increase fossil fuel cost but will increase greenhouse gas emissions and environmental pollution as well (Fazelpour et al. 2014). The fast growth in the quantity of Internal Combustion Engine Vehicles (ICEV) to cope with the enormous increase of inhabitants and transport sector demand all over the world is a frightening alarm of fuel cost increase and ecological pollution (Karmaker et al. 2018). Road transport is the largest source of environment pollution in urban areas (Grande et al. 2018). Studies showed that, transport sector consumed 62.3% of world's oil and produced 6,892 Mt emissions of CO_2 into the atmosphere (Dicks and Breedon 1988) and classified transport sector as second highest CO_2 emitter (Girard et al. 2019). This excessive reliance on oil generates various ecological problems such as natural resources depletion, pollution in cities and climate change, as well as economic challenges such as rise of oil prices (García-Villalobos et al. 2014). Since less fuel consumption contributes into carbon emissions reduction, which turns to be environmental friendly (Amjad et al. 2018), many countries have set their decision to replace ICEV with green vehicles (Amini et al. 2016) where plug-in electric vehicles (PEVs) are considered as ecological substitute for ICEV (García-Villalobos et al. 2014) and clean means of transportation (Zhang et al. 2019). Environmental issues and fossil-fuel exhaustion increase strongly motivated implementing PEVs throughout the world (Razipour et al. 2019). Sales of PEVs worldwide have increased from 320,000 in 2014 to 1.04 million in 2017 (Wu et al. 2019). PEVs compared with ICEV are environmentally friendly and from an economic point of view are more cost-effective especially where electricity production costs are very low (Mozafar et al. 2017). Among many benefits of PEVs compared to ICEVs, main ones are listed below:

- Dramatic fuel consumption reduction leading to energy efficiency enhancement (Amjad et al. 2018).
- Political and economic gains for less and non-oil producing countries by decreasing energy reliance on countries producing oil and decreasing imports of oil (García-Villalobos et al. 2014).
- Positive effect on climate through minimizing CO_2 emissions into the environment (García-Villalobos et al. 2014).
- Less pollution due to less CO_2 emissions leading to healthier atmosphere and improved overall health standards (García-Villalobos et al. 2014).
- PEVs technicalities are simpler than those of ICEVs since they include fewer mechanical components and in the long-term have much lower maintenance costs (Serra 2012).

1.1 Consequences of Plug-in Electric Vehicle Expansion on Electrical Grid and Energy System

For consumers and commercial businesses, PEV usage increase is promising for more energy-efficient and a cleaner transportation system (Kabli et al. 2019) and the growing pollution-free environment global awareness will increase the number of PEVs in the near future (Goli and Shireen 2014). To promote PEV growth, it is essential to expand fast-charging station (FCS) where PEVs' batteries could be charged within approximately 15 minutes (Domínguez-Navarro et al. 2019). Since PEVs have shorter driving distance and longer charging time, it is essential to establish sufficient quantity of charging stations (CS) in appropriate areas to meet PEVs charging demand (Zhang et al. 2019) which in turn initiates the need for an extensive amount of power load necessary to charge the PEVs (Goli and Shireen 2014). Due to their little market perception, PEVs are presently representing as relatively small demand (García-Villalobos et al. 2014) but to support PEV expansion, fast charging stations need to be expanded which requires power companies to confirm their possess of adequate energy in their network to fulfill the enlarged demand from these PEV charging stations (Kabli et al. 2019). PEV adoption increase promotes transport sector to convert less dependent on fossil fuels, however, as more and more PEVs take the road, the number of charging stations need to be increased (Kabli et al. 2019) and the use of electricity to fuel transportation needs will be increased (Hadley 2007) leading the power grid to face a challenging future due to the consequence of this extra load on the electric system (Colmenar-Santos et al. 2019). Supplying those charging stations by the network enforces an additional substantial load on the electrical network (Yang and Ribberink 2019) since one of the main fast-charging drawbacks is high electrical energy load and its effect on the network (Domínguez-Navarro et al. 2019). The interest towards PEVs is growing dramatically as it is expected that PEVs will grow largely in the next decade (Amjad et al. 2018), which would add significant load on the power grid (Wu et al. 2019). Deployment of huge quantity of PEVs imposes bulk demand to the electric system leading to substantial threats (Razipour et al. 2019) where greatly qualified electricity network is vital to supply PEVs' increasing charging load (Tulpule et al. 2013).

1.2 Impact of Supplying Charging Stations for Plug-in Electric Vehicle from Electrical Network

The effect of network-connected CS on the electrical network was analyzed by many authors (Domínguez-Navarro et al. 2019) where many researchers explored the impacts of charging demand (Tulpule et al. 2013) and emissions from power grid (Tulpule et al. 2013). Several studies have been done concerning the power quality problems caused by PEV chargers to the network (Khan et al. 2019). As per the analysis carried out in (Hadley 2007), PEV penetration to the vehicle market will significantly change the electric grid and vast pressure on the grid causes negative effect on the grid voltage profile (Karmaker et al. 2018). It will raise challenges such as transmission lines' thermal constraint violation due to overload and voltage drop in some sensitive network buses and demand uncertainty (Mozafar et al. 2017). Large number of PEVs charging can create many challenges for electrical grid such as transformers and lines overloads, and increase of power losses (García-Villalobos et al. 2014). PEV fast charging has harmful effects on the quality of the network causing DC offset, phase imbalance and voltage fluctuations (Khan et al. 2019).

PEV charging will destroy the capability of power network as well as will impact the overall distribution infrastructure (Amjad et al. 2018) as those PEVs will impose tension on the overloaded network generating new threats in the distribution infrastructure (Goli and Shireen 2014). PEVs will add major load to the network and studies concentrated on the problems that will be introduced to distribution networks (García-Villalobos et al. 2014). The PEVs fleet expansion will result in an extensive amount of power on the network, introducing negative impacts on distribution systems such as voltage violation unbalance and distribution transformers' lifespan reduction (Wu et al. 2019). Numerous articles were published addressing distribution transformers overload problems due to PEV charging (Goli and Shireen 2014). With increasing PEVs connected to the network for charging, profile of distribution network load will be severely altered causing network consistency concerns due to charging demand uncertainty (Rui et al. 2019). Also, many recent reports have assessed the increasing number of PEVs impact on transformers and underground cables within medium voltage distribution system where few have assessed the same through modeling (Farmer et al. 2010). Stress and negative impact on distribution networks such as saturations of transformers and lines, electrical losses increase and voltage fluctuations, and increase of will risk the consistency and safety of the network (García-Villalobos et al. 2014). The main technical challenge with PEVs is the huge uncertainties related to quantity and type of PEVs arriving at a specific time to a charging station (Eldeeb et al. 2018). PEVs arrive randomly to charging stations and connect to the distribution network for fast charging which seriously affects network safety and stability (Zhang et al. 2019). PEV charging demand is greatly unstable reliant on consumers' charging behavior and even if scattered charging periods are merged in unified periods, the charging load connected to the grid might be huge causing network instability (Lee and Hur 2019). Simultaneous numerous PEVs charging would increase total network demand which shall create a new peak time, lessening system voltage and frequency and lowering system stability and reliability (Lee and Hur 2019). It causes the peak-valley loads to alternate repeatedly and the grid power to vary significantly which is harmful to distribution network stability (Zhang et al. 2019). Night time PEVs charging represents a major load on present secondary and primary distribution infrastructure where numerous of these circuits do not have sufficient extra capacity (Tulpule et al. 2013). Night-charging would overload the distribution transformers designed to cool during night times (Goli and Shireen 2014). Installing transformers with higher capacities could resolve the issue but it is a costly and non-ecofriendly solution (Goli and Shireen 2014). Moreover, variable nature of PEV chargers results into higher order harmonics current in the network (Khan et al. 2019). These issues affect the operation and the reliability of the equipment used in the distribution network causing extra losses in cables and power transformers' windings (Khan et al. 2019).

The current PEVs implementation and projected rates are increasing and PEVs want to be recharged repeatedly, hence, the energy needed to secure that is enormous and increasing while several power networks do not possess adequate supply to fulfill this demand growth from PEV charging stations (Kabli et al. 2019). As per (García-Villalobos et al. 2014), the estimated energy load required for 30 million PEVs is 100 TWh per day which means 35 GW extra power is required. Power grid expansion takes time where power companies should take decisions based on inexact forecasts of PEVs acceptance rate per geographic region per year (Kabli et al. 2019). These tentative projections cause uncertainty in predicting the power demand by region and power utilities as they need to select where to enlarge the power network to supply the installation of charging stations (Kabli et al. 2019). Upgrade of transmission and distribution systems may assist in meeting the peak demand but may result into capacity surplus during normal operating conditions (Goli and Shireen 2014). Therefore, the integration of PEVs into electric grids is one of the biggest challenges of electric mobility technology (García-Villalobos et al. 2014).

1.3 Impact of Fossil Fuel-Based Energy Generation to Meet Plug-in Electric Vehicle Energy Demand

Although PEVs do not produce any emissions while moving on the roads, the electricity they use, if generated from fossil fuel based sources, produce emissions such as greenhouse gases (GHGs) (Grande et al. 2018). Energy is a necessity in an advanced world but expansions based on fossil fuel power generation are not sustainable solutions as they accelerate the depletion of existing natural reserves and polluting the environment (Colmenar-Santos et al. 2019). Since electricity is mostly produced from fossil fuels combustion such as petrol, gas, and coal (Girard et al. 2019), the huge growth in the global energy demand has raised up worries over the exhaustion of fossil fuel energy resources, heavy environmental impacts and deficiency in future energy supply (Ghenai and Bettayeb 2019). Electricity production is the most vital sectors accountable for high GHG emissions including CO_2, which affected numerous clear weather changes worldwide, causing numerous natural disasters affecting everyday millions of people's lives (Girard et al. 2019). Therefore, to obtain precise assessment results, PEVs' environmental impact assessments must evaluate the emissions produced by the energy generation sources supplying the charging stations (Galus et al. 2010).

1.4 Renewable Energy Integration in the Energy Mix to Fulfill Plug-in Electric Vehicle Charging Demand

The above discussion triggers the necessity for exploring alternative sustainable solutions to charge PEVs without affecting the power grid (Tulpule et al. 2013). Knowning that the current network is inadequate to fulfill the anticipated energy demand and the infeasibility of expanding the existing power grid, sustainable solutions are necessary to best meet the forecasted demand. In parallel with exploration for alternative sustainable resources to lessen yearly GHG emissions, climate change and global warming awareness have increased (Girard et al. 2019) demanding both transportation and power sectors (having the biggest potential for cutting CO_2 emissions) to increasingly contribute in the low carbon transition (Colmenar-Santos et al. 2019). The development of sustainable and renewable energy systems alternative to fossil fuels are needed to meet current and future increased energy demand (Ghenai and Bettayeb 2019).

Increasing the combination of renewable resources such as wind, solar, biomass, hydro and ocean in the energy mix and improving energy efficiency for transportation is among the 2030 sustainable energy targets (Ghenai and Bettayeb 2019) and to achieve those targets, renewable energy sources (RES) and PEVs have an vital part for a steady transition (Colmenar-Santos et al. 2019). To decrease the burden on the existing power sources and to promote a clean environment, RES are attracting more attention as alternative energy sources to conventional fossil fuel energy sources being the only unpolluted and uninterrupted energy solution to fulfill present and forthcoming energy demand (Ozlu and Dincer 2015). Accordingly, research is devoted to find renewable alternatives to meet the energy demand in a more ecological way (Girard et al. 2019) where RES in recent years are highly regarded as an eco-friendly alternative to fossil fuel power plants (Mozafar et al. 2017). Electricity from low emission RES such as PV, wind, water and ocean sources can almost eliminate CO_2 emissions (Colmenar-Santos et al. 2019). Investment in renewable energy has increased in recent years and renewable based generation has significantly expanded its market share (Abbasi et al. 2019) where many countries have fixed significant number of RES plants (Colmenar-Santos et al. 2019). This would reduce greenhouse gas emissions and the concentrations of pollution particles and smog in cities (Girard et al. 2019). RES play a key role in reducing the environmental impacts opposite to the conventional energy sources (Caliskan et al. 2013). Additionally, they do not deplete, and they support system decentralization and local solutions that are isolated from existing networks, which increases system's flexibility and provides economic benefits to small remote areas (Caliskan et al. 2013).

As per the International Renewable Energy Agency (IRENA) report, countries will increase their share of renewable energy sources dramatically by 2030 which would result in decreasing CO_2 emissions per kWh of produced electricity which sorts power production from renewable sources essential for enhancing PEVs ecological benefits (Grande et al. 2018). Subsequently, the popularity of PEVs is increasing due to their capability to rise the use of RES in transport sector (Karmaker et al. 2018). PEVs charging favors the use of RES (Amjad et al. 2018) since the ability of installing them near the load reduces technical losses, voltage fluctuations as well as investment costs (Mozafar et al. 2017). The integration of RES and PEV charging infrastructure shall reduce emissions from transport sector to more than 60% below 1990 levels (Colmenar-Santos et al. 2019) and maximize grid security by relieving the dependence on electricity network, which would efficiently decrease transmission line losses and improve the economic advantages of RES construction by selling the produced electricity to PEVs customers directly (Rui et al. 2019).

2 Previous Work

In this section, the recent work related to generation of electricity through RES which is used or could be used for charging PEVs is presented where the variability of PEV demand and RES power production is considered. Based on which, the optimal and novel design for this study shall be proposed in the subsequent section. The following sections shall cover proposed system's detailed description, thermodynamic analysis and obtained results. Integrated energy systems have the capability of utilizing the waste energy and achieving higher efficiencies (Ozlu and Dincer 2015). Multigeneration systems consisting of renewable sources merge high efficiency with unpolluted energy. Several researchers' interest was attracted by renewable generation and electric mobility (Mazzeo 2019), however, few combined renewable energy-based multigenerational electrical CS studies were found where PEV charging demand was segregated from the other energy demand as presented below.

The use of generated electricity by photovoltaic (PV) for PEV charging is well established and several studies investigated PV based PEV charging for both standalone and grid connected PV charging infrastructures (Esfandyari et al. 2019). A system that combines both solar system and PEVs was investigated by (Mazzeo 2019) considering diverse PV generation levels and diverse PEVs perceptions incorporating vehicle-to-grid and smart charging strategies (Mazzeo 2019). A smart charging infrastructure can use solar PV to provide CO_2 savings in both electricity generation and consumption for charging PEV (Esfandyari et al. 2019). Despite the obstacles of connecting MW range of solar system to the network, recent years PV based PEV parking lots became a popular subject (Eldeeb et al. 2018). PV based PEV charging stations will become dominant for dropping the effect of charging on the electric network and establishing ecological cities and societies (Islam and Mithulananthan 2018). The work in (Goli and Shireen 2014) proposed a PV powered charging station in a workplace to reduce the stress imposed on the distribution transformers. In (Tulpule et al. 2013), a day-time photovoltaic (PV) based PEV charging station located in a workplace parking garage was examined where the research highlighted many results such as reducing the effect of PEV charging on the distribution grid, charging load drop from the power grid, increasing RES utilization in transport sector. On the other hand, a study has examined wind-powered PEV charging station where access supply during no charging periods from 4 kW wind turbine is used to power the network (Lee and Hur 2019). The drawback of that proposal was that the intermittent nature of wind is reflected on the stability of generated power which in turn may effect negatively on PEV charging and the electric network (Lee and Hur 2019). Another study (Ozlu and Dincer 2015) proposed an entirely renewable solar-wind hybrid multi-generation system to generate both electricity and hydrogen where the two used renewable sources substitute each other such that if one is undersupplied or out of service the other system will back up.

Photovoltaic (PV) powered PEV CS is one method of integrating RES with PEV charging infrastructure and wind powered is another one, however, the energy supplied by both of them is intermittent (Colmenar-Santos et al. 2019) where PV or wind energy systems do not produce energy for a significant portion of time during the year mainly due to dependence on sunshine hours and wind speed which are variable (Ozlu and Dincer 2015). For PV based CS, energy produced from the solar system is generally unlike from PEVs' energy demand (Rui et al. 2019). Moreover, solar irradiation is absent during cloudy environments and in foggy days which would result into less or no energy generation during those circumstances. Therefore, relying completely on solar energy affects the system consistency (Karmaker et al. 2018). Same wise, the intermittent nature of wind generation can risk power system's reliable operation (Abbasi et al. 2019) due to comparatively high cut-in wind speeds ranging between 3.5 to 4.5 m/s resulting in capacity underutilization (Ozlu and Dincer 2015). Therefore, specific strategies should be developed to solve the variability of PEVs charging demand and renewable energy production simultaneously (Lee and Hur 2019). To resolve this vital issue, charging stations could be fixed with energy storage systems (Domínguez-Navarro et al. 2019). With the provision of energy storage systems, electricity produced from renewable sources can fulfill the required power loads despite the fluctuating and random nature of RESs (Colmenar-Santos et al. 2019).

The application of solar system can be additionally improved by integrating battery energy storage system (BESS) to overcome the intermittency of renewable sources and to rise the complete system consistency (Yang and Ribberink 2019). By storing the power that cannot be utilized directly and utilizing it when solar system cannot meet the demand, the BESS is used to ensure the security of supply (Grande et al. 2018). Many research studied the feasibility of applying either PV and/or BESS system with FCS (Yang and Ribberink 2019). In (Esfandyari et al. 2019), a research to maximize self-consumption and autonomy a photovoltaic (PV) array with battery storage for charging light weight PEV was studied. The work in (Lee and Hur 2019) investigates PEV charging with PV system and battery storage. During large PV power generation, PEVs are initially charged by solar system and surplus power used to charge the BESS. However, if the generated solar power is not satisfactory, PEVs are charged by the power discharged from the BESS (Lee and Hur 2019). The drawback of this system once again is the intermittent nature of solar is reflected on the stability of generated power which in turn may effect charging and discharging of BESS (Lee and Hur 2019). The results in (Girard et al. 2019) showed that PEV adaptation does not lead to real environmental benefits in comparison to its ICEV equivalent when it is charged with the grid, however, the results showed a considerable decrease in cost and CO_2 emissions per km travelled when using solar energy to charge the batteries.

Similarly, the application of wind turbines could be further improved by integrating BESS that can store the excess energy of wind generation during off-peak hours and feed electricity back to PEVs (Abbasi et al. 2019). Therefore, storage options make renewable sources more attractive as the stored energy can be used during power shortage periods (Caliskan et al. 2013) and it reduces the power fluctuation of random charging for fast charging stations (Zhang et al. 2019).

In order to reduce the required high power demand from the network and improve the profitability of the fast-charging stations, a study (Domínguez-Navarro et al.

2019) proposed a FCS consisting of solar and wind renewable sources and BESS. The attained outcomes showed that combination of renewable energy sources and energy storage systems attains the greatest results cost and efficient wise. That study assessed the optimal size and performed techno-economic optimization of an independent hybrid solar and wind turbine energy system with BESS, while other study (Ozlu and Dincer 2015) analyzed environmental and exergoeconomic impact of a hybrid PV and wind based hydrogen and electricity production system using energy, exergy and sustainability assessment methods. Based on the comparison among the different presented RES portfolios so far to generate power for FCS, it is concluded that the application RES such as solar and wind and storage systems would lessen the effect on the electrical network (Domínguez-Navarro et al. 2019). However, it is essential to note that because of the fast development in the solar system installations in MW capacities over the world where most of them are network connected, numerous worries raised about network steadiness and several challenges face the integration of RES to the network to perform similar to conventional production plants (Eldeeb et al. 2018). The intermittent nature of the irradiance of solar systems causes huge instabilities in the power output resulting undesirable effects on electric grid (Eldeeb et al. 2018). For those reasons, few authorities of electric power have set limits on the deviations rate of the power supplied from solar systems to the network (Eldeeb et al. 2018).

Henceforth, grid-independent generation units based on RES and BESS with the capability of generating domestic power may be applied to fulfill increasing energy demand (Razipour et al. 2019). The extensively applied generation units can be solar, wind turbines and fuel cells (FCs) (Razipour et al. 2019). For network-independency and pollution reduction purposes, the idea of integrating multigeneration system for vehicle charging is promising and numerous present studies have investigated the potential to improve vehicle charging stations in domestic zones (Ramadhani et al. 2019). Stand-alone systems have the benefit of working individually from the network with the aid of renewable sources leading to low emissions in greenhouse gases (Ghenai and Bettayeb 2019). A research studied an integrated multigeneration system generating heating, chilling space and electricity with a PEV charging station targeting for network-independent system (Ramadhani et al. 2019). Another study (Marino et al. 2019) proposed off-grid solar and storage system where outages were avoided, the quality and the security of the power supply was enhanced and the economic and environmental benefits were achieved. The obtained results from that research demonstrated that the off-grid PV-BESS are technically and economically viable and reliable, moreover, they are profitable while allowing a significant reduction of the air pollution. In (Karmaker et al. 2018), it was proposed a grid-independent PEV CS design, which decreases the emissions of CO_2 compared to the conventional network-based PEV CS by integrating a solar system and biogas generators to produce electricity for PEVs charging. FCS with optimal energy storage modeling was performed by (Zhang et al. 2019) which concluded that stand-alone designs can fulfill the electric energy demand of the PEVs relieving the effect of PEV charging on the network. Another study used PV and BESS system for PEV charging station for reducing the effect of PEV charging power on the network, which concludes that it is essential to use energy storage to meet PEV charging demand and optimize the PV consumption (Zhang et al. 2019).

Few research on hydrogen use was found in the literature where the produced electrical energy was from grid-independent renewable energy sources such as solar and wind (Caliskan et al. 2013). The fuel cell (FC) is used to supplement solar and wind, which are discontinuous energy sources, to meet the required energy load (Ghenai and Bettayeb 2019). A hybrid system consisting of solar and wind to produce electricity and hydrogen is investigated in (Caliskan et al. 2013) where solar and wind renewable energy resources were applied to generate electricity which can be stored by BESS. The electricity then was supplied to an electrolyzer to decompose water into hydrogen and oxygen. Hydrogen was stored in hydrogen tank and whenever required FC is used to reproduce electricity by utilizing oxygen from atmosphere (or from oxygen storage) and the stored hydrogen. Another study investigated a multi-generation system providing heating, cooling and electricity supply for buildings with a PEV charging station (Ramadhani et al. 2019). In (Colmenar-Santos et al. 2019), energetic and economic analysis of a stand-alone photovoltaic system with hydrogen storage was investigated. The objective of (Ghenai and Bettayeb 2019) research was to come up with an optimized stand-alone solar and hydrogen fuel cell system to supply the electrical load of a university building with less electricity cost and greenhouse gas emissions. In (Ramadhani et al. 2019), a system was proposed that delivers heat and cooling for houses, electricity for houses and PEV CS and hydrogen for hydrogen vehicle CS. FCs were applied for incorporating heating, cooling, electric power and hydrogen in a multigeneration system despite its high initial cost. The study concluded for combined residential supply, stand-alone grid-independent multi-generation system with solid oxide FC and electric vehicle charging station is the optimal arrangement. The literature analysis shows that hybrid electric renewable systems containing solar, wind and storage systems were investigated and enhanced individually for several years, however, the study of joint systems with energy storage is rather recent (Mazzeo 2019). Therefore, it is essential to strengthen experimental and theoretical research and studies to support more the fast perception of these integrated systems (solar, winds, BESS and PEVs). There are few studies found in literature that investigate these integrated systems in details. As per reviewed literature by the authors, the design and operation of off-grid charging stations is an important issue which still needs further investigation as off-grid charging stations have not been fully investigated and modeled (Mehrjerdi 2019).

3 Case Study

This case study is firstly proposing an optimal RES based PEV charging station having electro-chemical and chemical energy storage with no negative impact on the grid and secondly analyzing the designed system thermodynamically using proficient tools.

Solar energy as well as wind energy are deliberated as the chief sources of the designed charging station. Nevertheless, because of their intermittent nature, supplementary units are needed to store the excess energy and to generate the extra electricity needed in the times of night, peak periods or unpredicted weather conditions. Due to several environmental side effects of batteries for example use of toxic materials to human health (Bicer and Dincer 2018), the number of batteries in this case study are lowered, hence in addition to the battery storage, this study proposes chemical energy storage units that are used in FCs for further power generation.

This study proposes a unique case of standalone renewable energy-driven plug-in electric vehicle charging station by full combination of solar and wind energy resources along with battery, hydrogen and ammonia storage units. The specific objectives of this study are (i) designing a standalone hybrid renewable energy-based charging station consisting of solar PV plant and wind turbine plant, (ii) integrating sustainable energy storage alternatives such as chemical and electrochemical to ensure smooth operation of the charging stations during night times or unfavorable climate conditions, (iii) analyzing the proposed integrated system thermodynamically and assessing the energy and exergy efficiencies.

The schematic diagram of the designed integrated system for a fast charging station is shown in Fig. 1. The main inputs required to operate the integrated system are wind and sunlight as energy, and water and air as feedstock, while the main output is electricity. However, as intermediate products, there are hydrogen, ammonia and oxygen, which are later used for electricity generation through FCs.

The PV electricity is used for electrolyzer initially and then in battery units. Initially, the generated power from solar is supplied to H_2 production, because storing

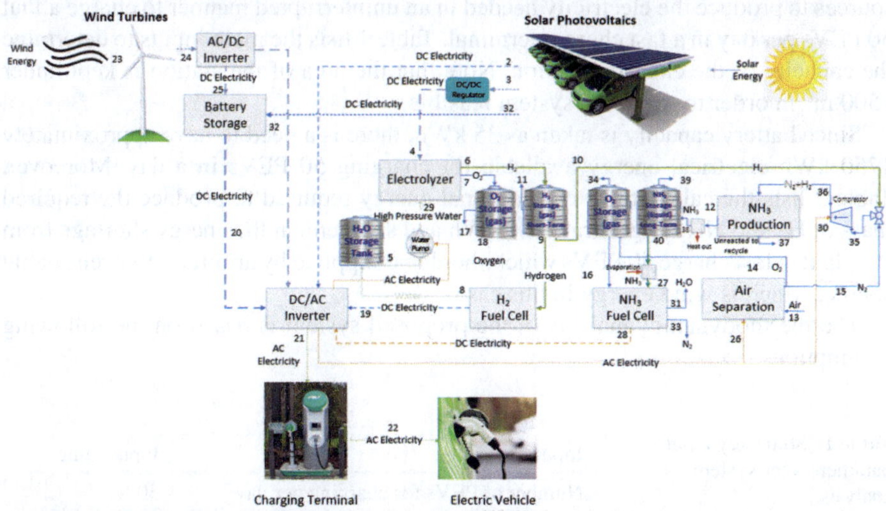

Fig. 1 The proposed system diagram for plug-in electric vehicle charging station working on 100% renewable energy

the solar energy is critical. Secondly, it is provided to PEV charging whereas the third priority is given to NH_3 and N_2 production. NH_3 can be stored longer periods of time compared to H_2 due to its thermo-physical characteristics. For wind turbine, the first priority is given to direct PEV charging and secondly, it is given to NH_3 and N_2 production and thirdly, the electricity is supplied to charge the batteries. After decomposition of water in the electrolyzer, the oxygen and hydrogen gases are stored separately in the storage tanks. On the other hand, a portion of the produced hydrogen is used as an input for H_2 FC and some portion is used for ammonia production. For ammonia production, there is a need of nitrogen, therefore, nitrogen is obtained through air separation unit and then supplied to ammonia production system along with hydrogen in the compressed form. Ammonia is condensed and stored in liquid form in the storage tank. The H_2 FC receives the oxygen and hydrogen, which are stored in tanks, and convert chemical energy into electrical energy. The product of FC reaction is water, hence, the produced water is sent to water tank to be used in the electrolyzer. Note that additional water input can be required for back-up purposes. Ammonia, which is stored in the tank, is converted into electricity through NH_3 FC for charging PEVs. Furthermore, the daily surplus power whether from solar plant or wind turbine is stored in the batteries for next day use, nighttime use or unfavorable days' times.

4 Analysis and Assessment

The key target of this study is to design a charging station using multi renewable sources to produce the electricity needed in an uninterrupted manner to charge about 50 PEVs per day in a fast charger terminal. Table 1 lists the main inputs to determine the capacity of the charging station. Note that the area of the station is kept under 1500 m^2 in order to make the system feasible.

Since battery capacity is taken as 35 kWh, there is a need to have approximately 1750 kWh electrical energy available for charging 50 PEVs in a day. Moreover, Table 2 lists the calculations of additional energy required to produce the required mass of H_2 and NH_3 to generate 200 kWh and supplement the energy shortage from PV plant to fast charge 50 PEVs which should be supplied by an alternative renewable source, which is wind energy in this case.

The thermodynamic analysis of the proposed system is based on the following assumptions:

Table 1 Some key input parameters for system analysis

Input parameters	Input value
Number of PEVs for charging per day	50
Number of FC slots at each station	4–6
Station area	1500 m^2
Battery capacity of PEVs	35 kWh

Table 2 Additional daily energy requirement of the integrated system due to chemicals production for chemical energy storage and PV unit limitation

Process	Total required energy (kWh/day)
Ammonia production	352
H_2 electrolyzer	520
Air separation	11
Energy shortage by PV plant	578
Energy produced by H_2 and NH_3 fuel cells	200
Total additional energy required from wind turbine	1261

- Steady state steady flow operations exist for all components (Rabbani et al. 2012).
- Reference pressure and temperature are 101.3 kPa and 25 °C, respectively.
- Kinetic and potential terms are neglected for all components except the wind turbine (Rabbani et al. 2012).
- Air is treated as ideal gas with 79% nitrogen and 21% oxygen for air separation plant only (Rabbani et al. 2012). The remaining processes employ real fluid characteristics.
- Average sun irradiance intensity is 700 W/m^2.
- Sun temperature is taken as 6000 °C.
- Average wind speed is taken as 5.6 m/s at the initial conditions.
- The average wind blowing hours per day is assumed to be 12 h.
- Wind turbine transmission and generator efficiencies are 95% (Cadu et al. 2012).
- AC-DC and DC-AC inverters efficiencies are taken as 95% (Notton 2010) and DC-DC converter efficiency is 100% (Fathabadi 2016).
- Battery charging and discharging efficiencies are taken as 95%.
- Water pump and ammonia production unit operate on average for 8 h per day.
- Hydrogen and oxygen gases have an average leakage of 10% from the storage tanks.
- At standard conditions, the PV panel efficiency is taken as 15%.
- Average sun hours is taken as 8 h.
- 40% of produced electricity from PV is used for H_2 production electrolyzer via DC–DC converter.
- 40% of produced electricity from PV is used for EV charging via DC-AC converter.
- 20% of produced electricity from PV is used for battery charging.
- Wind turbine rotor diameter is 54 m.
- Wind turbine lambda (λ) = 0.5.
- Pump isentropic efficiency is 85%.
- 30% of electricity produced by H_2 FC is equivalent to total heat rate generated by the FC.
- 30% of electricity produced by NH_3 FC is equivalent to total heat rate generated by the FC.

The thermodynamic analysis approach followed in this study is based on writing the balance equations for mass, energy, entropy and exergy for system's main components as well as their energy and exergy efficiencies equations. The codes are developed in Engineering Equation Solver (EES) (EES, 2018) for the overall system. Table 3 reflects the main thermodynamic balance equations utilized for the key components of the system.

5 Results and Discussion

In the analysis of the designed integrated system, using mass, energy, entropy and exergy balance equations, values of mass flow rate, temperature, pressure, specific enthalpy, specific entropy and specific exergy are determined for all the state points in the system at design conditions. Based on the fundamental balance equations and efficiency definitions given in Table 3, the energy and exergy efficiencies are calculated for subsystems and components as shown in Figs. 2 and 3. The H_2 and NH_3 FC energy efficiencies are in the range of 72–78%, respectively whereas the

Table 3 Main thermodynamic balance equations for the subsystems and components

Main component	Equation type	Balance and efficiency equations
PV solar	Energy	$\dot{Q}_{in,solar} = \dot{W}_{out,solar} + \dot{Q}_{out,solar}$
	Entropy	$\frac{\dot{Q}_{in,solar}}{T_{sun}} + \dot{S}_{gen} = \frac{\dot{Q}_{out,solar}}{T}$
	Exergy	$\dot{Q}_{in,\ solar}\left(1 - \frac{T_0}{T_{sun}}\right) = \dot{W}_{out,\ solar} + \dot{Q}_{out,\ solar}\left(1 - \frac{T_0}{T}\right) + \dot{E}x_{d,PV}$
	η_{energy}	$\frac{\dot{W}_{out,solar}}{\dot{Q}_{in,solar}}$
	η_{exergy}	$\frac{\dot{W}_{out,solar}}{\dot{Q}_{in,solar}\left(1 - \frac{T_0}{T_{sun}}\right)}$
Wind turbine	Mass	$\dot{m}_{in,turbine} = \dot{m}_{out,turbine}$
	Energy	$\dot{m}_{in,turbine}\cdot\left(h_{in,turbine} + \frac{v_1^2}{2}\right) =$ $\dot{m}_{out,turbine}\cdot\left(h_{out,turbine} + \frac{v_2^2}{2}\right) + \dot{Q}_{loss,turbine} + \dot{W}_{turbine}$
	Entropy	$\dot{m}_{in,turbine}\cdot s_{in,turbine} + \dot{S}_{gen} = \dot{m}_{out,turbine}\cdot s_{out,turbine} + \frac{\dot{Q}_{loss,turbine}}{T}$
	Exergy	$\dot{m}_{in,\ turbine}\cdot(ex_{in,\ turbine} + \frac{v_1^2}{2}) = \dot{m}_{out,\ turbine}\cdot\left(ex_{out,\ turbine} + \frac{v_2^2}{2}\right) +$ $\dot{Q}_{loss,\ turbine}\left(1 - \frac{T_0}{T}\right) + \dot{W}_{turbine} + \dot{E}x_{d,turbine}$
	η_{energy} and η_{exergy}	$\frac{\dot{W}_{turbine}}{\dot{m}_{turbine}\cdot\left(\frac{v_1^2}{2}\right)}$

(continued)

Table 3 (continued)

Main component	Equation type	Balance and efficiency equations
Electrolyzer	Mass	$\dot{m}_{H_2O,\text{in}} = \dot{m}_{H_2,\text{out}} + \dot{m}_{O_2,\text{out}}$
	Energy	$\dot{m}_{H_2O,\text{in}}.h_{H_2O,\text{in}} + \dot{W}_{\text{electrical,in}} =$ $\dot{m}_{H_2,\text{out}}.h_{H_2,\text{out}} + \dot{m}_{O_2,\text{out}}.h_{O_2,\text{out}} + \dot{Q}_{\text{loss,electrolysis}}$
	Entropy	$\dot{m}_{H_2O,\text{in}}.s_{H_2O,\text{in}} + \dot{S}_{\text{gen}} =$ $\dot{m}_{H_2,\text{out}}.s_{H_2,\text{out}} + \dot{m}_{O_2,\text{out}}.s_{O_2,\text{out}} + \frac{\dot{Q}_{\text{loss,electrolysis}}}{T}$
	Exergy	$\dot{m}_{H_2O,\text{ in}}.ex_{H_2O,\text{ in}} + \dot{W}_{\text{electrical, in}} = \dot{m}_{H_2,\text{out}}.ex_{H_2,\text{out}} +$ $\dot{m}_{O_2,\text{out}}.ex_{O_2,\text{out}} + \dot{Q}_{\text{loss, electrolysis}}\left(1 - \frac{T_0}{T}\right) + \dot{E}x_{d,\text{electrolysis}}$
	η_{energy}	$\frac{\dot{m}_{H_2,\text{out}}.h_{H_2,\text{out}} + \dot{m}_{O_2,\text{out}}.h_{O_2,\text{out}}}{\dot{W}_{\text{electrical,in}}}$
	η_{exergy}	$\frac{\dot{m}_{H_2,\text{out}}.ex_{H_2,\text{out}} + \dot{m}_{O_2,\text{out}}.ex_{O_2,\text{out}}}{\dot{W}_{\text{electrical,in}}}$
H_2 fuel cell	Mass	$\dot{m}_{H_2,\text{in}} + \dot{m}_{O_2,\text{in}} = \dot{m}_{H_2O,\text{out}}$
	Energy	$\dot{m}_{H_2,\text{in}}.h_{H_2,\text{in}} + \dot{m}_{O_2,\text{in}}.h_{O_2,\text{in}} =$ $\dot{m}_{H_2O,\text{out}}.h_{H_2O,\text{out}} + \dot{W}_{\text{el},H_2\text{FCout}} + \dot{Q}_{\text{loss},H_2\text{FC}}$
	Entropy	$\dot{m}_{H_2,\text{in}}.s_{H_2,\text{in}} + \dot{m}_{O_2,\text{in}}.s_{O_2,\text{in}} + \dot{S}_{\text{gen}} = \dot{m}_{H_2O,\text{out}}.s_{H_2O,\text{out}} + \frac{\dot{Q}_{loss,H_2\text{FC}}}{T}$
	Exergy	$\dot{m}_{H_2,in}.ex_{H_2,\text{in}} + \dot{m}_{O_2,\text{in}}.ex_{O_2,\text{in}} = \dot{m}_{H_2O,\text{ out}}.ex_{H_2O,\text{ out}} +$ $\dot{W}_{\text{el, }H_2\text{FC out}} + \dot{Q}_{\text{loss, }H_2\text{FC}}\left(1 - \frac{T_0}{T}\right) + \dot{E}x_{d,H_2\text{FC}}$
	η_{energy}	$\frac{\dot{W}_{\text{el},H_2\text{FCout}}}{\dot{m}_{H_2}.\text{LHV}_{H_2} + \dot{m}_{O_2}.h_{O_2,\text{in}}}$
	η_{exergy}	$\frac{\dot{W}_{\text{el},H_2\text{FCout}}}{\dot{m}_{H_2,\text{in}}.\text{ex}_{H_2,\text{in}} + \dot{m}_{O_2,\text{in}}.\text{ex}_{O_2,\text{in}}}$
NH_3 fuel cell	Mass	$\dot{m}_{NH_3,\text{in}} + \dot{m}_{O_2,\text{in}} = \dot{m}_{N_2,\text{out}} + m_{H_20,\text{out}}$
	Energy	$\dot{m}_{NH_3,\text{in}}.h_{NH_3,\text{in}} + \dot{m}_{O_2,\text{in}}.h_{O_2,\text{in}} =$ $\dot{m}_{N_2,\text{out}}.h_{N_2,\text{out}} + \dot{m}_{H_2O,\text{out}}.h_{H_2O,\text{out}} + \dot{W}_{\text{el},NH_3\text{FCout}} + \dot{Q}_{\text{loss},NH_3\text{FC}}$
	Entropy	$\dot{m}_{NH_3,\text{in}}.s_{NH_3,\text{in}} + \dot{m}_{O_2,\text{in}}.s_{O_2,\text{in}} + \dot{S}_{\text{gen}} =$ $\dot{m}_{N_2,\text{out}}.s_{N_2,\text{out}} + \dot{m}_{H_2O,\text{out}}.s_{H_2O,\text{out}} + \frac{\dot{Q}_{\text{loss},NH_3\text{FC}}}{T}$
	Exergy	$\dot{m}_{NH_3,\text{in}}.\text{ex}_{NH_3,\text{in}} + \dot{m}_{O_2,\text{in}}.\text{ex}_{O_2,\text{in}}$ $= \dot{m}_{N_2,\text{out}}.\text{ex}_{N_2,\text{out}} + m_{H_2O,\text{out}}.\text{ex}_{H_2O,\text{out}} + \dot{W}_{\text{el},NH_3\text{FCout}}$ $+ \dot{Q}_{\text{loss},NH_3\text{FC}}\left(1 - \frac{T_0}{T}\right) + \dot{E}x_{d,NH_3\text{FC}}$
	η_{energy}	$\frac{\dot{W}_{\text{el},NH_3\text{FCout}}}{\dot{m}_{NH_3}.\text{LHV}_{NH_3} + \dot{m}_{O_2,\text{in}}.h_{O_2,\text{in}}}$
	η_{exergy}	$\frac{\dot{W}_{\text{el},NH_3\text{FCout}}}{\dot{m}_{NH_3,\text{in}}.\text{ex}_{NH_3,\text{in}} + \dot{m}_{O_2,\text{in}}.\text{ex}_{O_2,\text{in}}}$

(continued)

Table 3 (continued)

Main component	Equation type	Balance and efficiency equations
Water pump	Mass	$\dot{m}_{in,pump} = \dot{m}_{out,pump}$
	Energy	$\dot{m}_{in,pump}.h_{in,pump} + \dot{W}_{pump,actual} = \dot{m}_{out,pump}.h_{out,pump}$
	Entropy	$\dot{m}_{in,pump}.s_{in,pump} + \dot{S}_{gen} = \dot{m}_{out,pump}.s_{out,pump}$
	Exergy	$\dot{m}_{in,pump}.ex_{in,\ pump} + \dot{W}_{pump,\ actual} =$ $\dot{m}_{out,\ pump}.ex_{out,\ pump} + \dot{E}x_{d,pump}$
	η_{energy}	$\frac{\dot{m}_{pump}.(h_{out,pump} - h_{in,pump})}{\dot{W}_{pump,actual}}$
	η_{exergy}	$\frac{\dot{m}_{pump}.(ex_{out,pump} - ex_{in,pump})}{\dot{W}_{pump,actual}}$
O_2 storage for H_2 fuel cell	Mass	$\dot{m}_{O_2,in} = \dot{m}_{O_2,out} + \dot{m}_{O_2,leak}$
	Energy	$\dot{m}_{O_2,in}.h_{O_2,in} = \dot{m}_{O_2,out}.h_{O_2,out} + \dot{m}_{O_2,leak}.h_{O_2,leak}$
	Entropy	$\dot{m}_{O_2,in}.s_{O_2,in} + \dot{S}_{gen} = \dot{m}_{O_2,out}.s_{O_2,out} + \dot{m}_{O_2,leak}.s_{O_2,leak}$
	Exergy	$\dot{m}_{O_2,in}.ex_{O_2,in} = \dot{m}_{O_2,out}.ex_{O_2,out} + \dot{m}_{O_2,leak}.ex_{O_2,leak} + \dot{E}x_{d,O_2}$
	η_{energy}	$\frac{\dot{m}_{O_2,out}.h_{O_2,out}}{\dot{m}_{O_2,in}.h_{O_2,in}}$
	η_{exergy}	$\frac{\dot{m}_{O_2,out}.ex_{O_2,out}}{\dot{m}_{O_2,in}.ex_{O_2,in}}$
O_2 storage for NH_3 fuel cell	Mass	$\dot{m}_{O_2,in} = \dot{m}_{O_2,out} + \dot{m}_{O_2,leak}$
	Energy	$\dot{m}_{O_2,in}.h_{O_2,in} = \dot{m}_{O_2,out}.h_{O_2,out} + \dot{m}_{O_2,leak}.h_{O_2,leak}$
	Entropy	$\dot{m}_{O_2,in}.s_{O_2,in} + \dot{S}_{gen} = \dot{m}_{O_2,out}.s_{O_2,out} + \dot{m}_{O_2,leak}.s_{O_2,leak}$
	Exergy	$\dot{m}_{O_2,in}.ex_{O_2,in} = \dot{m}_{O_2,out}.ex_{O_2,out} + \dot{m}_{O_2,leak}.ex_{O_2,leak} + \dot{E}x_{d,O_2}$
	η_{energy}	$\frac{\dot{m}_{O_2,out}.h_{O_2,out}}{\dot{m}_{O_2,in}.h_{O_2,in}}$
	η_{exergy}	$\frac{\dot{m}_{O_2,out}.ex_{O_2,out}}{\dot{m}_{O_2,in}.ex_{O_2,in}}$
H_2 storage	Mass	$\dot{m}_{H_2,in} = \dot{m}_{H_2,H_2FC} + \dot{m}_{H_2,NH_3Production} + \dot{m}_{H_2,leak}$
	Energy	$\dot{m}_{H_2,in}.h_{H_2,in} = \dot{m}_{H_2,H_2FC}.h_{H_2,H_2FC} +$ $\dot{m}_{H_2,NH_3Production}.h_{H_2,NH_3Production} + \dot{m}_{H_2,leak}.h_{H_2,leak}$
	Entropy	$\dot{m}_{H_2,in}.s_{H_2,in} + \dot{S}_{gen} = \dot{m}_{H_2,H_2FC}.s_{H_2,H_2FC} +$ $\dot{m}_{H_2,NH_3Production}.s_{H_2,NH_3Production} + \dot{m}_{H_2,leak}.s_{H_2,leak}$
	Exergy	$\dot{m}_{H_2,in}.ex_{H_2,in} = \dot{m}_{H_2,H_2FC}.ex_{H_2,H_2FC} +$ $\dot{m}_{H_2,NH_3Production}.ex_{H_2,NH_3Production} + \dot{m}_{H_2,leak}.ex_{H_2,leak} + \dot{E}x_{d,H_2}$
	η_{energy}	$\frac{\dot{m}_{H_2,H_2FC}.h_{H_2,H_2FC} + \dot{m}_{H_2,NH_3Production}.h_{H_2,NH_3Production}}{\dot{m}_{H_2,in}.h_{H_2,in}}$
	η_{exergy}	$\frac{\dot{m}_{H_2,H_2FC}.ex_{H_2,H_2FC} + \dot{m}_{H_2,NH_3Production}.ex_{H_2,NH_3Production}}{\dot{m}_{H_2,in}.ex_{H_2,in}}$

(continued)

Table 3 (continued)

Main component	Equation type	Balance and efficiency equations
NH_3 storage	Mass	$\dot{m}_{NH_3,in} = \dot{m}_{NH_3,out} + \dot{m}_{NH_3,leak}$
	Energy	$\dot{m}_{NH_3,in}.h_{NH_3,in} = \dot{m}_{NH_3,out}.h_{NH_3,out} + \dot{m}_{NH_3,leak}.h_{NH_3,leak}$
	Entropy	$\dot{m}_{NH_3,in}.s_{NH_3,in} + \dot{S}_{gen} = \dot{m}_{NH_3,out}.s_{NH_3,out} + \dot{m}_{NH_3,leak}.s_{NH_3,leak}$
	Exergy	$\dot{m}_{NH_3,in}.ex_{NH_3,in} =$ $\dot{m}_{NH_3,out}.ex_{NH_3,out} + \dot{m}_{NH_3,leak}.ex_{NH_3,leak} + \dot{E}x_{d,NH_3}$
	η_{energy}	$\frac{\dot{m}_{NH_3,out}.h_{NH_3,out}}{\dot{m}_{NH_3,in}.h_{NH_3,in}}$
	η_{exergy}	$\frac{\dot{m}_{NH_3,out}.ex_{NH_3,out}}{\dot{m}_{NH_3,in}.ex_{NH_3,in}}$
NH_3 production	Mass	$\dot{m}_{N_2,in} + \dot{m}_{H_2,in} = \dot{m}_{NH_3,out} + \dot{m}_{N_2,unreacted} + \dot{m}_{H_2,unreacted}$
	Energy ($\dot{W}_{elec}$: is for recycle compressor within NH_3 production unit)	$\dot{m}_{N_2,in}.h_{N_2,in} + \dot{m}_{H_2,in}.h_{H_2,in} + \dot{W}_{elec} =$ $\dot{m}_{NH_3,out}.h_{NH_3,out} + \dot{m}_{N_2,unreacted}.h_{N_2,unreacted} +$ $\dot{m}_{H_2,unreacted}.h_{H_2,unreacted}$
	Entropy	$\dot{m}_{N_2,in}.s_{N_2,in} + \dot{m}_{H_2,in}.s_{H_2,in} + \dot{S}_{gen} =$ $\dot{m}_{NH_3,out}.s_{NH_3,out} + \dot{m}_{N_2,unreacted}.s_{N_2,unreacted} +$ $\dot{m}_{H_2,unreacted}.s_{H_2,unreacted}$
	Exergy	$\dot{m}_{N_2,in}.ex_{N_2,in} + \dot{m}_{H_2,in}.ex_{H_2,in} + \dot{W}_{elec} =$ $\dot{m}_{NH_3,out}.ex_{NH_3,out} + \dot{m}_{N_2,unreacted}.ex_{N_2,unreacted} +$ $\dot{m}_{H_2,unreacted}.ex_{H_2,unreacted} + \dot{E}x_{d,NH_3}$
	η_{energy}	$\frac{\dot{m}_{NH_3,out}.h_{NH_3,out}}{\dot{m}_{N_2,in}.h_{N_2,in}+\dot{m}_{H_2,in}.h_{H_2,in}+\dot{W}_{elec}}$
	η_{exergy}	$\frac{\dot{m}_{NH_3,out}.ex_{NH_3,out}}{\dot{m}_{N_2,in}.ex_{N_2,in}+\dot{m}_{H_2,in}.ex_{H_2,in}+\dot{W}_{elec}}$
Air Separation	Mass	$\dot{m}_{air,in} = \dot{m}_{N_2,out} + \dot{m}_{O_2,out}$
	Energy	$\dot{m}_{air,in}.h_{air,in} + \dot{W}_{elec} = \dot{m}_{N_2,out}.h_{N_2,out} + \dot{m}_{O_2,out}.h_{O_2,out}$
	Entropy	$\dot{m}_{air,in}.s_{air,in} + \dot{S}_{gen} = \dot{m}_{N_2,out}.s_{N_2,out} + \dot{m}_{O_2,out}.s_{O_2,out}$
	Exergy	$\dot{m}_{air,in}.ex_{air,in} + \dot{W}_{elec} = \dot{m}_{N_2,out}.ex_{N_2,out} + \dot{m}_{O_2,out}.ex_{O_2,out} + \dot{E}x_d$
	η_{energy}	$\frac{\dot{m}_{N_2,out}.h_{N_2,out}+\dot{m}_{O_2,out}.h_{O_2,out}}{\dot{m}_{air,in}.h_{air,in}+\dot{W}_{elec}}$
	η_{exergy}	$\frac{\dot{m}_{N_2,out}.ex_{N_2,out}+\dot{m}_{O_2,out}.ex_{O_2,out}}{\dot{m}_{air,in}.ex_{air,in}+\dot{W}_{elec}}$

chemical storage efficiencies are between 90 and 85%. The overall energy efficiency of the integrated system is about 16.9% whereas the overall exergy efficiency is about 17.6%.

Figure 3 shows the summary of exergy efficiencies. The exergy efficiency of PV plant is about 16%. The exergy efficiency of the air separation plant is calculated to

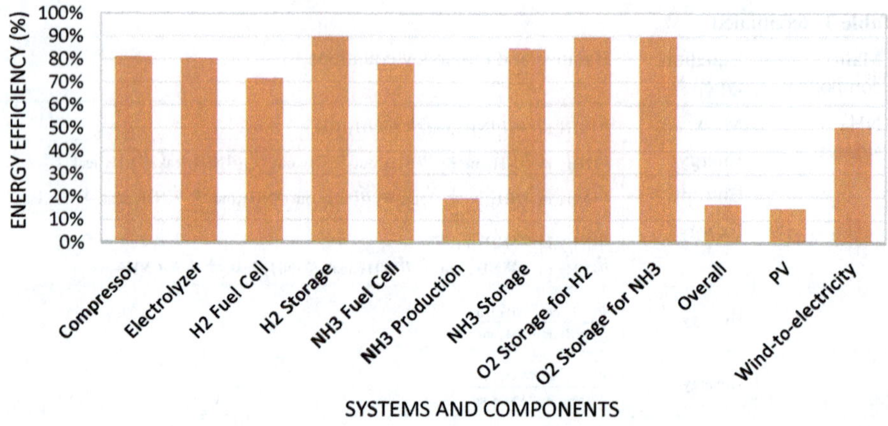

Fig. 2 Energy efficiencies of the subsystems and components within the integrated charging station

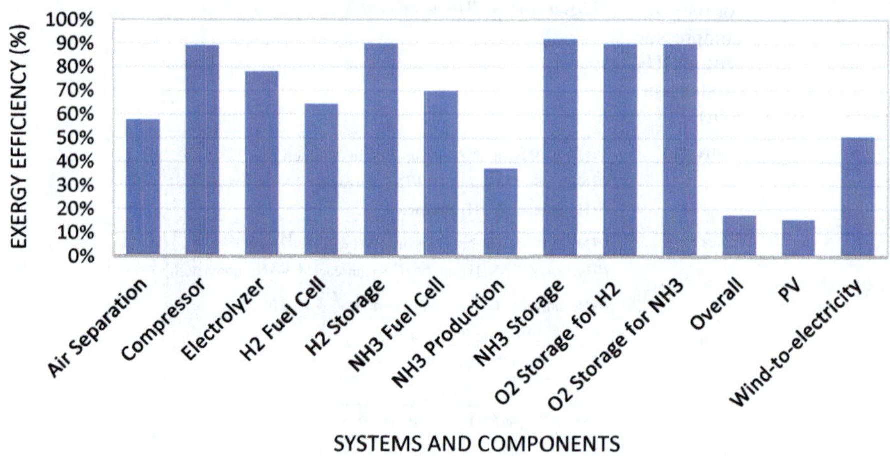

Fig. 3 Exergy efficiencies of the subsystems and components within the integrated charging station

be about 57%. Note that the ammonia production efficiency represents the single pass efficiency meaning that the unreacted gases (N_2 and H_2) are not recycled and taken into account in the efficiency calculations. The exergy efficiencies of the subsystems and components are different than energy efficiencies, implying that through exergy analysis, a deeper analysis is possible. Non-recovered part of exergy is denoted as exergy destruction for which the main exergy destruction values are given in Table 4.

The designed system is for a charging station; hence the size of the plants is not that huge. The highest exergy destruction (~840 kW) occurs in the solar PV power plant as expected due to huge heat losses and low energy conversion efficiency as listed in Table 4. The electrolyzer for hydrogen production and ammonia production plants

Table 4 Main exergy destruction rates within the integrated charging station

Component/subsystem	Exergy destruction rate (kW)
Air separation	0.1931
Compressor	0.2299
H_2 storage	0.08144
NH_3 production	8.37
NH_3 storage	0.00948
O_2 storage for H_2 fuel cell	0.04102
O_2 storage for NH_3 fuel cell	0.0203
Electrolyzer	14
H_2 fuel cell	2.168
NH_3 Fuel cell	2.074
Solar PV	840.3

have the second and third highest exergy destruction among other subsystems. The exergy destruction rate of H_2 and NH_3 FCs are about 2.2 kW and 2.1 kW, respectively.

Some of the subsystems have either heat inputs or outputs through their boundaries. The heat rates are calculated using energy balance equations are depicted in Table 5. Due to large size of PVs, the total heat input is quite high corresponding to about 1047 kW, where the respective losses are large as well (890 kW). The FCs have heat generation as well with the values of about 1.3 kW and 1.6 kW, respectively for H_2 and NH_3 FCs. Conceptually, the by-product heat from the FCs can also be utilized for several applications (hot water, heating etc.), however, in this case study they are not used because the focus is electricity required for the charging station.

The integrated system is mainly designed for work production. However, storage units will also require electricity, mainly for ammonia and hydrogen production. The produced electricity from solar energy is mostly consumed by electrolyzer for producing hydrogen as shown in Table 6. Solar energy is given priority for chemicals synthesis because it has more intermittent nature than wind energy, hence, requiring more storage. The NH_3 FC generated about 5.4 kW, whereas the H_2 FC generates about 4.5 kW electricity.

Table 5 Main heat rates within the integrated charging station

Description	Heat rate (kW)
Total heat input to PV	1047
Air separation	0.885
Compressor cooling	0.4
Electrolyzer	13.3
H_2 fuel cell	1.335
NH_3 fuel cell	1.627
NH_3 production	17.25
PV heat loss	890.1

Table 6 Main work rates within the integrated EV charging station

Description	Work rate (kW)
Air separation unit	0.4583
Compressor	2.137
Electrolyzer	62.83
Total PV output	157.1
PV output direct charging station (state point 2)	31.42
PV output for battery (state point 32)	62.83
PV output for electrolyzer (state point 3)	62.83
Wind turbine electrical output	113.6
H_2 fuel cell output	4.449
NH_3 fuel cell output	5.423

In order to observe the performance changes depending on the critical parameters, a parametric study is performed. Figure 4 depicts the changes of exergy destruction and exergy efficiency of air separation unit based on reference pressure change. This unit is more sensitive to reference pressure due to involving several gases. The exergy efficiency of the air separation plant decreases to about 56% from about 61% in case the reference pressure rises to 110 kPa.

Since exergy analysis takes into account the reference temperature, the effects of changing reference temperature will play a role in the system performance. In Qatar conditions, the ambient temperature may rise to 50 °C. Therefore, the parameters are changed from 10 to 50 °C. The exergy destruction of ammonia production unit increases from 7.5 to 9.8 kW whereas the exergy efficiency of the air separation plant rises to about 63% from 55% when the ambient temperature varies from 10 to 50 °C. Similarly, the overall system efficiency has a slight improvement due to

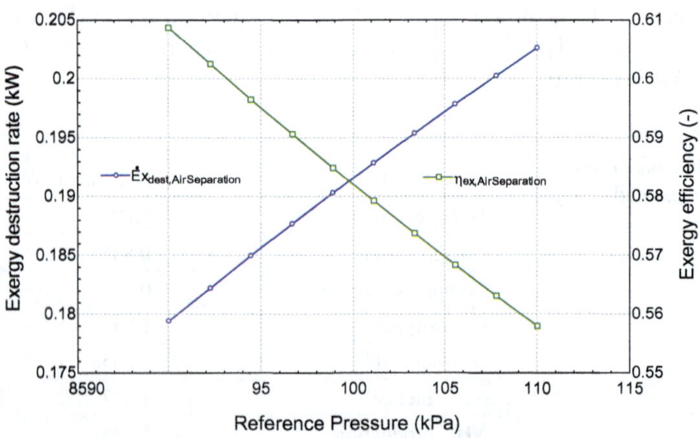

Fig. 4 The effect of reference pressure on the air separation exergy efficiency and exergy destruction

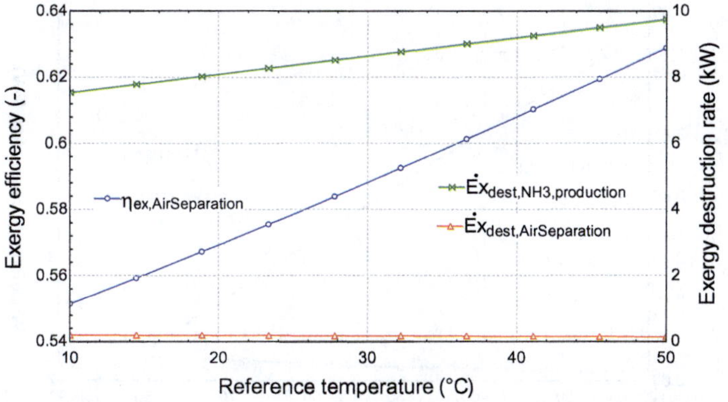

Fig. 5 The effects of reference temperature on the air separation and ammonia production systems

rising ambient temperature as shown in Fig. 6. On the other hand, the FC efficiencies decrease due to more internal heat generation and difficulties for heat rejection. This phenomenon is further explained in Fig. 7 where H_2 FC efficiency decreases from 68.3% to about 63.4%. The work production reduces to 4.8–4.4 kW with associated exergy destruction increase (Fig. 5).

There are mainly two inputs to the integrated system, (i) solar irradiation, (ii) wind. Therefore, the effects of these input parameters are also investigated as shown in Figs. 8 and 9. With the increase in the irradiance, the total power output from PVs rises, however, the exergy destruction increases as well. This causes a slight decrease in the overall efficiencies. At about 945 W/m^2 solar irradiance, the generated power from PV plant is about 211 kW as illustrated in Fig. 8. This can allow more hydrogen generation and storage capabilities. Note that the initial design of the system is for 700 W/m^2.

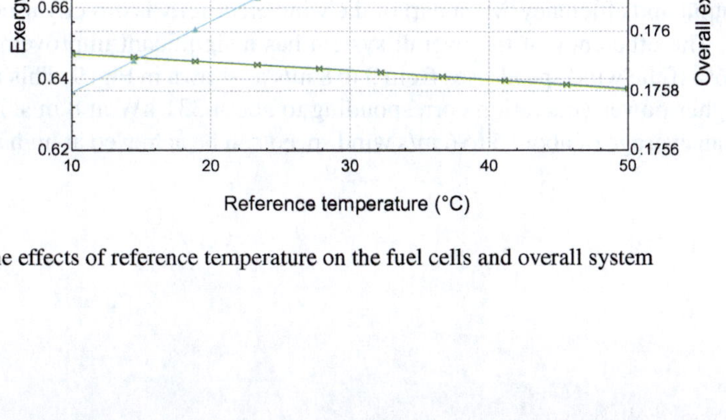

Fig. 6 The effects of reference temperature on the fuel cells and overall system

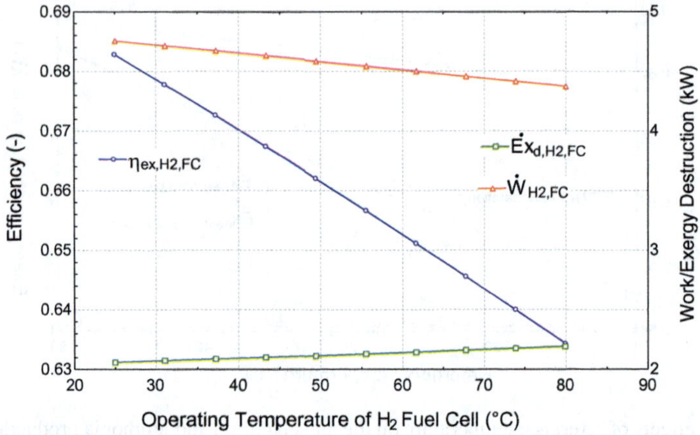

Fig. 7 The effects of operation temperature of H_2 fuel cell on the fuel cell work production, exergy destruction and exergy efficiency

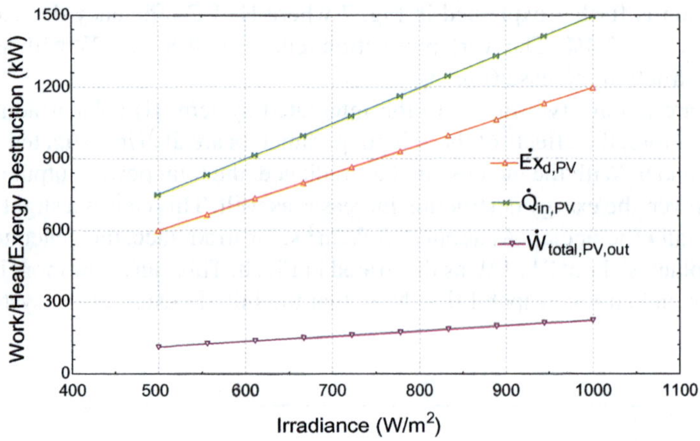

Fig. 8 The changes of total power input and output from PV system and exergy destruction based on solar irradiance

The wind velocity, on the other hand, has a more dominant effect on the total power output and efficiency. Most part of the wind electricity is directly used in PEV charging. The efficiency of the overall system has a significant improvement from 11.5 to 26% if the wind speed rises from 3 to 8 m/s as shown in Fig. 9. This is mainly due to higher power generation corresponding to about 331 kW at 8 m/s. Note that in Qatar, an average of about 5 to 6 m/s wind speed can be achieved at high altitudes.

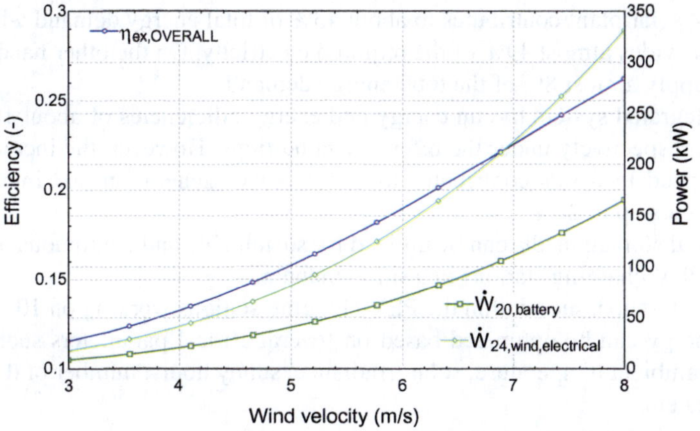

Fig. 9 The changes of total power output from wind turbine and exergy efficiency based on wind velocity

The designed system is a more reasonable and environmental alternative to fossil fuel-based PEV charging stations due to the fact that it enables continuous electricity supply through renewables independent from grid and with sustainable energy storage units.

6 Conclusions

In the context of the energy crisis and environment degradation, the energy sector has to accelerate the transition of low-carbon economy where the electrification of transportation is an important access to realize this transition. The main objective of this study is to propose ecofriendly stand-alone electrical car charging station. The contribution of this study is to propose a reliable and combined solar and wind systems along with battery, hydrogen and ammonia storage units capable of fast charging 50 number of PEVs per day in a country with arid weather condition such as State of Qatar. The results showed the below main findings:

- With 1500 m^2 land, 1200 kWh/day energy can be generated from a solar plant in addition to 1350 kWh/day from 250 kW wind turbine which are the two main sources of the proposed design. BESS with capacity of 650 kWh along with H_2 and NH_3 chemical storage systems are incorporated in the system to cater during insufficient power supply from the main two renewable sources.
- Although 1750 kWh/day is required to charge 50 PEVs per day with 35 kWh average battery capacity, additional 860 kWh/day is considered to produce the required chemicals to reproduce 200 kWh/day during night and unfavorable times. This enables a smooth operation of the charging station.

- The PV solar plant contributes to about 43% of total energy demand whereas the wind provides almost 49% of the required electricity. On the other hand, the fuel cells supply almost 8% of the total energy demand.
- The integrated system has an energy and exergy efficiencies of about 16.9% and 17.6%, respectively under the reference conditions. However, the increase in the wind speed to 8 m/s contributes to larger power generation and improves the efficiencies up to 26%.
- Chemical storage media can be utilized for sustainable and continuous operations of the PEV charging stations in a smart manner.
- Energy storage issues within the PEV charging stations working on 100% renewable energy can be optimized based on several critical parameters such as wind speed, ambient temperature, solar irradiance, sunny hours, number of demanding vehicles etc.

Acknowledgements The authors acknowledge the great support provided by Hamad Bin Khalifa University as well as Qatar Electricity and Water Corporation (KAHRAMAA).

References

Abbasi MH, Taki M, Rajabi A, Li L, Zhang J (2019) Coordinated operation of electric vehicle charging and wind power generation as a virtual power plant: a multi-stage risk constrained approach. Appl Energy 239:1294–1307. https://doi.org/10.1016/j.apenergy.2019.01.238

Amini MH, Boroojeni KG, Wang CJ, Nejadpak A, Iyengar SS, Karabasoglu O (2016) Effect of electric vehicle parking lots' charging demand as dispatchable loads on power systems loss. In: 2016 IEEE international conference on electronics information technology. IEEE, pp 499–503. https://doi.org/10.1109/eit.2016.7535291

Amjad M, Ahmad A, Rehmani MH, Umer T (2018) A review of EVs charging: from the perspective of energy optimization, optimization approaches, and charging techniques. Transp Res Part D Transp Environ 62:386–417. https://doi.org/10.1016/j.trd.2018.03.006

Bicer Y, Dincer I (2018) Life cycle environmental impact assessments and comparisons of alternative fuels for clean vehicles. Resour Conserv Recycl 132:141–157. https://doi.org/10.1016/j.resconrec.2018.01.036

Cadu M, Huijbregts MAJ, Althaus H, Koehler A, Hellweg S (2012) Wind power electricity : the bigger the Turbine. The Greener. https://doi.org/10.1021/es204108n

Caliskan H, Dincer I, Hepbasli A (2013) Exergoeconomic and environmental impact analyses of a renewable energy based hydrogen production system. Int J Hydrogen Energy 38:6104–6111. https://doi.org/10.1016/j.ijhydene.2013.01.069

Colmenar-Santos A, Muñoz-Gómez AM, Rosales-Asensio E, López-Rey Á (2019) Electric vehicle charging strategy to support renewable energy sources in Europe 2050 low-carbon scenario. Energy. 183:61–74. https://doi.org/10.1016/j.energy.2019.06.118

Dicks G, Breedon F (1988) World outlook, 2013th edn. International Energy Agency (IEA). https://doi.org/10.1111/j.1468-0319.1988.tb00400.x

Domínguez-Navarro JA, Dufo-López R, Yusta-Loyo JM, Artal-Sevil JS, Bernal-Agustín JL (2019) Design of an electric vehicle fast-charging station with integration of renewable energy and storage systems. Int J Electr Power Energy Syst 105:46–58. https://doi.org/10.1016/j.ijepes.2018.08.001

Eldeeb HH, Faddel S, Mohammed OA (2018) Multi-objective optimization technique for the operation of grid tied PV powered EV charging station. Electr Power Syst Res 164:201–211. https://doi.org/10.1016/J.EPSR.2018.08.004

Esfandyari A, Norton B, Conlon M, McCormack SJ (2019) Performance of a campus photovoltaic electric vehicle charging station in a temperate climate. Sol Energy 177:762–771. https://doi.org/10.1016/j.solener.2018.12.005

Farmer C, Hines P, Dowds J, Blumsack S (2010) Modeling the impact of increasing PHEV loads on the distribution infrastructure. In: 2010 43rd Hawaii international conference on system sciences. IEEE, pp 1–10. https://doi.org/10.1109/hicss.2010.277

Fathabadi H (2016) ScienceDirect novel high efficiency DC/DC boost converter for using in photovoltaic systems. Sol Energy 125:22–31. https://doi.org/10.1016/j.solener.2015.11.047

Fazelpour F, Vafaeipour M, Rahbari O, Rosen MA (2014) Intelligent optimization to integrate a plug-in hybrid electric vehicle smart parking lot with renewable energy resources and enhance grid characteristics. Energy Convers Manag 77:250–261. https://doi.org/10.1016/j.enconman.2013.09.006

Galus MD, Zima M, Andersson G (2010) On integration of plug-in hybrid electric vehicles into existing power system structures. Energy Policy 38:6736–6745. https://doi.org/10.1016/j.enpol.2010.06.043

García-Villalobos J, Zamora I, San Martín JI, Asensio FJ, Aperribay V (2014) Plug-in electric vehicles in electric distribution networks: a review of smart charging approaches. Renew Sustain Energy Rev 38:717–731. https://doi.org/10.1016/j.rser.2014.07.040

Ghenai C, Bettayeb M (2019) Modelling and performance analysis of a stand-alone hybrid solar PV/fuel cell/diesel generator power system for university building. Energy 171:180–189. https://doi.org/10.1016/j.energy.2019.01.019

Girard A, Roberts C, Simon F, Ordoñez J (2019) Solar electricity production and taxi electrical vehicle conversion in Chile. J Clean Prod 210:1261–1269. https://doi.org/10.1016/j.jclepro.2018.11.092

Goli P, Shireen W (2014) PV powered smart charging station for PHEVs. Renew Energy 66:280–287. https://doi.org/10.1016/j.renene.2013.11.066

Grande LSA, Yahyaoui I, Gómez SA (2018) Energetic, economic and environmental viability of off-grid PV-BESS for charging electric vehicles: case study of Spain. Sustain Cities Soc 37:519–529. https://doi.org/10.1016/J.SCS.2017.12.009

Hadley SW (2007) Evaluating the impact of plug-in hybrid electric vehicles on regional electricity supplies. In: 2007 IREP symposium on bulk power system dynamics control—VII, revitalizing operational reliability. https://doi.org/10.1109/irep.2007.4410538

Islam MS, Mithulananthan N (2018) PV based EV charging at universities using supplied historical PV output ramp. Renew Energy 118:306–327. https://doi.org/10.1016/j.renene.2017.11.009

Kabli M, Quddus MA, Nurre SG, Marufuzzaman M, Usher JM (2019) A stochastic programming approach for electric vehicle charging station expansion plans. Int J Prod Econ. https://doi.org/10.1016/j.ijpe.2019.07.034

Karmaker AK, Ahmed MR, Hossain MA, Sikder MM (2018) Feasibility assessment & design of hybrid renewable energy based electric vehicle charging station in Bangladesh. Sustain Cities Soc 39:189–202. https://doi.org/10.1016/j.scs.2018.02.035

Khan W, Ahmad F, Alam MS (2019) Fast EV charging station integration with grid ensuring optimal and quality power exchange. Eng Sci Technol Int J 22:143–152. https://doi.org/10.1016/j.jestch.2018.08.005

Lee Y, Hur J (2019) A simultaneous approach implementing wind-powered electric vehicle charging stations for charging demand dispersion. Renew Energy 144:172–179. https://doi.org/10.1016/j.renene.2018.11.023

Marino C, Nucara A, Panzera MF, Pietrafesa M, Varano V (2019) Energetic and economic analysis of a stand alone photovoltaic system with hydrogen storage. Renew Energy 142:316–329. https://doi.org/10.1016/j.renene.2019.04.079

Mazzeo D (2019) Solar and wind assisted heat pump to meet the building air conditioning and electric energy demand in the presence of an electric vehicle charging station and battery storage. J Clean Prod 213:1228–1250. https://doi.org/10.1016/j.jclepro.2018.12.212

Mehrjerdi H (2019) Off-grid solar powered charging station for electric and hydrogen vehicles including fuel cell and hydrogen storage. Int J Hydrogen Energy 44:11574–11583. https://doi.org/10.1016/j.ijhydene.2019.03.158

Mozafar MR, Moradi MH, Amini MH (2017) A simultaneous approach for optimal allocation of renewable energy sources and electric vehicle charging stations in smart grids based on improved GA-PSO algorithm. Sustain Cities Soc 32:627–637. https://doi.org/10.1016/J.SCS.2017.05.007

Notton G (2010) Hybrid wind-photovoltaic energy systems. Woodhead Publishing Limited. https://doi.org/10.1533/9781845699628.2.216

Ozlu S, Dincer I (2015) Development and analysis of a solar and wind energy based multigeneration system. Sol Energy 122:1279–1295. https://doi.org/10.1016/J.SOLENER.2015.10.035

Rabbani M, Dincer I, Naterer GF (2012) Thermodynamic assessment of a wind turbine based combined cycle. EGY. 44:321–328. https://doi.org/10.1016/j.energy.2012.06.027

Ramadhani F, Hussain MA, Mokhlis H, Fazly M, Ali JM (2019) Evaluation of solid oxide fuel cell based polygeneration system in residential areas integrating with electric charging and hydrogen fueling stations for vehicles. Appl Energy 238:1373–1388. https://doi.org/10.1016/j.apenergy.2019.01.150

Razipour R, Moghaddas-Tafreshi SM, Farhadi P (2019) Optimal management of electric vehicles in an intelligent parking lot in the presence of hydrogen storage system. J Energy Storage 22:144–152. https://doi.org/10.1016/j.est.2019.02.001

Rui T, Hu C, Li G, Tao J, Shen W (2019) A distributed charging strategy based on day ahead price model for PV-powered electric vehicle charging station. Appl Soft Comput J 76:638–648. https://doi.org/10.1016/j.asoc.2018.09.041

Serra JV (2012) Electric vehicles, 1st edn. Routledge, London. https://doi.org/10.4324/9780203125755

EES (2018) Engineering equations solver (EES)

Tulpule PJ, Marano V, Yurkovich S, Rizzoni G (2013) Economic and environmental impacts of a PV powered workplace parking garage charging station. Appl Energy 108:323–332. https://doi.org/10.1016/j.apenergy.2013.02.068

Wu D, Radhakrishnan N, Huang S (2019) A hierarchical charging control of plug-in electric vehicles with simple flexibility model. Appl Energy 253:113490. https://doi.org/10.1016/j.apenergy.2019.113490

Yang L, Ribberink H (2019) Investigation of the potential to improve DC fast charging station economics by integrating photovoltaic power generation and/or local battery energy storage system. Energy. 167:246–259. https://doi.org/10.1016/j.energy.2018.10.147

Zhang Y, He Y, Wang X, Wang Y, Fang C, Xue H, Fang C (2019) Modeling of fast charging station equipped with energy storage. Glob Energy Interconnect 1:145–152. https://doi.org/10.14171/j.2096-5117.gei.2018.02.006

Community Wind Under the Auctions Model: A Critical Appraisal

Stefan Gsänger and Timo Karl

Abstract As part of a 2.5-year study, the World Wind Energy Association (WWEA) and the Association for Renewable Energy of the German State North Rhine-Westphalia (LEE NRW) assessed the impact on the community wind sector of switching from fixed feed-in tariffs to auctions, with a particular focus on the German state of NRW. This study included a thorough analysis of the framework conditions, along with a series of interviews conducted with the affected community wind actors. In the first year of auctions (2017) the inadequacy of the community energy definition in the EEG became clear, but in that year the auction prices also fell sharply. After restricting the corresponding privileges for community energy in 2018, the share of community energy decreased significantly, and in 2019 actually fell to zero. Yet at the same time, the prices awarded through the auctions rose significantly, and had already risen above the level of the old EEG feed-in tariff (FIT) by the end of 2018. In 2019, the market for new installations collapsed completely, because now only projects that have been successful in an auction are eligible—until the end of 2018, most installed projects were still benefiting from the old fixed FIT. At the same time, participation in the auction rounds has dropped dramatically, to less than one third of the planned expansion volume. From early on, community wind actors have held a very negative assessment of the auction system. They clearly prefer a return to the old FIT system that was open to everyone. In particular, these community energy actors are still very critical of the additional risk and the increasing complexity of the auctions model. In the meantime, the NRW state government has been perceived very negatively, which is mainly related to the planning law-related deterioration in the state. The new state planning rules became effective only in July 2019, but prior anticipation had already led to high levels of uncertainty and dissatisfaction. Thus, from our perspective today, the German Federal Government has missed all three of the goals associated with the introduction of auctions. The set expansion corridor has not been achieved, nor have the auctions improved cost-efficiency, and the diversity of actors has declined greatly since the beginning of 2017, which calls into question the acceptance of the energy transformation as a whole. At the same time, significant obstacles in the field of obtaining permission, particularly from the

S. Gsänger (✉) · T. Karl
World Wind Energy Association (WWEA), Bonn, Germany
e-mail: sg@wwindea.org

T. S. Uyar (ed.), *Accelerating the Transition to a 100% Renewable Energy Era*, Lecture Notes in Energy 74, https://doi.org/10.1007/978-3-030-40738-4_11

areas of air traffic control, military airspace use and species protection, are preventing further expansion of wind energy throughout Germany. New planning restrictions at the state level, including in North Rhine-Westphalia, with its new flat-rate distance requirement of 1,500 m to residential areas and forest exclusion, will further slow the development of on-shore wind energy in the coming years. Given the dramatic nature of the situation, quick and clear countermeasures are urgently required. The following recommendations result from this study:

1. A clear commitment to the full transition to renewable energy with wind energy as a cornerstone and as a fundamental part of an effective climate change mitigation strategy.
2. In accordance with the principle of subsidiarity, a clear recognition of the importance of community energy and its many advantages, as well as a commitment to the creation of framework conditions conducive to the further development of community energy.
3. Including the prioritization of renewable energies in a national climate protection law or in constitutional law at state and federal level.
4. Creation of a non-discriminatory remuneration system beyond auctions, throughout Europe, in accordance with the decisions by the European Court of Justice.
5. Prompt and rapid reduction of bureaucratic barriers and hurdles under planning laws, such as general minimum distances.
6. Strengthening local energy schemes and promoting local and regional approaches to sector coupling.
7. Promotion and further development of prosumer models, as determined at the European level.
8. Promoting cooperation among regional, national and cross-border community energy actors.

1 Introduction

Community energy has been a major success factor and driving force behind the German energy transition ("Energiewende"), involving hundreds of thousands of dedicated citizens and communities across the country.

Although the term Energiewende was already coined in the 1980s, in practice implementation was initially only realized by individual pioneers, by citizens who worked on sustainable energy solutions on site. This was long before politicians recognized the opportunities in such an energy transition, then supporting it first through the Electricity Feed-in Act, and later through the significantly more impactful Renewable Energy Sources Act (EEG).

Community energy has been a cornerstone of the energy transition to date, with 42% of all German renewable energy system ownership being in the hands of citizens and farmers (WWEA 2019). But for some time now the energy transition has stalled, and the few renewable energy projects that are now being implemented are carried out by fewer and fewer players. Many community energy enterprises (CEE) are now facing major problems due to fundamentally changed general conditions, and many are wrestling with alternative business models in an attempt to further develop the energy transition locally.

Given the significance of community energy as a driving force for the growth of renewable energies, the World Wind Energy Association (WWEA) and the Association for Renewable Energy of North-Rhine Westphalia (LEE NRW) have been analysing the potential and opportunities for participation of community wind energy in Germany, with a focus on North Rhine-Westphalia, in a 2.5-year long-term study. This research refers specifically to the transition from the guaranteed feed-in tariff (FIT) of the original EEG to the auction model, which came into effect at the beginning of 2017 following amendments to the EEG. With these amendments, the German legislator explicitly sought to pursue the following goals:

1. Compliance with the approved expansion corridors (2,800 MW per year in the years 2017–2019, 2,900 MW from 2020 onwards)
2. Improved cost efficiency
3. Preservation of the diversity of actors.

Within the framework of the WWEA/LEE NRW project, the study analysed the effects of this paradigmatic shift with regard to participation opportunities, actual successes, and restrictions for community wind actors. In addition, the study examined the significance of planning law, including state regulations, in relation to their impact on the implementation of community wind projects. A previous study, "Headwind and Tailwind for Community Power" (WWEA 2016), showed how community energy actors were already expressing serious concerns in the run-up to the decision on the introduction of auctions. Accordingly, a key objective of this research project was to determine the extent to which subsequent developments showed these fears to have been well founded. In addition, an evaluation and systematisation was undertaken of alternative business models and market access for community energy. The entire analysis and evaluation took place on the basis of a thorough literature and document review. An integral part of the study was a survey undertaken among community wind actors in NRW. In total, more than 50 experts participated in the three rounds of interviews in the years 2017–2019.

The background to this approach is the assumption that the actual situation of community energy can only be captured in a synthesis of a comprehensive legal and planning law analysis on the one hand, and the individual market assessment of the community wind actors on the other hand. At the same time, in view of the size and importance of the state of North Rhine-Westphalia, a degree of transferability of the results to Germany and internationally is assumed, so that general conclusions can be drawn regarding the relationship between the success of community energy and the respective political framework conditions.

In addition to the expert survey and the general analysis, three international community wind symposia took place in November 2017, September 2018 and June

2019. Each of the symposia was attended by around 100 experts from some 20 countries. The main goal of these events was to discuss the research results with a national and international audience, and to improve networking among the community energy actors. The symposia in 2018 and 2019 each concluded with a declaration summarizing the main points of discussion and conclusions supported by numerous German and international community energy organizations (WWEA 2019 and Annex).

On the basis of the expert survey and extended analysis, a study was published at the beginning of 2018, titled "Community Wind in North Rhine-Westphalia" (WWEA 2018), and another in the spring of 2019, "Community Wind in the second year of tenders: A lot of shadow, little light" (WWEA 2019—only available in German), which provide essential information on the status of auctions at the time and their impact on community energy.

2 Development of the Wind Market in Germany and NRW 2017–2019

This chapter outlines the development of the wind market in Germany and North Rhine-Westphalia over the entire project period. As already noted, the development of community energy in Germany has taken place in parallel to the larger wind market development. On the basis of the original EEG and its statutory feed-in tariffs, new onshore wind energy installations in Germany reached a new record in 2017, with 1,792 new wind turbines (WTG) installed, with a capacity of 5,334 MW (Deutsche Windguard 2018). With these additions, wind energy overtook the electricity production of the three primary energy sources: nuclear energy, natural gas and hard coal. The record wind energy share of 20.4% in power generation, reached in 2018, is also entirely attributable to the original feed-in tariff (Fraunhofer ISE 2019). The record level of new installations in 2017 was largely based on proactive behaviour of investors to secure the old FIT before the impending switch to auctions.

In parallel with the new record federal level, NRW also achieved a new installation record in 2017, and at the end of 2017 some 3,630 wind turbines were in operation in the state, with a total installed capacity of 5,449 MW (Deutsche WindGuard 2017). Another 116 plants were added by the end of 2018, representing an additional capacity of 2,402 MW, again reflecting an imperative among investors to develop them prior to the introduction of auctions.

Yet these 2018 figures reflect a wind market that was already declining significantly in anticipated response to the impending auctions model. The gross additions, at 2,402 MW, were well below what had been expected by industry experts. The projects completed in 2018 had been permitted in 2016 under the "old" FIT regime (BWE 2019a). By the first half of 2019 it was clear that the German wind market had in fact collapsed, with only 287 MW of onshore wind capacity built—the lowest value of any year in this century. If we subtract the old WTGs dismantled in this period, the net addition was actually a mere 231 MW. In six federal states, not a single new plant was installed, and in North Rhine-Westphalia, the 43 MW of additional installations was only a fraction of the previous year (Deutsche WindGuard 2019). Since the beginning of 2019, only the projects that won with their auction bid are eligible to receive a remuneration. Accordingly, this result clearly illustrates the very negative impact of the auctions on the entire German wind market.

At the same time, there are significant obstacles in obtaining permissions that are preventing the further expansion of wind energy throughout Germany. For example, an industry survey conducted by FA Wind in July 2019 reveals that more than 1,000 prospective wind turbines (representing 4,800 MW) are currently not approved due to their direct influence on air traffic control, and another 900 turbines (3,600 MW) due to conflicts with military airspace. Another 300 plants (1,200 MW) nationwide are subject to legal actions, with "species protection" being the most common issue. In addition to these barriers related to permitting processes, a number of German states (including North Rhine-Westphalia) have set up new planning restrictions, such as flat-rate distance rules or extensive forest use bans, which will additionally and severely slow down wind energy expansion in the coming years.

3 Community Energy as Driver of the Energy Transformation

The overall circumstance of the energy transition is quite well reflected in the situation of community energy. There are good reasons to suppose that ultimately, the successful implementation of the energy transformation as a whole will depend on the successful integration of community energy. Community energy initiatives are responsible for numerous social, economic and technical innovations within the energy revolution. Overall, community energy has great potential in transforming a carbon-based energy system into a truly sustainable energy system based on decentralized renewable energies.

This is the case precisely because community energy contributes substantially to decentralized, that is to say local, value creation. Added value, the right to participate, and opportunities for participation in turn create the social acceptance that is so urgently needed for the shift towards renewables, since they bring with it practical advantages for local citizens. In addition, through this local added value, such as can be found in North Rhine-Westphalia, community energy makes an important contribution to coping with structural change, especially in the context of the phase out of coal power and coal mines.

3.1 What Is Community Energy?

In order to be clear about its subject matter, this study followed a stringent, internationally recognized definition of "community energy." A few years ago, representatives from all continents within the WWEA Community Energy Working Group agreed on three distinct criteria that define community energy: voting rights, capital shares, and profit distribution. The constituents are thus the majority economic participation of community actors, the voting sovereignty, which must lie with the citizens locally, and the value added, which must remain to a large extent in the region. Along with WWEA, the Community Energy Working Group of the Coalition for Action at IRENA also subscribes to this definition (IRENA 2018) (Fig. 1).

These criteria were repeatedly evaluated in the course of this project's citizen energy symposia, and found high approval among the participating experts and practitioners. Voting limits per shareholder were also discussed as a possible further criterion, but this is difficult to abstract on the international level, and has therefore not been included in the WWEA definition so far. The WWEA criteria by themselves do not constitute a directly legally implementable definition, but they do reflect what practitioners see as the essential elements of community energy, which then have to be interpreted and concretized according to the respective local conditions. So far, there are only a few countries that have adopted a legally valid definition. In Germany, a legal definition of community energy was not necessary in the past, because the feed-in tariff model did not disadvantage any of the participating actors, but gave equal access to all market participants. Before the switch to auctions was implemented, experts surveyed in a previous study (WWEA 2016) emphasized that the guaranteed feed-in tariffs, low bureaucratic requirements, and a wind energy-friendly

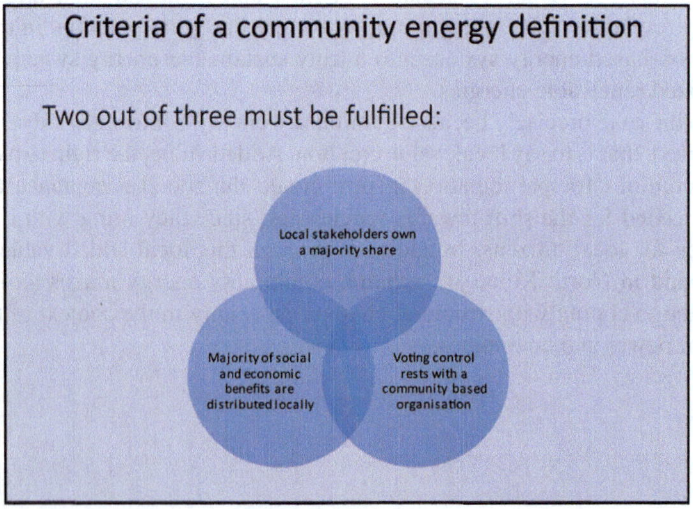

Fig. 1 Three main elements of community power

planning policy in the federal states were the main basis for extensive expansion of wind energy projects in the hands of citizens. The upcoming auctions were regarded with great concern at the time, as the CEE actors assumed that many of the above mentioned factors would be lost as a result of the changeover.

3.2 *Community Energy Definition in the EEG*

As a result of the switch to the auction system, the situation in Germany has changed fundamentally. In order to take account of the massive public concern, legislators decided to include special privileges for community energy in the Renewable Energy Act EEG, which accordingly required drafting a legal definition. This definition of a community energy actor, which remains valid, includes in particular:

- At least ten natural persons as voting shareholders
- At least 51% of the voting rights belong to natural persons resident in the project district
- No shareholder holds more than 10% of the voting rights.

However, as discussed below, this definition was inadequate to capture genuine community energy and has allowed widespread abuse of the associated privileges (see also WWEA 2018). The main reason is that the definition refers only to voting rights instead of shares, and so it neglects to consider who actually contributes the capital of the EEG "community energy" companies.

4 The Transition Towards Auctions: Three Phases

The study was divided into three phases according to the course of the observation period, which corresponds to three calendar years:

- Year 2017: auctions take place for the first time; "community energy" dominates; prices are moving down significantly
- Year 2018: stricter tendering conditions for community energy; substantial increase in awarded prices
- Year 2019: first year exclusive to tendering; collapse of the wind market with a further high surcharge level; community energy is marginalized.

4.1 *Year 2017—Phase 1: "Community Energy" Is Dominating the Auctions, Prices Go Down*

The introduction of the auction model in 2017, together with imprecise assigned privileges, confirmed the concerns expressed in the previous year. The previously balanced market conditions became heavily distorted in favor of financially strong players, as the smaller CEEs could not afford to bear the risks associated with participating in auctions. Indeed, due to the unfortunate EEG definition of community energy, companies that had submitted their projects under the label of community energy won a large number of bids in the first auction round. In the first auction round in 2017, the aggregate allotment for community projects was 96.1%; it was 94.6% in the second round, and 99.3% in the third round (see Table 1 in Chap. 5; detailed analysis can be found in WWEA 2018).

A closer look, however, shows that these projects were practically all initiated by larger developers, which did not meet the community energy criteria agreed by the community energy actors, described above. It is also noticeable that in many cases the bidding project companies only entered the commercial register shortly before the end of the submission deadline of the auction round, and in some cases even after that. In addition most were still at the very beginning of the approval process, which was facilitated by the use of the special rules for community energy according to which all investors fulfilling the legal community wind definition could claim some important privileges:

1. An extended implementation period of four instead of two years,
2. Permissions in accordance with the Federal Immission Control Act (BImSchG) did not have to be submitted when the bid was submitted,
3. Only half of the required financial security had to be provided with the bid, and
4. In the case of a successful participation, the price of the highest bid still awarded was allocated as a price for these projects (standard price procedure).

Despite or perhaps even because of these special regulations, the diversity of actors was reduced considerably under the new conditions. Only a few companies were successful, with the company UKA alone involved in sixty "community projects" with a total volume of more than 1,000 megawatts (MW). This corresponded to more than one third of all approved projects in 2017.

In addition, a large regional imbalance was apparent from the beginning. The southern federal states, particularly, in which, after the abolition of atomic energy, there is a great need to switch on wind energy in the grid, won almost zero bids in the 2017 auction rounds. Bavaria, the largest federal state in terms of area, had only four projects awarded; in Baden-Württemberg not a single bid was successful (WWEA 2018). Such an uneven distribution of electricity generation will ultimately lead to rising costs of the energy transition, since it leads to an additional demand for electricity transport capacities. A targeted expansion of renewable energies, combined with a high degree of system stability, can best be achieved by expanding wind energy generation as widely as possible.

In the state of North Rhine-Westphalia in 2017, development was similar to that seen at the federal level. Essentially five major project developers were behind the project bids in the area of community energy. Only two approved projects went to CEEs with a strong local connection (in the sense of the WWEA definition). Also, just as at the federal level, an unequal regional distribution could be observed. The administrative districts of Münster, Detmold and Arnsberg, which are located in the North of North Rhine-Westphalia, together won 25 bids, while the most densely populated districts of North Rhine-Westphalia, Cologne and Dusseldorf had only one project awarded (WWEA 2018). At the same time, the price level achieved in 2017 was far below what many experts in Germany had considered feasible.

Based on this evaluation of the experiences with the auctions made in 2017, WWEA and LEE NRW formulated several recommendations to the Federal Government and the NRW state government in early 2018:

1. The need for a clear commitment to the goals of the energy transition, the associated expansion of wind energy, and the preservation of the diversity of actors.
2. The introduction of a de minimis regulation in conformity with European law according to which up to three wind turbines are exempted from participating in tenders and continue to receive a guaranteed statutory feed-in tariff.
3. If, instead of the de minimis rule, the government wishes to continue to support community energy through a simplified framework, the definition of community energy should follow the criteria used in this study.
4. A significant increase in the auction volume is required in order to enable the many community wind projects that are far advanced in the project planning to be realized and to prevent "stranded investment".
5. To promote approaches that include a holistic approach to renewable energy, electromobility, energy efficiency and climate adaptation.

4.2 Year 2018—Phase 2: Stricter Tender Conditions for Community Energy, Significant Price Increase

Following on the experience of 2017, however, the legislature not only did not follow the recommendations set out here, but on the contrary, almost completely suspended the exemptions for community energy. As a result, since the beginning of 2018, CEEs can only participate in the auctions once their project has been fully approved under the Federal Immission Control Act. CEEs are now also bound to the standard project implementation period of 30 months. By contrast, CEE projects are still awarded based on the standard price procedure (implementation of the project based on the maximum successful price bid in the auction). Although these restrictions on the special rules for CEEs were initially intended to apply only to the first two rounds of 2018, the German parliament later extended these to 1 June 2020.

By switching to the auctions model, the German Government had been aiming for improvements in the area of compliance with the expansion corridors, the preservation of the diversity of actors and, in particular, cost efficiency. However the expansion corridors were only partially exploited in 2018 due to the lack of planning reliability. Approvals under the Federal Immission Control Act proved to be a particularly large hurdle, especially for community energy projects. This also means that awarded projects, with a total volume of 2,500 MW, are at risk of not being implemented (BWE 2019b). The uncertainties, the burgeoning bureaucracy and the removal of the community energy privileges had a negative effect on the preservation of the variety of actors. The proportion of projects submitted as community energy projects declined sharply in 2018; it was only 6.5% in the August auction, and slightly better, at 16.0%, in the October round (see Table 1).

Through the course of 2018, therefore, the question arose more and more frequently among CEEs as to whether the remaining privilege of the guaranteed maximum price of the auction round would justify the extra effort required to fulfill the legal definition as a "community energy" project—an effort that might not be worthwhile in the absence of significant competitive pressure. And low competitive pressure clearly marked the bidding rounds of the year 2018. Only the first auction of the year, in February, was oversubscribed, and not by a large margin. In the subsequent auctions in 2018, virtually all the eligible bids were awarded contracts (Table 1). At the same time, due to the low competitive pressure, the main economic argument for the introduction of the tendering model was dropped, and the awarded prices rose steadily in the course of the year.

The targeted cost reduction actually turned to an increase in costs. While the average additional value for a so-called reference location in November 2017 was still 3.82 ct/kWh, it increased from 4.73 to 6.26 ct/kWh between February and October 2018 (Bundesnetzagentur 2019). Since most of the projects, even in NRW, are rarely in a statistically average location with 100% of the wind yield, the reference yield model must be taken into account, since the actual tariffs per kWh are sometimes considerably higher depending on the wind yield. For the auctions in February and May, for example, for a 70% location this was 6.81 ct/kWh, and 8.10 ct/kWh, respectively (own calculations, based on data: Bundesnetzagentur 2019, see Fig. 2). In the course of 2018, this level was already well above the remuneration that would have been granted under the pre-2017 EEG feed-in tariffs. Thus it is evident that the goal of cost reduction was clearly missed.

The low competitive pressure that characterized the 2018 tender rounds can be attributed to the fact that many projects were undergoing a re-approval process, based on the experiences of the first auction rounds. That is, the participating companies hoped to be able to obtain better prices following the re-approval process. Also, the evaluation of the auctions showed that from January to August 2018 only 1,081 MW had received a permit according to the Federal Immission Control Act (BWE 2018).

In addition to the recommendations formulated in the 2017 report (see Chap. 4.1), WWEA and LEE NRW called on the federal and state legislators, on the basis of the findings of the year 2018 (WWEA 2019):

Table 1 Onshore wind power auctions in Germany 2017–2019

	May 17	Aug 17	Nov 17	Feb 18	May 18	Aug 18	Oct 18	Feb 19	May 19	Aug 19	Total/ø
Auction volume (MW)	800	1000	1000	700	670	670	670	700	650	650	7510
Submitted volume (MW)	2137	2927	2591	989	604	709	388	499	295	208	11347
Awarded volume (MW)	807	1013	1000	709	604	666	363	476	276	208	6122
Awarded volume for CEEs (MW)	776	958	993	155	113	43	58	92	12	0	3201
Share of CEEs (%)	96.1	94.6	99.3	21.9	18.8	6.5	16.0	19.3	4.2	0.0	52.3
Average awarded price (ct/kWh)	5.71	4.28	3.82	4.73	5.73	6.16	6.26	6.11	6.13	6.20	5.44
Lowest price (ct/kWh)	4.20	3.50	2.20	3.80	4.30	4.00	5.00	5.24	5.40	6.19	4.18
Highest price awarded (ct/kWh)	5.78	4.29	3.82	5.28	6.28	6.30	6.30	6.20	6.20	6.20	5.61

Source Bundesnetzagentur, FA Wind

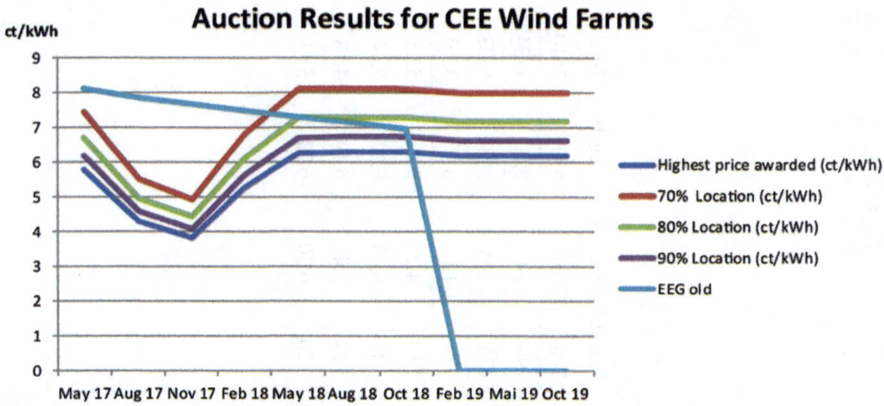

Fig. 2 Highest price awarded. *Source* Authors' calculations, Bundesnetzagentur

- to increase political support for community energy and, above all, to reduce planning restrictions;
- generally, to send confidence-building signals in favour of renewable energy and in favour of wind energy;
- to consider, for community energy in particular, a return to the old remuneration system, as the European Court of Justice has declared admissible.

4.3 *Year 2019—Phase 3: Collapse of the Wind Market, Auction Price Hike, Community Energy is Marginalized*

Thus, while the year 2018 was determined by a sharp decline in the share of community energy and, at the same time, significantly rising prices, the collapse of the German wind market really became clear in the first half of 2019. The first three auction rounds were dramatically undersubscribed; of the 2000 MW tendered, only 1002 MW were awarded, or only 50% of the tendered capacity. In fact, all those bidders who avoided making any formal mistakes can expect to be awarded a contract. While the settled auction price level remained high at 6.1 ct/kWh, or 6.2 ct/kWh for community energy, the formal share of community energy fell to 4.2% in the May auction. In August, not a single bid made use of the community energy privileges (Table 1).

At the same time, in the first half of 2019 there was a dramatic slump in the new installations, which for the first time ever took place only on the basis of auctions won in the previous two years: there was a gross increase of just 86 WTGs with a total of 287 MW, with the net addition only 231 MW (Deutsche WindGuard 2019).

The year-on-year trend that began in 2018, with a small number of approved projects not participating in the auctions, continued during the first round of the

2019 auctions. Projects totaling 499 MW participated, although a total of 1,840 MW would have been eligible (AEE 2019).

Accordingly, the poor situation of the wind industry and of community wind cannot be explained by under-compensation, nor by a lack of suitable land for wind energy use. Rather, the problems lie in the shortcomings of the auction model and the lack of planning security for the wind industry, in particular due to a complicated and increasingly restrictive planning law in many states—including in North Rhine-Westphalia.

5 Summary: All Goals Missed

All in all, it can be said that not one of the three goals issued stated by the Federal Government has been achieved in the nearly three years since the FIT was terminated. The auctions regime has not led to compliance with the expansion corridors, improved cost-efficiency, or preserved the variety of actors participating in wind energy development.

5.1 Compliance with the Installation Targets

Table 1 gives an overview of all auction rounds held between the beginning of 2017 and August 2019. It shows very clearly the significant decline in the bid volumes, from 2,137 MW and 2,972 MW in the first and second auction rounds (May and August 2017) to less than one tenth of that in the ninth round in May 2019.

A total of 6,122 MW have been awarded in ten rounds so far; this represents some 81.5% of the total tendered volume of 7,510 MW. However, particularly with regard to the high number of projects awarded in 2017 under the community energy label, the actual implementation is very uncertain, especially in view of the lack of a BImSchG approval and very low prices. Thus the actual implementation rate may be less than two-thirds.

With the number and volume of projects awarded so far, it is already certain that the installation targets aimed at by the Federal Government will be far undercut. The development of actual installation figures for 2019, presented in Chap. 2, further substantiates this expectation.

5.2 Cost Efficiency

In terms of cost-effectiveness, too, the results of the bidding rounds do not paint a positive picture. Indeed even in 2017, while the awarded prices were fairly moderate, they reached levels that many experts regarded as too low to provide a sound financial basis for development, given the equipment available on the market. However, this was mainly due to the fact that project developers had based their calculations and bids on future turbine technologies and anticipated leaps in efficiency, which made sense, given the extended implementation deadline for community energy projects—up to four and a half years (instead of the regular two and a half years).

With the elimination of most of the community energy privileges, including the introduction of a generally binding implementation period of a maximum of two and a half years after the auction, the price levels awarded increased steadily, reaching values in excess of the statutory tariffs of the old EEG. This only becomes apparent, however, when the theoretical auction prices are converted into actual rates in accordance with the reference yield model, as shown in Fig. 2. In the latest three rounds of 2019, a site with 70% of the average wind yield would receive remuneration of 8.0 ct/kWh over a period of 20 years. In the last three rounds of 2018 the figure was even higher, at 8.1 ct/kWh. By contrast, the highest statutory feed-in tariff for plants commissioned at the beginning of 2018 under the old EEG was 7.49 ct/kWh, and only 6.97 ct/kWh for plants installed between October and the end of 2018.

In addition, the implementation and administration of the auctions still entail considerable costs, above all with the Federal Network Agency, but these are not quantifiable in the context of this study.

5.3 *Preservation of Diversity of Actors*

Under the 2017 FIT rounds, the proportion of projects using the community energy privileges was in the range of 95–99%; this level dropped drastically to less than 20% in early 2018, and actually reached zero in the August 2019 round (Table 1). The original impression of the dominance of community energy has thus reversed, and there can be no talk of successfully preserving a diversity of actors.

The concentration of awarded projects on a few, especially northern, federal states is also detrimental to the variety of actors in terms of geography. After nine rounds of tendering, a corresponding evaluation revealed a strong geographic bias: the five leading states of Brandenburg, Lower Saxony, North Rhine-Westphalia, Mecklenburg-Vorpommern and Schleswig-Holstein could account for 72.9% of the aggregate capacity, while Bavaria, Baden-Württemberg Württemberg, Saxony, the Saarland and Bremen together secured only 7.6% of the total allocated capacity (FA Wind 2019b).

North Rhine-Westphalia received a total of 14.7% of the surcharges in 2017 and 2018, while representing 9.5% of Germany's area and 21.6% of its population. Beyond this, the results of the auctions in NRW have two main results:

- On the one hand, as on the federal level, regional disparities exist, and the awarded projects are centered in only a few districts in NRW. Of the 193 wind turbines contracted in 2017 and 2018, 100 were allocated to the rural districts of Hochsauerlandkreis, Paderborn, Minden-Lübbecke and Borken (40, 21, 20 and 19 wind turbines) (FA Wind 2019a).
- On the other hand, it is noteworthy that the bids that were successful in the bidding process can essentially be attributed to four companies that have already been active in NRW for a long time. These few North Rhine-Westphalian companies are joined by a few farmers and small engineering firms (EA.NRW 2017).

In NRW and nationwide, generally the small, new cooperatives and CEEs are no longer participating with new projects, or are not even being founded, since for them the market risks are too high. Nationwide there has been a dramatic decline in the number of newly founded energy cooperatives, from a peak of 167 start-ups in 2011, to just 14 in 2018 (DGRV 2018).

Relevant experts also confirmed the reasons given in this study at a symposium in 2019. The critical mass of different regional project sponsors, who are at the same time also pursuing smaller projects with lower investments on their own initiative, is therefore declining, which will have a negative impact on the sector's expansion in the medium to long term.

6 Planning Law

In addition to the problems caused by the auction system, the study also identified changed planning law requirements, especially at the state level, as a further major regressive barrier to new wind power installations. Of note is the so called 10 h regulation in Bavaria, according to which wind turbines must have a minimum distance of 10 times their height to residential buildings in areas with development plans. With current wind turbine heights often reaching 200 m, this means a minimum distance to the next residential area of 2,000 m. As a consequence of this, wind power expansion in Bavaria has practically come to a standstill. In Schleswig-Holstein, the state government even imposed a moratorium, so that with the exception of developers who are willing to undertake a complicated exemption procedure, the further expansion of wind power can be ruled out until the creation of the new regional plan in 2020.

In North Rhine-Westphalia, too, the state government has created a great deal of uncertainty with its changes in the State Development Plan (LEP), which in principle should include a flat minimum distance of 1,500 m to residential areas and a widespread ban on wind turbines in forests. Permitting for wind energy in commercial forests is not strictly speaking impossible, but it has become immensely more difficult. However, since municipalities must provide sufficient space for wind turbines in accordance with federal law, the risk of incorrect planning, lawsuits or legal disputes also increases. According to LEE NRW's forecasts in North Rhine-Westphalia, more than 80% of wind energy potential in the state will be excluded in the future due to the general minimum distances and the extensive ban on the use of commercial forests for wind energy. The results of a potential analysis by the state government, which has sought to investigate the effects of the area restrictions and to measure the remaining area, are still pending.

In the present study, it was possible to show that the intended changes to the LEP led to considerable uncertainty and confusion among planners and municipalities. While some municipalities impose planning moratoria for fear of lawsuits, others proactively intervene in the proposed applications in order to not be ignored in the planning process. These inconsistencies are difficult to convey to citizens who are not fully knowledgeable. In addition, the proposed LEP could be the basis for a large number of legal actions, as investors may be denied building rights only in the case of higher-ranking interests such as species protection or air traffic control, but not because of a flat-rate distance regulation. Instead of installing new wind turbines, the NRW Ministry of Economic Affairs plans in the future to replace older turbines with more powerful wind turbines, what is called "repowering"—regardless of the fact that for the repowering of old plants, new permits under the Federal Immission Control Act still need to be obtained.

The LEP regulations also appear contradictory against the background that the NRW state government recently adopted. This entails an energy strategy in which the stated goal is to almost double the wind energy capacity in the country, from today's 5,800 MW to 10,500 MW by 2030. This target does not accord with the limitations of the LEP in the least, since the necessary planning process of the municipalities—including substantial allocation of land with the new specifications of the LEP—is hardly possible.

7 Results of the Survey

The survey of key actors in the field of community wind energy gives a dramatic picture of the developments outlined here. In total, more than 50 community energy experts and practitioners from North Rhine-Westphalia were interviewed by means of an online questionnaire in each of the three years of the study, and in personal interviews in 2017 and 2018. A strong majority of the community energy experts assessed the switch to the auction system negatively—together with increasing problems related to permission processes, this policy shift has been identified as the main cause of the wind energy crisis in Germany.

7.1 General Assessment of Auctions

At the end of the observation period in the first half of 2019, 45.5% of respondents assessed the changeover to auctions as somewhat negative, and 27.3% as very negative. When asked about the key hurdles for community wind players in the auction

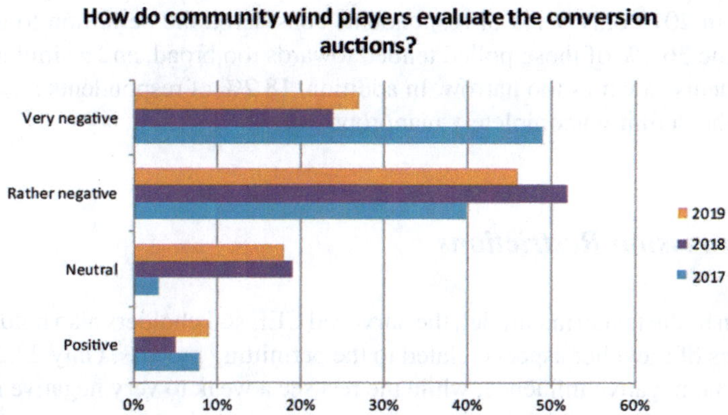

Fig. 3 Auctions evaluation by community wind players

process, 100% cited concerns about the general risk of not winning the bid, while 89% saw the increasing complexity of the process as negative or very negative. This shows that after the auctions were launched, the fears expressed in advance appeared to hold true: in 2015, 95% of respondents considered increasing complexity and 90% saw the additional risk of a failed bid as negative or very negative (WWEA 2016).

The fear of penalties and withdrawal of the awarded contract also played a role, but the negativity was not perceived as widely as with the previous factors. In 2019, 44% of respondents assessed this as negative or very negative, compared with 70% in 2016.

In the survey it became also clear that those projects that are planned at all, are all 3–4 MW or larger. Small projects are no longer lucrative since the implementation in the tendering system.

In the overall review of the auctions it must also be said that the actual situation of the industry is worse than the official figures indicate at first glance. The number of visible permits is deceptive, because all of the approved projects are listed. However, a considerable portion of those permits are either in a reassignment procedure or delayed in a legal case, and thus cannot be implemented.

Meanwhile, 72.7% of the CEEs surveyed now see themselves in increased competition with project developers, while only 18.2% expect more cooperation between CEE and project developers (Fig. 3).

7.2 About the Community Energy Definition

Concerning the community energy definition in the EEG, the expressed opinions remain very divided, although in summary they are equally negative. In 2016, 36% of respondents considered the definition appropriate, while 24% saw it as too broad and 40% understood it too narrow. This mixed picture remained through to 2019.

However, in 2019 only 9.1% of respondents considered the definition to be appropriate, while 36.4% of those polled tended towards too broad, and a similar number of respondents saw it as too narrow. In addition, 18.2% of respondents now say that they find the definition completely inappropriate.

7.3 Permission Restrictions

In addition to the tendering model, the surveyed CEE stakeholders also hold negative perceptions of the other aspects related to the permitting process. Only 22.2% of the actors see no negative influence, while the rest see a weak to very negative influence and understand these permitting issues as genuine obstacles.

7.4 Importance of New Business Fields

As already suggested by the document analysis in the study, the complexity of the tender process has encouraged many actors to look into the idea of becoming active in business areas other than wind energy. This was being considered by some 36% of the interviewed stakeholders, while another 27% had not yet decided on it. A clear development can be observed over time. In the first survey in 2017, shortly after the launch of the auction model, 80% of respondents said that they could not imagine being active in other business areas (Fig. 4).

The business segments that were rated most interesting were the self-supply of renewable energies, regenerative heat supply, and electric mobility. Regarding self-supply, 60% of respondents said they were already active in this area, while 40% considered it to be potentially of interest. This was followed closely by the energy

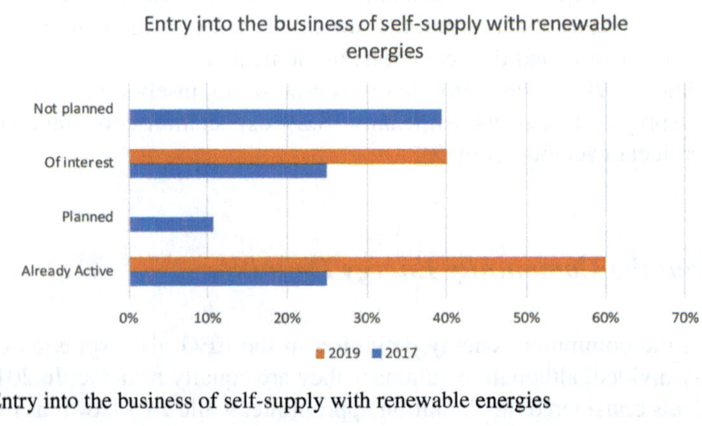

Fig. 4 Entry into the business of self-supply with renewable energies

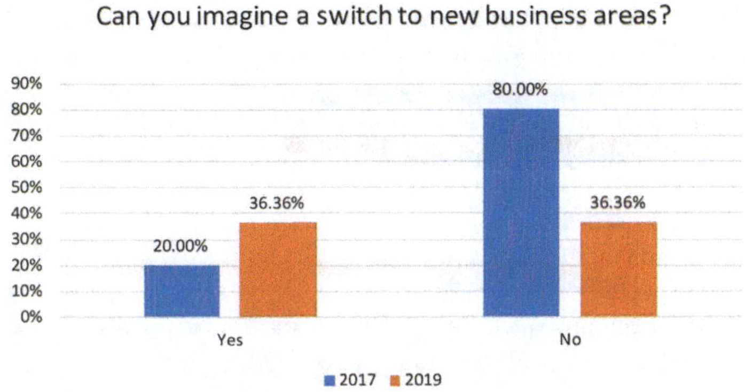

Fig. 5 Switch to new business areas

efficiency market segment, in which 50% of the respondents are already active and the other 50% described the market segment as being of interest.

Comparing the data from 2017 and 2019, it is notable that the CEE players are now increasingly turning to alternative business areas, due to the difficult conditions on the wind market. The strongest growth is in the area of electromobility: While in 2017 only 22% of respondents said they would be active in the segment, by 2019 that portion had reached 60%. In the area of self-supply, the increase from 25 to 60% is somewhat lower, but still substantial, as the level of activity is more than twice as high as two years earlier (Fig. 5).

7.5 Assessment of NRW State Policies

The consequence of the permitting-related deterioration in North Rhine-Westphalia described above is a massive dissatisfaction among wind energy stakeholders with the state government. In 2018 and 2019, the community wind experts were asked for an assessment of the state government—in 2018, in relation to the state's wind energy policy, and in 2019 to energy policy in general. Since the two topics are very closely related to each other, the two survey results can be compared directly (Fig. 6).

While in the year 2018 5% of the respondents were satisfied, at least 40% expressed dissatisfaction and 50% were very dissatisfied with the wind energy policy in NRW. Only one year later, 91% of the respondents were very dissatisfied with the energy policy of the state government, while a further 9% of the respondents were dissatisfied; none gave a positive or neutral assessment.

Here, too, the development of opinion over time must be considered. In a previous survey in 2017, 47.8% stated they were satisfied with the support offered by the state of North Rhine-Westphalia; at that time only 4.6%, and 14.3%, respectively, were very dissatisfied or dissatisfied with the state development plan and the guidelines on wind energy in forests (Fig. 7).

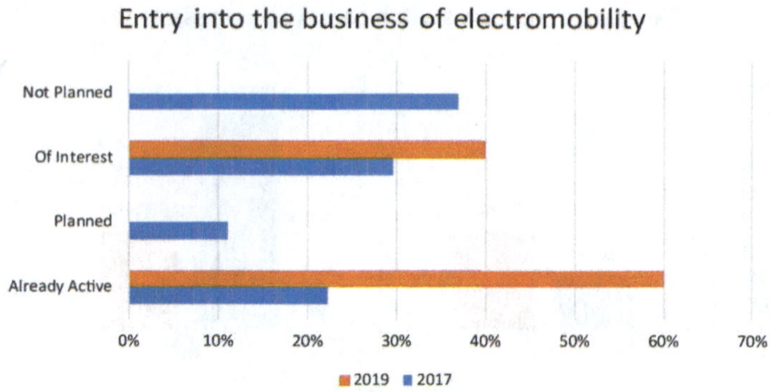

Fig. 6 Entry into the business of electromobility

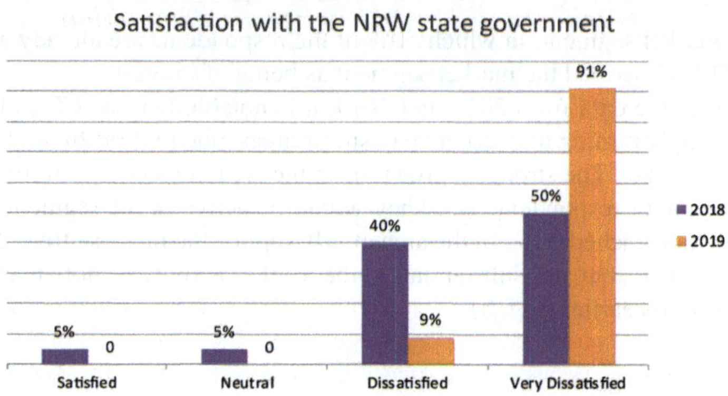

Fig. 7 Satisfaction with the NRW state government

In contrast to this development, however, it is worth mentioning that in 2019, 54% of the respondents rated the support from the district administration and the municipality as very positive or positive, which shows that the mood toward public actors and authorities is by no means consistently negative.

8 International Community Wind Symposium and Community Power Forum

Another instrument used to verify the results of the research and to capture the sector's sentiment was the annual international symposia. The symposia held in 2015, 2017 and 2018 have been described in some previous publications (WWEA 2016, 2018, 2019).

At the fourth International Community Wind Symposium and Community Power Forum in May 2019, the preliminary results of this research project were presented and initial conclusions were drawn. Participants at the symposium again showed an almost unanimous rejection of the current auctions model. The discussion brought out a range of proposals, from reforming the tendering system to returning to the previous EEG, as well as the establishment of a new market model outside the EEG remuneration system. Representatives of small CEEs especially emphasized that under the prevailing conditions it would be virtually impossible to launch new community energy enterprises such as cooperatives. There was a very broad consensus about the urgent need to remove the current bureaucratic obstacles to CEE activities.

The current problems around the acceptance of wind energy on land were another point of discussion. It was stated that acceptance is usually lower where wind turbines are built for the first time, as there are many biases against wind energy. A new campaign for acceptance involving the entire wind industry was proposed, especially given that the expansion of decentralized renewable energies could make major parts of the planned network extension obsolete. At the same time, it was repeatedly stated that federal as well as state policies should give renewable energies much more support and thus increase public acceptance.

However, as there are currently massive problems in the wind sector that cannot be solved in the short-term, many participants at the symposia deliberated on which market segments and which business models CEEs can profitably and meaningfully enter beyond public power purchase agreements. Such community projects make sense not only in the electricity sector, as these approaches combine participatory possibilities for citizens with environmental benefits. In addition, rural structures can be strengthened in case of successful implementation. In addition to the indisputable importance of legislation supporting a decentralized energy transition, then, the symposia stressed the importance of a social vision and a guiding positive narrative, one that highlights the opportunities and benefits of energy in the hands of citizens.

Much hope was therefore expressed regarding the Clean Energy Package of the European Union, which also contains a new Renewable Energy Directive. In this new legal framework, community energy and the role of the "prosumer" should be enhanced, for example through protection against discriminatory levies that prevent participation in the energy transition. The right to generate, store, consume and sell renewable energy without disproportionately large burdens is thus intended to benefit all EU citizens. Numerous CEEs are hoping for a better starting position in terms of self-generation and the associated direct marketing following national implementation of the Directive.

Many panelists and participants also expressed hope for greater public support from the ever-growing public protest movement for climate protection, notably through Fridays for Future. The discussions triggered by these regular protests have the potential to spur meaningful legislative decisions, such as a Climate Protection Act, or the inclusion of climate protection or prioritization of renewable energy in constitutional law.

Moreover, and beyond national borders, the possibilities and prospects of international cooperation among community energy actors were also discussed with international representatives. Concrete examples were presented, and participants discussed the conditions that would need to be fulfilled so that cooperation, especially between developing and industrialized countries, can be fruitful.

The "Bonn Community Power Declaration", adopted during the 4th symposium in 2019, can be found in the annex.

9 Conclusions and Recommendations

The wind energy industry and community energy especially are in a deep, politically induced crisis. There is an urgent need for confidence-building measures for climate protection, for renewable energy and for wind energy—as happened in the aftermath of the Fukushima nuclear disaster. The public discussion surrounding the urgency of effective measures against climate change offers a hitherto unprecedented opportunity, in which the importance of renewable energies and also of community energy has not yet been adequately addressed.

Building on the recommendations already made in the previous study, WWEA and LEE NRW recommend the following measures:

1. A clear commitment to the full transition to renewable energy with wind energy as a cornerstone and as a fundamental part of an effective climate change mitigation strategy.
2. In accordance with the principle of subsidiarity, a clear recognition of the importance of community energy and its many advantages, as well as a commitment to the creation of framework conditions conducive to the further development of community energy.
3. Including the prioritization of renewable energies in a national climate protection law or in constitutional law at state and federal level.
4. Creation of a non-discriminatory remuneration system beyond auctions, throughout Europe, in accordance with the decisions by the European Court of Justice.
5. Prompt and rapid reduction of bureaucratic barriers and hurdles under planning laws, such as general minimum distances.
6. Strengthening local energy schemes and promoting local and regional approaches to sectoral coupling.
7. Promotion and further development of prosumer models as determined at the European level.
8. Promoting cooperation among community energy actors, regional, national and cross-border.

Annex

Bonn Community Power Declaration

We, the participants of the 4th International Community Wind Symposium and Community Power Forum, have gathered in Bonn on 28 and 29 May under the theme "Shaping the Energy Transition – Strength through Alliances".

Building on the predecessor events and in particular on the Fukushima Community Power Declaration, the Community Power for All! resolution and the Bamako Community Power Declaration, we underline the urgency of a rapid switch to a renewable energy future, based on a fair and equal distribution of wealth and prosperity.

Every day, we understand more the urging pressure to act on climate change and to counter the growing disparity amongst people around the world. A 100% renewable energy supply and community power are the primary answers to these two challenges.

At the same time, we notice with growing concern that there are increasing barriers against the rapid growth of renewable energy globally, e.g. through the introduction of auctions in the renewable power sector which represents an insurmountable hurdle, together with additional barriers in form of restrictive permission rules and manifold other forms of discrimination and exclusion from fair market access.

In spite of the growing difficulties, we notice an encouraging growth of the renewable energy and community power movement around the world, we highly appreciate that community power has become a topic on national and international agenda and we welcome the growing international community power networking and international solidarity of the community power community.

In light of the urgency of the situation, we call on decision makers on all levels of society:

- to prioritise renewable energy and community power as cornerstones of a sustainable world
- to recognise the utilisation of renewable energy in all its forms, including for self-consumption, as a basic human right
- accordingly, to remove all barriers against the utilisation of renewable energy and against community ownership models, in particular counterproductive policies such as auctions
- to enable communities around the world to play an active role and to work together
- to amend basic legislation and include the basic right to use renewable energy and the priority for renewable energy in the overarching legal frameworks, including climate change legislation and constitutional law.

Bonn, 29 May 2019

- World Wind Energy Association
- Landesverband Erneuerbare Energien LEE NRW
- Bündnis BürgerEnergie

- Global100RE
- International Geothermal Association
- ICLEI
- Institute for Sustainable Energy Policies ISEP Japan
- Rescoop
- Bergische BürgerEnergie Genossenschaft
- Danish Small Wind Energy Association
- DIEM25
- Energiegenossenschaft Starkenburg
- EnergiE zum Leben, auch in Wuppertal-Nord
- EW-Nord
- FridaysForFuture Dithmarschen
- Nordic Folkecenter for Renewable Energy
- Solarverein Goldene Meile.

References

AEE (Agentur für Erneuerbare Energien) (15 Feb 2019) Mangelnde Teilnahme an Ausschreibungen zeigt hohe Verunsicherung der Branche, online abrufbar unter: https://unendlich-viel-energie.de/erneuerbare-energie/wind/onshore/mangelndeteilnahme-an-ausschreibungen-zeigt-hohe-verunsicherung-der-branche (11 Jul 2019)

Bundesnetzagentur (2019) Beendete Ausschreibungen Windenergieanlagen an Land, online: https://www.bundesnetzagentur.de/DE/Sachgebiete/ElektrizitaetundGas/Unternehmen_Institutionen/Ausschreibungen/Wind_Onshore/BeendeteAusschreibungen/BeendeteAusschreibungen_node.html (11 Jul 2019)

BWE (Bundesverband WindEnergie) (11 Oct 2018): Genehmigungsstau blockiert Energiewende, online: https://www.wind-energie.de/presse/pressemitteilungen/detail/genehmigungsstau-blockiert-energiewende/ (11 Jul 2019)

BWE (Bundesverband WindEnergie) (29 Jan 2019a) Ausbauzahlen für das Gesamtjahr 2018 in Deutschland: Windenergie an Land – Zubau bricht stark ein, Mittel- und Langfristperspektive muss jetzt gesetzlich fixiert werden, online abrufbar unter: https://www.wind-energie.de/presse/pressemitteilungen/detail/ausbauzahlen-fuer-das-gesamtjahr-2018-in-deutschland-windenergie-an-land-zubau-bricht-stark-ein-m/ (11 Jul 2019)

BWE (Bundesverband WindEnergie) (30 Apr 2019b) Quartalszahlen für Windenergiezubau an Land bedrohlich – Politische Entscheidungen dringend erforderlich, online abrufbar unter: https://www.wind-energie.de/presse/pressemitteilungen/detail/quartalszahlen-fuer-windenergiezubau-an-land-bedrohlich-politische-entscheidungen-dringend-erforder/ (11 Jul 2019)

Deutsche WindGuard GmbH (2017) Status des Windenergieausbaus an Land in Deutschland, online: https://www.wind-energie.de/fileadmin/redaktion/dokumente/publikationen-oeffentlich/themen/06-zahlen-und-fakten/20180125_factsheet_status_windenergieausbau_an_land_2017.pdf (11 Jul 2019)

Deutsche WindGuard GmbH (2018) Status des Windenergieausbaus an Land in Deutschland 1. Halbjahr 2018, online: https://www.windguard.de/veroeffentlichungen.html?file=files/cto_layout/img/unternehmen/veroeffentlichungen/2018/Status%20des%20Offshore%20Windenergieausbaus%20in%20Deutschland%2C%201.%20Halbjahr%202018.pdf (11 Jul 2019)

Deutsche WindGuard GmbH (2019) Status des Windenergieausbaus an Land in Deutschland 1. Halbjahr 2019, online: https://www.wind-energie.de/fileadmin/redaktion/dokumente/publikationen-oeffentlich/themen/06-zahlen-und-fakten/20190725_Factsheet_Status_des_Windenergieausbaus_an_Land_-_Halbjahr_2019.pdf (7 Aug 2019)

DGRV (Deutscher Genossenschafts- und Raiffeisenverband) (2018) Energy cooperatives. Results of the DGRV-Survey 2018. Umfrageergebnisse, online https://www.genossenschaften.de/dgrv-annual-survey-energy-cooperatives

EA.NRW (EnergieAgentur.NRW) (2017) Bürgerenergie Atlas, online abrufbar unter: https://www.energieagentur.nrw/tool/buergerenergie/liste.php? (11 Jul 2019)

FA Wind (Fachagentur Windenergie an Land) (2017a) 1. Ausschreibung für Windenergieanlagen an Land (Juni 2017), online abrufbar unter: https://www.fachagentur-windenergie.de/fileadmin/files/Veroeffentlichungen/FA_Wind_Analyse_1_Ausschreibung_Wind_an_Land_2017.pdf (11 Jul 2019)

FA Wind (2017b) 2. Ausschreibung für Windenergieanlagen an Land (September 2017), online abrufbar unter: https://www.fachagentur-windenergie.de/fileadmin/files/Veroeffentlichungen/FA_Wind_Analyse_2_Ausschreibung_Wind_an_Land_2017.pdf (11 Jul 2019)

FA Wind (2017c) 3. Ausschreibung für Windenergieanlagen an Land (Dezember 2017), online abrufbar unter: https://www.fachagentur-windenergie.de/fileadmin/files/Veroeffentlichungen/FA_Wind_Analyse_3_Ausschreibung_Wind_an_Land_2017.pdf (11 Jul 2019)

FA Wind (2018) 6. Ausschreibung für Windenergieanlagen an Land (September 2018), online abrufbar unter: https://www.fachagentur-windenergie.de/fileadmin/files/Veroeffentlichungen/FA_Wind_Analyse_6_Ausschreibung_Wind_an_Land.pdf (11 Jul 2019)

FA Wind (2019a) Ausbausituation der Windenergie an Land im Jahr 2018, online abrufbar unter: https://www.fachagentur-windenergie.de/fileadmin/files/Veroeffentlichungen/FA_Wind_Zubauanalyse_Wind-an-Land_2018.pdf (11 Jul 2019)

FA Wind (2019b) 9. Ausschreibung für Windenergieanlagen an Land (Mai 2019), online abrufbar unter: https://www.fachagentur-windenergie.de/fileadmin/files/Veroeffentlichungen/Analysen/FA_Wind_Analyse_9_Ausschreibung_Wind_an_Land.pdf (7 Aug 2019)

Fraunhofer ISE (13 Mar 2019) Energy Charts - Nettostromerzeugung in Deutschland in 2018, online abrufbar unter: https://www.energy-charts.de/energy_pie_de.htm?year=2018 (2 Apr 2019)

IRENA Coalition for Action (2018) Community energy. Broadening the Ownership of Renewable Energy, online abrufbar unter: https://irena.org/-/media/Files/IRENA/Agency/Articles/2018/Jan/Coalition-for-Action_Community-Energy_2018.pdf

Landtag NRW (17 Apr 2019) Zukunft von Windkraftanlagen – Wie plant die Landesregierung?, Antwort der Landesregierung auf die Kleine Anfrage 2172 vom 15. März 2019 des Abgeordneten Andreas Keith, AfD, Drucksache 17/5497, online abrufbar unter: https://www.landtag.nrw.de/portal/WWW/dokumentenarchiv/Dokument/MMD17-5857.pdf (11 Jul 2019)

LEE NRW (o.D.) Reden und Handeln der NRW-Landesregierung passen nicht zusammen, online: https://www.lee-nrw.de/reden-und-handeln-der-nrw-landesregierung-passen-nicht-zusammen/ (11 Jul 2019)

WWEA (World Wind Energy Association) (2016) Headwind and Tailwind for Community Power. Community Wind Perspectives from North Rhine-Westphalia and the World. WWEA Policy Paper Series (PP-01-16), online: https://www.wwindea.org/download/community_power/Community_Wind_NRW.pdf

WWEA (2018) Community Wind in North Rhine-Westphalia. Perspectives from State, Federal and Global Level. WWEA Policy Paper Series (PP-01-18), online: https://www.wwindea.org/wp-content/uploads/2018/02/CP_Study_English_reduced.pdf

WWEA (2019) Bürgerwind im zweiten Jahr der Ausschreibungen: Viel Schatten, wenig Licht. WWEA Policy Paper Series (PP-01-19), online: https://wwindea.org/blog/2019/05/27/new-study-proves-community-power-is-increasingly-being-marginalised/#

100% Renewable Energy Generation with Integrated Solar Energy Systems

Ayhan Atiz and Mehmet Karakilcik

Abstract In this study, heat, electricity and hydrogen production performances of integrated solar energy systems producing 100% renewable energy were investigated. For this purpose, energy analyzes of the systems were performed to evaluate the systems in terms of usability. Two types of integrated systems were used in this analyzes. The first system is consisted of a solar pond (SP) integrated with an Organic Rankine Cycle (ORC). The second is comprised of a solar pond integrated with evacuated tubes collectors (EVTCs) and the ORC that works with organic liquids. The analysis of these systems was performed by Engineering Equation Solver (EES) under solar energy, which is known as one of the 100% renewable energy sources. With ORC, low and medium temperature hot water sources have been used to generate electricity in important areas. When the low temperature of the water is increased with solar collectors, electricity production efficiency is increased. Furthermore, the performance of ORC in the system was evaluated in terms of thermodynamics. As a result, it was found that integrated systems can produce thermal energy, electricity and hydrogen efficiently from one hundred percent renewable sources.

Keywords Solar energy · Solar pond · Solar collectors · Organic Rankine Cycle · Energy analyses

1 Solar Pond

Due to the increasing electricity consumption of countries, electricity is produced from many different sources in the world. Most of the electricity produced today is produced using fossil fuels. However, fossil fuels are damaging the environment. Therefore, electricity generation should be produced by both clean and renewable

A. Atiz
Department of Mathematics and Science Education, Faculty of Education, Alanya Alaaddin Keykubat University, Alanya, Antalya 07400, Turkey

M. Karakilcik (✉)
Department of Physics, Faculty of Sciences and Letters, Cukurova University, Adana 01330, Turkey
e-mail: kkilcik@cu.edu.tr

T. S. Uyar (ed.), *Accelerating the Transition to a 100% Renewable Energy Era*, Lecture Notes in Energy 74, https://doi.org/10.1007/978-3-030-40738-4_12

energy sources. Solar energy is one of the largest clean and renewable energy sources (Şahin and Okumuş 2013). Solar ponds (SPs) are one of the important systems that use the solar energy to convert as thermal energy. SPs are one of the best heat storage systems. The temperature of each layer of the solar pond depends on the penetrating solar radiation, layer thicknesses, shading areas of the zones and heat losses. Changing layer thickness increases the performance and stability of the solar pond (Karakilcik et al. 2006). Due to the heat storage capacity of SPs, it is possible to extract the heat from this system. The SPs can be used as a sustainable source of heat energy. This aim is possible to carry out as experimentally and theoretically (Abdullah et al. 2016). The SP occurs from three layers; upper convective zone (UCZ), non-convective zone (NCZ) and heat storage zone (HSZ). If thermal energy is extracted from the NCZ and HSZ, the stored thermal energy in the SP is used more efficiently (Date et al. 2013). Many studies have shown that electricity is produced by utilizing heat capacity in the HSZ. In one of these studies, the amount of electricity generation has increased up to 5000 MW (Ding et al. 2018).

The best time to use the storage area of the SP is in the summer. Because the storage temperature of the SP is the highest in this season than other seasons (Karakilcik et al. 2013). A Solar pond not only generate heat, but also generate electricity. Figure 1 shows the components of a system that generates electricity from a SP. This system is simulated for June, July, August and September. The system includes SP, heat exchanger (HE), pumps and ORC. One of the most important components of this system is the SP. Because the SP stores a significant part of the reaching solar energy in the HSZ. The dimensions of the solar pond in the system are (1.5 m × 8m × 8 m) 96 m^3. The SP consists of UCZ, NCZ and HSZ. UCZ is a homogeneous layer that occurs from fresh water. The thickness of this zone is 0.1 m. The density of the NCZ is increasing towards the HSZ. This zone is a thermal insulation zone which prevents the heat in the storage zone from solar pond to the environment. The thickness of the NCZ is 0.5 m. The HSZ is the inner region of the solar pond where solar energy is stored as thermal energy and its thickness is 0.9 m (Bozkurt and Karakilcik 2012).

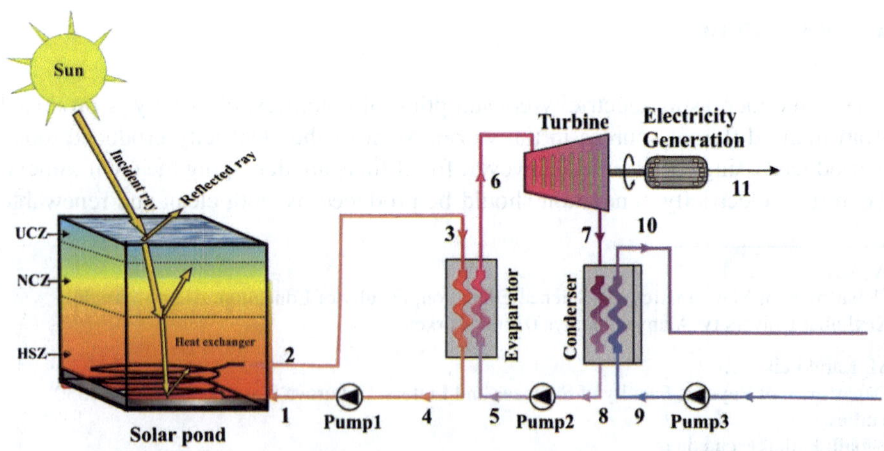

Fig. 1 Solar pond integrated with ORC

In this system, cold water is pumped to the storage area of the SP via pump1. This water is heated by the HSZ via the HE and pumped to the ORC. Isobutane is used as working fluid in ORC. The isobutene fluid, which uses the hot water coming to the ORC, is heated in the evaporator by means of pump2 and transferred to the turbine. Thus, the turbine rotates with steam pressure and electricityis generated by the generator. The cycle in the system continues like this, (Atiz and Karakilcik 2018).

Some assumptions were accepted in this system:

- For each month, 90% of the thermal energy in the HSZ was transferred to the water by utilizing heat exchanger.
- It is assumed that the SP is very well insulated and that the temperature of the HSZ does not change when the sun is available.
- Environment temperature is accepted from the meteorology station.
- In this study, the shading effect on the yield of the SP is ignored. The yields of the HSZ were taken as 29.03%, 31.65%, 37.25% and 35.07% for June, July, August and September, respectively (Karakilcik et al. 2006).

The solar pond converts the solar radiation into thermal energy and stores in the HSZ. The stored thermal energy can be transformed to the electrical energy by different methods. As shown in Fig. 1, the solar pond occurs three zones: UCZ, NCZ and HSZ. Density of the UCZ almost equals to fresh water. NCZ is occurred from different density salty layers. The density of the layers in the NCZ is increased from the end of the UCZ to starting of the HSZ. These saline layers prevent to pass the thermal energy from the SP to the environment. Thanks to this system, a significant amount of thermal energy remains in the HSZ. The stored thermal in the HSZ is obtained from solar radiation that is attenuated by the layers of the UCZ and NCZ and shading effect of sidewalls (Karakilcik et al. 2013). The remaining of the attenuated solar radiation in the HSZ is found as follows:

$$Q_{s,HSZ} = \dot{Q}_{s,sp}\beta[(1 - F)h(\Delta x - \delta)] \tag{1}$$

where δ is the thickness of the layer of the UCZ that absorbs long-wave solar radiation, F is the absorbed solar radiation fraction at a zone of δ.

When energy analyses of SP are found, the solar radiation reaching on the surface of the SP surface must be computed necessarily. The solar radiation can be given as:

$$\dot{Q}_{s,sp} = \dot{E}_{net} A_{HSZ} \tag{2}$$

where $\dot{E}_{net}$ is the incident solar radiation and A_{HSZ} is the total surface area of the SP. Some of the solar radiation reaching the surface of SP is reflected from surface of the SP and another remaining part of solar radiation is absorbed by the zone of the SP. Thus, the solar radiation passes from inside to the zones and decreases and arrives to the HSZ. β is the part of the incident solar radiation and is found by (Ding et al. 2016):

$$\beta = 1 - \frac{1}{2}\left[\frac{\sin^2(\theta_i - \theta_r)}{\sin^2(\theta_i + \theta_r)} + \frac{\tan^2(\theta_i - \theta_r)}{\tan^2(\theta_i + \theta_r)}\right] \quad (3)$$

where θ_i and θ_r are the incidence and refraction angles and h is the rate of the solar radiation at the deepness of x and is calculated by (Bryant and Colbeck 1977):

$$h = 0.36 - 0.08 \ln\left(\frac{X}{\cos\theta_r}\right) \quad (4)$$

1.1 Heat Exchanger

Heat exchanger (HE) is the device that operates for heat transferring between two fluids at different temperature without mixing. The HE is generally worked for heat transferring systems, air conditioner, power plants and chemical processes. The mixing of the two fluids is usually prevented by a pipe that have good thermal conduction during the heat transferring (Cengel 2006). The HE is used in the SPs for extracting thermal energy in the HSZ and NCZ. As shown in Fig. 1, a lot of thermal energy was pumped from the HSZ to ORC by employing the HE in the SP. The amount of thermal energy is found as given:

$$\dot{Q} = UA_{sur}\Delta T_{lm} \quad (5)$$

where U is the thermal conduction coefficient of the HE, A_{sur} is the total surface of the HE and ΔT_{lm} is the logarithmic mean temperature difference given as follows:

$$\Delta T_{lm} = \frac{\Delta T_1 - \Delta T_2}{\ln\left(\frac{\Delta T_1}{\Delta T_2}\right)} \quad (6)$$

where ΔT_1 and ΔT_2 are the temperature differences between the HSZ and input and output temperature of water of the HE that are found as follows:

$$\Delta T_1 = T_{HSZ} - T_{exc,out} \quad (7)$$

$$\Delta T_2 = T_{HSZ} - T_{exc,in} \quad (8)$$

where T_{HSZ} is the temperature of the HSZ. $T_{exc,in}$ and $T_{exc,out}$ are the temperature of the input and output water of the HE, respectively. Thus, the thermal energy extracted from the HSZ is found as follows:

$$\dot{Q} = \dot{m}_w c_p\left[T_{exc,out} - T_{exc,in}\right] \quad (10)$$

In June, July, August and September, temperature of the HSZ can rise above 50 °C even if shading is high (Bozkurt and Karakilcik 2015).

1.2 Energy Analysis of ORC integrated with Solar Pond

Generally an ORC consists of evaporator, turbine and condenser that is working by the thermal energy. ORC contains an organic fluid which can be vaporized at low temperature in the evaporator and produces a mechanical power by expanding it. Electricity is generated in the turbine generator thanks to this cycle (Quoilin et al. 2013). ORC has been the focus of great research in recent years because of its distinctive advantages in using low-grade heat sources such as industrial waste heat, solar energy and geothermal energy (Wang et al. 2013). If temperature of the water entering the ORC is higher than 50 °C, the ORC can be more efficient (Tchanche et al. 2009). Therefore, a temperature higher than 50 °C was preferred in this study. The net power obtained from the ORC is expressed as follows:

$$\dot{W}_{net} = \dot{W}_T - \left(\dot{W}_{pump1} + \dot{W}_{pump2} + \dot{W}_{pump3}\right) \tag{11}$$

where $\dot{W}_T$ is the generated electricity by turbine generator, $\dot{W}_{pump1}$,$\dot{W}_{pump2}$ and $\dot{W}_{pump3}$ are pumping power of the pumps.

$$\dot{W}_T = \eta_T \eta_G \dot{m}_6 (h_6 - h_7) \tag{12}$$

where $\eta_T \eta_G = 0.4$ the total performance of the turbine generator, $\dot{m}_6$, h_6 and h_7 are the mass flow rate, enthalpies at the points 6 and 7, respectively. The energy efficiency of ORC is obtained as the ratio of the net electricity produced to input energy of the evaporator. It can be obtained as given (Long et al. 2014).

$$\eta_{ORC} = \frac{\dot{W}_{net}}{\dot{Q}_{ev}} \tag{13}$$

where $\dot{Q}_{ev}$ is obtained as follows:

$$\dot{Q}_{ev} = \dot{m}_5 (h_6 - h_5) \tag{14}$$

where $\dot{m}_5, h_5$ and h_6 are the mass flow rate, enthalpies at the points 5 and 6, respectively.

Table 1, 2, 3 and 4 shows the temperatures, pressures and mass flow rates of the fluids used in the system for four months. These values are used in EES to make thermodynamic calculations. In this system, inlet temperature of the ORC and air temperature vary from month to month. Other values are given for each month as follows.

Table 1 Thermodynamic parameters of the system for June

Num.	Fluid	Temp. (°C)	Pressure (kPa)	Mass flow rate (kg/s)
0	Air	25.5	101.3	–
1	H_2O	35	150	0.040
2	H_2O	53.48	150	0.040
3	H_2O	53.48	150	0.040
4	H_2O	36	150	0.040
6	Isobutene	53	2000	0.010
7	Isobutene	35	1750	0.010
8	Isobutene	32	1750	0.010
5	Isobutene	30	2000	0.010

Table 2 Thermodynamic parameters of the system for July

Num.	Fluid	Temp. (°C)	Pressure (kPa)	Mass flow rate (kg/s)
0	Air	28	101.3	–
1	H_2O	35	150	0.040
2	H_2O	54.08	150	0.040
3	H_2O	54.08	150	0.040
4	H_2O	36	150	0.040
6	Isobutene	53.5	2000	0.010
7	Isobutene	35	1750	0.010
8	Isobutene	32	1750	0.010
5	Isobutene	30	2000	0.010

Table 3 Thermodynamic parameters of the system for August

Num.	Fluid	Temp. (°C)	Pressure (kPa)	Mass flow rate (kg/s)
0	Air	28.4	101.3	–
1	H_2O	35	150	0.040
2	H_2O	54.75	150	0.040
3	H_2O	54.75	150	0.040
4	H_2O	36	150	0.040
6	Isobutene	54.25	2000	0.010
7	Isobutene	35	1750	0.010
8	Isobutene	32	1750	0.010
5	Isobutene	30	2000	0.010

Table 4 Thermodynamic parameters of the system for September

Num.	Fluid	Temp. (°C)	Pressure (kPa)	Mass flow rate (kg/s)
0	Air	25.8	101.3	–
1	H_2O	35	150	0.040
2	H_2O	52.70	150	0.040
3	H_2O	52.70	150	0.040
4	H_2O	36	150	0.040
6	Isobutene	52.2	2000	0.010
7	Isobutene	35	1750	0.010
8	Isobutene	32	1750	0.010
5	Isobutene	30	2000	0.010

To make the energy analyses of SP, the amount of solar radiation reaching the surface of the SP must be calculated initially. This is a very important parameter to evaluate the performance of the integrated system. Therefore, monthly solar energy was obtained from Adana Regional Meteorology Station. Monthly average of the solar energy distribution reaching to the horizontal surface can be seen in Fig. 2. The maximum solar energy was found as 756 MJ/m^2, 792 MJ/m^2, 735 MJ/m^2 and 513 MJ/m^2 for June, July, August and September and the minimum solar energy was recorded as 226 MJ/m^2, 269 MJ/m^2 and 240 MJ/m^2 for January, November and December, respectively. The minimum and maximum solar energy reached on the horizontal surface in January and July. In summer, the amount of stored thermal energy in the solar technologies is increased by solar radiation . Depending on the

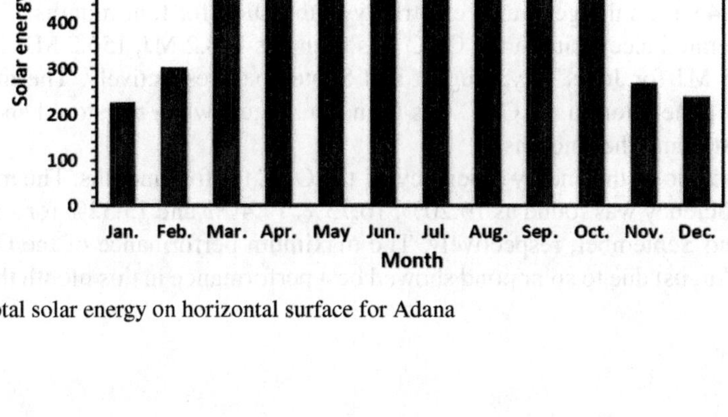

Fig. 2 Total solar energy on horizontal surface for Adana

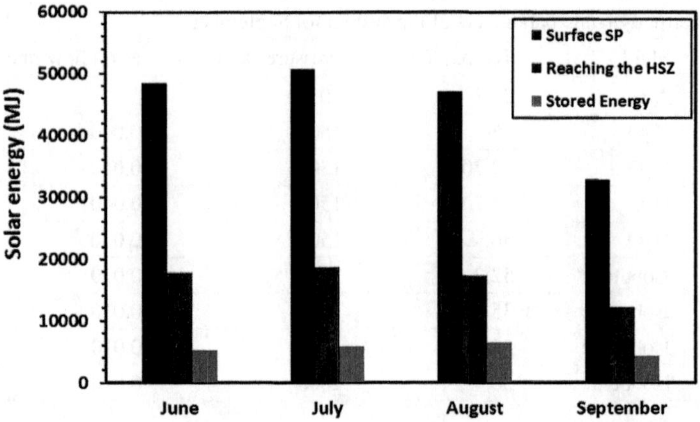

Fig. 3 Distribution of solar energy from surface to bottom of the SP in different months (Karakilcik at al. 2006)

amount of reaching solar energy, the storage area of the solar pond has also reached the highest temperatures (Karakilcik et al. 2006) (Fig. 3).

The solar energy enters from the surface of the SP and reaches the HSZ. Some energy is transferred to the environment due to heat losses and the remaining solar energy is stored in the HSZ. Monthly averages of the solar energy reaching on the horizontal surface of the solar pond. The maximum solar energy was found as 48,410 MJ, 50,730 MJ, 47,071 MJ and 32,889 MJ for June, July, August and September, respectively. The maximum solar energy reaching the HSZ was found as 17,911 MJ, 18,770 MJ, 17,416 MJ and 12,168 MJ for June, July, August and September, respectively. Eventually, some heat loss occurs due to environmental conditions and transmission losses, and the solar energy reaching the HSZ converts into heat energy for storing in it. The maximum stored solar energy in the HSZ was found as 5199 MJ, 5865 MJ, 6487 MJ and 4267 MJ for June, July, August and September, respectively. The reason for August to store higher heat energy than other months: More solar energy is stored by the solar pond due to its highest efficiency in this month than other months.

Figure 4 shows the generated electricity in the ORC for four months. The maximum generated electricity in the ORC was found as 143.2 MJ, 156.2 MJ, 168.5 MJ and 130.5 MJ for June, July, August and September, respectively. The maximum electricity generation in the ORC was found in August when the stored energy was the highest than other months.

Figure 5 shows the energy efficiency of the ORC for four months. The maximum energy efficiency was found as 16.20%, 16.73%, 17.47% and 15.32% for June, July, August and September, respectively. The maximum performance of the ORC was found in August due to solar pond showed best performance in this month than other months.

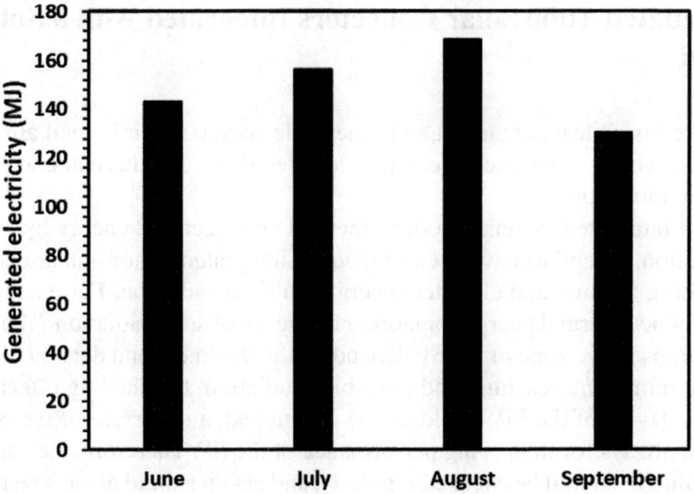

Fig. 4 Generated electricity in the ORC for four months

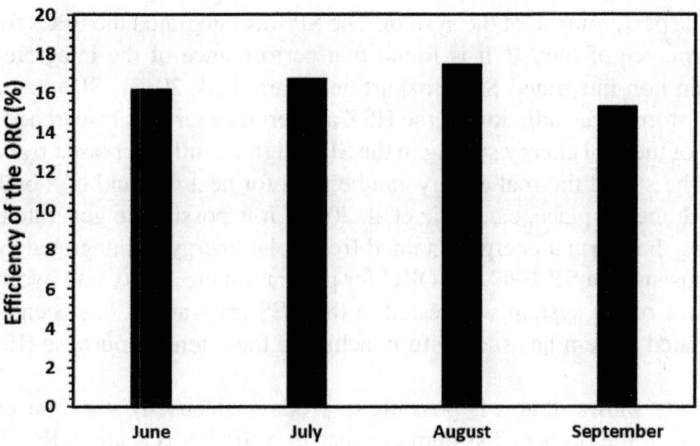

Fig. 5 The energy efficiency of ORC for four months

In this study, a rectangular prism model SP (the dimensions of 1.5 m × 8 m × 8 m) integrated with ORC has been successfully analyzed. This system is operated by pumping the thermal energy from the HSZ to the ORC by using a pump. Thus, electricity is produced at low temperatures. The EES simulation program was successfully carried out for the analyses of the system. Analyzes were conducted separately for June, July, August and September when the storage temperature of the solar pond was highest. As a result, the maximum amount of electricity was produced in August when the solar pond showed the highest performance.

2 Evacuated Tube Solar Collectors Integrated with a Solar Pond

Solar energy is a clean, abundant and renewable source and it is seen almost every place in the world. Solar energy can provide electricity and thermal energy needed for a lot of applications.

A lot of integrated systems produce thermal and electrical energy by employing solar radiation. Scientists have been studied the integrated or non-integrated systems for producing thermal and electrical energy via solar radiation. For example, SP is used to produce thermal energy for storing by means of solar radiation. The temperature variations of the zone of the SP depend on the thickness and density of the brine layers, the remaining reaching and absorbing radiation, the shading effect and heat losses. The layers of the SP should not be intermixed, and surface of the pond must be cleaned always for improving performance of the SP. Therefore, the salt concentration of the SP should be checked regularly, and also it should always be kept clean (Karakilcik et al. 2006). One of the important systems is EVTCs for storing solar energy. It is found that EVTCs have advantageous thermal efficiency as compared to flat-plate collectors (FPCs) (Chow et al. 2011). Integrating solar technologies can improve the performance of the system. The SP was integrated the FPCs for improving performance of the SP. It is found that performance of the integrated SP was higher than non-integrated SP (Bozkurt and Karakilcik 2012). SP integrated with EVTCs to store solar radiation in the HSZ as thermal energy is researched. The performance of thermal energy storing in the SP is significantly increased by employing EVTCs. The stored thermal energy can be used for heat demand of a building, hot water for domestic place, etc. (Atiz et al. 2015). It is possible to generate electricity by utilizing the thermal energy obtained from solar energy in integrated systems. A system consists of a SP, FPC and ORC for generation electricity and hydrogen. The performance of the system was tested in the EES program. It has been found that this integrated system has successfully achieved the intended purpose (Erden et al. 2017).

This study shows that it is possible to produce electricity via solar energy. As seen in Fig. 6, the proposed system consists of a SP, EVTCs and ORC that works with isobutene, a cooling tower and an electrolysis system. Enough amount of hot water is transferred to the ORC for electricity production. Thus, the integrating of the SP with EVTCs provide a lot of hot water for working of the system properly. The aim of this system is to generate thermal and electrical energy. The act of the SP in the integrated system is to store thermal energy by utilizing solar radiance for pumping to the EVTCs. The EVTCs upgrade the preheated water that obtained from the HSZ by the heat exchanger (HE). Thus, the temperature of the water entering the ORC is increased to higher temperatures. Thanks to the hot water that is coming from the integrated system, considerable electricity is generated by the ORC. The inlet water at 39 °C enters the HE from point 1 and exits the HE at 56 °C at point 2. Thus, the hot water temperature is upgraded to 96 °C by EVTCs for entering the evaporator at point 3. The temperature isobutene is reached 90 °C at point 6 and

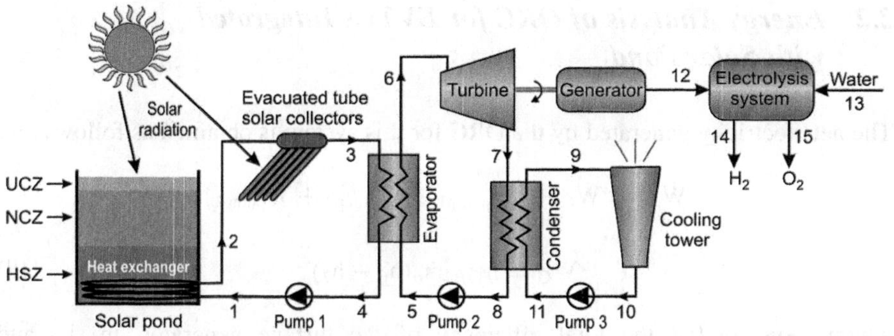

Fig. 6 Solar pond integrated with EVTCs and ORC (Atiz et al. 2019)

produces electricity by expanding to the turbine. Then isobutene leaves the turbine at 70 °C for entering condenser from point 7 and is cooled by the condenser. Thus, some of the energy of the fluid is transferred to the cooling tower. The fluids are circulated by pumps 1, 2 and 3 in the integrated system. The electricity produced in this system can be converted to hydrogen energy if desired.

2.1 Evacuated Tube Collectors

The EVTC contains 24 cylindrical glass tube that have very good light transmittance. The surface of the interior tube is covered with a good light absorption material. In this system, water moves in the natural cycle. The solar radiation is absorbed by the water to rise the reservoir tank for transforming to thermal energy (Budihardjo and Morrison 2009). Thus, useful heat is obtained in this system for using whenever needed. Useful energy of the EVTCs is found by (Bryant and Colbeck 1977):

$$\dot{Q}_U = \eta_E A_{EVTC} I_{EVTC} \tag{15}$$

where η_E is the energetic performance of the EVTC, A_{EVTC} is the total aperture area of the EVTCs and I_{EVTC} is the incident solar radiation for per square meter of the EVTCs. The useful energy is transferred to the water in the EVTCs and it is found as given:

$$\dot{Q}_U = \dot{m}_w C_p (T_3 - T_2) \tag{16}$$

where $\dot{m}_w$ is mass flow rate water leaving the EVTCs, T_2 is the inlet water temperature to the EVTCs and T_3 is the outlet water temperature of EVTCs.

2.2 Energy Analysis of ORC for EVTCs Integrated with Solar Pond

The net electricity generated by the ORC for this system is obtained as follows:

$$\dot{W}_{net} = \dot{W}_{G} - \left(\dot{W}_{pump1} + \dot{W}_{pump2} + \dot{W}_{pump3}\right) \tag{17}$$

$$\dot{W}_{G} = \eta_T \eta_G \dot{m}_6 (h_6 - h_7) \tag{18}$$

where $\eta_T \eta_G = 0.8$ the total efficiency of the turbine generator, $\dot{m}_6, h_6$ and h_7 are the mass flow rate and enthalpies at the points 6 and 7 respectively. Besides,$\dot{W}_{pump1}\,\dot{W}_{pump2}$ and $\dot{W}_{pump3}$ are the pumping power for circulating the fluids.

$$\dot{W}_{pump1} = \dot{m}_1 (h_4 - h_1) \tag{19}$$

$$\dot{W}_{pump2} = \dot{m}_5 (h_5 - h_8) \tag{20}$$

$$\dot{W}_{pump3} = \dot{m}_{10} (h_{11} - h_{10}) \tag{21}$$

Thus, utilizing the above equations, energetic performance of the ORC (η_{ORC}) is found as given:

$$\eta_{ORC} = \frac{\dot{W}_{net}}{\dot{Q}_{ev}} \tag{22}$$

where

$$\dot{Q}_{ev} = \dot{m}_5 (h_6 - h_5) \tag{23}$$

where $\dot{m}_5$, h_6 and h_5 are the mass flow rate, enthalpies at the points 5 and 6 fluid, respectively.

In this study, to find the thermodynamic performance of the integrated system, some parameter such as air temperature, pressure, mass flow rate is given in Table 5. This parameter plays very important role to evaluate thermodynamic analyses of the system. Firstly, the temperature of the HSZ of the SP is accepted as 55 °C while the temperature of the water at the input and output of the HE in the SP is assumed as 39 °C and 53 °C, respectively. These temperatures are acceptable values that is taken in the section of the solar pond for this study. Thus, the temperature of the water entering the evaporator is upgraded from 53 to 96 °C by the EVTCs. In Fig. 7, the daily distributions the solar energy reaching to the surface of the EVTCs was found from 5 a.m. to 19 p.m. for 20th July. The maximum solar energy was found as 947 MJ at noon. The total solar energy reaching on the surface of the EVTCs was found as 7899 MJ for area of 300 m^2.

Table 5 The most important parameters for calculations in the system

Num.	Fluid	Temp. (°C)	Pressure (kPa)	Mass flow rate (kg/s)
0	Air	28	101.3	–
1	H_2O	39	200	0.5
2	H_2O	53	200	0.5
3	H_2O	96	200	0.5
6	Isobutene	90	2000	0.2
7	Isobutene	70	1750	0.2
5	Isobutene	40	1750	0.2
8	Isobutene	39	2000	0.2

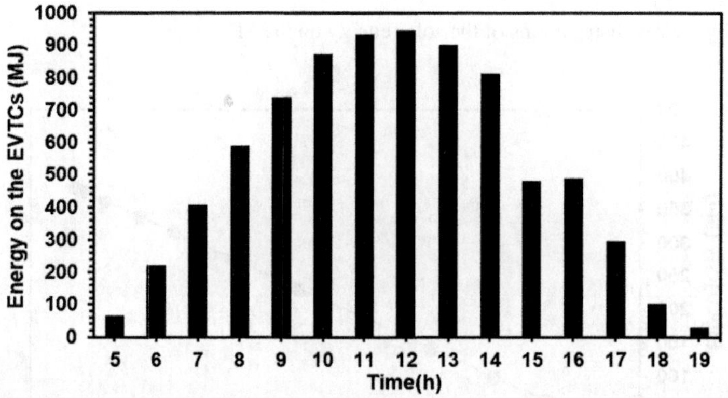

Fig. 7 The daily solar energy variations for the EVTCs

In Fig. 8, the daily energy distributions of the solar energy reached on the surface of the SP was found from 5 a.m. to 19 p.m. for 20th July. The maximum solar energy was found as 696 MJ at noon. Since the surface area of the solar pond is 217 m^2, a total of 5806 MJ of solar energy has been obtained from morning to evening. Due to the surface area of SP is smaller the EVTCs, total solar energy reaching on SP is found smaller than the EVTCs. Besides, total surface area of the SP is sufficient for producing the warm water to pump to the EVTCs. The solar energy on surface of the SP is reached the highest at noon and then it decreased from noon to evening.

As seen in Fig. 9, there is a direct relationship between electricity generation and thermal energy transferred to the evaporator. Thus, the thermal energy in the evaporator directly affects electricity production of the ORC. Thus, turbine generator produced 377 MJ electricity by using 1461 MJ thermal energy in the evaporator. Thanks to a lot of solar energy reached the surface on the SP and EVTCs, much more electricity is produced by the ORC. The energetic performance of the ORC was found as 25.40% with respect to Eq. 22. The electricity generated in this system can be converted to hydrogen for later use.

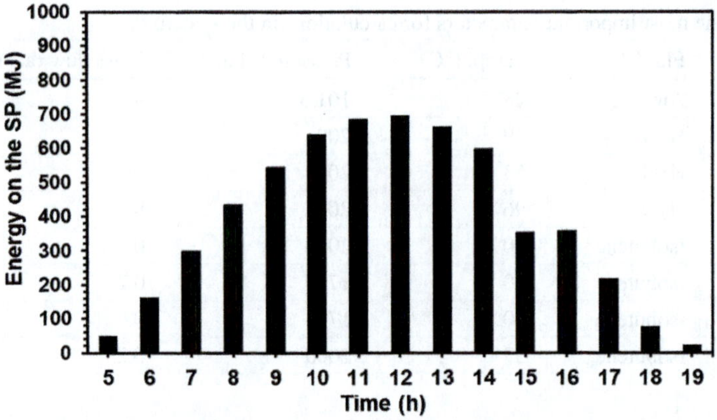

Fig. 8 Daily energy distributions of the solar energy on the SP

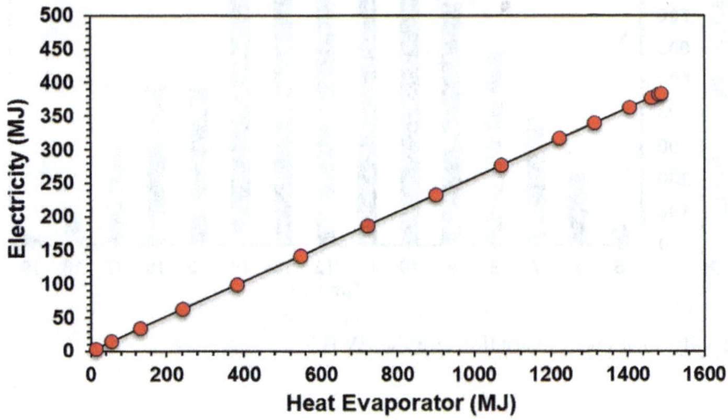

Fig. 9 The produced electricity and thermal energy in the evaporator for a day

The converted thermal energy of the SP and EVTCs plays very important duty for electricity generation. The SP is a significant part of the system to improve the thermal energy performance of the EVTCs in the integrated system. The daily thermal energy generation of the integrated system was converted to electricity by the ORC for 20th of July. It is observed that as the performance of each component in the system increases, the produced amount of electricity in the system increases. With such an integrated system, a lot of thermal and electrical energy can be produced and excess electricity can be converted into hydrogen. As a result, this system operates with a 100% renewable energy source and does not harm the environment by producing 100% renewable energy. As these systems increase, the world will continue to be a livable place for longer.

3 Conclusion

Solar energy is a 100% renewable energy source. In this context, solar energy plays a key role in the production of various types of clean energy such as thermal, electrical and hydrogen energy. By using integrated energy systems, 100% of our energy needs can be met with these energy sources. This is only possible with the development of integrated systems capable of generating multiple energy based on solar energy. In this study, the integrated system consisting of solar pond (SP), evacuated tube collectors (EVTCs), organic Rankine Cycle (ORC) and electrolysis system was successfully analyzed. In particular, thermal efficiency has been significantly improved with a solar pond integrated with EVTCs. Thus, a significant amount of thermal, electricity and hydrogen were produced with this system. It has been determined that if system components are improved, thermal, electrical and hydrogen energy can be produced more efficiently than solar energy. As a result, it was concluded that the use of renewable energy sources in homes, buildings and industry can meet 100% of the thermal, electrical and hydrogen energy needs. Therefore, more attention should be paid to scientific studies on 100% renewable energy systems. Because these systems produce 100% clean and renewable energy and are environmentally friendly.

References

Abdullah AA, Lindsay KA, AbdelGawad AF (2016) Construction of sustainable heat extraction system and a new scheme of temperature measurement in an experimental solar pond for performance enhancement. Sol Energy 130:10–24

Atiz A, Karakilcik M (2018) The Theoretical investigation of performance of the hydrogen attaination system with electricity from a solar pond. Çukurova Univ J Fac Eng Arch 33(4):1–8

Atiz A, Bozkurt I, Karakilcik M, Dincer I (2015) Investigation of effect of using evacuated tube solar collector on solar pond performance. Prog Clean Energy 2:261–273 (Springer International Publishing, Switzerland)

Atiz A, Karakilcik H, Erden M, Karakilcik M (2019) Assessment of electricity and hydrogen production performance of evacuated tube solar collectors. Int J Hydrogen Energy 44(27):14137–14144

Bozkurt I, Karakilcik M (2012) The daily performance of a solar pond integrated with solar collectors. Sol Energy 86:1611–1620

Bozkurt I, Karakilcik M (2015) The effect of sunny area ratios on the thermal performance of solar ponds. Energy Convers Manag 91:323–332

Bryant HC, Colbeck I (1977) A solar pond for London. Sol Energy 19:321–322

Budihardjo I, Morrison GL (2009) Performance of water-in-glass evacuated tube solar water heaters. Sol Energy 83:49–56

Cengel YA (2006) Heat and mass transfer, 3rd edn. Mc Graw Hill

Chow TT, Dong Z, Chan LS, Fong KF, Bai Y (2011) Performance evaluation of evacuated tube solar domestic hot water systems in Hong Kong. Energy Build 43:3467–3474

Date A, Yaakob Y, Date A, Krishnapillai S, Akbarzadeh A (2013) Heat extraction from non-convective and lower convective zones of the solar pond: a transient study. Sol Energy 97:517–528

Ding LC, Akbarzadeh A, Date A (2016) Transient model to predict the performance of thermoelectric generators coupled with solar pond. Energy 103:271–289

Ding LC, Akbarzadeh A, Tan L (2018) a review of power generation with thermoelectric system and its alternative with solar ponds. Renew Sustain Energy Rev 81(1):799–812

Erden M, Karakilcik M, Dincer I (2017) Performance investigation of hydrogen production by the flat-pate collectors assisted by a solar pond. Int J Hydrogen Energy 42:2522–2529

Karakilcik M, Dincer I, Rosen M (2006) Performance investigation of a solar pond. Appl Therm Eng 26:727–735

Karakilcik M, Dincer I, Bozkurt I, Atiz A (2013a) Performance assessment of a solar pond with and without shading effect. Energy Convers Manag 65:98–107

Karakilcik M, Bozkurt I, Dincer I (2013) Dynamic exergetic performance assessment of an integrated solar pond. Int J Exergy 12:70–85

Long R, Bao YJ, Huang XM, Liu W (2014) Exergy analysis and working fluid selection of Organic Rankine Cycle for low grade waste heat recovery. Energy 73:475–483

Quoilin S, Den Broek MV, Declaye S, Dewallef P, Lemort V (2013) Techno-economic survey of Organic Rankine Cycle (ORC) systems. Renew Sustain Energy Rev 22:168–186

Şahin ME, Okumuş Hİ (2013) Güneş Pili Modülünün Matlab/Simulink ile modellenmesi. EMO Bilimsel Dergi 3:5

Tchanche BF, Papadakis G, Lambrinos G, Frangoudakis A (2009) Fluid selection for a low-temperature solar Organic Rankine Cycle. Appl Therm Eng 29:2468–2476

Wang J, Yan Z, Wang M, Maa S, Dai Y (2013) Thermodynamic analysis and optimization of an (Organic Rankine Cycle) ORC using low grade heat source. Energy 49:356–365

The Role of Hydrogen in Global Transition to 100% Renewable Energy

Haris Ishaq and Ibrahim Dincer

Abstract This chapter discusses how hydrogen can replace the traditional energy sources and it can make global transition possible to 100% renewable energy. It focuses on three different aspects namely; renewable energy sources, hydrogen production system using renewable energy sources in terms of electricity and thermal energy and the services and applications where hydrogen can be employed. A classification is made showing the types of the renewable energy sources and their applications for production of heat, electricity or fuel extracted from the described sources, hydrogen production methods from these commodities and usage of hydrogen in different industries. In the renewable hydrogen production methods section, a system for each renewable source is presented which can be integrated with the renewable energy source for hydrogen production. A case study is presented which uses the wind energy source for electricity production. A part of electricity is supplied to the community while remaining is employed to the proton exchange membrane electrolyser to produce hydrogen. The produced hydrogen is stored by passing through multistage compression system and stored in the storage tank. During the intervals of low wind speed, the stored hydrogen is fed to the proton exchange membrane fuel cell which produces electrical power and heat for the community. Numerous parametric studies are conducted to explore how designed case study operates under different conditions. This chapter covers all the renewable energy sources and presents the potential of implementing hydrogen as a source for the global transition to 100% renewable energy.

Keywords Hydrogen · Renewable energy · Sustainability · Energy · Exergy · Efficiency

---

H. Ishaq · I. Dincer (✉)
Faculty of Engineering and Applied Science, University of Ontario Institute of Technology, Oshawa, ON, Canada
e-mail: Ibrahim.Dincer@uoit.ca

H. Ishaq
e-mail: haris.ishaq@uoit.net

T. S. Uyar (ed.), *Accelerating the Transition to a 100% Renewable Energy Era*, Lecture Notes in Energy 74, https://doi.org/10.1007/978-3-030-40738-4_13

1 Introduction

Energy is a key resource in wealth generation and economic development of any country. Major growth in the population and economic development is occurring around the world. Energy demand and availability are the two key factors which play an important part in bringing economic development. The resources which covers this increase in energy demand are facing excessive challenges and also resulting in quick depletion. The major part of the energy demand is set to be covered by the fossil fuels which is the particular reason for their faster depletion. The main drawback while using fossil fuels is the CO_2 emissions produced which result in numerous environmental problems and global warming (Dincer 2010). With the deficiency of efficiently available fossil fuel resources and the environmental problems accompanies with them, renewable energy sources are the best candidate to replace these traditional sources and to produce the energy through environmentally benign way. The main reason behind industrial and academic research efforts is global warming (Chiari and Zecca 2011). This has much importance regarding environmental impacts, which is somehow connected to social sustainability and economic sustainability (Dincer and Zamfirescu 2011). The renewable energy sources such as solar, wind, hydro, geothermal, biomass and ocean thermal energy conversion (OTEC) have the capability to replace the traditional fossil fuels and eradicate the global warming issues through viable and environmentally benign energy systems. The energy storage mediums which are clean, cheap, abundant and efficient have been and still being discovered due to the intermittent nature of the renewable energy sources in order to take the full advantage from these renewable energy sources and hydrogen has the capability to be used as storage media. Figure 1 displays a consumption comparison of traditional and renewable energy sources.

Hydrogen is not new but getting more and more recognized currently and globally as a potential fuel and a unique energy solution due to the fact that it offers advantages carried out by its usage and the availability of carbon-free solutions. Furthermore, such advantages are well known as hydrogen can use the currently available transportation and fuel storage infrastructures which is used for other chemical fuels and it can also be stored for extended periods. In comparison with hydrogen, electricity cannot be transferred or stored by employing currently available fuel storage systems and it also faces transmission and heat losses because of the high voltages and electrical resistance.

Hydrogen is the lightest, abundant and simplest among all other chemical elements by nature, but it is rare to find it in pure form but in combination with different other elements. The simplest example of hydrogen where it is found in combination with oxygen is water and it is also found in combination with different other elements such as nitrogen, carbon, chlorine and sulfides. Hydrogen becomes a striking energy carrier when its combination splits with different other elements using an energy source as it is not a prime energy source. When used as fuel or for combustion, hydrogen carries zero harmful emissions. In fuel cells, hydrogen is used as energy carrier and produces water by combining with oxygen emitting zero carbon emissions.

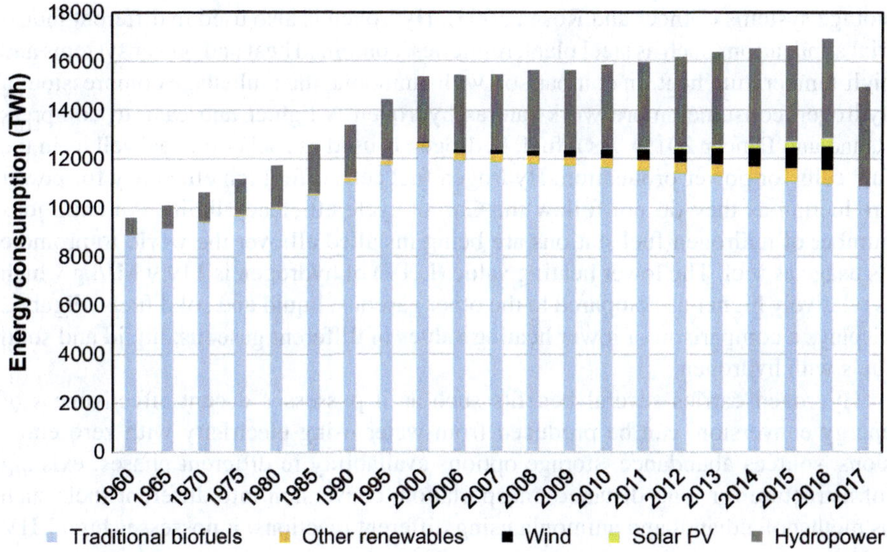

Fig. 1 Energy consumption comparison of traditional and renewable energy sources

The objective of this study chapter is to present the potential of implementing hydrogen as a source for the global transition to 100% renewable energy. The commercial plants are using traditional sources for hydrogen production and these sources result in numerous environmental impacts. This chapter also provides a complete route including the selection of the renewable energy source, the selection of the system which needs to be employed for hydrogen production to generate optimized results and the hydrogen applications. The necessity is to replace these traditional sources with renewable energy sources to provide a clean pathway for producing hydrogen and presenting hydrogen as the potential source for global transition to 100% renewable energy.

2 Hydrogen and Its Importance

Hydrogen is a promising energy carrier which can be employed for multiple purposes. The hydrogen demand does not only limit to one department but expands to the various sectors. It is used for different chemical processes such as hydrodealkylation, hydrodesulfurization, hydrocracking, metallic ores reduction, welding, dehydrogenation of oils and fats and also for the formation of different chemicals such as methanol, hydrochloric acid and ammonia in the commercial Haber Bosch process. The advantage of being lighter also extends its usage for filling balloons. It has the potential of being used as energy carrier and it is also feasible to be used in energy storage systems (Dincer and Rosen 2003). Hydrogen is also used in different industrial applications such as steel plant, refineries, combined heat and power systems and high-temperature heat. In comparison with ammonia, the multistage compression of hydrogen consumes more work rate as hydrogen is lighter and easy to compress (Ishaq and Dincer 2019). As a fuel, hydrogen is used as rocket fuel as well as in the fuel cells for power production. Hydrogen fuel cells offer high efficiency for power production as they do not follow the Carnot cycle efficiency limits. An enormous number of hydrogen fuel stations are being installed all over the world to promote its usage as fuel. The lower heating value (LHV) of hydrogen is 119.9 MJ/kg which is relatively higher as compared to the other gaseous, liquid and solid fuels. Figure 2 displays a comparison of lower heating values of different gaseous, liquid and solid fuels with hydrogen.

Hydrogen carries several benefits such as it possesses decent effectiveness of energy conversion, can be produced from water using electricity with zero emissions, sources abundance, storage options availability in different phases, existing infrastructure for long-distance transportation, conversion into different fuels such as methanol, ethanol and ammonia using different reactions, it possesses high LHV

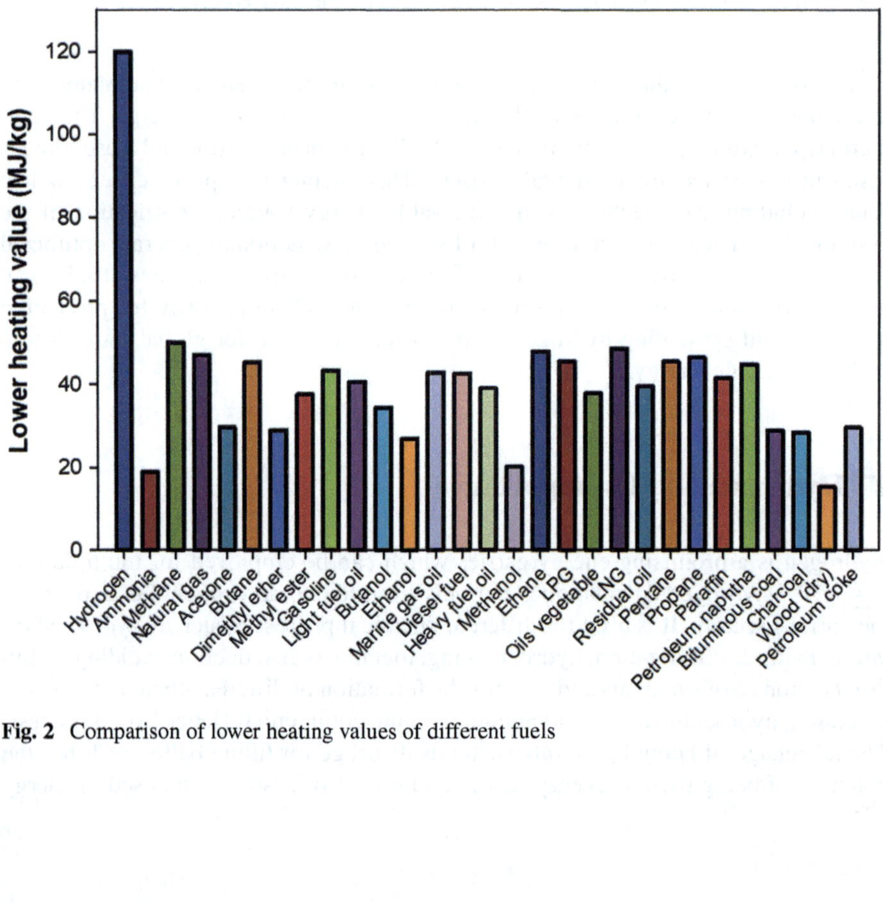

Fig. 2 Comparison of lower heating values of different fuels

and HHV in comparison with many traditional fossil fuels and it also provides possibilities to abolish the environmental effects from energy sector if produced from renewable energy sources.

Hydrogen can be produced from numerous conventional and renewable energy sources. In the conventional energy sources, nuclear energy source and the chemical processes of gasification, reforming and hydrocarbon reforming of fossil fuels are leading in the list but these sources are the reason of numerous environmental impacts. This conventional energy sources can be replaced with renewable energy sources such as solar, wind, geothermal, hydro and biomass. Each renewable energy sources have different categorizations which can be used for hydrogen production. For instance, solar energy source can be used in photovoltaic electrolysis, photolysis, photocatalysis, artificial photosynthesis and photoelectrochemical water splitting methods for hydrogen production. All of the renewable energy sources may result in one of the following types of energies; thermal energy, biochemical energy, electrical energy and photonic energy. Green hydrogen can be produced from all of these types of energies.

Figure 3 displays the global hydrogen demand in the major sectors which are refining, ammonia synthesis and others from 1975 till 2018. In the graphical representation, first five years from 1975 till 1995 are represented as bars and plotted according to the primary Y-axis while from years 2000 to 2018, the hydrogen demand is presented as lines which refers to the secondary Y-axis. The legends provide the

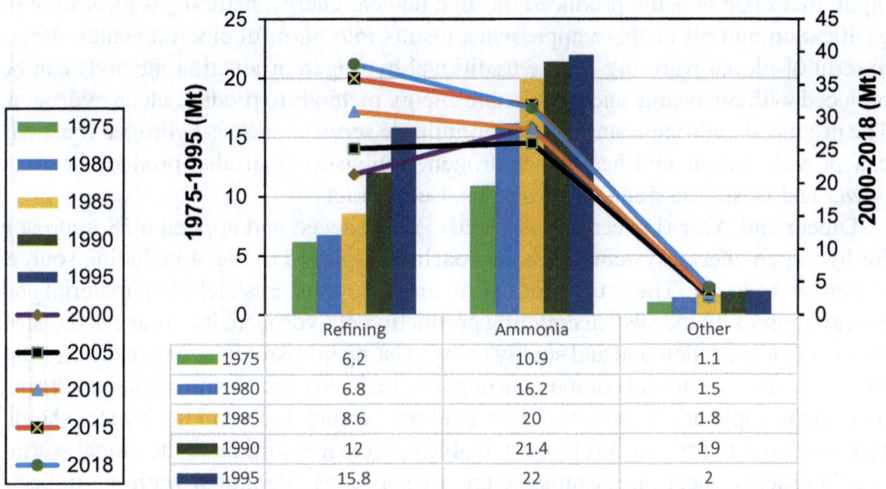

	Refining	Ammonia	Other
1975	6.2	10.9	1.1
1980	6.8	16.2	1.5
1985	8.6	20	1.8
1990	12	21.4	1.9
1995	15.8	22	2
2000	21.4	28.6	2.5
2005	25.3	26.1	2.7
2010	31	28.3	3.1
2015	36	31.9	3.8
2018	38.2	31.5	4.2

Fig. 3 Global demand of pure hydrogen (data from Agency IE 2019)

representation of each year. It can be observed that the global hydrogen demand is increasing gradually in all the relevant sectors. The global hydrogen demand from 1975 to 2018 increases from 6.2 to 38.2 Mt in the refining sector and it jumps from 10.9 to 31.5 Mt in the ammonia sector and ammonia industry is considered as the backbone of fertilizer industry and hydrogen demand rises from 1.1 to 4.2 Mt in all other sectors. As completely supplied by fossil fuels including 2% from coal and 6% from natural gas, this quantity of hydrogen is responsible for 830 million tonnes of CO_2 emissions per year which can be eliminated by using renewable energy sources. By replacing these traditional energy sources with renewable and sustainable energy system, the global hydrogen demand can be met through environmentally benign solutions.

In the ammonia synthesis sector, hydrogen requirement is three times the nitrogen and this requirement is fulfilled by producing hydrogen from catalytic natural gas reforming which emits carbon emissions. This whole process can be switched to the environmentally benign system if hydrogen source will be replaced from natural gas reforming to the renewable energy-based water electrolysis for hydrogen production.

3 Renewables and Hydrogen Production

In most of the working commercial plants where hydrogen is required as a key input, hydrogen is being produced through nuclear energy, natural gas reforming or gasification and all of these approaches results into harmful environmental effects in term of global warming. These traditional hydrogen production methods can be replaced with renewable and sustainable energy methods to produce clean hydrogen. The produced hydrogen can be used for multiple services such as hydrogen fuel cells can provide power and heating, hydrogen combustion can also produce heating, power and fresh water and it can also be used as fuel.

Dincer and Acar (Dincer and Acar 2015) introduced and applied a 3S approach for hydrogen energy systems. This approach is displayed in Fig. 4 including source, system and service. The sources of hydrogen energy systems including material and energy sources, to the system counting production of hydrogen, its storage and distribution to meet the demand and supply using existing infrastructure and to the end-use services namely; internal combustion engines, fuel cells and chemical manufacturing to meet the applications such as power, cooling, heating, fuels and fresh water should be considered to provide a fully sustainable approach to eliminate the global warming. In order to meet the sustainable energy solutions, the global warming impacts should be eliminated completely.

The source selection for hydrogen production should be the first step which should be abundant, clean, affordable and reliable. Considering some of the renewable energy sources which are intermittent by nature, reliable and trusted storage systems needs to be integrated with the hydrogen production systems. With the advancements in renewable energy sources, the cost for the renewable energy has been dropped down, highly efficient and large-scale renewable energy (solar, wind

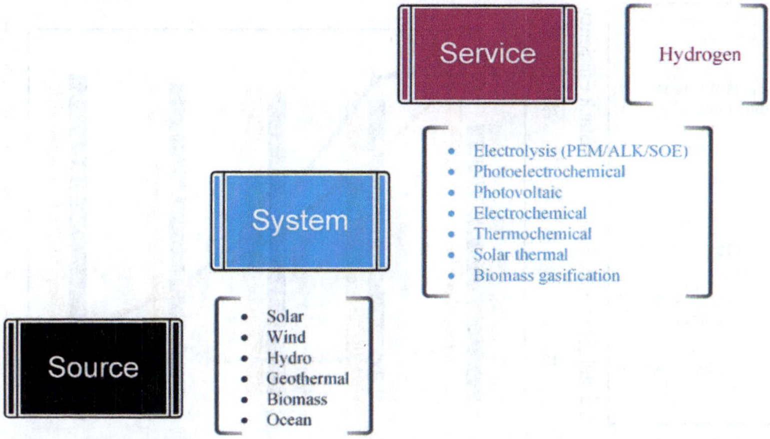

Fig. 4 3S approach for renewable energy-based hydrogen production systems (Dincer and Acar 2015)

and hydro) based systems have been developed to provide the transition from traditional sources (fossil fuels) to the renewable energy sources. If the selected source is suitable but the accompanied system is not proper, it might not generate optimum results. The system accompanied with the sustainable source should be efficient, viable, abundant and reliable and it should be possible to scale-up the system. This section can be subdivided into three subsystems of hydrogen production, storage and distribution. These systems can be categorized on the basis of energy type such as electrical, thermal, biological, photonic, electrochemical photochemical and thermochemical which further connects to the end-users. Such systems can be used for different applications such as power, cooling, heating, fresh water, space heating, fuels and formation of different chemicals. Figure 5 displays the future predictions for renewable energy-based hydrogen production.

The renewable energy sources such as wind, hydro, OTEC and tidal produces electrical power which can be used for water splitting into hydrogen using proton exchange membrane (PEM)/alkaline/solid oxide electrolyser (SOE). The geothermal energy source can be subdivided into high-temperature and geothermal heat pump categories. The high temperature can be converted into power and the simplest example is thermoelectric generator while geothermal heat pump results into heat which can be used in different thermochemical cycles for clean hydrogen production. These thermochemical cycles depend more on the thermal energy than electric power. The solar energy source can also be subdivided into photovoltaic (PV) and solar thermal. The PV system produces the electricity while solar thermal source produces heat and these two energy carriers can be converted to the hydrogen using electrolyser and thermochemical cycles.

The biomass source is further divided into three major types of wood, pulp and industrial wastes and these sources can be used to produce electricity, heat or fuels. The fuels can be used in different sectors such as natural gas reforming, oil refineries,

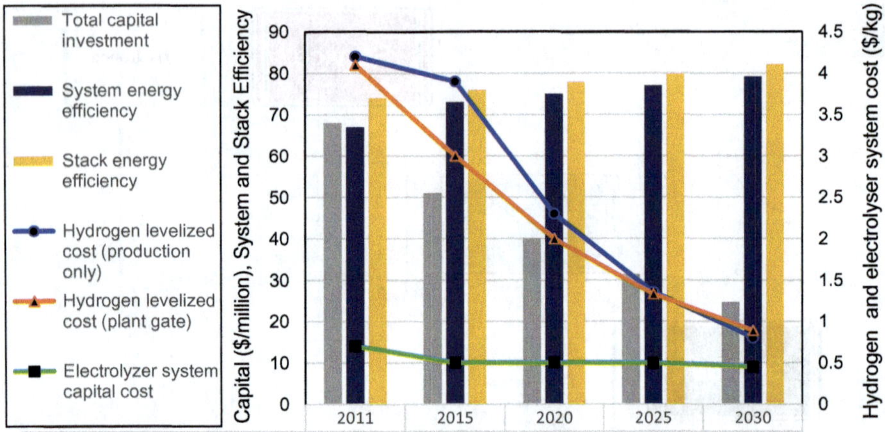

	2011	2015	2020	2025	2030
Total capital investment	68	51	40	31.4	24.6
System energy efficiency	67	73	75	77.1	79.2
Stack energy efficiency	74	76	78	80.1	82.2
Hydrogen levelized cost (production only)	4.2	3.9	2.3	1.4	0.8
Hydrogen levelized cost (plant gate)	4.1	3	2	1.3	0.9
Electrolyzer system capital cost	0.7	0.5	0.5	0.5	0.5

Fig. 5 Future predictions for renewable energy-based hydrogen production (data from U.S. Department of Energy. https://www.energy.gov/eere/fuelcells/doe-technical-targets-hydrogen-production-electrolysis)

steel plants and for methanol, hydrochloric acid and ammonia production. All of the described renewable energy sources are used for clean hydrogen production. The sources with the intermittent nature can be integrated with energy storage systems to meet the optimum outcomes. The produced hydrogen can be used for different purposes such as oil refineries, methanol, hydrochloric acid and ammonia synthesis, steel plants, combined heat and power systems, as fuel for rocket jets and road transport vehicles and in the fuel cells to produce heat and electricity. Figure 6 displays the distribution of different renewable energy sources, their outcomes as electricity, heat and fuels, their conversion processes to the hydrogen and the usage of hydrogen for different purposes. All of these sources are displayed with the systems leading to the end-use commodities.

The renewable energy-based hydrogen production systems can be categorized into the following sections.

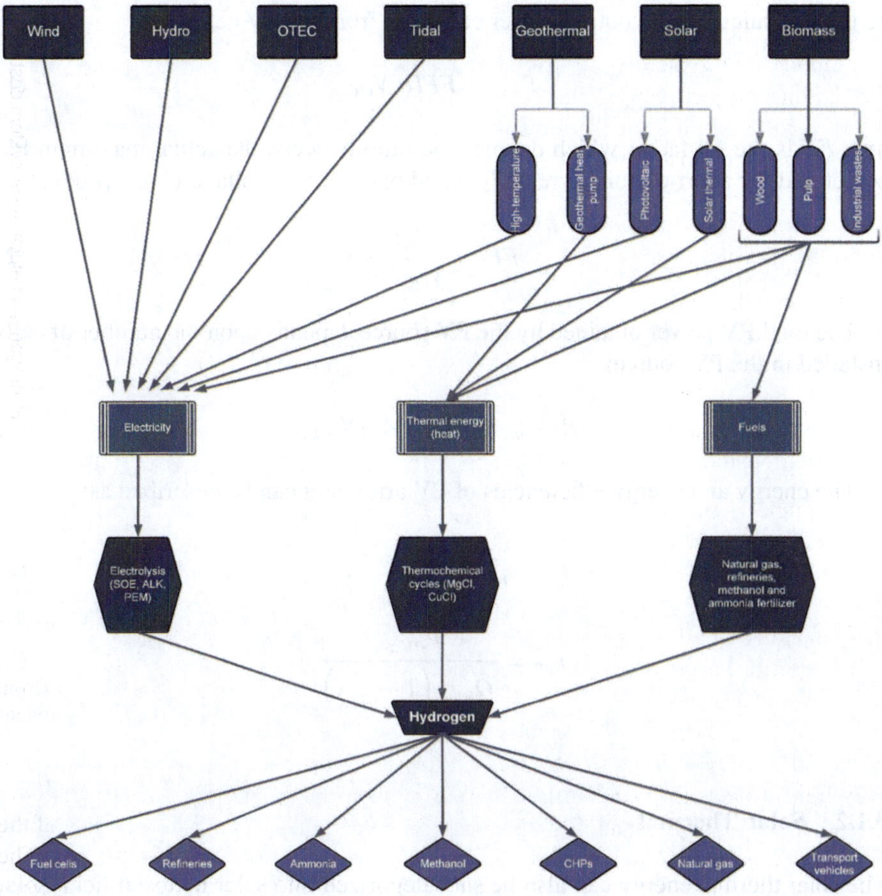

Fig. 6 Hydrogen production and utilization for the global transition to 100% renewable energy

3.1 Solar Energy Source

The solar energy source can be subdivided into three different categories of photovoltaic, solar thermal and photoelectrochemical. Because of the intermittent nature of the solar energy source, a storage system is needed to be integrated with the renewable energy source to achieve the optimum outcomes.

3.1.1 Photovoltaic

The energy input through the solar radiation is absorbed by the solar PV array and delivers electrical output. The maximum power which is possible to extract from the PV array is evaluated through the I-V characteristics. The following correlation can be used to calculate the output power extracted from the PV cell:

$$\dot{P}_{PV} = FFI_{SC}V_{OC} \tag{1}$$

here FF is the fill factor which denotes the ratio between the actual maximum PV power and the short-circuit current (I_{SC}) and open-circuit voltage (V_{OC}) product.

$$FF = \frac{V_m I_m}{V_{OC} I_{SC}} \tag{2}$$

The total PV power obtained by the PV source depends upon the number of cells installed in the PV source:

$$\dot{P}_{PV,tot} = FFI_{SC}V_{OC}N_{cells} \tag{3}$$

The energy and exergy efficiencies of PV array unit can be described as:

$$\eta_{en} = \frac{\dot{P}_{PV,tot}}{\dot{Q}_{solar}} \tag{4}$$

$$\eta_{en} = \frac{\dot{P}_{PV,tot}}{\dot{Q}_{solar}\left(1 - \frac{T_0}{T_{sun}}\right)} \tag{5}$$

3.1.2 Solar Thermal

The solar thermal energy can also be subcategorized into solar heliostat field, solar parabolic trough and solar dish collector. In the solar tower, the total amount of the solar heat depends upon the solar tower efficiency η_{he}, direct normal irradiance $\dot{I}_b$, area of the heliostat A_{he} and number of heliostats N_{he}. The thermal energy extracted from the solar tower is represented by $\dot{Q}_s$ and correlation can be written as:

$$\dot{Q}_{solar} = \eta_{he}\dot{I}_b A_{he} N_{he} \tag{6}$$

3.1.3 Photoelectrochemical

Photoelectrochemical cell uses solar light and transforms it to hydrogen. In this photoelectrochemical process, the photoelectrodes absorb the solar light and one of the electrodes works as semiconductor. The material bandgap helps in detecting the light utilization ability and photoelectrochemical cells have the capability to generate either electrical or chemical energy according to the user requirement. Some natural losses are linked with the solar energy transformation methods accompanied with materials. The conversion efficiencies of photoelectrochemical cell can be defined in terms of the photocurrent density "J" (mA cm^{-2}), water splitting reaction potential pH = 0 E^0_{rev}, light intensity "I_o" (mW cm^{-2}) and applied external voltage V_{bias} as follows:

$$\eta_{PEC} = \frac{J\left(E^0_{rev} - V_{bias}\right)}{I_o} \tag{7}$$

3.2 Wind Energy Source

Wind energy source utilizes the wind through the wind turbine to deliver the mechanical power. This produced mechanical power turns the electric generators and electric work is employed for different purposes. Wind energy source is considered as a renewable and sustainable energy source having minimum environment impact as compared to the fossil fuels. The produced electric work can be employed to the electrolyser (PEM/ALK/SOE) to produce hydrogen. The generated power can be calculated with the help of wind turbine efficiency η_{wt}, wind turbine area A_{wt}, air density ρ_{air}, wind speed V and specific heat C_p. The correlation used to evaluate the wind power is as follows:

$$P_{wt} = \frac{1}{2}\eta_{wt}A_{wt}\rho_{air}V^3C_{p,wt}n_T \tag{8}$$

3.3 Hydro Energy Source

The hydro energy source can be subcategorized into two different types of ocean thermal energy conversion and tidal energy. In the ocean thermal energy conversion system, the water is used from the different depths having different temperature and electrical power is generated on the basis of temperature difference and water is reinjected to the source while tidal energy uses the tides to rotate the turbine for mechanical energy which is further converted into the electrical energy.

3.3.1 Ocean Thermal Energy Conversion

Ocean thermal energy conversion cycle is known as a cycle that produces electrical power by utilizing the temperature difference in the warm surface water and deep cold ocean water. The OTEC cycle pumps great measures of the surface and deep cold seawater to operate a power cycle to generate electricity. This cycle consists of a condenser, boiler, turbine and a pump which circulates the water. The boiler converts the ammonia from saturated liquid to the saturated vapors and turbine expands the boiler output vapors to generate the electrical power. The condenser removes the additional heat available in the seawater and pump circulates the working fluid to the boiler.

3.3.2 Tidal

The tidal energy uses the tides to rotate the turbine for mechanical energy which is further converted into the electrical energy. Due to the intermittent nature of the tidal energy source, a storage system is needed to be integrated to the system source to achieve the optimum outcomes. Tides associated with the tidal energy source are more predictable as compared to the solar and wind sources. Tidal energy has conventionally suffered with limited sites availability having comparatively high flow velocities or tidal ranges and relatively high cost among other renewable energy sources. Though, the recent technological advancements, improvements and developments in the design and turbine technology considering axial and crossflow turbines indicated the total tidal power availability comparatively higher than previously assumed and environmental and economic costs can be fetched down to the viable levels.

3.4 Geothermal Energy Source

Geothermal energy source utilizes the thermal energy stored and generated in the earth. This source contains the thermal energy confined in the rocks present in the earth crust and fluid which seals the pores and fractures in the rocks. Geothermal energy sources are the hot water reservoirs existing at different depths and temperatures under the earth surface. The geothermal energy source is subcategorized into two different types of high-temperature that produces power and geothermal heat pump that produces thermal energy.

3.4.1 High-Temperature

The high-temperature geothermal energy source is associated with the water temperature extracted from the earth crust. If the extracted water is at the temperature of 160 °C or higher, it can be directly employed to the steam turbine to generate electricity and produced electricity can be further used to produce hydrogen. In the high-temperature geothermal sources, the heat flux remains 3–4 times higher as compared to the normal heat flux and it is usually found in 3–15 km depth.

3.4.2 Geothermal Heat Pump

A ground-source or geothermal heat pump is a fundamental heating and/or cooling unit which has the capability to transfer the heat to or from the earth. Unlike the solar and wind energy sources, geothermal heat pump does not require any storage system as it remains available all the time without any intermittency. This source works as the heat source during winter and as heat sink during summer. This geothermal heat pump system receives the benefit of the reasonable temperatures available in the ground to drop down the operational costs for heating and cooling and to boost the system efficiency.

3.5 Biomass Energy Source

Biomass energy source utilizes the organic material known as biomass which is derived from animals and plants. This biomass comprises of the energy stored in the plants from the sun during the photosynthesis process and biomass releases heat when it is burnt. The chemical energy in biomass is released as heat. Biomass-derived from the animal and plants is used for electrical or thermal energy production and they are also employed to the numerous industrial processes as raw material. The burning of the biomass releases CO_2 emissions but it is classified as one of the renewable energy sources by the American and European Union because plant absorb CO_2 during the photosynthesis process. The amount of heat or electricity generated by the biomass energy source can be used for hydrogen production using electrolysis or thermochemical cycles.

3.5.1 Biomass Combustion

In the combustion process, two ingredients of oxygen and biomass are reacted under high-temperature conditions to produce heat. The estimated chemical equation of combustible biomass portion is $CH_{1.44}O_{0.66}$. The biomass combustion basically represents the burning of the organic material. The heat generated by the biomass combustion can be further used to generate electricity. Direct combustion process is the most common and established technology used to convert biomass into heat. During the combustion process, the excess air is supplied to the combustion process to burn the biomass fuel and produce heat. The contact between the biomass fuel and the oxygen available in the air effects the efficiency of the combustion process. In an efficient combustion process, the CO_2 and water are the major products which are also accompanied with alkaline ash particles and tar which are needed to be minimized to meet the standards of an environmentally acceptable design for biomass combustion.

The biomass combustion reaction and the correlation for the biomass LHV calculation and exergy-to-energy ratio (Φ) which is used to evaluate the chemical exergy of the fuel are as follows:

$$C_xH_yO_z + \lambda(O_2 + 3.76N_2) \rightarrow aCO_2 + BH_2O + cN_2 \tag{9}$$

$$LHV_{Biomass} = \frac{400{,}000 + 100{,}600y - \frac{z}{1+0.5y}(117{,}600 + 100{,}600y)}{12 + y + 16z} \tag{10}$$

$$\Phi = \frac{ex_f}{LHV_{dry}} \tag{11}$$

$$\Phi = 1.0401 + 0.1728\frac{\beta}{\alpha} + 0.0432\frac{\delta}{\alpha} + 0.2169\frac{\gamma}{\delta}\left(1 - 0.2062\frac{\beta}{\alpha}\right) \tag{12}$$

3.5.2 Biomass Gasification

Gasification is a process used to convert solid fuel into a combustible producer gas including carbon dioxide, carbon monoxide and hydrogen (synthesis gas) through a series of thermo-chemical reactions. This producer gas is a fuel with low-heating value, calorific value in the range of 1000 and 1200 kcal N^{-1} m^{-3}. This is established by achieving the high-temperatures condition without combustion and providing a controlled oxygen and steam amount. The power or thermal energy derived from the biomass combustion or gasification of the producer gas is considered as a renewable energy source. The biomass gasification carries an advantage that consuming the syngas is much more effective as compared to the direct combustion because of the reason of high-temperature combustion and also in fuel cells which are not bound to follow the thermodynamic upper limit of Carnot efficiencies.

The overall biomass gasification reaction and the chemical reactions of char decomposition, volatile matter combustion and char decomposition into chemical species can be written as follows:

$$Biomass \rightarrow Char + (C_6H_6 + CO + H_2 + CO_2 + N_2 + CH_4 + H_2O + H_2S) \tag{13}$$

$$C_6H_6 + 7.5O_2 \rightarrow 3H_2O + 6CO_2 \tag{14}$$

$$CH_4 + 2O_2 \rightarrow 2H_2O + CO_2 \tag{15}$$

$$char \rightarrow C + O_2 + N_2 + H_2 + S + Ash \tag{16}$$

$$H_2 + 0.5O_2 \rightarrow H_2O \tag{17}$$

$$CO + 0.5O_2 \rightarrow CO_2 \tag{18}$$

$$S + H_2 \rightarrow H_2S \tag{19}$$

$$C + 2H_2 \rightarrow CH_4 \tag{20}$$

$$C + O_2 \rightarrow CO_2 \tag{21}$$

$$CO + H_2O \rightarrow CO_2 + H_2 \tag{22}$$

$$CH_4 + H_2O \rightarrow CO_2 + H_2 \tag{23}$$

$$C + H_2O \rightarrow CO + H_2 \tag{24}$$

4 Renewable Hydrogen Production Methods

The source selection for hydrogen production is the first step in the hydrogen production system and source should be abundant, clean, affordable and reliable. The storage system needs to be integrated with some of the renewable energy sources which are intermittent by nature. Renewable energy cost has been dropped down, large-scale and highly efficient renewable energy-based systems have been developed with the advancements in renewable energy sources to replace the traditional fossil fuels. The renewable hydrogen production methods can be categorized into different section depending upon the renewable energy source and hydrogen production methods. To produce the hydrogen from electricity, electrolyser technology (PEM/ALK/SOE) is used while thermochemical cycles are used to employ the heat for hydrogen production. In this section, different renewable energy sources-based hydrogen production methods are presented and discussed in detail.

4.1 Solar Energy-Based Hydrogen Production Methods

The solar energy source is subdivided into three different categories of solar thermal, photoelectrochemical and photovoltaic. A storage unit needs to be integrated with the system to achieve the optimum results and to overcome the intermittency of the solar energy source. Each of these sources uses different hydrogen production method.

4.1.1 Solar Thermal Energy-Based Hydrogen Production

The thermal energy extracted from the solar heliostat field is fed to the thermochemical Cu–Cl cycle for clean hydrogen production. Figure 7 displays the system schematic of the solar thermal energy-based hydrogen production. The molten salt is used as working fluid which extracts the heat from the solar tower and passes through hot and cold storage tanks which stores the additional available heat and molten salts recycles back to the solar tower using pump. The water is fed to the heat exchanger which converts it into steam which is introduced in the thermochemical Cu–Cl cycle. This cycle comprises of four key steps. In the first step, copper (II) chloride reacts with steam to at high temperature of 400–450 °C and produces hydrogen chloride gas (HCl) and copper oxychloride (Cu_2OCl_2) and this step is known as hydrolysis. In the second step, Cu_2OCl_2 decomposes at very high temperature of 500–550 °C into oxygen and copper (I) chloride and this step is known as thermolysis. In the third step, copper (I) chloride reacts HCl at ambient temperature of 25 °C in the presence of water and produces aqueous copper (II) chloride and hydrogen gas and this step is known as electrolysis. In the fourth step, aqueous copper (II) chloride passes through a dryer at high temperature of 100–120 °C which converts the water into steam solid

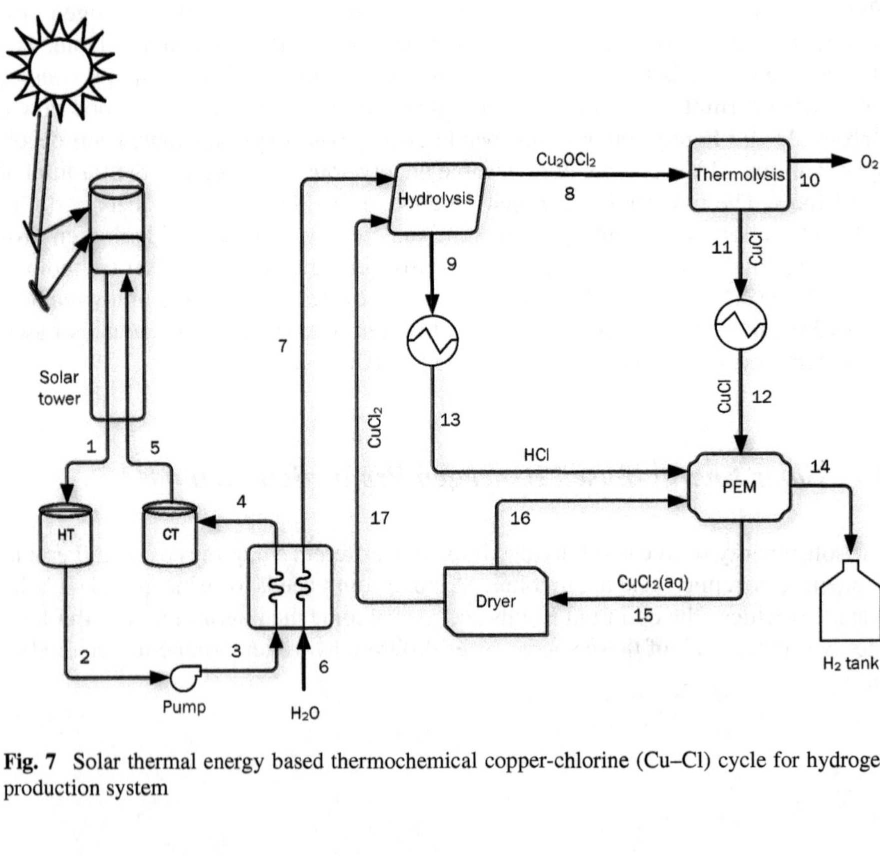

Fig. 7 Solar thermal energy based thermochemical copper-chlorine (Cu–Cl) cycle for hydrogen production system

Table 1 Four key steps of the thermochemical Cu–Cl cycle

Process	Chemical reaction	Temperature (°C)	Reaction nature
Hydrolysis	$2CuCl_2(s) + H_2O(g) \rightarrow Cu_2OCl_2(s) + 2HCl(g)$	400	Endothermic
Thermolysis	$Cu_2OCl_2(s) \rightarrow 0.5O_2(g) + 2CuCl(l)$	500	Endothermic
Electrolysis	$2CuCl(aq) + 2HCl(aq) \rightarrow H_2(g) + 2CuCl_2(aq)$	25	Electrochemical
Drying	$CuCl_2(aq) \rightarrow CuCl_2(s)$	100	Drying

copper (II) chloride is fed to the hydrolysis reactor to continue the cycle and this step is known as drying. Table 1 displays the chemical reaction occurring on each step of the thermochemical cycle along with the reaction temperatures and the nature of the reaction.

4.1.2 Photoelectrochemical Hydrogen Production

In the photoelectrochemical energy-based hydrogen production system, the photoelectrode existing in the photoelectrochemical cell absorbs the solar irradiance while electricity is also inputted to the photoelectrochemical cell. The water is fed to the photoelectrochemical cell which splits into positive (H^+) and negative (O^{-2}) ions. When two protons react with each other, hydrogen gas is produced. Figure 8 displays the schematic of the photoelectrochemical based hydrogen production system.

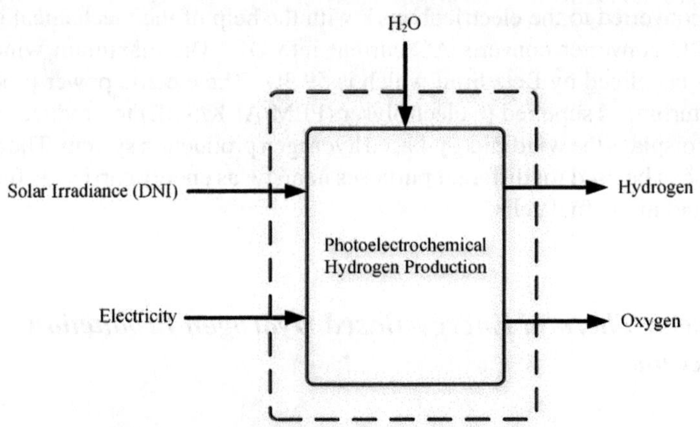

Fig. 8 Photoelectrochemical hydrogen production

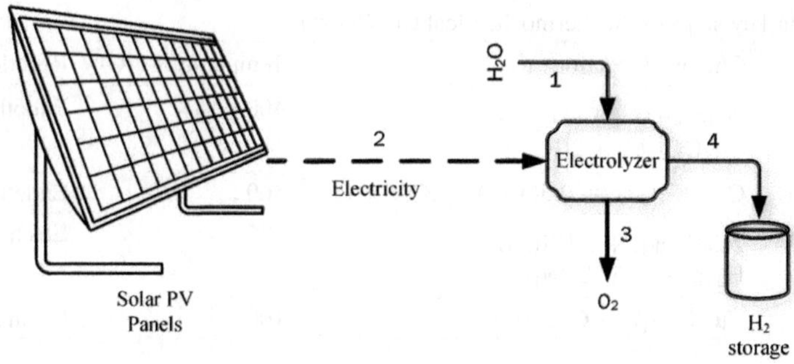

Fig. 9 Solar photovoltaic based hydrogen production system

4.1.3 Photovoltaic Electricity-Based Hydrogen Production

The solar PV panel absorbs the solar radiation and delivers the electrical output. The I-V curves indicate the maximum possible power which can be extracted from the PV panel. Figure 9 displays the solar photovoltaic based hydrogen production system. The electricity produced by the solar PV panel is fed to the electrolyser which splits water into oxygen and hydrogen.

4.2 Wind Energy-Based Hydrogen Production Method

In the wind energy source, wind turbine rotates and generates mechanical work which is converted to the electrical work with the help of the mechanical generator and AC/DC converter converts AC current into DC. The maximum wind turbine efficiency is defined by Betz limit which is 59.3%. The electric power produced by the wind turbine is supplied to electrolyser (PEM/ALK/SOE) to produce hydrogen. Figure 10 displays the wind energy-based hydrogen production system. The produced hydrogen can be used for different purposes namely; as energy carrier, as fuel in road vehicles and in the fuel cells.

4.3 Ocean Thermal Energy-Based Hydrogen Production Method

Ocean thermal energy conversion cycle uses the temperature difference of the warm surface water and deep cold ocean water to generate electrical power. Ammonia is used as the working fluid in the OTEC cycle. The cold-water temperature is usually in the temperature range of 0–3 °C and the average sea surface temperature is 17 °C. The OTEC cycle works on the principle of using working fluid with low boiling point and pump pressurizes the working fluid at high-pressure, warm water supports NH_3

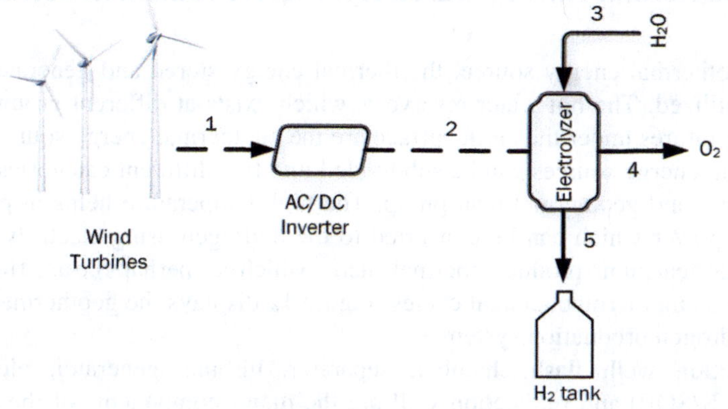

Fig. 10 Wind energy-based hydrogen production system

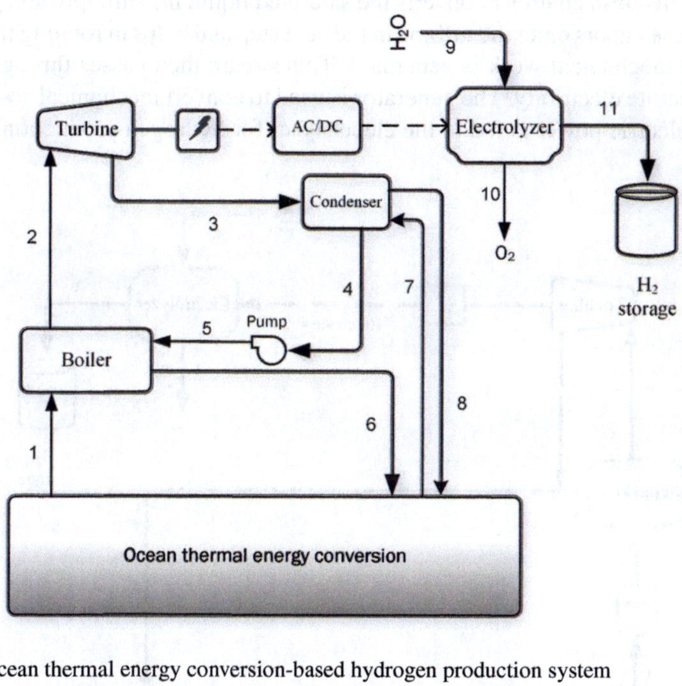

Fig. 11 Ocean thermal energy conversion-based hydrogen production system

to transform into saturated vapors and these high-pressure vapors drive the turbine which generates electricity while deep-seawater is used to condense the working fluid again into liquid. The OTEC cycle pumps great measures of the surface and deep cold seawater to operate a power cycle to generate electricity. Figure 11 displays the OTEC energy-based hydrogen production system.

4.4 Geothermal Energy-Based Hydrogen Production Method

In the geothermal energy source, the thermal energy stored and generated by the earth is utilized. The hot water reservoirs which exists at different ground depths and temperatures under the earth surface are the geothermal energy sources. These geothermal energy sources can be subdivided into two different categories of high-temperature and geothermal heat pump. The high temperature helps in producing produces power which can be converted to the hydrogen using electrolyser while geothermal heat pump produces thermal energy which can perhaps be used to produce hydrogen using thermochemical cycles. Figure 12 displays the geothermal energy-based hydrogen production system.

Production well, flash chamber, separator, turbine, generator, electrolyser (PEM/ALK/SOE) and reinjection well are the major components of the designed geothermal energy-based hydrogen production system. The hot geothermal fluid is extracted from the production well at high temperature and enters to the flash chamber. This flash chamber converts the saturated liquid into mixture and generates vapors. These vapors enter the turbine in the next step and helps in rotating the turbine blades and mechanical work is generated. This stream then passes through the turbine to generate electricity. The generator is used to convert mechanical-to-electrical work and electric power is fed to the electrolyser for hydrogen production.

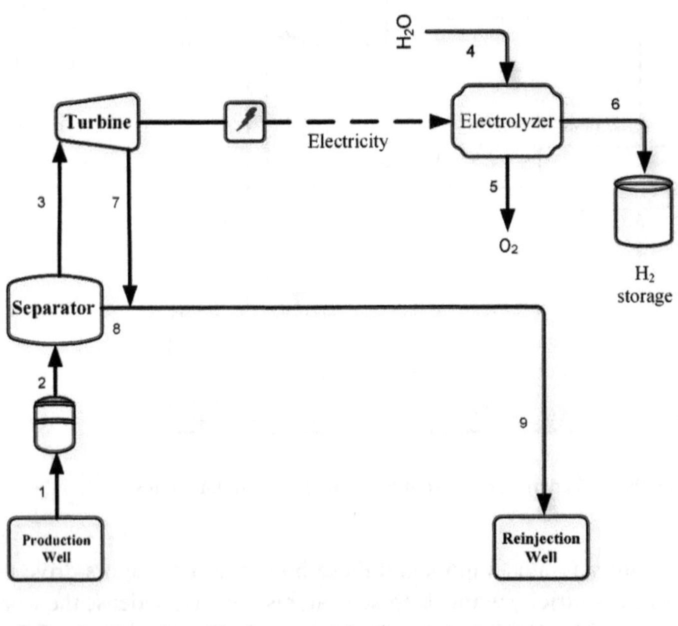

Fig. 12 Geothermal energy-based hydrogen production system

4.5 Biomass Gasification-Based Hydrogen Production Method

Biomass gasification is a process which is used to convert solid biomass fuel (wet/dry) into a combustible producer gas including carbon dioxide, carbon monoxide and hydrogen. This process is carried out by establishing the high-temperatures condition by providing controlled oxygen and steam amount without combustion. Although, biomass gasification produces some carbon emissions, but the power or thermal energy derived from the biomass combustion or gasification of the producer gas is considered as a renewable energy source because of the reason that plants consume CO_2 during photosynthesis. The biomass gasification receives the benefit of consuming the syngas is much more effective as compared to the direct combustion. This biomass gasification occurs at very high-temperature of 700–900 °C. Figure 13 shows the schematic of the biomass energy-based hydrogen production method. In the first step, air is fed to the air separation unit (ASU) which separates oxygen and supplies it to the gasifier. The steam and biomass fuel enter the gasifier from the top. A water gas shift reactor (WGSR) is employed after the gasifier which is used to convert the carbon monoxide from the gasifier into carbon dioxide by consuming steam and produces more hydrogen. A pressure swing adsorption unit is installed after WGSR which separates hydrogen from the waste gas.

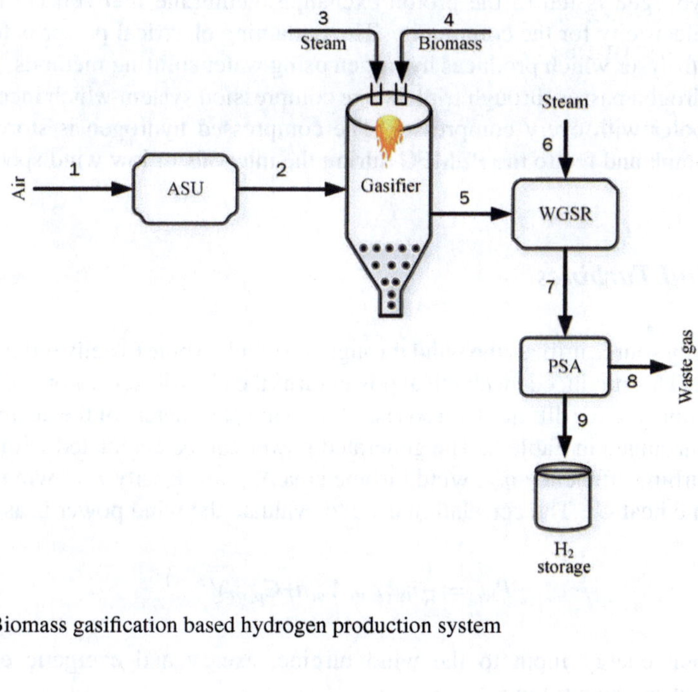

Fig. 13 Biomass gasification based hydrogen production system

5 Case Study

A case study using the wind energy source is presented in this section. In the first step of the source selection for hydrogen production, the wind energy source is found clean, abundant, affordable and reliable. Due to the intermittent nature of the wind energy source, a part of electric power generated by the wind turbines is supplied to the community instead of integrating a storage unit. During the intervals of low wind speed, the storage hydrogen is fed to the proton exchange membrane fuel cell (PEMFC) to generate electricity for the community. A part of electricity from the wind turbines is employed to the PEM electrolyser which produces hydrogen via water splitting method. With the advancements in renewable energy sources, the cost for the renewable energy has been dropped down. After deciding the appropriate source, the efficient, viable, abundant and reliable system utilization is necessary to generate optimized results. The designed case study is analyzed using Engineering Equation Solver (EES). The last step is to decide the services which the developed system is expected to provide, and present system is designed to provide the community with electricity, heat and hydrogen.

In this designed case study, the wind turbines are self-employed by the generator which convert mechanical work into electrical and further connected to an AC/DC converter. The system schematic diagram is shown in Fig. 14. A part of electric power produced is supplied to the community. During the intervals of low wind speed, the storage hydrogen is fed to the proton exchange membrane fuel cell (PEMFC) to generate electricity for the community. The remaining electrical power is fed to the PEM electrolyser which produces hydrogen using water splitting methods. The produced hydrogen passes through triple-stage compression system which incorporates an intercooler with every compressor. The compressed hydrogen is stored in the hydrogen tank and fed to the PEMFC during the intervals of low wind speed.

5.1 *Wind Turbines*

Wind energy source utilizes the wind through the wind turbine to deliver the mechanical power. This produced mechanical power turns the electric generators and electric work in employed for different purposes. The design parameters of the designed case study are arranged in Table 2. The generated power can be calculated with the help of wind turbine efficiency η_{wt}, wind turbine area A_{wt}, air density ρ_{air}, wind speed V and specific heat C_p. The correlation used to evaluate the wind power is as follows:

$$P_{wt} = \frac{1}{2}\eta_{wt}\rho_{air}A_{wt}n_T C_{p,wt}V^3 \quad (25)$$

The total energy input to the wind turbine, exergy and exergetic efficiency correlations can be written as:

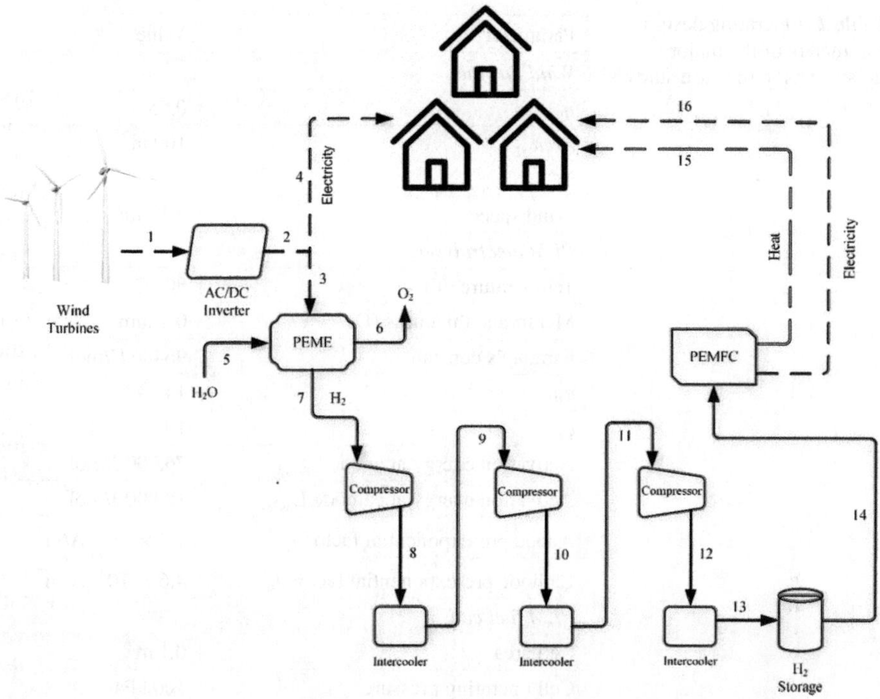

Fig. 14 Wind energy-based system for electricity, heat and hydrogen production

$$Input = \frac{1}{2}\rho_{air}A_{wt}n_T V^3 \tag{26}$$

$$\dot{E}x_{wt} = \frac{1}{2}\rho_{air}A_{wt}V^3 \tag{27}$$

$$\eta_{wt} = \frac{P_{wt}}{\dot{E}x_{wt}} \tag{28}$$

5.2 *PEM Electrolyser*

In the PEM electrolyser integrated with the wind turbines, hydrogen is produced via water splitting method and reaction is as follows:

$$\mathrm{H_2O} + \Delta\mathrm{H} \rightarrow \mathrm{H_2} + \frac{1}{2}\mathrm{O_2} \tag{29}$$

Table 2 Operating design parameters of the major subsystems in the case study

Parameter	Value
Wind turbine	
η_{wt}	0.55
$Area_{wt}$	100 m^2
n_T	1
Wind speed	4.17 m/s
PEM electrolyser	
Temperature (T)	80 °C
Membrane thickness (D)	0.1 mm
Faraday's constant	96486 C/mol
φ_a	14
φ_a	10
Activation energy at anode $E_{act,a}$	76,000 J/mol
Activation energy at cathode $E_{act,c}$	18,000 J/mol
Anode pre-exponential factor J_a^{ref}	1.7×10^5 A/m^2
Cathode pre-exponential factor J_a^{ref}	4.6×10^3 A/m^2
PEM fuel cell	
Cell area	0.1 m^2
Cell operating pressure	100 kPa
Cell operating temperature	80 °C
No of cells	2

here ΔH represents change in enthalpy. The energy required by the electrolyser can be written in terms of thermal and Gibbs energy.

$$\Delta H = \Delta G + T\Delta S \tag{30}$$

The production rate of hydrogen carried out by the PEM electrolyser can be written in terms of current density and Faraday's constant as follows:

$$\dot{N}_{H_2} = \frac{J_{el}}{2F} \tag{31}$$

The correlations for the excess electricity and the distribution of the cell voltage distribution into reversible cell potential, activation and Ohmic and concentration overpotentials and power density are as follows:

$$\dot{W}_{PEM} = J_{el}V \tag{32}$$

$$V = V_0 + V_{act,a} + V_{act,c} + V_{con,a} + V_{con,c} + V_{Ohmic} \tag{33}$$

$$V_0 = \frac{\Delta G}{nF} + \frac{RT}{nF}\ln\left(\frac{(p_{H_2})(p_{O_2})^{0.5}}{p_{H_2O}}\right) \tag{34}$$

$$V_{act,i} = \left(\frac{RT}{F}\right) sinh^{-1}\left(\frac{J_{el}}{2J_{o,i}}\right) \tag{35}$$

$$J_{o,i} = J_i^{ref}\exp\left(\frac{E_{act,i}}{RT}\right) \tag{36}$$

The Ohmic over-potential can be written in terms of current density J_{el}.

$$V_{Ohmic} = R_{PEM} J_{el} \tag{37}$$

5.3 Compressor

The balance equations for the first compressor can be written as:

$$\dot{m}_7 = \dot{m}_8 \tag{38}$$

$$\dot{m}_7 h_7 + \dot{W}_C = \dot{m}_8 h_8 \tag{39}$$

$$\dot{m}_7 s_7 + \dot{S}_{gen} = \dot{m}_8 s_8 \tag{40}$$

$$\dot{m}_7 ex_7 + \dot{W}_C = \dot{m}_8 ex_8 + \dot{E}x_d \tag{41}$$

5.4 Intercooler

The thermodynamic balance equations for the first intercooler can be represented as:

$$\dot{m}_8 = \dot{m}_9 \tag{42}$$

$$\dot{m}_8 h_8 = \dot{m}_9 h_9 + \dot{Q}_{out} \tag{43}$$

$$\dot{m}_8 s_8 + \dot{S}_{gen} = \dot{m}_9 s_9 + \frac{\dot{Q}_{out}}{T_0} \tag{44}$$

$$\dot{m}_8 ex_8 = \dot{m}_9 ex_9 + \dot{Q}_{out}\left(1 - \frac{T_0}{T_{IC}}\right) + \dot{E}x_d \tag{45}$$

5.5 Fuel Cell Subsystem

The correlation used to determine the PEMFC performance can be written in terms of the reversible cell potential, ohmic, activation and concentration overpotential.

$$E = E_r - E_{act} - E_{Ohm} - E_{conc} \tag{46}$$

The cathodic and anodic and total activation potential expression can be expressed with the help of universal gas constant, fuel cell temperature, electron transfer coefficient, number of fuel cells, Faraday's constant and exchange current density.

$$E_{act} = E_{act,an} + E_{act,ca} \tag{47}$$

$$E_{act,i} = \frac{RT_{FC}}{\alpha_i nF} \ln\left(\frac{J}{J_0}\right) \tag{48}$$

$$J_0(T) = 1.08 \times 10^{-21} \times \exp(0.086 \times T_{FC}) \tag{49}$$

$$E_r(T, P_i) = 1.482T - 0.000845T + 4.31 \times 10^{-5} \ln\left(p_{H_2} P_{O_2}^{0.5}\right) \tag{50}$$

The expressions for the concentration polarization, membrane water content λ_{mem} and membrane conductivity σ_{mem} can be represented as:

$$\sigma_{mem} = (0.005139\lambda_{mem} - 0.00326) \exp\left(1268\frac{1}{303} - \frac{1}{T_{FC}}\right) \tag{51}$$

$$\lambda_{mem} = 0.043 + 17.81a - 39.85a^2 + 39.85a^3,\ 0 < a \le 1 \tag{52}$$

$$E_{conc} = \frac{RT}{2F} \ln\left(1 - \frac{J}{J_{L,an}}\right) + \frac{RT}{2F} \ln\left(1 + \frac{P_{H_2} J}{P_{H_2O} J_{L,an}}\right) - \frac{RT}{4F} \ln\left(1 - \frac{J}{J_{L,ca}}\right) \tag{53}$$

The power generated by one cell and the cell stack can be calculated with the help of cell area, number of cells and electric work generated by each cell.

$$\dot{W}_{cell} = E(I) \times J \times A_{cell} \tag{54}$$

$$\dot{W}_{Stack} = n \times \dot{W}_{cell} \tag{55}$$

The fuel cell exergy destruction can be expressed as:

$$\dot{E}x_{dest,fc} = \dot{E}x_{H_2} + \dot{E}x_{O_2} - \dot{Q}_{fc} - \dot{W}_{stack} \tag{56}$$

5.6 Efficiencies

The system is designed in a way that a part of electricity extracted from the wind turbine is fed to the electrolyser and which produces hydrogen and stored hydrogen is used for power production through PEMFC during the period of low wind speed. Thus, the system efficiency is defined considering three different scenarios of low, average and high wind speed. The following equation represent the efficiency correlations for low, average and high wind speed respectively. Equation (57) represents the net electrical power output. For the high wind speed, the system efficiency can be defined as:

$$\dot{P}_{net} = \dot{P}_{wt} - \dot{W}_{comp1} - \dot{W}_{comp2} - \dot{W}_{comp3} \tag{57}$$

$$\eta_{ov} = \frac{\dot{P}_{net} + \dot{m}_{H_2} LHV_{H_2}}{Input} \tag{58}$$

The system efficiency when the wind speed is equal to the average wind speed:

$$\eta_{ov} = \frac{\dot{P}_{net}}{Input} \tag{59}$$

The energetic and exergetic efficiencies when wind speed is lower than the average speed and hydrogen is fed to the PEMFC to meet the electrical load:

$$\eta_{ov} = \frac{W_{stack} + \dot{Q}_{FC}}{\dot{m}_{H_2} LHV_{H_2}} \tag{60}$$

$$\psi_{ov} = \frac{W_{stack} + \dot{Q}_{FC}\left(1 - \frac{T_0}{T_{boundary}}\right)}{\dot{m}_{H_2} LHV_{H_2}} \tag{61}$$

5.7 Results and Discussions

The presented case study is important to be investigated under different operating conditions. The system efficiency is defined considering three different scenarios of low, average and high wind speed. The design investigated under different current densities, number of cells in fuel cell and wind speed to explore these effects on the system performance. Figure 15 exhibits the effect on current density of the ohmic, activation and concentration overpotential of the fuel cell. Three different types of lines are used to represent the ohmic, activation and concentration overpotential. The range of the current density is taken from 0 to 1600 mA/cm^2. It can be observed that increase

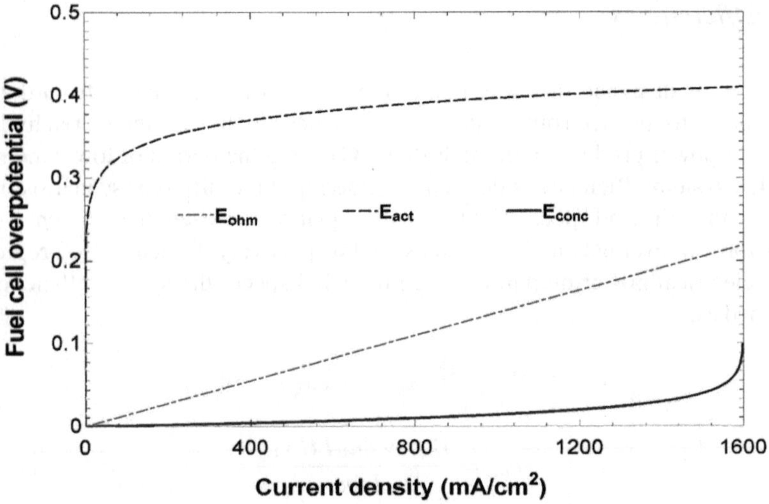

Fig. 15 Effect of current density on fuel cell overpotential

Table 3 Significant results of the designed case study

Parameter	Value
Wind turbine	
Wind turbine power	1157 kW
Wind turbine input	4292 kW
$\eta_{forlowV}$	0.2896
$\psi_{forlowV}$	0.2927
$\eta_{foreverageV}$	0.2695
$\eta_{forhighV}$	0.4425

in current density causes the ohmic, activation and concentration overpotential to increase gradually.

The significant results such as wind turbine input, power generated by wind turbine and system efficiencies under low, average and high wind speed are arranged in Table 3. The effect of the wind speed on the different operating parameters is important to be investigated for the proposed case study. Figure 16 displays the effect of the wind speed on the power generated by the wind turbine and the exergy destruction rates of the fuel cell and PEM electrolyser. The primary Y-axis represent wind power and secondary Y-axis is used to display the exergy destruction rates of the fuel cell and PEM electrolyser. Wind speed is considered in the range of 3.5–10 m/s. The different type of lines is used to represents three different parameters. It can be observed that with the rise in wind speed, the wind turbine power increases from 684 to 15953 kW, the exergy destruction of the PEM electrolyser rise from 12.82

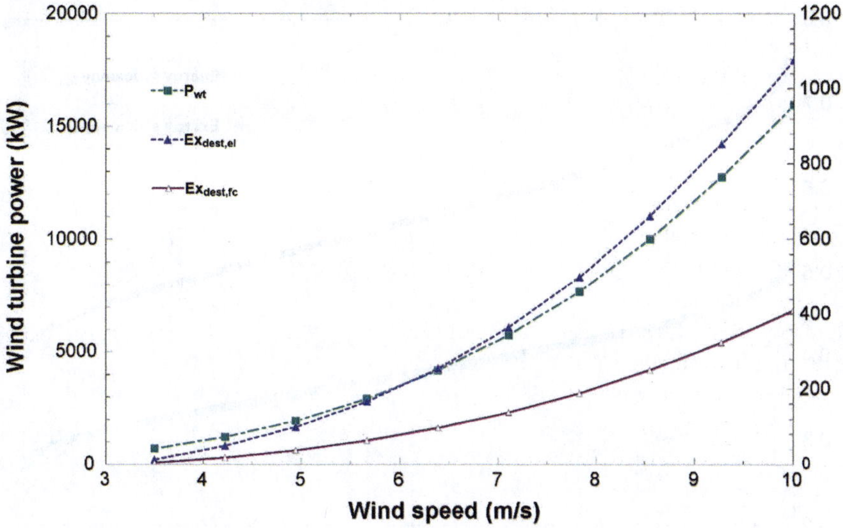

Fig. 16 Wind speed effect on the wind turbine power and exergy destruction rates

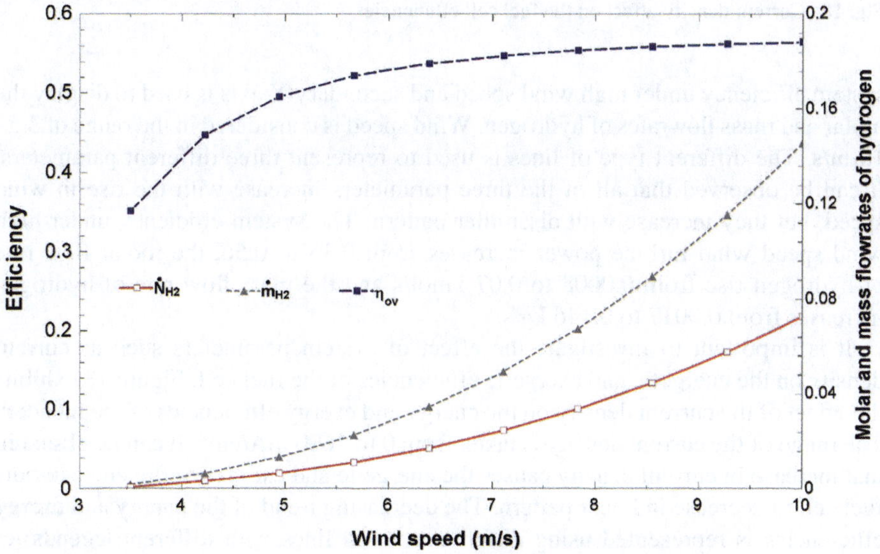

Fig. 17 Wind speed effect on the system efficiency during high wind speed and H_2 flowrate

to 1077 kW and the exergy destruction of the PEM fuel cell increases from 3.63 to 409.5 kW.

Figure 17 shows the effect of wind speed on the system efficiency under high wind speed and molar and mass flow rate of hydrogen. The primary Y-axis represent the

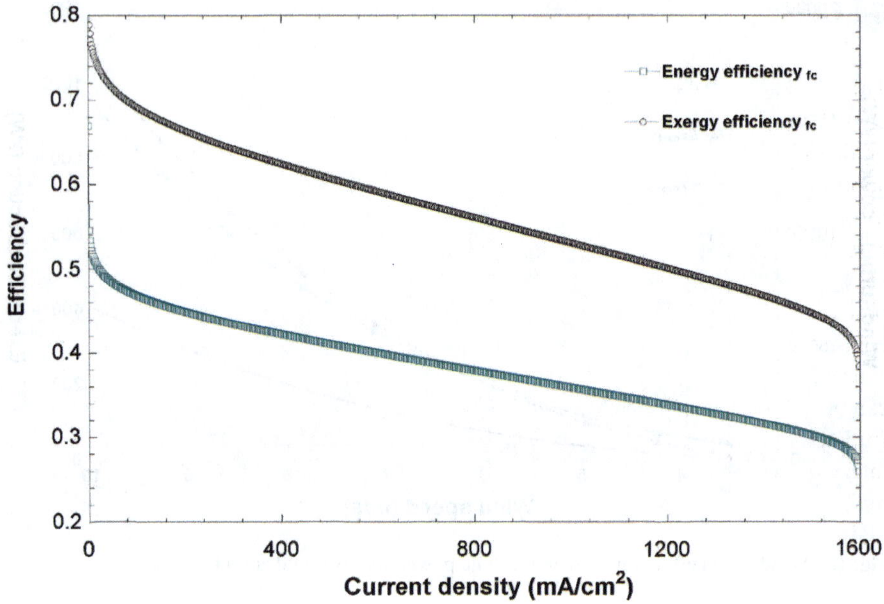

Fig. 18 Current density effect on the fuel cell efficiencies

system efficiency under high wind speed and secondary Y-axis is used to display the molar and mass flowrates of hydrogen. Wind speed is considered in the range of 3.5–10 m/s. The different type of lines is used to represent three different parameters. It can be observed that all of the three parameters increase with the rise in wind speed, but they increase with dissimilar pattern. The system efficiency under high wind speed wind turbine power increases from 0.35 to 0.56, the molar flow rate of hydrogen rise from 0.0008 to 0.07 kmol/s and the mass flow rate of hydrogen increases from 0.0017 to 0.146 kg/s.

It is important to investigate the effect of system parameters such as current density on the energetic and exergetic efficiencies of the fuel cell. Figure 18 exhibits the effect of the current density on the energy and exergy efficiencies of the fuel cell. The range of the current density is taken from 0 to 1600 mA/cm^2. It can be observed that increase in current density causes the energetic and exergetic efficiencies of the fuel cell to decrease in linear pattern. The decreasing trend of the energy and exergy efficiencies is represented using different colored lines with different legends. In the legends, symbol η signifies the energy efficiency while symbol ψ represents the exergy efficiency.

6 Conclusions

This chapter discusses the transition of the traditional energy sources into renewable energy sources using hydrogen as a source. The main focus of this study is derived on the three significant steps of selecting the renewable energy source, choosing the appropriate hydrogen production system to generate optimized results using electricity and heat and the services which can be provided by using hydrogen as energy carrier or storage medium or fuel. A classification has been made considering different types of the renewable energy sources for the production of heat, electricity or fuel extracted from the defined sources, methods for producing hydrogen from these commodities and hydrogen usage for different purposes. A system for each renewable source is presented in the renewable hydrogen production methods section, which can be integrated with the renewable energy source for hydrogen production. A case study using the wind energy source is presented and discussed in detail. During the intervals of high wind speed, a part of electricity is supplied to the community while remaining is employed to the proton exchange membrane electrolyser to produce hydrogen. The produced hydrogen is stored by passing through multistage compression system and stored in the storage tank. The generated electrical power is supplied directly to the community during average wind speed and during the intervals of low wind speed, the stored hydrogen is employed to the PEM fuel cell which produces electrical power and heat for the community. The designed case study is investigated under different operating conditions using numerous parametric studies. All the renewable energy sources are covered in detail and the potential implementation hydrogen as a source for the global transition to 100% renewable energy is discussed in detail.

Nomenclature

A	area (m^2)
E	actual voltage of cell (V)
$E_{act,a}$	activation overpotential at anode (V)
$E_{act,c}$	activation overpotential at cathode (V)
E_{conc}	concentration overpotential (V)
E_{ohmic}	ohmic overpotential (V)
$\dot{E}x$	exergy rate (kW)
ex	specific exergy, kJ/kg
$\dot{E}x_{dest}$	exergy destruction (kW)
F	Faraday constant, C/mol
H	specific enthalpy, kJ/kg
J	current density, A/m^2
J_{oa}	anodic exchange current density (A/m^2)
J_{oc}	cathodic exchange current density (A/m^2)

LHV	lower heating value (kJ/kg)
$\dot{m}$	mass flow rate, kg/s
N	molar flow rate, mole/s
n	number of moles
Q	heat rate (kW)
R	gas constant, kJ/kmol K
Re	Reynolds number
s	specific entropy, kJ/kg K
$\dot{S}_{gen}$	Entropy generation rate (kW/K)
T	temperature (°C)
V	wind speed, m/s
$\dot{W}$	work rate (kW)

Greek Letters

η	energy efficiency
ψ	exergy efficiency
ϕ	fuel exergy to energy ratio
σ_{mem}	membrane conductivity, 1/(Ω cm)
λ_{mem}	membrane water content
λ	stoichiometric constant
α	symmetry factor
$\alpha, \beta, \delta, \gamma$	atoms of C, H, N and O

Subscripts

0	ambient conditions
act	activation
a	anode
c	cathode
comp	compressor
el	electrolyser
FC	fuel cell
PEM	proton exchange membrane
PV	photovoltaic

Acronyms

Cu-Cl	copper chlorine

EES	engineering equation solver
LHV	lower heating value
LNG	liquified natural gas
LPG	liquified petroleum gas
OTEC	ocean thermal energy conversion
PEMFC	proton exchange membrane fuel cell
PEM	proton exchanger membrane
PV	photovoltaic
SHF	solar heliostat field
SOE	solid oxide electrolyser
WGSR	water gas shift reactor

References

Agency IE (2019) The future of hydrogen executive summary and recommendations. Japan

Chiari L, Zecca A (2011) Constraints of fossil fuels depletion on global warming projections. Energy Policy. 39:5026–5034

Dincer I (2010) Global warming: engineering solutions. Springer, London, New York

Dincer I, Acar C (2015) A review on clean energy solutions for better sustainability. Int J Energy Res 39:585–606

Dincer I, Rosen M (2003) Thermal energy storage systems and applications. Wiley, Hoboken, NJ

Dincer I, Zamfirescu C (2011) Sustainable energy systems and applications. Springer, New York, Dordrecht, Heidelberg, London. https://doi.org/10.1007/978-0-387-95861-3

Ishaq H, Dincer I (2019) A comparative evaluation of OTEC, solar and wind energy based systems for clean hydrogen production. J Clean Prod 118736. https://doi.org/10.1016/j.jclepro.2019.118736

U.S. Department of Energy. DOE technical targets for hydrogen production from electrolysis| Department of Energy. Fuel cell Technol. Off. https://www.energy.gov/eere/fuelcells/doe-technical-targets-hydrogen-production-electrolysis

Solar Hydrogen's Role for a Sustainable Future

Canan Acar

Abstract In this study, hydrogen's role during the transition to 100% renewable energy systems is discussed thoroughly, and the importance of sustainable hydrogen production is highlighted. For a successful transition to hydrogen-based renewable energy systems, hydrogen has to be produced in a clean, reliable, affordable, efficient, and safe manner. Therefore, in the second part of this study, a comprehensive life cycle assessment of solar hydrogen production options is conducted. The selected clean hydrogen production options are steam methane reforming, conventional electrolysis, photoelectrochemical cells, PV electrolysis, and photocatalysis. A complete source to service approach is taken when evaluating the environmental and technical performance of the selected hydrogen production options. Greenhouse gas (GHG) emissions, resource use, fossil fuel use, water use, energy and exergy efficiencies, and cost of hydrogen are the selected sustainability performance criteria. The selected hydrogen production methods are compared based on these performance criteria. In the next part, the performance evaluation results of each option are normalized and ranked in the 0–10 range where 0 gives the least sustainable manner, and 10 is the hypothetical ideal case where there is no damage to the environment, zero resource and water use, and 100% energy and exergy efficiencies, and zero cost. The GHG emissions, resource use, fossil fuel use, and water use results indicate that photoelectrochemical cells (PEC) is the most advantageous. Steam methane reforming has the highest efficiencies and the lowest. When all of the selected performance criteria are considered together, PEC has the highest sustainability rankings (5.24/10), and steam methane reforming has the lowest (3.24/10).

Keywords Hydrogen · Energy · Exergy · Solar · Sustainability

C. Acar (✉)
Faculty of Engineering and Natural Sciences, Bahcesehir University, Istanbul, Turkey
e-mail: Canan.Acar@eng.bau.edu.tr

T. S. Uyar (ed.), *Accelerating the Transition to a 100% Renewable Energy Era*, Lecture Notes in Energy 74, https://doi.org/10.1007/978-3-030-40738-4_14

1 Introduction

Hydrogen has a key role during the transition to clean energy systems for a sustainable future. When produced from clean and renewable resources like solar energy and water, hydrogen can provide safe, reliable, affordable, efficient, clean, and efficient energy to a wide variety of end users. Hydrogen plays an especially important role during the transition to 100% renewable energy systems in terms of energy security in supply and demand. Therefore, the recent literature shows that there have been numerous studies towards more efficient, clean, and cheap solar hydrogen production.

The first step towards truly sustainable energy systems via hydrogen is the production. For that reason, solar hydrogen has to be produced in affordable, reliable, environmentally benign, and efficient manners with zero or minimal resource depletion (Acar and Dincer 2019). In the literature, different studies are focusing on different aspects of solar hydrogen production. For instance, Wang et al. (2017) have focused on photocatalysis and photoelectrochemical cells (PEC). The authors have reviewed the recent progress of black TiO_2 for photocatalytic hydrogen evolution and PEC water splitting, along with a detailed introduction to its unique structural features, optical property, charge carrier transfer property, and related theoretical calculations. Lee et al. (2018) have provided a review of advanced hydrogen passivation applied on p-type, n-type, and upgraded metallurgical grade crystalline silicon solar cells, respectively. Acar et al. (2016) have examined photocatalytic hydrogen generation as a key to solve climate crisis issues by enabling the transition to 100% renewable energy. The authors have considered social, environmental, and economic characteristics of hydrogen production while evaluating different types of photocatalysts. De Crisci et al. (2018) have highlighted some of the methods of eliminating hydrogen sulfide pollution via partial oxidation, reformation, and decomposition techniques and approaches. The authors have proposed an approach to convert hydrogen sulfide to sulfur, water and, more importantly, hydrogen. With their approach, hydrogen is produced with zero GHG emissions, and the proposed method also helps to lower and eventually eliminate hydrogen sulfide.

In the literature, there have been numerous examples of innovative approaches to sustainable hydrogen production. For instance, Khetkorn et al. (2017) have reviewed the recent technological progress, enzymes involved, and genetic as well as metabolic engineering approaches towards sustainable hydrogen production from microalgae. Research and development activities in the field of energy and cost-effective solar and wind hydrogen have been summarized by Saeedmanesh et al. (2018). Im-orb et al. (2018) have developed a user-friendly solid oxide electrolysis cell (SOEC) model in a flowsheet simulator. The authors have developed a model to investigate the effects of key process parameters such as operating temperature, current density, steam concentration, sweep gas type and several cells, on the environmental, energetic, and economic performance of the SOEC for sustainable hydrogen production. Sharma (2019) have presented hydrogen production from biomass carbohydrates by using different types of catalysis such as chemical catalysis, biocatalysis, and their combinations. Zhang and Wang (2015) have investigated semiconductor photocatalysts for solar water splitting at 600 nm wavelength for sustainable hydrogen production.

Kadier et al. (2016) have provided a brief overview of recent advances in research on scalable microbial electrolysis cells (MEC) reactor designs and configurations for sustainable hydrogen gas production. Bolatkhan et al. (2019) have investigated the sustainable methods of hydrogen production with the help of two large groups of phototrophic microorganisms, which are microalgae and cyanobacteria. Santos et al. (2018) have designed a new family of highly effective catalysts for low-temperature water gas shift reaction based on gold modified copper-zinc mixed oxides. Nikolaidis and Poullikkas (2017) have provided a comparative overview of the major hydrogen production methods, including an overall comparison of conventional and alternative methods from fossil fuels and renewables. Wang and Yin (2018) have reviewed various biomass as feedstock, including waste activated sludge produced from the wastewater treatment plant, algae, agricultural residuals, and municipal wastes used for biological hydrogen production.

Dincer and Acar (2017) have introduced the critical perspectives of innovation for specifically for hydrogen production under a new concept (so-called: 18S concept), covering source, system, service, scope, staff, scale-up, safety, scheme, sector, solution, stakeholder, standardization, subsidy, stimulation, structure, strategy, support and sustainability. Jiang et al. (2018) have demonstrated an innovative approach to biological hydrogen production and introduced a tandem inorganic-biological hybrid by combining $AgInS_2/In_2S_3$ and a facultative anaerobic bacterium, *Escherichia coli*, for sustainable biological hydrogen production. Show et al. (2018) have presented an elucidation on development in biohydrogen encompassing innovative biological pathways, bioreactor designs and operation, and techno-economic evaluation. The authors have also outlined the challenges and prospects of biohydrogen production. Esmieu et al. (2018) have investigated alternative methods from protein engineering to artificial enzymes and utilized innovative biological and biomimetic approaches towards sustainable hydrogen production. He et al. (2017) have taken an innovative approach to photocatalysis by introducing novel metal-free catalysts for highly efficient and stable photocatalytic hydrogen production from water splitting.

For sustainable hydrogen production, it is essential to investigate the system performance, including all system components thoroughly. For instance, Shinagawa and Takanabe (2017) have investigated the impact of electrolyte engineering on the performance and sustainability of hydrogen production systems. Ahmed and Dincer (2018) have reviewed the engineering design principles for different hydrogen production system configurations, including single, dual/tandem photoelectrodes, tandem PEC-PV, and multi-junction designs. All of these studies have pointed out the need for considerable efforts from both technical and managing aspects to achieve a full-scale application of sustainable hydrogen production from affordable, reliable, clean, and abundant sources.

In the literature, there is a need for studies comprehensively investigating solar hydrogen production systems from environmental, technical, and economic perspectives quantitatively based on their life cycle performances. For this reason, in this study, the impact of solar hydrogen production systems on the environment is quantitatively investigated by comparing their life cycle GHG emissions, fossil fuel use, water use, and resource use together with their energy and exergy efficiencies and cost. The life cycle term used in this study follows the source-system-service approach and includes all energy and material source harvesting and processing into account as well. Steam methane reforming (SMR), conventional electrolysis (CE), PEC, PV electrolysis (PE), and photocatalysis (PC) is the selected hydrogen production options for comparison purposes. For a better and clearer insight on the true sustainability of the selected options, the environmental, economic, and technical performance results of each option are normalized and ranked to highlight their strengths and weaknesses and provide future directions.

2 How Hydrogen Empowers the Transition to 100% Renewable Energy

The need for an energy transition towards 100% renewable sources is extensively recognized, and the demand for renewables has been increasing steadily. On the other hand, there are several implications and challenges which must be resolved before a complete switch from fossil fuels to renewables. These issues require a concerted, collaborative, and interdisciplinary effort to come up with innovative solutions. Among the possible solutions towards the 100% renewable energy systems, hydrogen has the potential to be an effective enabler of this transition, since it provides a clean, sustainable, reliable and safe alternative for resolving many problems that stand in the way of resilient sustainable and renewable energy systems.

Hydrogen is the candidate to replace the fossil-based energy systems with renewables, which is the key to a sustainable future. What is more, hydrogen has the potential to enable long-term energy independence and security for all types of end-users in the energy supply and distribution. With the advancements in technology and new developments in materials science through vigorous R&D efforts, most of the challenges associated with hydrogen energy systems have been tackled, and the costs have been decreasing steadily as the scale of the system keep increasing.

The energy transition requires an interdisciplinary approach and crosscutting technologies. And it should be noted that there is no single technology alone which can enable full decarbonization of energy systems and 100% renewable energy. Therefore, a strong collaboration of the industry, academia, and governments is needed to design, develop, build, test, and commercialize large-scale renewable energy systems. Hydrogen seems to be a promising candidate for truly sustainable energy systems: it is a multipurpose energy carrier, it can be stored in a wide variety of options, and it can be used to meet the energy demands of all kind of end-users in the market. When it is produced from 100% renewable energy, hydrogen gives the industry, buildings, transportation, and all other industries and sectors a secure and reliable access to clean energy at all times without the burden of excess renewable

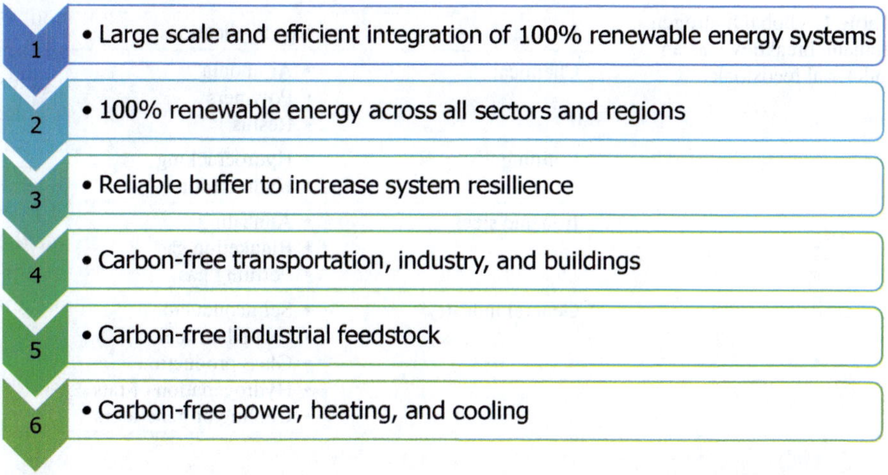

Fig. 1 Hydrogen's role towards renewabilization and decarbonization of energy systems for a sustainable future

electricity. Hydrogen's substantial benefits for all types of end users are summarized in Fig. 1.

Hydrogen can be used to decarbonize the global power and heat consumption with 100% renewable energy systems. Hydrogen is capable of providing high-grade heat which could be used to meet a wide range of energy requirements which is very challenging to address via electricity. As a chemical fuel, hydrogen can be transported to longer distances than electricity with considerably lower transmission losses. Besides, hydrogen can be used in mobile applications such as cars, buses, trains, trucks, planes, ships, etc. Hydrogen is also extensively used in different industrial processes including ammonia and methanol synthesis, fertilizer production, and so on. All of these advantages make hydrogen the key solution in the transformation of the global energy system towards renewables.

Currently, 95% of the global hydrogen supply is produced from fossil fuels. And the largest share of hydrogen demand comes from the industry for the synthesis of various chemicals such as ammonia. Hydrogen is also mainly used in refining for hydrocracking and desulfurization of fuels. In addition, a large portion of the global fertilizer industry is produced from ammonia synthesized in the Haber-Bosh process. Haber-Bosh process utilizes the hydrogen which is the product of natural gas steam reformation. The breakdown of global hydrogen demand by the industry is given in Table 1. Currently, hydrogen is produced mainly from fossil fuels, which is largely contributing to a wide range of issues including climate change. For that reason, it is very important for the major consumers of hydrogen to switch to renewable sources and completely eliminate the use of fossil fuels for hydrogen production. By doing so, they would not only reduce their CO_2 emissions and costs in the long-term in an effective manner but also fasten the transition to 100% renewables for a truly sustainable future.

Table 1 Global hydrogen demand breakdown as an industrial feedstock

Sector	Key applications
Chemical	• Ammonia • Polymers • Resins
Refining	• Hydrocracking • Hydrotreating
Iron and steel	• Annealing • Blanketing gas • Forming gas
General industry	• Semiconductor • Propellant fuel • Glass production • Hydrogenation of fats • Cooling of generators

It is important to note that fossil fuels cause dependence on import and/or export of these sources since they are found only in the limited regions around the world. Importing fossil fuels needs transportation and the entire supply chain infrastructure and this process emits CO_2. In addition to the social, political, and environmental impacts of the whole fossil fuel supply chain, there are noise and pollution issues related to fossil fuel consumption at the end-user site. In addition, remote locations often rely on diesel generators, which are noisy and polluting the environment with GHG emissions. With hydrogen being the storage medium for 100% renewable energy systems, these regions' dependence on fossil fuel extraction, processing, transportation, and use, as well as import can be eliminated, which would essentially increase energy security in these regions in a clean and reliable manner.

Global coal, oil, and natural gas reserves are not distributed homogeneously around the world. For that reason, almost every country in the world relies on fossil fuel imports and/or exports, Cutting the reliance on these imports and exports enhances the resilience of these countries by making them more energy independent with energy security through 100% renewable energy systems. For this manner, microgrids can be developed with hydrogen as a storage medium for renewables. These microgrids can effectively support all types of end users have secure and uninterrupted access to reliable renewable energy. Having reliable and uninterrupted access to energy and material resources at all times is considered as the first step towards sustainable communities. Decentralization of energy systems with locally available renewables not only eliminates the transmission and distribution losses but also the negative environmental impact of the entire fossil fuel supply chain and empowers local communities.

In addition to their significant energetic, economic, and environmental benefits, 100% renewable energy based microgrids are also smarter than traditional centralized grid structures. By including hydrogen as a storage medium for renewables, new collaboration and business model opportunities are generated for cleaner communities to solve climate change issues. In these systems, hydrogen is used to provide

Table 2 Potential benefits of hydrogen for different sectors

Sectors		Hydrogen use
Industry	Cement	Kiln heat source
	Iron and steel	Reduction agent and heat source
	Ammonia	Input to Haber-Bosch process
	Ethylene	Heat source Plastics feedstock
Transportation	Aviation	Fuel cells Internal combustion engines
	Land (e.g., cars, trucks, buses, etc.)	
	Sea (e.g., ships)	
Buildings	Heating and cooling	As a substitute for natural gas
Electricity	Renewable energy storage	Storage medium for surplus renewable electricity to match supply and demand

heating, cooling, and power and as a transportation fuel. Therefore, all types of industries, from residential services to transportation, chemicals, buildings, and so on are expected to have substantial benefits from hydrogen. For that reason, hydrogen is seen as the driver for a clean, independent, and energy secure future. With renewables, hydrogen has the potential to enhance energy security and system resilience of different sectors in the following ways (Table 2):

- In the transportation sector, decarbonizing short- and long-distance driving, trucking, shipping, and aviation requires greater energy densities than batteries currently offer. And the current grid system is not capable of delivering renewable electricity to longer distances. Besides, electrification of the entire transportation sector would generate a load that cannot be met with the current grid infrastructure. On the other hand, hydrogen has the potential to address the issues related to electrification.
- In the steel sector, blast furnace emissions arise from the iron reduction process as well as the energy input requirements. Replacing coking coal with hydrogen as the reduction agent may be a key route to decarbonization, which has already been embraced by some major companies within the industry.
- In other industrial processes (e.g., cement and ethylene production), the use of a decarbonized heat source such as hydrogen may prove more cost-effective than direct electrification. Because further developments are still required before electric furnaces can be commercially deployed at large-scales.
- Last, but not least, in residential heating, it is possible to use hydrogen piped by converting the existing natural gas grid.

In this study, the importance of hydrogen and its key role towards the decarbonization of the whole global energy chain is discussed up to this point. For the hydrogen economy to successfully mature and reach larger scales in all over the world, the first challenge to tackle is hydrogen production. Sustainability requires hydrogen to be produced in a safe, affordable, reliable, clean, and effective manner. And with this motivation in mind, in this study, a comprehensive environmental, economic, and thermodynamic performance evaluation of different solar hydrogen production options is conducted.

3 Solar Hydrogen Production

PEC has become a very promising solution for the efficient, reliable, clean, and affordable hydrogen production challenge. On the other hand, PEC technologies have several challenges which can be summarized as low photon-to-hydrogen energy conversion efficiency, slow electron-hole separation rate, and rapid electron-hole recombination issue. For this reason, there have been many studies focusing on different aspects of PEC to enhance their performance by lowering costs and increasing efficiencies. When the challenges mentioned above are addressed via advancements in materials science and technologies, sustainable hydrogen production could be attainable via PEC. The chemical reactions in PEC water splitting are as follows:

$$\text{Anode:}\quad 2H_2O \rightarrow 4H^+ + O_2 + 4e^- \tag{1}$$

$$\text{Cathode:}\quad 4H^+ + 4e^- \rightarrow 2H_2 \tag{2}$$

$$\text{Overall:}\quad 2H_2O \rightarrow 2H_2 + O_2 \tag{3}$$

Using earth-abundant catalysts for efficient and affordable hydrogen production is a common strategy used in the literature. For instance, Morales-Guio et al. (2015) have reported PEC-based hydrogen production in alkaline solutions, which are more favorable than acidic solutions for the complementary oxygen evolution half-reaction. The authors have shown that amorphous molybdenum sulfide is a highly active hydrogen evolution catalyst in basic medium. The authors have developed catalyst coated Cu_2O photoelectrodes which exhibit high PEC activity for hydrogen evolution in alkaline solutions. Ding et al. (2016) have focused on molybdenum disulfide (MoS_2) and related compounds as inexpensive alternatives for hydrogen evolution reaction catalysis and PEC water splitting. In their studies, the authors have considered key approaches to improving the intrinsic catalytic activity and overall catalytic performance and the developments in combining MoS_2 with semiconductors to realize solar-to-fuel conversion. The authors have also discussed different design approaches for efficient PEC water-splitting systems and some important challenges and future directions for earth abundant hydrogen evolution reaction electrocatalysis and PEC water splitting.

Kwon et al. (2016) have developed wafer-scale, transferable, and transparent thin-film catalysts based on MoS_2, which consists of cheap and earth-abundant elements for highly efficient and cheaper PEC-based hydrogen production. Zhang et al. (2016)

have reported the rational design of a novel 3D p-Si/NiCoSe$_x$ core/shell nanopillar (NP) array photocathode via uniform photo-assisted electrodeposition of NiCoSe$_x$ electrocatalyst on bamboo shoot-like Si NP array backbones. The authors have shown that the design of p-Si/NiCoSe$_x$ core/shell NP arrays offers a new strategy for preparing highly efficient PEC-based solar energy conversion devices. Zhang et al. (2015) have reported amorphous MoS$_x$Cl$_y$ as a high-performance electrocatalyst for both electrochemical and PEC-based hydrogen generation. The authors have shown that the MoS$_x$Cl$_y$ electrocatalysts exhibit stable and high catalytic activity toward the hydrogen evolution reaction.

Fan et al. (2016) have synthesized and characterized TiO_2/reduced graphene oxide/Cu_2O heterostructure constituted by TiO_2 nanowires, reduced graphene nanostructures, and Cu_2O. The authors have shown that their complex exhibits significantly enhanced PEC performance, including high photocurrent density and hydrogen production at higher efficiencies. Zheng et al. (2016) have fabricated strongly coupled Nafion molecules and ordered porous CdS networks for efficient visible-light PEC hydrogen evolution. The authors have shown that the Nafion layer coating shifts the band position of CdS upward and accelerates charge transfer in the photoelectrode/electrolyte interface. Tong et al. (2017) have reported an environmentally friendly, high-efficiency PEC device in which the photoanode consists of a mesoporous TiO_2 film sensitized with heavy metal-free, near-infrared (NIR) colloidal $CuInSe_xS_{2-x}$ (CISeS) quantum dots. The authors have reported that their PEC device is environmentally friendly with outstanding stability, cost-effectiveness, and high efficiency.

Gross et al. (2016) have reported the improvement of a dye-sensitised p-type nickel oxide (NiO) photocathode with a hexaphosphonated Ru(2,2′-bipyridine)$_3$ based dye (RuP3) and a tetraphosphonated molecular $[Ni(P_2N_2)_2]^{2+}$ type proton reduction catalyst (NiP) for the photoreduction of aqueous protons to H_2. With this method, the authors have concluded that the PEC could achieve higher solar-to-hydrogen efficiencies in the visible light region. Tsai and Hsu (2015) have established the use of CdSe/graphene quantum dot (QD) nanoheterostructures as the photoanode for outstanding PEC-based hydrogen production. Zhang et al. (2015) have synthesized a dense array of CdS–ZnS core-shell nanorods. This photocatalyst array has been reported to have high photocorrosion resistance, charge separation and transportation efficiencies, photocurrent density, photon to electron conversion efficiency and stability, which are essential for sustainable PEC-based hydrogen production.

Hydrogen can be produced from solar electricity as well via electrolysis, which is usually referred to as PV-electrolysis. PV-electrolysis is an effective approach to utilize excess (surplus) electricity from the PV panels. In the literature, there are numerous studies focusing on the performance enhancement of solar electrolysis. These studies focus mainly on efficiency enhancement and cost reduction of hydrogen production. For instance, Tebibel et al. (2017) have investigated the energetic, economic, and environmental aspects of PV electrolysis. The authors focused on the optimal management strategy to achieve high system efficiency and safe operation. Shaner et al. (2016) have performed a techno-economic analysis of PEC and PV electrolysis hydrogen production of 10,000 kg H_2/day (3.65 kilotons per year) to assess the economics of each technology and to provide a basis for comparison between these technologies as well as within the broader energy landscape. Dahbi et al. (2016) have presented a new way for hydrogen production by adapting the electrolysis to a renewable source of energy such as PV to generate the maximum quantity of hydrogen.

Bhattacharyya et al. (2017) have addressed the design of a standalone solar PV energy system that meets the energy requirements of the electrolysis process, followed by the performance analysis under different environmental conditions. The authors have also presented a step-by-step simplified approach for the preliminary PV power system design and analysis for an electrolysis-based hydrogen production unit. Kumari et al. (2016) have demonstrated a new method to produce hydrogen fuel from solar energy by splitting seawater vapor from ambient humidity at near-surface ocean conditions. The authors have used a proton exchange membrane electrolyzer with seawater-humidified air at 80% relative humidity at the anode, and dry N_2 at the cathode and maintained a sufficient electrolysis current density to operate near the maximum energy-conversion point. Huang et al. (2016) have simulated hydrogen production scenarios for fuel cell electric vehicle (FCEV) hydrogen refueling stations by examining an electrolysis hydrogen production system powered by small wind turbines and a PV system.

4 Sustainability Investigation of Hydrogen Production Options

The sustainability investigation includes GHG emissions, resource use, fossil fuel use, water use, energy and exergy efficiencies, and cost. All of the sustainability performance criteria values are calculated based on the life cycle performances, from the source to the end user, by using the GaBi (Life Cycle Assessment (LCA) modelling and reporting software). It should be noted that life cycle results sum up all of the energy and emissions associated with the process inputs, including upstream. Energy and emission adjustments associated with by-products are not included. In the GHG emissions category, the following are considered:

- CO_2 emissions
- SO_x emissions
- NO_x emissions
- CO emissions
- CH_4 emissions
- Particulate matter (PM_{10} and $PM_{2.5}$) emissions
- Volatile organic compound (VOC) emissions
- Primary organic carbon emissions
- Black carbon emissions.

All emissions are given in units as kg emissions per kg of hydrogen produced. Resource use indicates the amount of resource energy used (MJ) per kg of hydrogen produced. This resource energy can be natural gas, grid electricity (conventional electrolysis), or solar energy (PEC, photocatalysis, or PV electrolysis). Fossil fuel use indicates the amount of fossil energy used (MJ) per kg of hydrogen produced. And water use is the total amount of water usage per kg of hydrogen production. It should be noted that the life cycle includes the preparation of resources, distribution, transportation, processing, waste handling, and the hydrogen production process itself. That is why the selected options have fossil fuel use despite the fact that the production processes alone do not involve any fossil fuel use. Cost indicates the life cycle cost of the selected options. The energy efficiency of the hydrogen production options are calculated based on the following equation:

$$\eta = \frac{HHV_{H_2}}{Resource\,use\left(\frac{MJ}{kg\,H_2}\right) + Fossil\,fuel\,use\left(\frac{MJ}{kg\,H_2}\right)} \quad (4)$$

Here, η is the energy efficiency and HHV_{H_2} is the higher heating value (also known gross calorific value or gross energy) of hydrogen which is defined as the amount of heat released by one kg hydrogen (initially at 25 °C) once it is combusted and the products have returned to a temperature of 25 °C. HHV_{H_2} takes into account the latent heat of vaporization of water in the combustion products. And the exergy efficiency of the selected hydrogen production options are calculated as

$$\psi = \frac{ex_{H_2}^{ch}}{Exergy\,content\,of\,the\,resource\left(\frac{MJ}{kg\,H_2}\right) + Exergy\,content\,of\,the\,fossil\,fuel\left(\frac{MJ}{kg\,H_2}\right)} \quad (5)$$

In this equation, ψ is the exergy efficiency and $ex_{H_2}^{ch}$ is the chemical exergy content of one kg hydrogen. For comparison purposes, all chemical exergies are calculated at standard state, which is 25 °C and 1 atm. And the hydrogen cost is gathered from the recent literature.

In the next step, for a more comprehensive investigation, the environmental, economic, and technical performance results are normalized and ranked within the range of 0–10 where 0 indicates the least desired performance and ten is given to a hypothetical ideal case. 0 is given to the highest GHG emissions, resource use, fossil fuel use, water use, and cost and 0% energy and exergy efficiencies. On the other hand, ten is assigned to a non-existing ideal situation where the hydrogen production option has zero emissions, cost, resource use, fossil fuel use, and water use and 100% energy and exergy efficiencies. Given these conditions, the normalized rankings of the GHG emissions, resource use, fossil fuel use, water use, and cost are calculated based on:

$$Rank_i = \frac{max - i}{max} \times 10 \quad (6)$$

Here, $Rank_i$ is the rank of the selected hydrogen production option (i.e., SMR, CE, PEC, PVE, or PC). And max is the maximum GHG emissions, resource use, fossil fuel use, water use, or cost among the selected options while "*i*" is the GHG emissions, resource use, fossil fuel use, water use, or cost of the selected hydrogen production option. And the energy and exergy efficiency rankings of the selected hydrogen production options are calculated from the following equation:

$$Rank_i = efficiency_i \times 10 \tag{7}$$

Here, since the energy and exergy efficiencies of each option are within the range of 0–1, the efficiency data are normalized by simply multiplying them with 10. The comprehensive performance investigation results are presented and discussed in detail in the next section with some possible future directions.

5 Results and Discussion

In this part of the present study, comprehensive sustainability examination outcomes of the carefully chosen hydrogen production methods are presented, and the GHG emissions, resource, fossil fuel, and water use, energy and exergy efficiencies, and the hydrogen production costs of the selected options are discussed in detail.

Achieving the objectives of the Paris Agreement, substantial greenhouse gas emission reductions are required across all sectors and all over the world. Currently, the consensus is to substantially increase the share of renewable energy in the global final energy consumption. But this study highlights the importance of 100% renewable energy systems for a carbon-free society. Despite that, one-third of global energy-related emissions come from the industry, where currently no viable economic alternative to fossil fuels exists. This study shows how hydrogen from renewable sources could be a critical element of a strategy to tackle this issue.

The second section of this study explains the advantages of hydrogen as a fuel, including providing high-grade heat; addressing a range of energy needs that direct electrification cannot meet; and replacing fossil fuel-based feedstocks, such as natural gas, in high-emission applications of the industry sector.

Hydrogen has zero GHG emissions during the storage and end-use steps. However, for a completely clean and renewable energy supply chain, hydrogen must be produced from renewables in the cleanest possible way. Therefore, it is essential to reduce, and possibly eliminate, GHG emissions of hydrogen production. Making the current hydrogen production completely renewable and emissions-free is challenging but can have a positive impact on reaching lowering global CO_2 emissions target and can play an important role in realizing cost declines. For this reason, the first investigation in this study is to comparatively evaluate the lifecycle GHG emissions of the selected hydrogen production options.

In Fig. 2, GHG emissions are compared, and the results show that steam methane reforming has the highest emissions (about 9 kg GHG/kg H_2 produced), followed by conventional electrolysis, which is around 7 kg GHG/kg H_2 produced. The lowest

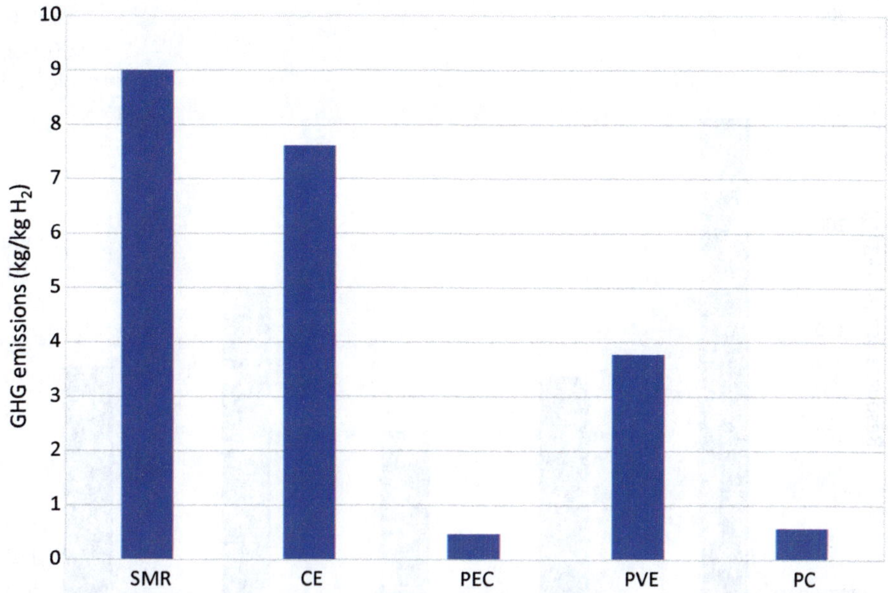

Fig. 2 GHG emissions of the selected hydrogen production options

GHG emissions come from PEC (less than 0.5 kg GHG/kg H_2 produced). PV electrolysis has about 4 kg GHG emissions and photocatalysis has 0.6 kg GHG emissions per kg hydrogen production. During the operation part of the hydrogen generation processes, there is no GHG emissions emitted from the conventional electrolysis, PEC, PV electrolysis, and photocatalysis. The high emissions from PV electrolysis are because of the PV manufacturing process. In conventional electrolysis, grid electricity is used, and since almost 80% of the grid electricity comes from fossil fuel combustion, there are relatively high emissions associated with this method. In all options, a thorough life cycle approach is considered, and all transportation, distribution, and operational emissions are calculated from the source to the very end user. In solar hydrogen production options, construction and transportation of the PEC reactors, photocatalysts, or PV panels are taken into consideration which causes the emissions.

The ultimate objective of this study is to show the potential of hydrogen to make integration of renewables to the existing energy systems happen and eventually eliminating fossil fuel use throughout the world. In this step of the present study, the selected hydrogen production methods are compared based on their resource use (MJ) to produce one kg of hydrogen. By doing so, it is aimed to find the method that uses the least amount of resource energy to produce the maximum amount of hydrogen. It should be noted that fossil-based hydrogen production methods are given here for comparison purposes.

Figure 3 shows the resource use calculation results. In steam methane reforming, the results show the amount of natural gas energy (MJ) required to produce one kg

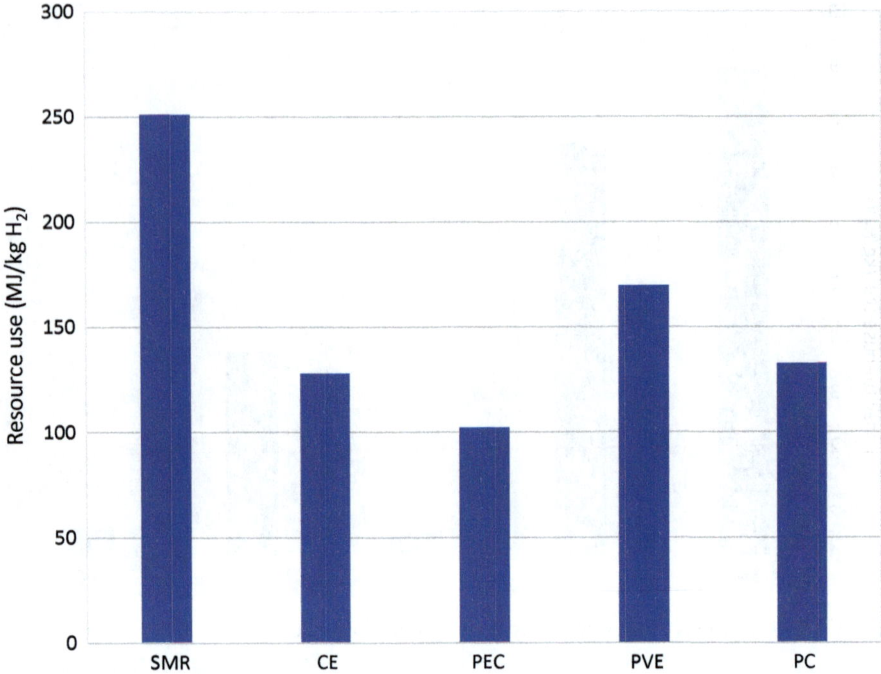

Fig. 3 Resource use comparison of the selected hydrogen production options

hydrogen. Similarly, in conventional electrolysis, this option indicates the amount of energy resource, which is mainly natural gas or coal (MJ) required to produce one kg hydrogen. In PEC, photocatalysis, and PV electrolysis, the results present MJ solar energy to produce one kg hydrogen. Here, it can be seen that steam methane reforming has the highest resource use (about 251 MJ/kg H_2), followed by conventional electrolysis (around 128 MJ/kg H_2). PEC has the lowest resource use, which is around 100 MJ/kg H_2.

In addition to the resource use, the lifecycle fossil fuel use performance of the selected options must be compared. Even though the solar hydrogen production options do not use fossil fuels at the hydrogen production site, the construction of materials, reactors, synthesis of catalysts, transportation of the components, construction of the process units, etc. might consume large amounts of fossil fuels which would cause significant GHG emissions. As the transition to fully renewabilized energy systems require the elimination of fossil fuel use, the auxiliary steps of each hydrogen production process must be carefully investigated to evaluate the actual fossil fuel consumption in the entire lifecycle. In this study, the entire lifecycles of the selected five hydrogen production methods are thoroughly investigated to accurately estimate each option's true fossil fuel use.

The life cycle fossil fuel use performance of the carefully chosen hydrogen production options is shown in Fig. 4. Since all upstream processes and additional steps

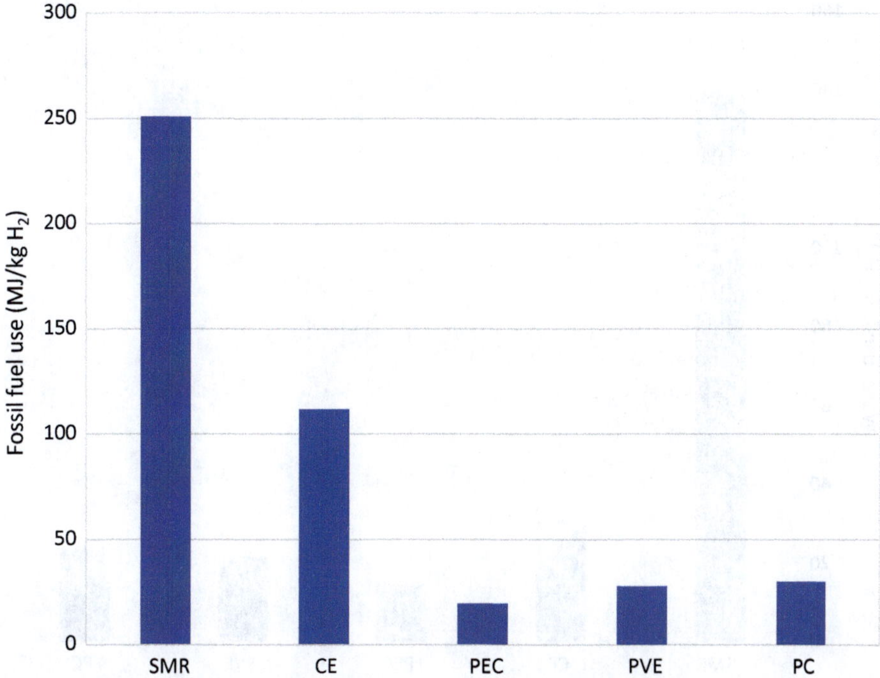

Fig. 4 Fossil fuel use comparison results of the selected hydrogen production options

like transportation are included in this study, the highest fossil fuel use happens to be in steam methane reforming (251 MJ/kg H_2). The second highest fossil fuel use belongs to conventional electrolysis (110 MJ/kg H_2), which is because of the high fossil fuel contribution in the grid electricity. PEC has the lowest fossil fuel use of about 20 MJ/kg H_2. PV electrolysis and photocatalysis have similar fossil fuel use of around 30 MJ/kg H_2. The fossil fuel use of PEC, photocatalysis, and PV electrolysis is mostly because of the processing of the reactors, panels, photocatalysts, membranes, etc.

Even though this study's primary aim is to highlight the possibility of the full renewabilization of energy systems through hydrogen, one cannot deny that water, as a scarce source, must also be considered while choosing the most sustainable hydrogen production option. Water is becoming more and more important every other day because of the issues related to fossil fuel use, pollution, deforestation, and so on. While evaluating the alternatives, water consumption performance has to be considered as well. The ultimate goal must be to find a method with the least water consumption performance. Ideally, the options with water recycling or wastewater usage possibilities should be selected to secure our global fresh water supplies.

In Fig. 5, the water use performances are comparatively assessed. Since clean water is an essential resource which is needed for all natural and industrial processes, it is essentially one aim of sustainable energy systems to minimize the waste of clean

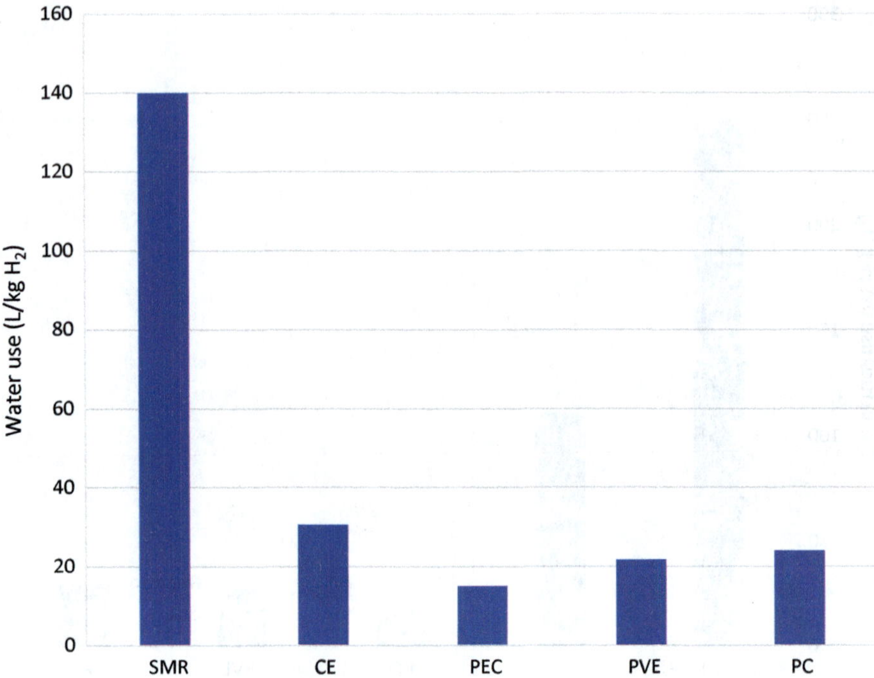

Fig. 5 Water use comparison results of the selected hydrogen production options

water. Water use includes two factors: water consumption and water degradation, such as water pollution (L/kg H_2). The results show that steam methane reforming has the highest water use, which is around 140 L/kg H_2, followed by conventional electrolysis, which is almost 30 L/kg H_2. One reason for the considerably high water consumption of the conventional electrolysis is the grid electricity generation process. PEC has the lowest water consumption of around 15 L/kg H_2, followed by PV electrolysis and photocatalysis (about 20 L/kg H_2). Another advantage of PEC is the fact that it can be integrated into wastewater treatment and desalination processes. As a result, these integrated systems do not only produce hydrogen, but also provides fresh water, which is a scarce resource around the world.

While the cost and performance of hydrogen energy systems have improved in the recent years (e.g., fuel cell cost fell more than 50%), performance improvement is not capturing its full potential as industry standards have been set for specific applications but remain limited overall. Advancing the energy transition requires harmonized regional and sector-specific hydrogen standards that will allow for economies of scale in research, development, and deployment (R, D & D) and manufacturing.

In this study, hydrogen production costs are calculated by taking the life cycle approach, and the results are presented in Fig. 6. Here, it can be seen that PEC, due to its early R&D stage and smaller scales, has the highest cost, which is around 9 USD/kg H_2. Following PEC, PV electrolysis and photocatalysis have the second

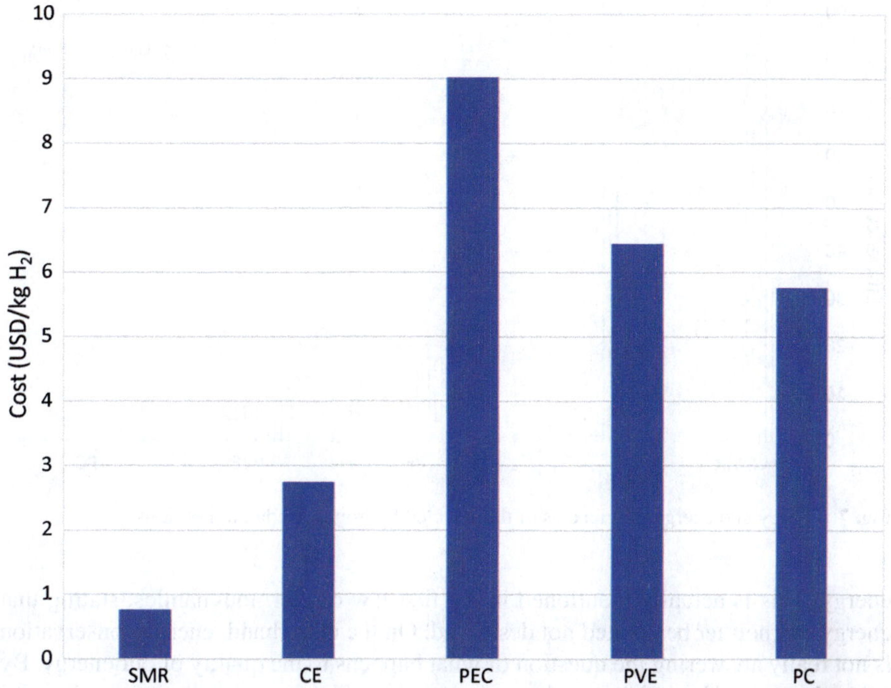

Fig. 6 Hydrogen production costs of the selected hydrogen production options

highest production cost of about 6 USD/kg H_2. The lowest production cost belongs to steam methane reforming (less than 1 USD/kg H_2), followed by conventional electrolysis (almost 3 USD/kg H_2). PEC, photocatalysis, and PV electrolysis have higher costs compared to steam methane reforming and conventional electrolysis because of a couple of reasons. Firstly, PEC, PV, and photocatalysis are in smaller scales than steam methane reforming, and conventional electrolysis, which increases their production costs significantly. In addition, most of the materials used in PEC, PV, and photocatalysis are still expensive, and the high initial cost of these options make hydrogen production more expensive than the more traditional options. However, it should be noted that the cost of the renewable-based hydrogen production options is expected to decrease significantly in the future. The target is to make renewable hydrogen cost-competitive with fossil fuel-based alternatives.

The last category is energy and exergy efficiencies of the selected hydrogen production options, and the results are presented in Fig. 7. In this study, in addition to energy efficiency, exergy efficiency of each hydrogen production method is calculated to provide a deeper understanding of the system performances from lifecycle perspective. Energy efficiency is the ratio of the energy content of the product hydrogen, divided by the energy content of the input. Energy efficiency is very useful to lower any waste and enhance the system performance of a process. However, it does not reflect the whole picture since it does not focus on the useful work portion of

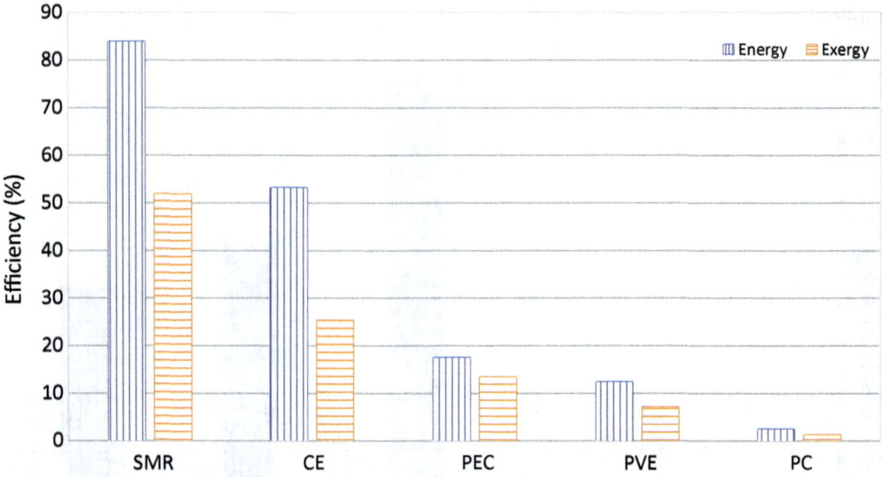

Fig. 7 Energy and exergy efficiencies of the selected hydrogen production options

energy. This is actually mentioned in the first law of thermodynamics, stating that energy can neither be created not destroyed. On the other hand, energy conservation is not really answering the question of what happens to the quality of our energy. By using the second law of thermodynamics, exergy efficiency actually states what part of the useful exergy is conserved in a process. For instance, in a hydrogen production process, exergy efficiency is the ratio of exergetic content of the product hydrogen to the exergetic content of all the input. By doing so, exergy efficiency gives a better perspective when selecting the most sustainable hydrogen production option.

From Fig. 7, it can be seen that steam methane reforming and conventional electrolysis have significant advantages due to their larger scale operation. Steam methane reforming has about 84% energy efficiency and 52% exergy efficiency. The energy and exergy efficiencies of conventional electrolysis are 53% and 25%, respectively. Photocatalysis has the lowest efficiencies among the selected options with 3% energy efficiency and 2% exergy efficiency. With 18% energy and 12% exergy efficiency, PEC has a higher performance than PV electrolysis, which has 12% energy and 7% exergy efficiency.

After performing the environmental, energetic, exergetic, and economic performance of the selected hydrogen production methods, the results are normalized and ranked based on the procedure explained in the previous section. The results are given in detail in Table 3 and presented in Fig. 8.

The overall comparison shows that PEC has advantages in terms of its low emissions and the use of energy and material resources. Steam methane reforming has the highest results based on the energy and exergy efficiencies and cost. In contrast, steam methane reforming has the lowest performance in terms of emissions, resource use, fossil fuel use, and water use. PEC has the lowest performance in cost criteria while photocatalysis has the weakest performance in energy and exergy efficiency

Table 3 Normalized performance rankings of the selected hydrogen production options

	GHG emissions	Resource use	Fossil fuel use	Water use	Efficiency		Cost	Average ranking
					Energy	Exergy		
Steam methane reforming (SMR)	0.00	0.00	0.00	0.00	8.38	5.18	9.15	3.24
Conventional electrolysis (CE)	1.53	4.90	5.54	7.81	5.31	2.53	6.96	4.94
Photo-electrochemical calls (PEC)	9.48	5.92	9.22	8.93	0.75	0.35	0.00	4.95
PV electrolysis (PVE)	5.81	3.23	8.88	8.43	1.24	0.72	2.86	4.45
Photocatalysis (PC)	9.37	4.71	8.79	8.28	0.24	0.13	3.62	5.02
Ideal	10	10	10	10	10	10	10	

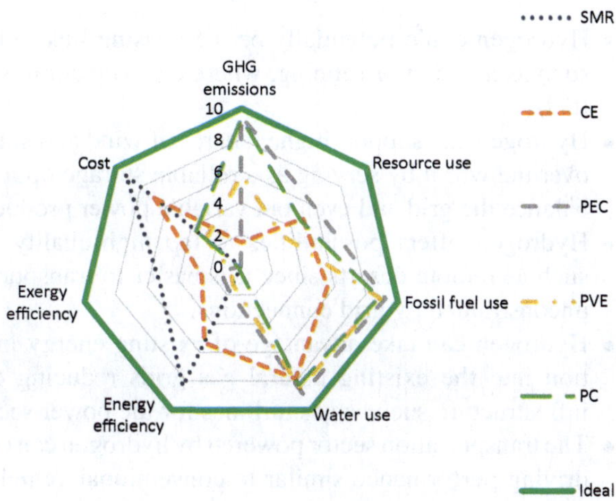

Fig. 8 Normalized performance rankings of the selected hydrogen production options

categories. When the averages of the normalized performance rankings are taken, it is seen that PEC has the highest normalized average ranking of 5.24 out of 10, which is immediately followed by photocatalysis with 5.02. The third highest normalized average ranking belongs to conventional electrolysis, that is 4.94. Among the selected options, steam methane reforming has the lowest average normalized ranking (3.24), and PV electrolysis has 4.45.

In this study, for the first time in the literature, a comprehensive quantitative comparison approach is taken to evaluate the environmental, economic, thermodynamic, and technical performances of the solar, conventional, and fossil fuel-based hydrogen production options. This study takes GHG emissions, resource use, fossil fuel use, water use, energy, and exergy efficiencies, and cost into account to provide a better insight into the cleaner hydrogen production methods. The results not only show a clear picture of the current status of these options, but also provides potential future directions to guide researchers, policymakers, and industries in the field of renewable energy systems for a sustainable future.

6 Conclusions

This study is conducted to explain the advantages of hydrogen as a fuel to support the transition to 100% renewable energy systems. The benefits of hydrogen include providing high-grade heat; addressing a range of energy needs that direct electrification cannot meet; and replacing fossil fuel-based feedstocks, such as natural gas, in high-emission applications of different sectors and the industry as a whole. According to the present study, renewables-based hydrogen could be critical for deeper energy transition for the following reasons:

- Hydrogen could potentially be the missing link in the transformation of sectors, such as aviation or refining, where electrification is not suitable to replace fossil fuels.
- Hydrogen can support higher shares of wind and solar energy in power sectors all over the world by serving as a reliable storage option for renewable electricity to balance the grid and even out variable power production.
- Hydrogen offers possibilities to tap high-quality renewable energy resources, such as remote deserts since it is easier to transport compared to electricity and unconstrained by grid connections.
- Hydrogen can take advantage of existing energy infrastructure, including injection into the existing natural gas grids reducing emissions of the current gas infrastructure such as gas turbines for the power sector.
- The transportation sector powered by hydrogen can offer consumers low emissions driving performance similar to conventional vehicles using internal combustion engines.

In this study, the GHG emissions, resource use, fossil fuel use, water use, energy and exergy efficiencies, and cost of hydrogen via five different production methods are investigated in detail from a life cycle perspective. The results are then comparatively evaluated by normalizing and ranking the environmental, energetic and exergetic, and economic performances of the selected hydrogen production options. Steam methane reforming, conventional electrolysis, PEC, PV electrolysis, and photocatalysis are the selected hydrogen production options. This study is one of the first attempts to thoroughly investigate the life-cycle environmental and economic impacts of sustainable hydrogen production. And the results indicate that:

- In terms of GHG emissions, steam methane reforming has the highest emissions (9 kg GHG/kg H_2), and PEC has the lowest emissions (0.47 kg GHG/kg H_2).
- Steam methane reforming has the highest resource consumption (251 MJ natural gas energy/kg H_2), and PEC has the lowest resource consumption (103 MJ solar energy/kg H_2).

- Fossil fuel consumption data shows that steam methane reforming has the highest fossil fuel consumption (251 MJ fossil fuel energy/kg H_2) and PEC has the lowest resource consumption (20 MJ fossil fuel energy/kg H_2).
- Steam methane reforming has the highest water consumption (140 L water/kg H_2), and PEC has the lowest water consumption (15 L water/kg H_2).
- Steam methane reforming has the highest energy and exergy efficiencies, which are 83% and 52%, respectively and photocatalysis has the lowest energy and exergy efficiencies, which are 2% and 1%.
- Cost comparison shows that steam methane reforming has the lowest cost (0.76 USD/kg H_2) while PEC has the highest cost (9.02 USD/kg H_2).
- The overall normalized performance ranking comparison shows that PEC has the highest average normalized ranking (5.24/10) and steam methane reforming has the lowest average normalized ranking (3.24/10).

References

Acar C, Dincer I (2019) Review and evaluation of hydrogen production options for better environment. J Clean Prod 218:835–849

Acar C, Dincer I, Naterer GF (2016) Review of photocatalytic water-splitting methods for sustainable hydrogen production. Int J Energy Res 40(11):1449–1473

Ahmed M, Dincer I (2018) A review on photoelectrochemical hydrogen production systems: challenges and future directions. Int J Hydrogen Energy 44(5):2474–2507

Bhattacharyya R, Misra A, Sandeep KC (2017) Photovoltaic solar energy conversion for hydrogen production by alkaline water electrolysis: conceptual design and analysis. Energy Convers Manag 133:1–13

Bolatkhan K, Kossalbayev BD, Zayadan BK, Tomo T, Veziroglu TN, Allakhverdiev SI (2019) Hydrogen production from phototrophic microorganisms: reality and perspectives. Int J Hydrogen Energy 44(12):5799–5811

Dahbi S, Aboutni R, Aziz A, Benazzi N, Elhafyani M, Kassmi K (2016) Optimised hydrogen production by a photovoltaic-electrolysis system DC/DC converter and water flow controller. Int J Hydrogen Energy 41(45):20858–20866

De Crisci AG, Moniri A, Xu Y (2018) Hydrogen from hydrogen sulfide: towards a more sustainable hydrogen economy. Int J Hydrogen Energy 44(3):1299–1327

Dincer I, Acar C (2017) Innovation in hydrogen production. Int J Hydrogen Energy 42(22):14843–14864

Ding Q, Song B, Xu P, Jin S (2016) Efficient electrocatalytic and photoelectrochemical hydrogen generation using MoS_2 and related compounds. Chem 1(5):699–726

Esmieu C, Raleiras P, Berggren G (2018) From protein engineering to artificial enzymes–biological and biomimetic approaches towards sustainable hydrogen production. Sustain Energy Fuels 2(4):724–750

Fan W, Yu X, Lu HC, Bai H, Zhang C, Shi W (2016) Fabrication of TiO_2/RGO/Cu_2O heterostructure for photoelectrochemical hydrogen production. Appl Catal B 181:7–15

Gross MA, Creissen CE, Orchard KL, Reisner E (2016) Photoelectrochemical hydrogen production in water using a layer-by-layer assembly of a Ru dye and Ni catalyst on NiO. Chem Sci 7(8):5537–5546

He Z, Fu J, Cheng B, Yu J, Cao S (2017) $Cu_2(OH)_2CO_3$ clusters: novel noble-metal-free cocatalysts for efficient photocatalytic hydrogen production from water splitting. Appl Catal B 205:104–111

Huang PH, Kuo JK, Wu ZD (2016) Applying small wind turbines and a photovoltaic system to facilitate electrolysis hydrogen production. Int J Hydrogen Energy 41(20):8514–8524
Im-orb K, Visitdumrongkul N, Saebea D, Patcharavorachot Y, Arpornwichanop A (2018) Flowsheet-based model and exergy analysis of solid oxide electrolysis cells for clean hydrogen production. J Clean Prod 170:1–13
Jiang Z, Wang B, Jimmy CY, Wang J, An T, Zhao H, Li H, Yuan S, Wong PK (2018) $AgInS_2/In_2S_3$ heterostructure sensitization of Escherichia coli for sustainable hydrogen production. Nano Energy 46:234–240
Kadier A, Simayi Y, Abdeshahian P, Azman NF, Chandrasekhar K, Kalil MS (2016) A comprehensive review of microbial electrolysis cells (MEC) reactor designs and configurations for sustainable hydrogen gas production. Alex Eng J 55(1):427–443
Khetkorn W, Rastogi RP, Incharoensakdi A, Lindblad P, Madamwar D, Pandey A, Larroche C (2017) Microalgal hydrogen production—a review. Biores Technol 243:1194–1206
Kumari S, White RT, Kumar B, Spurgeon JM (2016) Solar hydrogen production from seawater vapor electrolysis. Energy Environ Sci 9(5):1725–1733
Kwon KC, Choi S, Hong K, Moon CW, Shim YS, Kim DH, Kim T, Sohn W, Jeon JM, Lee CH, Nam KT (2016) Wafer-scale transferable molybdenum disulfide thin-film catalysts for photoelectrochemical hydrogen production. Energy Environ Sci 9(7):2240–2248
Lee SH, Bhopal MF, Lee DW, Lee SH (2018) Review of advanced hydrogen passivation for high efficient crystalline silicon solar cells. Mater Sci Semicond Process 79:66–73
Morales-Guio CG, Liardet L, Mayer MT, Tilley SD, Grätzel M, Hu X (2015) Photoelectrochemical hydrogen production in alkaline solutions using Cu_2O coated with earth-abundant hydrogen evolution catalysts. Angew Chem 127(2):674–677
Nikolaidis P, Poullikkas A (2017) A comparative overview of hydrogen production processes. Renew Sustain Energy Rev 67:597–611
Saeedmanesh A, Mac Kinnon MA, Brouwer J (2018) Hydrogen is essential for sustainability. Curr Opin Electrochem 12:166–181
Santos JL, Reina TR, Ivanov I, Penkova A, Ivanova S, Tabakova T, Centeno MA, Idakiev V, Odriozola JA (2018) Multicomponent Au/Cu-ZnO-Al_2O_3 catalysts: robust materials for clean hydrogen production. Appl Catal A 558:91–98
Shaner MR, Atwater HA, Lewis NS, McFarland EW (2016) A comparative technoeconomic analysis of renewable hydrogen production using solar energy. Energy Environ Sci 9(7):2354–2371
Sharma K (2019) Carbohydrate-to-hydrogen production technologies: a mini-review. Renew Sustain Energy Rev 105:138–143
Shinagawa T, Takanabe K (2017) Towards versatile and sustainable hydrogen production through electrocatalytic water splitting: electrolyte engineering. ChemSusChem 10(7):1318–1336
Show KY, Yan Y, Ling M, Ye G, Li T, Lee DJ (2018) Hydrogen production from algal biomass–advances, challenges and prospects. Biores Technol 257:290–300
Tebibel H, Khellaf A, Menia S, Nouicer I (2017) Design, modelling and optimal power and hydrogen management strategy of an off grid PV system for hydrogen production using methanol electrolysis. Int J Hydrogen Energy 42(22):14950–14967
Tong X, Zhou Y, Jin L, Basu K, Adhikari R, Selopal GS, Zhao H, Sun S, Vomiero A, Wang ZM, Rosei F (2017) Heavy metal-free, near-infrared colloidal quantum dots for efficient photoelectrochemical hydrogen generation. Nano Energy 31:441–449
Tsai KA, Hsu YJ (2015) Graphene quantum dots mediated charge transfer of CdSe nanocrystals for enhancing photoelectrochemical hydrogen production. Appl Catal B 164:271–278
Wang J, Yin Y (2018) Fermentative hydrogen production using various biomass-based materials as feedstock. Renew Sustain Energy Rev 92:284–306
Wang B, Shen S, Mao SS (2017) Black TiO_2 for solar hydrogen conversion. J Mater 3(2):96–111
Zhang J, Wang X (2015) Solar water splitting at $\lambda = 600$ nm: a step closer to sustainable hydrogen production. Angew Chem Int Ed 54(25):7230–7232

Zhang X, Meng F, Mao S, Ding Q, Shearer MJ, Faber MS, Chen J, Hamers RJ, Jin S (2015a) Amorphous MoS_xCl_y electrocatalyst supported by vertical graphene for efficient electrochemical and photoelectrochemical hydrogen generation. Energy Environ Sci 8(3):862–868

Zhang J, Wang L, Liu X, Li XA, Huang W (2015b) High-performance CdS–ZnS core–shell nanorod array photoelectrode for photoelectrochemical hydrogen generation. J Mater Chem A 3(2):535–541

Zhang H, Ding Q, He D, Liu H, Liu W, Li Z, Yang B, Zhang X, Lei L, Jin S (2016) A p-Si/$NiCoSe_x$ core/shell nanopillar array photocathode for enhanced photoelectrochemical hydrogen production. Energy Environ Sci 9(10):3113–3119

Zheng XL, Song JP, Ling T, Hu ZP, Yin PF, Davey K, Du XW, Qiao SZ (2016) Strongly coupled nafion molecules and ordered porous CdS networks for enhanced visible-light photoelectrochemical hydrogen evolution. Adv Mater 28(24):4935–4942

Design and Analysis of a New Environmentally Benign Ammonia-Based Solar Thermochemical Integrated Plant

Yunus Emre Yüksel, Fatih Yilmaz and Murat Ozturk

Abstract Supplying the remarkably increasing global power needs with no or minimal ecological damage and carbon-based sources dependence can only be achieved by utilizing the clean power plants. For this aim, a novel environmentally benign ammonia-based solar thermochemical combined plant for heating, cooling and electricity production is proposed, and thermodynamic analysis is done for performance evaluation. The examined plant includes an ammonia-based solar thermochemical process, a two turbine Rankine cycle, and a single effect absorption cooling plant with ejector. In the proposed system, ammonia is utilized as a working fluid in the solar tower subsystem, and ammonia is converted into hydrogen and nitrogen as a result of the thermochemical reaction, also the hydrogen and nitrogen can be stored. Also, the changes in energetic efficiency, exergetic efficiency, and useful products of the examined plant are investigated, with respect to varying solar radiation, dead state temperature and pinch point temperature of ammonia reactor. The energy and exergy efficiency of suggested plant are calculated as 49.03 and 38.94% for charging time, 52.17 and 43.45% for storing time, and 50.79 and 34.36% for discharging time.

Keywords Solar energy · Thermodynamic · Ammonia thermochemical cycle · Integrated plant · Environmentally benign viewpoint

1 Introduction

Energy is a necessary indicator for life of humans, societies, countries and the world. Efficient utilization of energy resources in applications such as production and utilization is a critical engineering problem in achieving the world's power sustainability aims (Hogerwaard et al. 2017). Due to factors such as industrialization and the

Y. E. Yüksel
Education Faculty, Math and Science Education, Afyon Kocatepe University, ANS Campus, Afyonkarahisar 03200, Turkey

F. Yilmaz · M. Ozturk (✉)
Department of Mechatronics Engineering, Faculty of Technology, Applied Science University of Isparta, Cunur West Campus, Isparta 32200, Turkey
e-mail: muratozturk@isparta.edu.tr

T. S. Uyar (ed.), *Accelerating the Transition to a 100% Renewable Energy Era*, Lecture Notes in Energy 74, https://doi.org/10.1007/978-3-030-40738-4_15

increase in the human population, the energy need has increased in direct proportion in recent years, and also, nowadays, about 80% of this energy requirement is obtained from fossil sources (Ozturk and Yuksel 2016). Global warming and climate change are seriously discussed topics mainly because of carbon-based source consumption. Depending on the use of fossil fuels, environmental impacts such as greenhouse gas impact, climate change, and global warming have begun to emerge, and thus, renewable energy sources have become a major role in the sustainability benign future. Therefore, based on this extensive utilization of fossil energy sources, power sustainability is becoming the worldwide necessities, and also, is directly linked to the general opinion of sustainability that affects all of human civilization. These ideas bring significance to renewable power resources and efficient generation and use of energy. In addition to renewable energy sources, multigeneration systems have come to the fore in terms of efficient use of energy sources. In other words, the performance of conventional electricity generation systems is generally less than 40% due to only one energy output (Ahmadi et al. 2012). Furthermore, the electricity and heat power generated by cogeneration plants, tri-generation plants offer one more beneficial exit from the combined plant, such as freshwater, cooling, hydrogen, drying or synthetic fuels. On the other hand, it should be noted that the benefit achieved from the further exit should be weighed against the rise in overall cost and emissions of the integrated plant. Hence, it should be helpful to consider these criteria in the final decision-making viewpoints.

Renewable energy source based integrated plants are expected to supply some useful outputs, in addition to power generation purpose. In other words, with the integrated combined power plant, high efficiency can be reached by obtaining many outlets such as power, heating, cooling, hydrogen, and oxygen with single power inlet (Agency 2008). Moreover, the combined multi-generation plant has many advantages such as higher system operation, low energetic and exergetic losses, decreased carbon emissions, low maintenance cost and improvements in resource usage (Ozturk and Dincer 2013; Dincer and Zamfirescu 2012). In recent years, there are many studies conducted by researchers about multigeneration systems, especially solar energy based. Solar power is a pure, abundant and endless source that could be found almost anywhere in the globe. Assisted by solar power, the needs of power, heating-cooling, hydrogen, and many different applications could be met by combined plants (Atiz et al. 2019).

A lot of research has been done in the literature about environmentally benign renewable energy supported multigeneration systems. Ozlu and Dincer (2016) have proposed a performance evaluation of the novel poly-generation process driven by solar power. The outputs of their work are namely hydrogen, heat, clean water and power. The maximum energy and exergy performances are computed as 36 and 44%, based on the 24 m^2 collector area. Kasaeian et al. (2018) have investigated a review of the integrated heat and electricity generation with solar energy. In their review study, they examined solar system and photovoltaics as heat and energy plants. As a result of their study, one of their recommendations is the more comprehensive investigation of energetic, exergetic and economic aspects of large-scale combined heat and electricity generation systems. Behzadi et al. (2019) have investigated a multiobjective optimization of power, hydrogen and cooling production with solar power assisted. The main difference of their work is the proposition of a new solar-based combined power plant with a thermoelectric generator (TEG) to supply cooling and hydrogen generation. At the optimum point, the general exergy efficiency of the study is 12.01%, and also, the total cost rate is 0.1762 $/h.

Luqman et al. (2019) have reported an energy and exergy assessments of the oxy-hydrogen combustor based solar and wind power integrated multigeneration process for useful products. Their suggested study including a solar power, a wind turbine, an oxy-hydrogen combustor, a Rankine cycle, a desalination plant, an electrolyzer, a refrigeration cycle, a drying process, a water heater as well as oxygen and hydrogen storages. They found that the overall energetic and exergetic performances are 50 and 34%. El-Emam and Dincer (2018) have given an investigation and evaluation of the integrated combined plant driven by solar energy. They used a concentrated solar collector as an energy source, also the maximum and minimum exergetic efficiency of their work are 39 and 21.7%, respectively.

Luzzi and Lovegrove (1997) have performed an evaluation of the solar thermochemical power plant for greenhouse gas reduction. In their proposed study, it is an ammonia-based, closed-loop, solar thermochemical electricity process concept utilizing the principle of direct power output through a Brayton subsystem or conventional steam Rankine cycle. Their experimental and theoretical research has shown that solar energy systems, including an ammonia-based closed-circuit thermochemical power storage, are technically feasible. Liu et al. (2019) have examined a design and modelling of the100 kW solar thermochemical electricity production process with mid-and-low temperature. In their suggested plant, the solar power is used to the chemical energy of the solar fuel (H_2 and CO) by the solar thermochemical cycle of the methanol decomposition reaction. The energy conversion efficiency from solar to the chemical is calculated in the range of 37.12–45.51%.

Kreetz and Lovegrove (1999) have explored a theoretical and experimental assessment of a closed-loop solar thermochemical power storage plant utilizing ammonia synthesis at Australian National University (ANU). 1 kW ammonia synthesis reactor was designed experimentally on a laboratory scale for stable and reproducible operation. According to their study conclusions, at the mass flow rate of 0.3 g/s and the pressure of 15 MPa, the maximum reactor outer wall temperature was able to reach about 450 °C. Again, same authors, Kreetz and Lovegrove (2002), have evaluated an exergetic analysis of the ammonia-based thermochemical power storage plant under the Australian National University (ANU) conditions. According to their results, as in the industry, it is necessary to use either a very small diameter reactor or adiabatic reactor to approach the maximum exergy curve.

Chen et al. (2017) have suggested a designing of the ammonia synthesis to generate supercritical steam for thermochemical power storage. In their examined study, solar energy is stored by ammonia decomposition by the generation of hydrogen (H_2) and nitrogen (N_2). The results illustrate that the process in their study reactor is limited to heat transfer and is most sensitive to the activation power. Again, same authors, Chen et al. (2018), have proposed a design and optimization of ammonia-based solar thermochemical cycle for power storage. They conducted the parametric studies for the purpose of the impacts of the diameter, mass flow rate and inlet reactor temperature on the plant. According to their results, they stated that pre-conditioning plant plays a significant aim in the required wall volume.

Siddiqui and Dincer (2019) have modelled a design and examination of the new solar-wind assisted combined plant with ammonia for energy storage. They found that whole energy and exergy performances vary between 46.1–53.3% and 34–41.5%, respectively. Chen et al. (2019) have investigated an efficiency assessment of the solar thermochemical power storage system with conical ammonia reactors. They examined the effect of the geometry and configuration of conventional ammonia reactors, and the results also showed that a higher conversion of converging reactors can be achieved with a preheating heat exchanger.

The depletion of carbon-based sources and the harmful emissions that accompany a classic conversion technic create the demand for different environmentally friendly energy conversion plants that can produce several useful outputs. For this aim, the solar thermochemical cycle for multigeneration is expected to play a significant indicator in the near future with the chemical power storage option. In this study, a thermodynamic performance evaluation of the new environmentally benign ammonia assisted solar thermochemical cycle is examined for power, heating, and cooling generation. In this regard, a comprehensive energetic and exergetic performances of the examined plant are performed based on various indicators such as dead state temperature, solar irradiation and pinch point temperature of ammonia synthesis reactor. Furthermore, this plant is designed for the generation of electricity, heating, and cooling applications in a completely clean and sustainable way. The main difference of this paper is the ammonia-based solar thermochemical cycle which is used as an energy source. In addition, this system offers the possibility to store energy in cases where solar irradiation is insufficient. Furthermore, two turbine Rankine cycle sub-system is also used for a power generation. Finally, this book chapter examines multigeneration from ammonia-based solar cycle and aims to reach the following objectives:

- To describe the proposed thermochemical plant and its components to perform thermodynamic assessment.
- To conduct a comprehensive energy and exergy modeling of ammonia-based solar thermochemical combined plant.
- To apply comprehensive thermodynamic analysis utilizing the energy and exergy to investigate the feasibility to generate H_2/N_2 gas mixture from liquid ammonia by using a solar tower power plant and use the exothermic reaction heat energy to produce useful outputs.
- To present plant parts which have a higher irreversibility rate for the multigeneration purpose.
- To investigate the impact of reference and working conditions on the three statuses such as charging, storing and discharging times.

2 System Design

Power inlet to integrated plants can be supplied by one primary resource or more. In this study, the solar power is chosen as the primary power source. The aim is to minimize waste power and increase the sustainability of power production. The waste heat energy taken from a medium-temperature plant in the ammonia synthesis reactor is converted into mechanical energy by using the power plants, which have a maximum performance equal to the Carnot system performance. Power plants can reject heat energy, which should be done beneficial by combining with another plant to reach a higher integrated plant performance. As given in this book chapter, tri-generation plants combine different cycles for efficient plant design aims. The co-generation and tri-generation plants supply better utilization of power sources while decreasing greenhouse gas emissions compared to a single generation of electricity. This also allows the cost-effectiveness and independence for different communities that import power sources.

In this part of this book chapter, a detailed description of the proposed system is provided. This plant constitutes several units. The main units are the solar tower, ammonia synthesis reactor, and liquid ammonia storage tank. The analysis conducted on the system units is used to examine how competitive the thermochemical plant is to produce useful outputs. Figure 1 illustrate the flow diagram of ammonia-based solar thermochemical integrated plant.

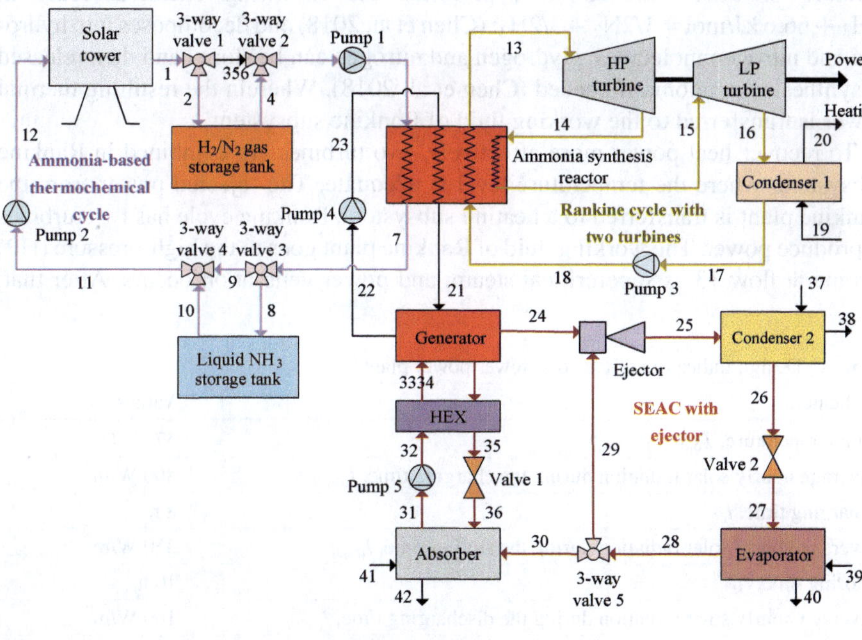

Fig. 1 Schematic illustration of the investigated plant

As illustrated in Fig. 1, the examined plant consists of three main sub-systems which are an ammonia-based solar thermochemical cycle, a Rankine cycle with two turbines and a single effect absorption cycle (SEAC) with ejector. The ammonia-based solar thermochemical cycle consists of a solar tower, four 3-way valves, two pumps, a H_2/N_2 gas storage tank, a liquid NH_3 storage tank, and an ammonia synthesis reactor. To increase the plant performance, heat energy losses from the plant should be minimized through a recovery subsystem to make the heat energy a beneficial commodity for other subsystems, such as SEAC with ejector which can use the low-grade heat energy. For the proposed plant, the solar thermal energy is concentrated by using the solar power tower subsystem. Also, some design indicators of the solar tower power process are given in Table 1. Additionally, in this subchapter, the key subsystems addressed in the theoretical assessment part are explained. These subsystems are namely Rankine subsystem with two turbines and solar tower process. Supercritical heated water is the most common working fluid employed in the Rankine systems for different integrated systems due to its exergy destruction rate, lower cost and higher enthalpy. The primary aim of the solar tower power is to minimize the area of the receiver-area with respect to the aperture-area. This method can be utilized to decrease the heat energy losses through irradiation with respect to the beneficial heat energy flux. In addition to that, Table 2 shows the data utilization for the examined system.

The liquid ammonia in the storage tank passes through points 10-11-12 respectively and comes to the solar tower. Ammonia, which reaches a temperature of about 550–650 °C, performs the following chemical reaction; $NH_3 + 66.6\,kJ/mol = 1/2N_2 + 3/2H_2$; (Chen et al. 2018) and decomposes into hydrogen and nitrogen molecules. Hydrogen and nitrogen can be stored and then released to synthesize ammonia if needed (Chen et al. 2018). Wherein the resulting thermal power is transferred to the working fluid of Rankine subsystem.

To recover heat power more effectively, two turbines are combined in Rankine subsystem, where the temperature level is adequate. The rejected power from the Rankine plant is transferred to a heating subsystem. Rankine cycle has two turbines to produce power. The working fluid of Rankine plant goes to the high-pressure (HP) turbine at flow 13 as supercritical steam, and power generation occurs. After that,

Table 1 Design indicators of the solar tower power plant

Indicators	Values
Sun temperature, T_{Sun}	5727 °C
Average hourly solar radiation during the charging time, $I_{av,ct}$	800 W/m^2
Charging time, t_{ct}	8 h
Average hourly solar radiation during the storing time, $I_{av,st}$	350 W/m^2
Storing time, t_{st}	10 h
Average hourly solar radiation during the discharging time, $I_{av,dt}$	100 W/m^2
Discharging time, t_{dt}	6 h

Table 2 The data utilized for investigated plant

Parameters	Values
Solar power tower exit temperature, T_1	625 °C
Solar power tower exit pressure, P_1	4500 kPa
Solar power tower inlet temperature, T_{12}	340 °C
Pinch point temperature of ammonia synthesis reactor, $T_{PP,ASR}$	12 °C
Mass flow rate of solar plant	8.61 kg/s
HP turbine input temperature, T_{13}	590 °C
HP turbine input pressure, P_{13}	3108 kPa
LP turbine input temperature, T_{15}	374 °C
LP turbine input pressure, P_{15}	2095 kPa
Evaporator temperature, T_{eva}	9 °C
Energetic coefficient of performance, COP_{en}	1.047
Exergetic coefficient of performance, COP_{ex}	0.276

the working fluid entering the low-pressure (LP) turbine at flow 15 then expands and enters the condenser at point 16. In addition, the heating application occurs in the condenser component.

The thermal energy exiting the ammonia synthesis reactor at point 21 provides the energy required for the operation of the SEAC plant. The SEAC plant comprises the condenser, evaporator, absorber, heat exchanger (HEX), pump, valve, generator, and ejector. The cooling application occurs in the evaporator part. The whole proposed system and sub-systems work together simultaneously.

3 Thermodynamic Assessment

For the integrated plant offered in Fig. 1, a whole thermodynamic assessment can be carried out by utilizing the principles of energetic and exergetic analyses viewpoints. To build the thermodynamic section of assessment, the assumptions can be done as given below:

- The whole plant and all of its elements are working at steady-state conditions.
- Changes in the kinetic and potential energetic and exergetic variables are not important.
- There are no pressure drops in the pipelines and HEXs.
- The pumps and turbines are considered as adiabatic.
- The refrigerant working fluid of absorption cooling plant is ammonia-water.

The exergetic assessment is a useful methodology for the estimation of the performance of components and processes, and also includes examining the exergetic efficiencies at different points in a series of energetic transformation points. For all thermal systems, heat energy transfer within the plant, or between the plant and reference conditions, happens at a finite temperature variation, which is an important contributor to exergetic losses for the investigated plant. The entropy generation evaluates the impacts of these exergy destruction rates in the considered plant during a cycle and assists compare each part in the plant based on how much they contribute to the operational inefficiencies of the whole plant. Hence, the entropy generation connected with each cycle requires to be examined to determine the whole plant performance. Although, the energetic assessment is the mostly utilized procedure for investigating thermal plants, it is only concerned with the conservation of energy, which neither takes the corresponding reference status into calculation, nor supplies how, where and why the plant efficiency degrades. In the general, any balance equality for a steady-state condition in a plant should be described as given below:

$$\text{Input} + \text{Generation} = \text{Output} + \text{Consumption} \tag{1}$$

The ammonia-based solar thermochemical integrated plant is investigated thermodynamically and the common mass, energetic, entropy, and exergetic balance equalities can be given as (Dincer and Rosen 2012; Cengel and Boles 2015)

$$\sum_{in} \dot{m} = \sum_{out} \dot{m} \tag{2}$$

$$\sum_{in} \dot{m}h + \sum_{in} \dot{Q} + \sum_{in} \dot{W} = \sum_{out} \dot{m}h + \sum_{out} \dot{Q} + \sum_{out} \dot{W} \tag{3}$$

$$\sum_{in} \dot{m}s + \sum_{in} \frac{\dot{Q}}{T} + \dot{S}_{gen} = \sum_{out} \dot{m}s + \sum_{out} \frac{\dot{Q}}{T} \tag{4}$$

$$\sum_{in} \dot{m}ex + \sum_{in} \dot{E}x^{Q} + \sum_{in} \dot{E}x^{W} = \sum_{out} \dot{m}ex + \sum_{out} \dot{E}x^{Q} + \sum_{out} \dot{E}x^{W} + \dot{E}x_{d} \tag{5}$$

Here, in shows the inlet condition while out denotes the exit condition. In addition to that, $\dot{Q}$ and $\dot{W}$ show the heat transfer and work rate, $\dot{m}$ denotes the mass flow rate, h and s show the enthalpy and entropy, $\dot{S}_{gen}$ and $\dot{E}x_d$ signify the entropy generation and exergy destruction rate. Both the physical and chemical energies and exergies should be considered for the gaseous elements in the chemical reactions. The correlation equations can be written as Kotas (1985);

$$ex_{ph} = (h - h_o) - T_o(s - s_o) \tag{6}$$

$$ex_{ch} = \sum y_j ex_{ch}^{0} + RT_o \sum y_j ln(y_j) \tag{7}$$

Here, y_j shows the mole fraction and ex_{ch}^{0} gives the standard specific chemical exergy. The net specific exergy is defined as:

$$ex = ex_{ph} + ex_{ch} \tag{8}$$

Finally, in the exergy balance equation, $\dot{E}x^{Q}$ and $\dot{E}x_{W}$ show the exergetic rate of heat power transfer and associated with shaft energy.

$$\dot{E}x_{Q} = \left(1 - \frac{T_o}{T}\right)\dot{Q} \tag{9}$$

$$\dot{E}x_{W} = \dot{W} \tag{10}$$

$$\dot{E}x_{d} = T_0\dot{S}_{gen} \tag{11}$$

A potential to increase the exergetic performance of the generation of the useful output from the ammonia-based solar thermochemical integrated plant can be investigated by utilizing the concept of potential improvement viewpoint. This approach examines how much available energetic rate can be redirected towards useful outputs generation. The potential improvement level in the exergy should be computed from (Kalinci et al. 2009):

$$PI = \dot{E}x_{d}(1 - \psi) \tag{12}$$

Here, ψ is the exergetic efficiency. The mass, energetic, entropy and exergetic balance equalities for different important plant parts are defined below:
Solar tower;

$$\dot{m}_{12} = \dot{m}_{1} \tag{13a}$$

$$\dot{m}_{12}h_{12} + \dot{Q}_{st} = \dot{m}_{1}h_{1} \tag{13b}$$

$$\dot{m}_{12}s_{12} + \dot{Q}_{st}/T_{st} + \dot{S}_{g,st} = \dot{m}_{1}s_{1} \tag{13c}$$

$$\dot{m}_{12}ex_{12} + \dot{E}x_{st}^{Q} = \dot{m}_{1}ex_{1} + \dot{E}x_{D,st} \tag{13d}$$

H_2/N_2 gas storage tank;

$$\dot{m}_{2} = \dot{m}_{4} \tag{14a}$$

$$\dot{m}_{2}h_{2} = \dot{m}_{4}h_{4} + \dot{Q}_{L}^{gst} \tag{14b}$$

$$\dot{m}_{2}s_{2} + \dot{S}_{g,gst} = \dot{m}_{4}s_{4} + \dot{Q}_{L}^{gst}/T_{gst} \tag{14c}$$

$$\dot{m}_{2}ex_{2} = \dot{m}_{4}ex_{4} + \dot{E}x_{L,gst}^{Q} + \dot{E}x_{D,gst} \tag{14d}$$

Ammonia synthesis reactor;

$$\dot{m}_6 = \dot{m}_7;\ \dot{m}_{23} = \dot{m}_{21},\ \dot{m}_{18} = \dot{m}_{13};\ \dot{m}_{14} = \dot{m}_{15} \tag{15a}$$

$$\dot{m}_6 h_6 + \dot{m}_{14} h_{14} + \dot{m}_{18} h_{18} + \dot{m}_{23} h_{23} = \dot{m}_7 h_7 + \dot{m}_{13} h_{13} + \dot{m}_{15} h_{15} + \dot{m}_{21} h_{21} \tag{15b}$$

$$\dot{m}_6 s_6 + \dot{m}_{14} s_{14} + \dot{m}_{18} s_{18} + \dot{m}_{23} s_{23} + \dot{S}_{g,asr} = \dot{m}_7 s_7 + \dot{m}_{13} s_{13} + \dot{m}_{15} s_{15} + \dot{m}_{21} s_{21} \tag{15c}$$

$$\dot{m}_6 ex_6 + \dot{m}_{14} ex_{14} + \dot{m}_{18} ex_{18} + \dot{m}_{23} ex_{23} = \dot{m}_7 ex_7 + \dot{m}_{13} ex_{13} + \dot{m}_{15} ex_{15} + \dot{m}_{21} ex_{21} + \dot{E}x_{D,asr} \tag{15d}$$

High pressure turbine;

$$\dot{m}_{13} = \dot{m}_{14} \tag{16a}$$

$$\dot{m}_{16} h_{16} = \dot{m}_{17} h_{17} + \dot{W}_{HPT} \tag{16b}$$

$$\dot{m}_{16} s_{16} + \dot{S}_{g,HPT} = \dot{m}_{17} s_{17} \tag{16c}$$

$$\dot{m}_{16} ex_{16} = \dot{m}_{17} ex_{17} + \dot{E}x^{W}_{HPT} + \dot{E}x_{D,HPT} \tag{16d}$$

3.1 Solar Power Tower Sub-Plant

Thermal power transport between the solar plant receiver surface and H_2/N_2 gas mixture of concentrating solar plant should be defined as given below (Xu et al. 2011):

$$\dot{Q}_{str} = \dot{Q}_{shp} - \dot{Q}_{L,str} \tag{17}$$

where $\dot{Q}_{shp}$ shows heat power received utilizing solar heliostat plant, $\dot{Q}_{L,str}$ gives solar plant receiver surface losses by convection and radiation. Given in Eq. (17), $\dot{Q}_{shp}$ is calculated by

$$\dot{Q}_{shp} = \eta_{sp} I_{sd} A_{pra} n_{pr} \tag{18}$$

Here, η_{sp} is solar plant performance, I_{sd} shows direct solar radiation, A_{pra} illustrates plant reflector area, n_{pr} defines plant reflector number. Finally, $\dot{Q}_{L,str}$ is calculated by:

Table 3 Design variables for solar tower plant

Design variables	Values
plant reflector area, A_{pra} (Siddiqui and Dincer 2017)	11 m × 11 m
Direct solar radiation, I_{sd}	825 W/m^2.
Plant reflector number, n_{pr} (Dincer and Rosen 2012)	510
Solar plant performance, η_{sp}	0.88

$$\dot{Q}_{L,str} = h_{air}A_r(T_1 - T_0) - \sigma\varepsilon\left(T_1^4 - T_0^4\right) \tag{19}$$

Here, h_{air}, A_r, σ and ε give air convective heat transfer coefficient, plant receiver area, Stefan-Boltzmann constant, and absorber emissivity, respectively. In Eq. (19), h_{air} should be computed as given below:

$$h_{air} = 10.45 - \upsilon_a + 10\sqrt{\upsilon_a} \tag{20}$$

Here, υ_a shows wind speed of reference air. Different significant design variables for solar tower plant are defined in Table 3.

3.2 Performances of Integrated Plant

Performing exergetic assessment is a valid procedure utilizing the conservation of both mass and energy with the exergetic analysis viewpoints to design and investigate the conversion of integrated plant. To evaluate the efficiency of the overall process and its sub-processes, the corresponding energetic and exergetic performances should be determined. The exergetic performance for a plant under-investigated should be defined as the ratio between useful exergetic outputs from the plant to the necessary exergetic inlet to the plant.

The energetic performance for any process is

$$\eta = \frac{\text{Energy in output products}}{\text{Energy in inputs}} \tag{21}$$

Similar to the energetic performance term, the exergetic performance for any process is

$$\psi = \frac{\text{Exergy in output products}}{\text{Exergy in inputs}} \tag{22}$$

The energetic and exergetic efficiency equations of the solar based ammonia thermochemical sub-plant are:

$$\eta_{SAT} = \frac{(\dot{m}_{13}h_{13} - \dot{m}_{18}h_{18}) + (\dot{m}_{15}h_{15} - \dot{m}_{14}h_{14}) + (\dot{m}_{21}h_{21} - \dot{m}_{23}h_{23})}{\dot{Q}_{solar} + \dot{W}_{P1} + \dot{W}_{P2}} \tag{23}$$

$$\psi_{SAT} = \frac{(\dot{m}_{13}ex_{13} - \dot{m}_{18}ex_{18}) + (\dot{m}_{15}ex_{15} - \dot{m}_{14}ex_{14}) + (\dot{m}_{21}ex_{21} - \dot{m}_{23}ex_{23})}{\dot{E}x^{Q}_{solar} + \dot{E}x^{W}_{P1} + \dot{E}x^{W}_{P2}} \tag{24}$$

The energetic and exergetic efficiencies of Rankine sub-plant can be written as:

$$\eta_{RC} = \frac{\dot{W}_{HPT} + \dot{W}_{LPT} + (\dot{m}_{20}h_{20} - \dot{m}_{19}h_{19})}{\dot{W}_{P3} + (\dot{m}_{13}h_{13} - \dot{m}_{18}h_{18}) + (\dot{m}_{15}h_{15} - \dot{m}_{14}h_{14})} \tag{25}$$

$$\psi_{RC} = \frac{\dot{E}x^{W}_{HPT} + \dot{E}x^{W}_{LPT} + (\dot{m}_{20}ex_{20} - \dot{m}_{19}ex_{19})}{\dot{E}x^{W}_{P3} + (\dot{m}_{13}ex_{13} - \dot{m}_{18}ex_{18}) + (\dot{m}_{15}ex_{15} - \dot{m}_{14}ex_{14})} \tag{26}$$

The energetic and exergetic performance equalities of single effect absorption cooling process with ejector should be given as:

$$\eta_{SEACE} = \frac{\dot{Q}_{Cooling}}{(\dot{m}_{21}h_{21} - \dot{m}_{22}h_{22}) + \dot{W}_{P5}} \tag{27}$$

$$\psi_{SEACE} = \frac{\dot{E}x^{Q}_{Cooling}}{(\dot{m}_{21}ex_{21} - \dot{m}_{22}ex_{22}) + \dot{E}x^{W}_{P5}} \tag{28}$$

The overall energetic performance of the designed plant which includes the ammonia-based solar thermochemical cycle and Rankine cycle with two turbines can be written as given below:

$$\eta_{WS} = \frac{\dot{W}_{net} + \dot{Q}_{Cooling} + \dot{Q}_{Heating}}{\dot{Q}_{Solar}} \tag{29}$$

For the ammonia-based solar thermochemical plant, the net produced power can be calculated as follows:

$$\dot{W}_{net} = \dot{W}_{HPT} + \dot{W}_{LPT} - \sum \dot{W}_{p} \tag{30}$$

Also, the overall exergetic performance of the designed plant can be given as:

$$\psi_{WS} = \frac{\dot{E}x^{W}_{net} + \dot{E}x^{Q}_{Cooling} + \dot{E}x^{Q}_{Heating}}{\dot{E}x^{Q}_{solar}} \tag{31}$$

In addition to that, the energetic and exergetic performance coefficient of cooling plant should be written as given below;

$$COP_{en} = \frac{\dot{Q}_{Eva}}{\dot{Q}_{Gen} + \dot{W}_{P5}} \tag{32}$$

and

$$COP_{ex} = \frac{\dot{E}x^{Q}_{Eva}}{\dot{E}x^{Q}_{Gen} + \dot{E}x^{W}_{P5}} \tag{33}$$

4 Results and Discussion

In this sub-chapter, several outlet key indicators are investigated in detail by using the thermodynamic assessment. The Engineering Equation Solver (EES) code for the Microsoft Windows working process is developed in order to analyze the method improved to simulate the ammonia-based solar thermochemical integrated plant and present the parametric studies. In this part, thermodynamic analysis results and parametric analyses as well are presented for ammonia-based solar thermochemical integrated plant. Table 4 indicates the power outputs of turbines, heating and cooling outputs of integrated plant and overall energetic and exergetic efficiency of the investigated system for three different working time which are charging, storing and discharging time.

As seen from Table 4, total power outputs are about 11400 kW for charging and storing time while it is about 7300 kW for discharging time. However, energy and exergy performance comparison show that energetic and exergetic efficiency values for charging, storing and discharging time are close to each other, expectantly energetic and exergetic efficiency values are a little bit higher for storing time with 52.17% and 43.45%, respectively.

After this point of the study, parametric analyses result in which dead state temperature, solar irradiation and pinch point temperature of ammonia synthesis reactor are

Table 4 Computed outputs utilized to evaluate the plant

Parameters	Charging time	Storing time	Discharging time
HP turbine electricity output (kW)	8426	8424	6107
LP turbine electricity output (kW)	3048	3012	1253
Inlet solar energy, $\dot{Q}_{solar}$ (kW)	18652	12364	3407
Heating output, $\dot{Q}_{Heating}$ (kW)	484	857	1025
Cooling output, $\dot{Q}_{Cooling}$ (kW)	2125	1872	1334
Overall energetic performance (%)	49.03	52.17	50.79
Overall exergetic performance (%)	38.94	43.45	34.36

given in detail. The effect of dead state temperature from 0 °C to 40 °C on the energetic performance of the whole plant for three different periods of time is presented in Fig. 2. The following figure reveals that as dead state temperature increases up to a certain point energy efficiency of the whole system rises for each period of time. It can be inferred that the increase in dead state temperature causes a rise in system performance. The energy efficiencies of integrated system for charging, storing and discharging periods can reach up to 51.42%, 54.39%, and 52.64%, respectively, at 40 °C.

Figure 3 similar to the previous one presents the exergetic efficiency change with respect to rising dead state temperature. As always, an increase in dead state temperature makes the temperature difference between the plant and reference decreases, so that, exergy efficiencies of the integrated system for charging, storing and discharging time increase about 19.01%, 17.15%, and 15.34%, respectively. Supporting the previous results for performances, the highest exergetic performance is obtained for storing period. On the other hand, the discharging period has lower exergetic performance compared to the other two periods.

Some energetic and exergetic quantities depend on the intensive characteristics of the dead condition, such as temperature and pressure. Thus, the energetic and exergetic analyses outputs generally are sensitive to variations in these characteristics. In this book chapter, it is seen that reference pressure change has a very limited effect on the energetic and exergetic values. On the other hand, the reference temperature has an important effect on both efficiencies, especially on the exergetic performance of the plant.

The impacts of dead state temperature on total electricity, cooling and heating production rates are shown in Fig. 4 for charging, storing and discharging periods. Figure 4 not only shows the useful outputs change of integrated plant for varying

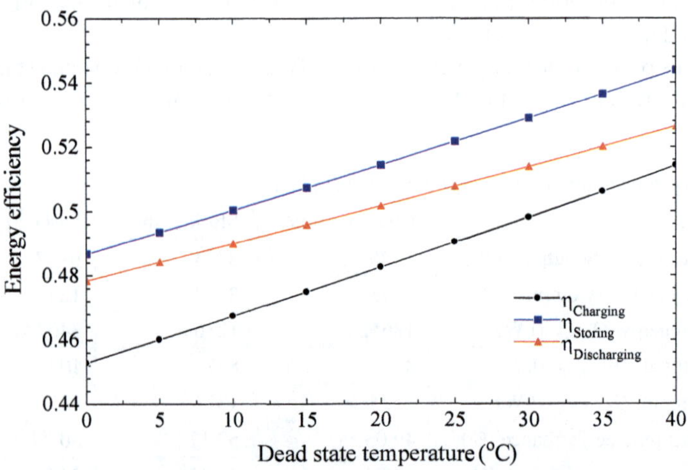

Fig. 2 Effect of dead state temperature on the whole system energetic efficiency at charging, storing and discharging time

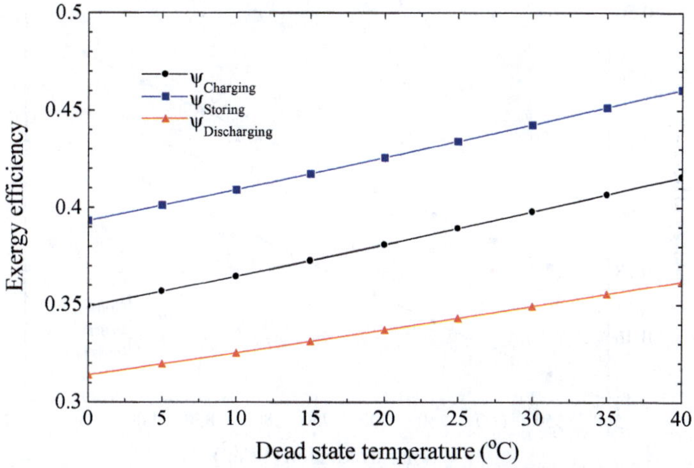

Fig. 3 Effect of dead state temperature on the whole system exergetic performance at charging, storing and discharging time

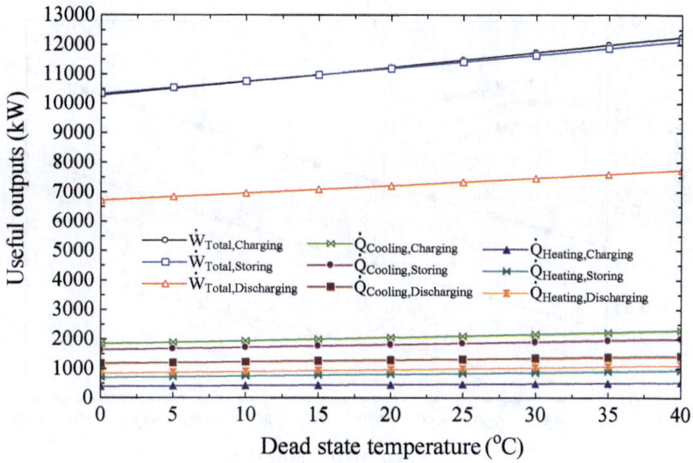

Fig. 4 Impact of dead state temperature on the useful outlets rates from integrated plant at charging, storing and discharging time

dead state temperature but also explains why energetic and exergetic efficiency values increase with increasing dead state temperature. As seen from the figure, the most distinctive increases occur in total work output for storing time from about 10500–12000 kW and for discharging time from about 6700 to nearly 7700 kW.

Solar irradiance is altered in Figs. 5, 6 and 7 in the range of 0.5–1 kW/m2, which has a very significant effect on the plant performance due to the large area of reflector surface. Therefore, the main parameter for this integrated plant is obviously solar

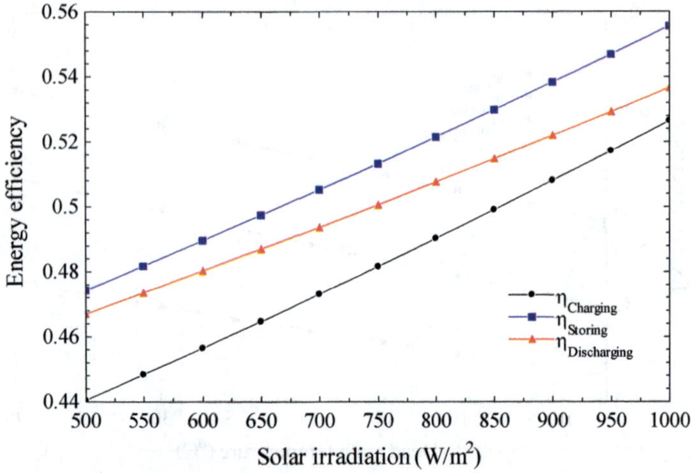

Fig. 5 Effect of solar radiation on the whole plant energetic performance at charging, storing and discharging time

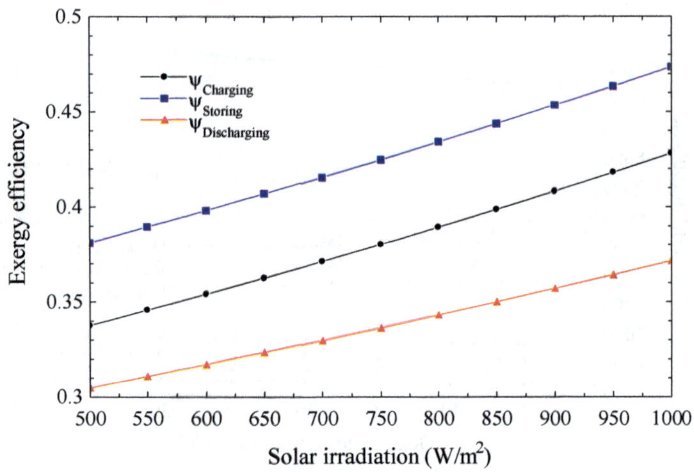

Fig. 6 Effect of solar radiation on the whole plant exergetic performance at charging, storing and discharging time

irradiation due it is the only source of the proposed system. While solar irradiation is 500 W/m^2, energy efficiencies of charging, storing and discharging time are 44%, 46.7% and 46.4%, respectively as given in Fig. 5. These values are 52.4%, 54.8% and 53.8% for the same working period of time at 1000 W/m^2. So as solar irradiation increases from 0.5 kW/m2 to 1 kW/m^2 energy efficiencies increase by about 8% for all working time of the integrated system. Hence it can be concluded that as solar radiation increases, energy efficiencies increase almost linearly up to a certain point.

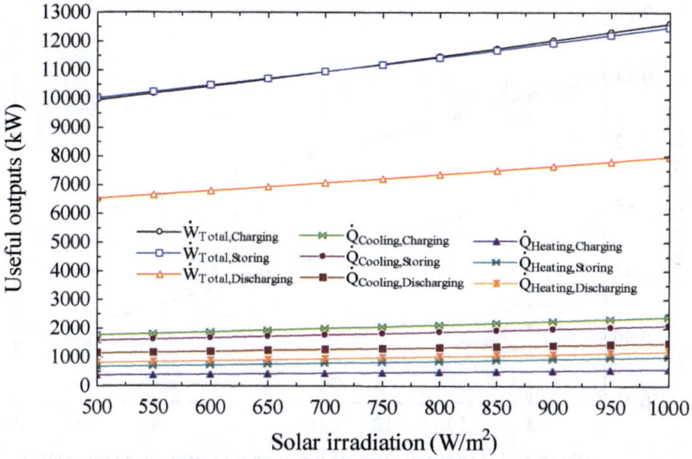

Fig. 7 Impact of solar radiation on the useful outputs rates from integrated plant at charging, storing and discharging time

Figure 6 illustrates the impact of solar irradiation rise in exergetic efficiencies of the proposed process. As solar irradiation rises the input of the integrated system increases thus proposed system can produce more useful products at the same time. Exergy efficiency values similar to energy efficiency values go up about 9% for each period of time with increasing solar radiation from 0.5 kW/m^2 to 1 kW/m^2.

Figure 7 presents the impact of solar radiation ranging from 0.5 kW/m^2 to 1 kW/m^2 on useful outputs of the combined plant. As seen from the figure all useful product generation raises with increasing solar irradiation different than increasing dead state temperature. Again, total work done during the storing period is the most affected element with rising solar irradiation. Resultantly it can be concluded that solar irradiation is the most effective parameter among others for these types of integrated systems feeding by solar energy.

Another important parameter is the pinch point temperature of ammonia synthesis reactor of which effects on the energy efficiency, exergy efficiency and useful products are presented in Figs. 8, 9 and 10, respectively. In this book chapter, the base case pinch point temperature of ammonia synthesis reactor is chosen as 12 °C, and the pinch point temperature of ammonia synthesis reactor is varied between 6 and 22 °C. It is evaluated that the pinch point temperature differences of the ammonia synthesis reactor have an important effect on energy and exergy performances. In the plots in Figs. 8 and 9, it can be seen that a rise in the pinch point temperature of ammonia synthesis reactor results in a decrease in plant energetic and exergetic efficiencies for charging, storing and discharging times.

Figure 8 shows how energy efficiencies of integrated plant for different working periods of times decrease with rising pinch point temperature. As pinch point temperature rises from 6 to 22 °C, energy efficiency of combined process decreases 51.11

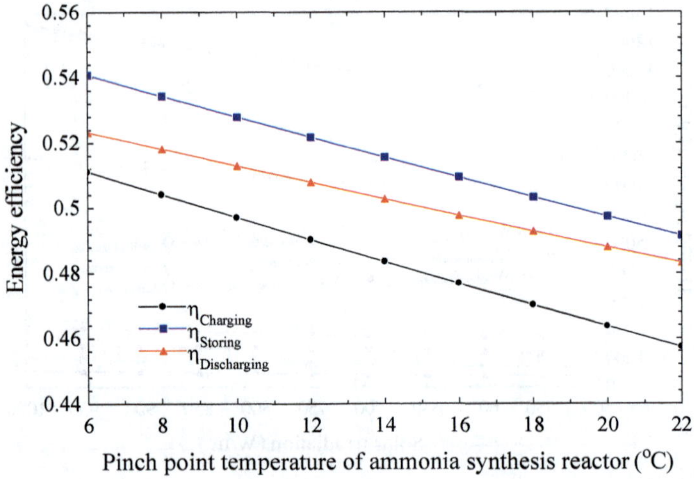

Fig. 8 Impact of pinch point temperature of ammonia synthesis reactor on the whole plant energetic performance at charging, storing and discharging time

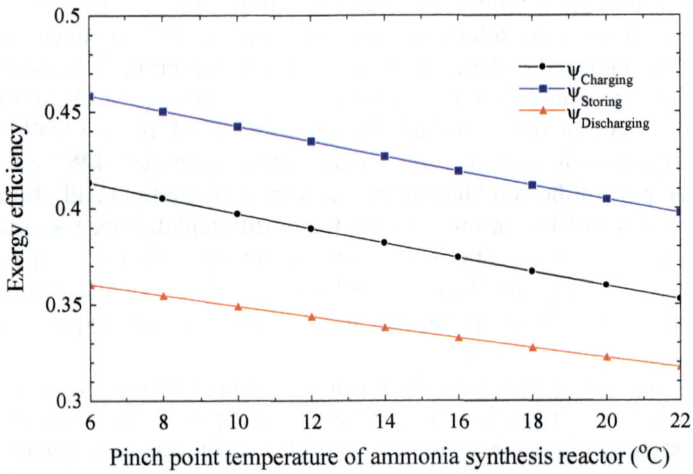

Fig. 9 Effect of pinch point temperature of ammonia synthesis reactor on the whole system exergetic performance at charging, storing and discharging time

to 45.73% for charging time, from 54.07 to 49.14% for storing time, and finally from about 52.33 to 48.31% for discharging time.

Exergetic performance changes with rising pinch point temperature are presented in Fig. 9. As pointed out by the figure, as pinch point temperature rises from 6 °C to 22 °C, the exergetic efficiency values of combined plant for charging time from 41.32 to 35.26%, from 45.83 to 39.74% for storing time, and finally from about 36.03 to 31.73% for discharging time.

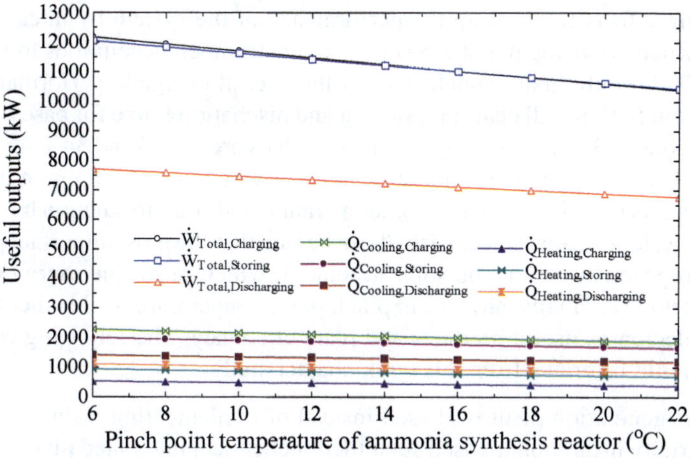

Fig. 10 Effect of pinch point temperature of ammonia synthesis reactor on the useful outlets rates from integrated plant at charging, storing and discharging time

The last figure of parametric analysis illustrates how useful products generation falls out with increasing pinch point temperature of ammonia synthesis reactor. When pinch point temperature is 6 °C, total power production for storing time is about 12000 kW however when pinch point temperature boosts up to 22 °C, total work done during storing comes to about 11000 kW. It can be the reason for this that the higher the pinch point temperature for ammonia synthesis reactor, the lower the power being used in this subcomponent which leads to a decrease of the Rankine cycle turbines electricity exit and the cooling effect of SEAC for the charging, storing and discharging periods.

5 Conclusions

Nowadays ecological and economic challenges in the earth require the utilization of power as efficiently as possible. To make utilization of low-grade heat resources such as industrial waste heat and sustainable resources such as the geothermal, solar or biomass combustion, there are different thermodynamic process ways. Also, decreasing the dependency on carbon-based resources and declining the potentially damaging emissions should be achieved by utilizing renewable and sustainable power technologies. Tri-generation plants present a promising alternative to traditional single and co-generation plants. In this study, a design of a novel ammonia-based solar thermochemical integrated plant is proposed, and it is analyzed both thermodynamically and parametrically in order to see the impacts of some parameters such as dead state temperature, solar irradiation and pinch point temperature of ammonia synthesis reactor. The integrated system produces power from Rankine turbines with heating and cooling effects to boost up the performance of the system by integrating subsystems which are being popular recently for environmental solutions to the energy problem. This study firstly concludes that the overall energetic performance of the plant is about 50% for all charging, storing and discharging time for base parameters given in Table 4. However, exergy efficiency values are 39, 43 and 34% for charging, storing and discharging time. Parametric analysis which is very important for energy system designs reveals that dead state temperature and solar irradiation have positive effects on system performance. Also, it can be said that the most important parameter for this proposed system is the solar irradiation which raises the exergy efficiency of the system 9%. Finally, any rise in pinch point temperature while other indicators are fixed decreases useful products and plant efficiency. The following concluding outputs should be drawn from this book-chapter study:

- If the tri-generation plant is chosen instead of implementing a single generation of electricity in ammonia-based solar thermochemical integrated plant, the overall energetic performance rises from 41.83 to 50.24%, where the overall exergetic performance will increase from 33.19 to 39.24%.
- A rise in solar irradiation from 0.5 kW/m^2 to 1 kW/m^2 rises the overall plant energetic performance by 19.24% and exergetic efficiency by 18.59%.
- In addition to that, when the reference-state temperature rises from 0 to 40°C, the overall energetic performance rises by 9.24% and the overall exergetic efficiency rises by approximately 11.53%.
- A rise in pinch point temperature of the ammonia synthesis reactor reduces the plant energetic and exergetic performance.

References

Agency IE (2008) Combined heat and power: evaluating the benefits of greater global investment report. International Energy Agency (IEA), Paris

Ahmadi P, Dincer I, Rosen MA (2012) Exergo-environmental analysis of an integrated organic Rankine cycle for trigeneration. Energy Convers Manag 64:447–453

Atiz A, Karakilcik H, Erden M, Karakilcik M (2019) Assessment of electricity and hydrogen production performance of evacuated tube solar collectors. Int J Hydrog Energy 44:14137–14144

Behzadi A, Habibollahzade A, Ahmadi P, Gholamian E, Houshfari E (2019) Multi-objective design optimization of a solar based system for electricity, cooling, and hydrogen production. Energy 169:696–709

Cengel YA, Boles MA (2015) Thermodynamics: an engineering approach, 8th edn. McGraw-Hill Education, New York

Chen C, Aryafar H, Lovegrove KM, Lavine AS (2017) Modeling of ammonia synthesis to produce supercritical steam for solar thermochemical energy storage. Sol Energy 155:363–371

Chen C, Lovegrove KM, Sepulveda A, Lavine AS (2018) Design and optimization of an ammonia synthesis system for ammonia-based solar thermochemical energy storage. Sol Energy 159:992–1002

Chen C, Liu Y, Aryafar H, Wen T, Lavine AS (2019) Performance of conical ammonia dissociation reactors for solar thermochemical energy storage. Appl Energy 255:1–10

Dincer I, Rosen MA (2012) Exergy: energy, environment and sustainable development, 2nd edn. Elsevier, Oxford

Dincer I, Zamfirescu C (2012) Renewable-energy-based multigeneration systems. Int J Energy Res 36:1403–1415

El-Emam RS, Dincer I (2018) Investigation and assessment of a novel solar-driven integrated energy system. Energy Convers Manag 158:246–255

Hogerwaard J, Dincer I, Naterer GF (2017) Solar energy based integrated system for power generation, refrigeration and desalination. Appl Therm Eng 121:1059–1069

Kalinci Y, Hepbasli A, Dincer I (2009) Exergetic assessment of an integrated gasifier/boiler system for hydrogen production with different biomass types. In: Proceedings of the international conference on hydrogen production, May 03–06, Oshawa, Canada

Kasaeian A, Nouri G, Ranjbaran P, Wen D (2018) Solar collectors and photovoltaics as combined heat and power systems: a critical review. Energy Convers Manag 156:688–705

Kotas TJ (1985) The exergy method of thermal plant analysis, 1st edn. Butterworths, New York

Kreetz H, Lovegrove K (1999) Theoretical analysis and experimental results of a 1 kW_{chem} ammonia synthesis reactor for a solar thermochemical energy storage system. Sol Energy 67(4–6):287–296

Kreetz H, Lovegrove K (2002) Exergy analysis of an ammonia synthesis reactor in a solar thermochemical power system. Sol Energy 73(3):187–194

Liu T, Bai Z, Zheng Z, Liu Q, Lei J, Sui J, Jin H (2019) 100 kWe power generation pilot plant with a solar thermochemical process: design, modeling, construction, and testing. Appl Energy 251:1–11

Luqman M, Bicer Y, Al-Ansari T (2019) Thermodynamic analysis of an oxy-hydrogen combustor supported solar and wind energy-based sustainable polygeneration system for remote locations. Int J Hydrog Energy (In Press)

Luzzi A, Lovegrove K (1997) A solar thermochemical power plant using ammonia as an attractive option for greenhouse-gas abatement. Energy 22(2–3):317–325

Ozlu S, Dincer I (2016) Performance assessment of a new solar energy-based multigeneration system. Energy 112(1):164–178

Ozturk M, Dincer I (2013) Thermodynamic analysis of a solar-based multi-generation system with hydrogen production. Appl Therm Eng 51:1235–1244

Ozturk M, Yuksel YE (2016) Energy structure of turkey for sustainable development. Renew Sustain Energy Rev 53:1259–1272

Siddiqui O, Dincer I (2017) Comparative assessment of the environmental impacts of nuclear, wind and hydro-electric power plants in Ontario: a life cycle assessment. J Clean Prod 164:848–860

Siddiqui O, Dincer I (2019) Design and analysis of a novel solar-wind based integrated energy system utilizing ammonia for energy storage. Energy Convers Manag 195:866–884

Xu C, Wang Z, Li X, Sun F (2011) Energy and exergy analysis of solar power tower plants. Appl Therm Eng 31:3904–3913

An Overview of Hydrogen Production from Biogas

Yagmur Nalbant and C. Ozgur Colpan

Abstract Today, hydrogen is mainly produced by natural gas through thermochemical processes. Renewable hydrogen production has increased its importance due to the increase in the pure hydrogen demand, depletion in the fossil fuel reserves, and global environmental and energy issues. In this regard, biogas obtained from biomass by anaerobic digestion or organic matter fermentation processes could be an alternative resource for hydrogen production. Different reforming processes can be selected for this purpose depending on the composition of biogas, availability of investment, required hydrogen purity, and the amount of desired hydrogen. In this chapter, different chemical reforming processes that can be used to produce hydrogen are first presented in detail. Then, the considerations for producing hydrogen from biogas reforming technologies are discussed. The literature review related to hydrogen production from biogas is also presented. Finally, membrane reactor, which provides both hydrogen production and hydrogen separation in the same reactor, is introduced as an alternative route for hydrogen production from biogas.

Keywords Hydrogen production · Hydrogen · Biogas · Reforming process · Membrane reactor

1 Introduction

The energy demand in the world rises with the increase of global population and economy, and this also gradually increases the using of fossil fuels (e.g., coal, natural gas and oil) and leads to the negative impact for environment such as global warming, air pollution, and water pollution. Therefore, the energy crisis has become one of the most major problems for humanity and a sustainable world. The environment issues have led to the utilization of several renewable and alternative energy resources.

Y. Nalbant · C. O. Colpan (✉)
Faculty of Engineering, Mechanical Engineering Department, Dokuz Eylul University, Buca, Izmir, Turkey
e-mail: ozgur.colpan@deu.edu.tr

T. S. Uyar (ed.), *Accelerating the Transition to a 100% Renewable Energy Era*, Lecture Notes in Energy 74, https://doi.org/10.1007/978-3-030-40738-4_16

Hydrogen is one of the most important energy carriers since it is storable, highly reactive, light, and has higher gravimetric energy density than other fuels. The contribution of hydrogen to the global energy supply can be two types: (1) hydrogen can be directly utilized in the useful energy production (e.g., electricity and heat); and (2) hydrogen can be converted into storable gaseous or liquid hydrogen-based fuels (e.g., synthetic methane and methanol) (IEA 2019). Today, around 76% of hydrogen is produced from natural gas; and the majority of the remaining is produced from coal. These production processes are achieved through thermo-chemical methods. Among these methods, steam-methane reforming process is the most widely used one for natural gas. For coal and petroleum, partial oxidation and autothermal reforming are generally preferred. Electricity produced using renewable energy resources (e.g., wind, solar, geothermal, hydro, and biomass) can also be used to produce hydrogen through electrolysis of water. Among different types of electrolyzers, alkali and proton exchange membranes types are the most widely preferred ones for this purpose. There is also ongoing research on the development of ceramic material based solid oxide electrolyzers. However, it is expected that, the production of hydrogen by chemical reforming processes will be more preferred in the near future as the electrolysis process is expensive and still under development (Minh et al. 2018). Taking into account the current situation, the majority of hydrogen, which is produced by chemical reforming of various hydrocarbons such as natural gas, causes greenhouse gas generation. To reduce this environmental problem and produce hydrogen through renewable resources, biogas obtained from biomass could be an alternative (Gao et al. 2018).

Biogas is a gas mixture that mainly consists of methane (CH_4) and carbon dioxide (CO_2), in addition to other gases such as hydrogen sulfide (H_2S), hydrogen (H_2), nitrogen (N_2), oxygen (O_2), ammonia (NH_3), and water vapor (H_2O). Biogas is produced by anaerobic digestion or organic matter fermentation processes of several feedstocks in the presence of microorganisms that separate organic material. The different feedstocks are generally divided into three generations. The first generation of feedstocks includes seeds, sugars, and grains, and the biogas obtained from this generation is used in some applications such as production of heat and steam, injection into the networks of natural gas, and the generation of the combined heat and electricity. The second generation of feedstocks is given as lignocellulosic biomass such as woody crops and crop residues, and the second-generation biogas is used in the production of syngas, hydrogen, bio-methanol, hydrocarbons, and alcohols. The promising third generation of feedstocks is the algae because they have high growth rate and carbohydrate content (Gao et al. 2018; Verma and Samanta 2016). Table 1 shows the typical range of chemical composition of biogas obtained from different feedstocks.

Biogas contains some undesirable compounds that can damage the quality of biogas fuel. For example, H_2S is a corrosive and toxic gas that can damage the equipment in the overall processes. CO_2 and H_2O are known as impurity matter that reduces the energy value of biogas. Therefore, the purification and upgrading technologies of biogas are essential to control the impurity levels that depend on the several usage technologies. Today, the main traditional and new technologies

Table 1 Chemical composition of biogas (Verma and Samanta 2016; Alves et al. 2013)

Compounds	Formula	Units	Percentage
Methane	CH_4	vol. %	55–70
Carbon dioxide	CO_2	vol. %	30–45
Hydrogen sulfide	H_2S	ppm	500–4000
Ammonia	NH_3	ppm	100–800
Hydrogen	H_2	vol. %	<1
Nitrogen	N_2	vol. %	<1
Oxygen	O_2	vol. %	<1
Water vapor	H_2O	vol. %	<1

for purification include the adsorption, the pressure swing adsorption (PSA), the membrane technologies, the biological methods, and the cryogenic technologies (Gao et al. 2018; Verma and Samanta 2016).

In general, the reforming of biogas is a promising and attractive technique to produce clean hydrogen. The advantages of using biogas for this purpose are the reduction of greenhouses gas emissions, its reliability, and its local availability and therefore, it is better environmentally and economically.

The objective of this chapter is to present the different biogas reforming processes for hydrogen generation. In addition, the membrane reactors that are used to produce and purify hydrogen are emphasized.

2 Reforming Processes Used for Hydrogen Production from Biogas

The reforming processes for hydrogen production from biogas or methane can be split into two as the conventional and non-conventional reforming processes. The conventional reforming processes are mainly known as steam reforming (SR), partial oxidation reforming (POR), auto-thermal reforming (ATR), dry reforming (DR), and dry oxidation reforming (DOR). The non-conventional reforming processes are known as solar reforming, catalytic decomposition, and thermal plasma reforming (Verma and Samanta 2016).

Hydrogen can be generated by reforming of biogas or methane in a wide range of temperature (600–1000 °C) through endothermic or exothermic and reversible reactions. These reforming reactions generally occur at low pressure. In various reforming processes, biogas that is mainly a mixture of methane and carbon dioxide is reacted with agents such as steam, air or oxygen to produce syngas including hydrogen and other gases (Galvagno et al. 2013). Figure 1 shows the main steps of hydrogen production from biogas.

Before the biogas reforming process, undesirable and toxic substances (H_2S) in the biogas must be removed using some purification techniques; (1) physicochemical treatment such as the production of metal sulfide with chemical adsorption of

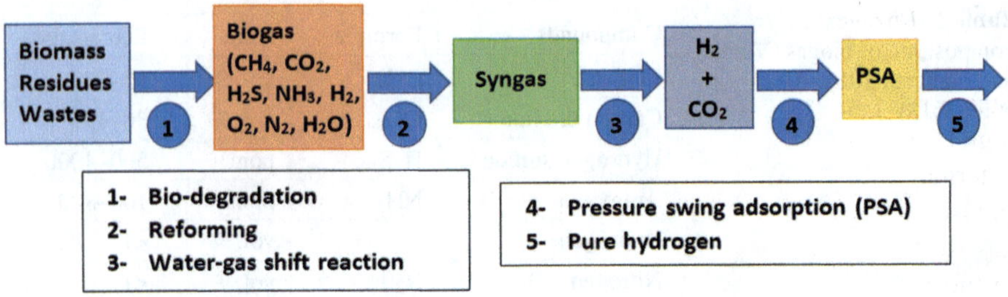

Fig. 1 Main steps of hydrogen production from biogas (Adapted from Minh et al. 2018)

H_2S on the solid adsorbent and absorption in hydrated solutions, and (2) biological treatment, for example, the living microorganisms like bacteria eliminates impurities with converting harmful toxic substances into less harmful forms (Verma and Samanta 2016; Alves et al. 2013; Holladay et al. 2009). Consequently, reactions for the biogas reforming are similar to those for the methane reforming. Table 2 shows the chemical reactions formed in reforming processes and its enthalpy of reaction at 25 °C and 1 atm.

Table 2 Chemical reactions formed in reforming processes (Verma and Samanta 2016; Alves et al. 2013)

Equations no	Reaction name	Chemical reaction	ΔH_{298} (kJ/mol)
Equation (1)	Steam reforming	$CH_4 + H_2O \rightarrow CO + 3H_2$	206
Equation (2)	Water-gas shift reaction	$CO + H_2O \rightarrow CO_2 + H_2$	−41.2
Equation (3)	Combined reaction	$CH_4 + 2H_2O \rightarrow CO_2 + 4H_2$	165
Equation (4)	Methane cracking	$CH_4 \rightarrow C + 2H_2$	74.9
Equation (5)	Boudouard reaction	$2CO \rightarrow C + CO_2$	−172.4
Equation (6)	Reduction of CO	$CO + H_2 \rightarrow C + H_2O$	−131.3
Equation (7)	Partial oxidation reforming	$CH_4 + 1/2O_2 \rightarrow CO + 2H_2$	−35.6
Equation (8)	Complete oxidation	$CH_4 + 2O_2 \rightarrow CO_2 + 2H_2O$	−801.7
Equation (9)	Auto-thermal reforming	$CH_4 + x/2O_2 + yCO_2 + (1\text{-}x\text{-}y)H_2O \rightarrow (y + 1)CO + (3\text{-}x\text{-}y)H_2$	~0
Equation (10)	Dry reforming	$CH_4 + CO_2 \rightarrow 2CO + 2H_2$	247.4
Equation (11)	Dry oxidation reforming	$CH_4 + \beta CO_2 + (1-\beta)/2O_2 \rightarrow (1 + \beta)CO + 2H_2$	$(285\beta–38), 0 < \beta < 1$

2.1 Steam Reforming (SR)

Steam reforming reaction is highly endothermic and the required temperature for high H_2 yields (60–70%) is in between 650 and 850 °C. Although it needs significant external energy input, it is the most widely used method for producing hydrogen in industrial applications. Steam reforming process involves steam reforming, water-gas shift, and direct steam reforming reactions. Firstly, methane reacts with water in the SR, and the syngas including carbon monoxide and hydrogen is produced in the presence of suitable catalyst (Eq. (1) in Table 2). After that, the syngas is cooled down to a temperature between 300 and 500 °C and fed to the WGS reactor where carbon monoxide reacts with water to reduce CO content in the syngas, and carbon dioxide and hydrogen are produced (Eq. (2) in Table 2). WGS reactor includes high-temperature (HT) and low-temperature (LT) water-gas shift reactors. HT-WGS reactor has high CO conversion and low reaction kinetics; whereas LT-WGS reactor has low CO conversion and high reaction kinetics (Minh et al. 2018). The used catalysts in the WGS are Fe, Mo, Cu, and Fe-Pd alloys (Verma and Samanta 2016). In addition, direct steam reforming reaction (Eq. (3)) or combined reaction, which is the sum of Eqs. (1) and (2), occurs. To produce highly pure hydrogen in SR, the separation unit such as PSA or membrane is needed to remove CO_2, CO, and other gases (except hydrogen) from syngas. Figure 2 shows the schematic diagram of the steam reforming process including reactants and products.

In some cases, parallel reactions that cause carbon formation on the catalytic surface can occur such as methane cracking (Eq. (4)), Boudouard (Eq. (5)), and CO reduction reactions (Eq. (6)). When the carbon formation appears as the form of nanotube, it provides lower dispersion of carbon over the catalyst surface, consequently, the catalyst activity is preserved for a longer time.

In SR, the reaction temperature, the molar rate of H_2O/CH_4, and reaction pressure are the critical operating parameters. Let's consider two cases; (i) Case A: 1 mol of H_2O and 1 mol of CH_4, and (ii) Case B: 3 mol of H_2O and 1 mol of CH_4. When the case A is considered, the production of hydrogen occurs when the temperature exceeds 180 °C; the CO_2 and solid carbon that are undesirable byproducts are observed at 250–800 and 450–800 °C, respectively. When the Case B is considered, solid carbon formation is eliminated and CH_4 is completely consumed above 720 °C.

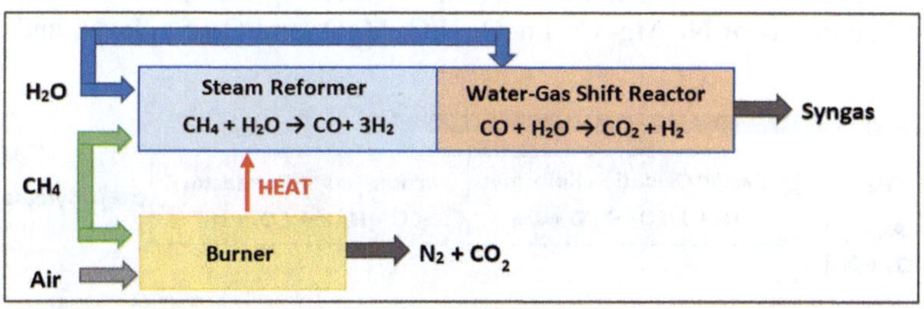

Fig. 2 Schematic diagram of steam reforming process including reactants and products

Considering this temperature level, Case B seems to be more advantageous than Case A. Thus, when the reaction temperature and the molar rate of H_2O/CH_4 increase, the conversion of CH_4 increases and the carbon formation decreases. In addition, when the reaction pressure increases, the molar flow rate of hydrogen decreases and the molar flow rate of CH_4 increases, thus, the conversion of CH_4 decreases (Minh et al. 2018).

Another critical parameter is the type of catalyst. Catalyst can be some transition metals (Ni, Fe, Co, Mo etc.) and noble metals (Pd, Ru, Pt, Ir, Rh etc.). For example, Kikuchi et al. (1974) reported that reactivity of metals is in the following order for steam reforming of methane: Fe = Co < Pt = Pd < Ir < Ni < Ru = Rh (Kikuchi et al. 1974). The most used metal is Ni due to its lower cost. However, Ni is less resistant to the formation of coke, therefore, Pt and Pd are more attractive for the stability of catalyst. Rh is used in the event of mass transfer limitations, since it has better activity and provides lower formation of coke (Alves et al. 2013; Iulianelli et al. 2010).

2.2 *Partial Oxidation Reforming (POR)*

The partial oxidation reforming for producing hydrogen decreases the cost of energy because of its moderately exothermic reaction. At the atmospheric pressure, methane is partially oxidized with oxygen, and the syngas including hydrogen and carbon monoxide is produced (Eq. (7) in Table 2). Figure 3 shows the schematic diagram of partial oxidation reforming process including reactants and products.

To provide highly CH_4 conversion and to reduce the formation of coke, the required temperature should be between 700 and 900 °C. In addition, if the selectivity of carbon monoxide decreases slightly, methane can react with oxygen to produce CO_2, which is called as complete combustion reaction (Eq. (8) in Table 2). This reaction is a strong exothermic reaction that can cause hot spots in the reactor bed and the formation of coke on the catalytic surface (Alves et al. 2013; Dantas et al. 2012).

The oxidation reforming process can be realized with or without catalyst (Ahmed and Krumpelt 2001; Rohendi et al. 2018). Recently, researchers have investigated stability and activity of several catalysts for POR. At high temperatures (>800 °C), the solid solutions of Ni–Mg–Cr–La–O, NiO–MgO, and Ca–Sr–Ti–Ni and mixed

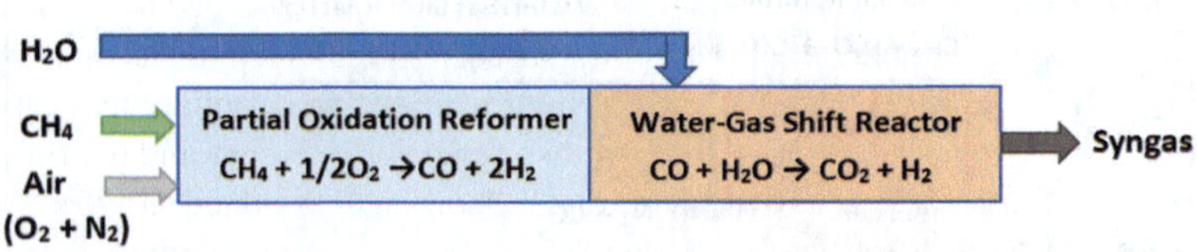

Fig. 3 Schematic diagram of partial oxidation reforming process including reactants and products

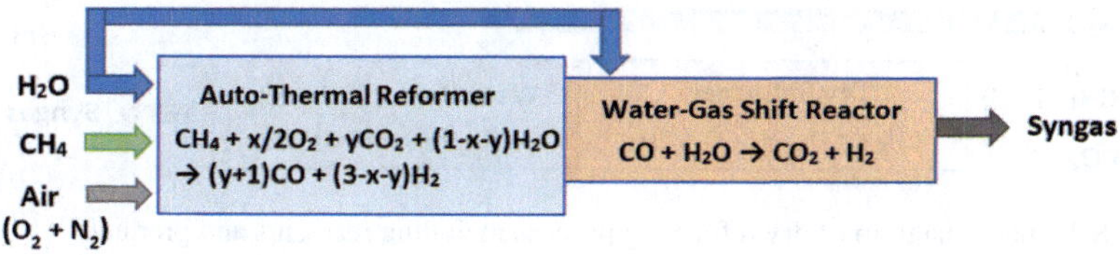

Fig. 4 Schematic diagram of auto-thermal reforming process including reactants and products

metal oxides, which are used as catalysts, ensure higher coke formation resistance (Verma and Samanta 2016; Alves et al. 2013).

2.3 Auto-thermal Reforming (ATR)

Auto-thermal reforming is defined as the combination of partial oxidation reforming (Eqs. (7) and (8) in Table 2) and steam reforming (Eq. (3) in Table 2). In ATR, the heat generated from partial oxidation reaction is used for the needed heat in the highly endothermic steam reforming reaction. The external heating is not required; therefore, ATR provides more efficiently and economical hydrogen production. Figure 4 shows the schematic diagram of auto-thermal reforming process including reactants and products.

ATR has some advantages such as faster reforming processing and higher H_2/CO ratios in comparison to POR. The generation of desired product is provided by different feed ratio combinations of CH_4–H_2O–O_2. Moreover, the control of temperature during the process can be advanced by the combination of steam reforming and partial oxidation reforming reactions. Consequently, hot-spots formation causing the coke formation on the catalyst surface is also reduced (Verma and Samanta 2016; Alves et al. 2013). To produce hydrogen from biogas, the range of ratios for H_2O/CH_4, O_2/CH_4, and H_2/CO can be taken as 1–2.5, 0.25–0.55, and 2–3.5, respectively (Mosayebi et al. 2012).

2.4 Dry Reforming (DR)

In DR, methane reacts with carbon dioxide to produce syngas consisting of hydrogen and carbon monoxide (Eq. (10) in Table 2). Figure 5 shows the schematic diagram of dry reforming process including reactants and products. This reaction type can be attractive considering the environmental issues, because two main greenhouse gas types (CH_4 and CO_2) are consumed in DR. Contrarily, the needed heat for the endothermic DR reaction is generated with the burned fuels, which causes the CO_2 emissions to the atmosphere.

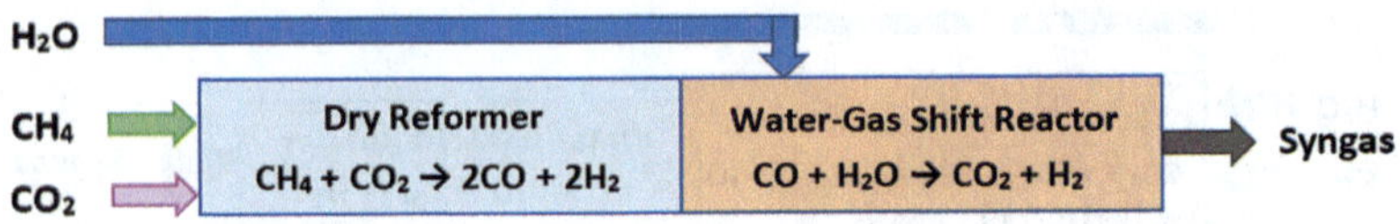

Fig. 5 Schematic diagram of dry reforming process including reactants and products

The other parallel reactions such as reverse water-gas shift reaction (Eq. (2) in Table 2), methane cracking reaction (Eq. (4) in Table 2), and Boudouard reaction (Eq. (5) in Table 2) can change the equilibrium conversion in the major reaction. Generally, the temperature of DR process is between 700 and 900 °C and the molar ratio of CH_4/CO_2 is between 1 and 1.5 (Alves et al. 2013).

The major problem in DR process is the carbon accumulation or coke formation on the catalyst surface. The parallel reactions can cause tendency of coke formation that affects the activation of catalyst. For example, when the removal rate of carbon is lower than the methane cracking, the carbon formed as a result of the methane cracking reaction accumulates and causes an important problem like coke formation. Moreover, the Boudouard reaction that is favorable at low temperatures promotes the methane cracking reaction in order to produce additional carbon, since this reaction is exothermic in nature (Verma and Samanta 2016; Alves et al. 2013).

The most effective catalysts used in DR process have been investigated by some researchers, and they report several metals that belong to 8, 9, and 10 groups in the Periodic Table, such as Ru, Pt, and Rh. These catalysts are very expensive, and their availability is limited in nature; therefore, they are not appropriate for industrial applications. Other catalysts that have cheaper and higher resistance of coke formation have been investigated, such as Ni and Co (Verma and Samanta 2016; Bereketidou and Goula 2012).

2.5 *Dry Oxidation Reforming (DOR)*

Dry oxidation reforming is defined as the combination of partial oxidation reforming (Eqs. (7) and (8) in Table 2) and dry reforming (Eq. (10) in Table 2). Furthermore, DOR provides the control of carbon formation on the catalyst surface, because the carbon accumulated in DR reactions can react with oxygen, and carbon including gases (CO or CO_2) can be produced. Figure 6 shows the schematic diagram of dry oxidation reforming process including reactants and products. β is defined as the stoichiometric ratio of CO_2 fed with conventional DR. If β is taken as 1, this reaction will be non-oxidative DR reaction, whereas if β is taken as 0, this reaction will be a partial oxidation reaction. Therefore, the value of β should be taken between 0 and 1 (Alves et al. 2013).

Additional advantages of DOR where CH_4 is fed with O_2 and CO_2 are as follows: improved resistance of deactivation, increased stability of catalyst, advanced CH_4

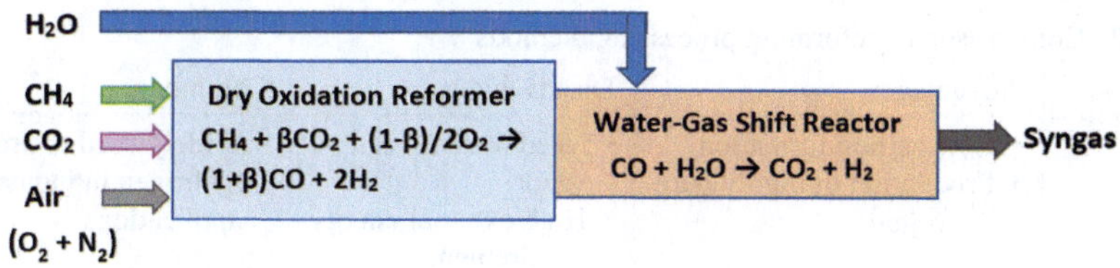

Fig. 6 Schematic diagram of dry oxidation reforming process including reactants and products

conversion, decreased overall energy related to the process, and enhanced hydrogen yield at low temperatures.

In the DOR, some critical parameters such as the temperature of reaction and/or the concentration of oxygen feed cause a change in the ratio of H_2/CO and reaction nature as endothermic or exothermic. As known, POR reaction (Eq. (7) in Table 2) is exothermic while DR reaction (Eq. (10) in Table 2) is endothermic. Hence, the process of DOR, in which oxygen with different molar ratios can be fed at various temperatures, could be an alternative to the conventional DR process. In addition, oxygen feed decreases the formation of coke on the catalyst surface and provides better control for the partial oxidation to produce CO or CO_2 formations. Consequently, the syngas production from biogas from this process has higher energy efficiency requiring less external energy (Alves et al. 2013; Avraam et al. 2010). Recent studies have focused on catalysts (e.g., Ni) that includes a second metal support to increase the stability of catalyst (Verma and Samanta 2016; Lau et al. 2011). Table 3 presents the summary of the comparison of the reforming processing methods discussed above.

3 Considerations for Producing Hydrogen from Biogas Reforming Technologies

Reforming processing method used to produce hydrogen can be selected depending on the composition of biogas, availability of investment, required hydrogen purity, and the amount of desired hydrogen (Alves et al. 2013; Scarlat et al. 2018). Biogas as a fuel cannot be directly used in conventional reforming processes. H_2S content in biogas causes a rapid reduction in the activity of catalyst due to its poisoning effect. Therefore, some purification methods such as adsorption, membrane technologies, and biological methods are used to remove H_2S.

The chemical composition of treated biogas involves mainly CH_4 and considerable CO_2. Therefore, CO_2 naturally present in the biogas can be used as an oxidant to react with CH_4 in dry reforming. Nevertheless, if the ratio of CO_2/CH_4 in biogas is lower than 1 (as it generally occurs), the addition of alternative oxidant is required to produce syngas that consists of hydrogen and carbon monoxide. This situation leads to the addition of oxygen through dry oxidation reforming, which could be

Table 3 Comparison of reforming processing methods

Methods	Advantages	Disadvantages	Comments
SR	• Low carbon formation • Production of highly pure hydrogen	• Need to add water vapor • High external energy requirement • High operating temperature • Expensive catalyst	• Widely used to produce hydrogen in industrial applications
POR	• Relatively low operating temperature • High energy efficiency	• Formation of hot spots in the catalytic bed with increasing temperature • May completely oxidize methane if CO selectivity is decreased	• Can be combined with other reforming methods • Limited industrial applications
ATR	• Production of highly pure hydrogen • High energy efficiency • No need for external heat • Lower cost • Can be stopped and restarted	• Complex control process • Requires multiple catalysts • Can be unstable	• Combination of POR and SR • ATR improves the temperature control and reduces the formation of hot spots • Self-sustaining process
DR	• High conversion efficiency • Environmentally friendly process (CH_4 and CO_2 are consumed)	• Carbon formation • Parallel reactions consume hydrogen • High operating temperature • High energy demand • Expensive catalyst	• Can utilize CH_4 and CO_2 that are in biogas
DOR	• Less carbon formation • Reduced overall energy involved • High conversion efficiency • Increased stability of catalyst and deactivation resistance	• Complex control process • High CO_2 production • A second metal is required for the stability of the catalyst	• Combination of POR and DR • Can be an endothermic or exothermic reaction according to O_2 feeding

considered as a more suitable method than DR for hydrogen production from biogas. This reforming process also slightly increases the ratio of H_2/CO (Alves et al. 2013).

The main challenges in all reforming processes are to choose suitable catalyst that prevents the carbon formation on its active surface and to increase the useful life of catalyst. The carbon formation problem is critical for DR process, since the parallel reactions occurred in DR cause the carbon formation to increase. Therefore, the hydrogen production from biogas in DR is unfeasible because of the fast deactivation of the catalyst. On the other hand, DOR process can be suitable for the elimination of carbon formation, because additional O_2 feed helps the gasification of carbon, thus the availability of catalytic active surface is enhanced. However, the ratio of H_2/CO in both DR and DOR processes is approximately equal to 1, thus it limits the high purity potential of hydrogen produced by WGS reactor, since the volume of produced CO is high. In addition, the part of hydrogen produced using dry reforming process can react with carbon dioxide via the reverse water-gas shift reaction (reverse Eq. (2) in Table 2), consequently, water and carbon monoxide are produced, and the hydrogen yield is reduced. The above discussion shows that both DR and DOR processes are not quite suitable and feasible for producing hydrogen from biogas (Alves et al. 2013).

Before almost all reforming processes, biogas should be purified from H_2S, and the chemical composition of biogas becomes 93–96% of methane, 4–7% of carbon dioxide, and <20 ppm of H_2S. Low CO_2 content in the biogas enhances CH_4 conversion and provides higher hydrogen purity. However, for both DR and DOR processes, CO_2 that is the reactant matter should be added. Therefore, SR, POR, and ATR processes are more suitable than DR and DOR processes (Alves et al. 2013).

ATR process can be appeared as an attractive path for hydrogen production from biogas, since this process provides high energy efficiency and high hydrogen yield. Moreover, ATR can be rapidly stopped and restarted due to self-sustaining process, resulting in the lowest ratio of expense/utility compared to all the other reforming processes. ATR has complex control process since it can be unstable (Alves et al. 2013).

SR process could be seen to be a less advantageous method due to the need of high external heat, but it can be a more feasible solution if appropriate conditions are selected. In SR reactor, syngas is produced from biogas, and followed by WGS reactor to minimize the CO content in the syngas. This technology is better known and controlled. Therefore, SR is the most widely used method for producing hydrogen in industrial applications (Alves et al. 2013).

Generally, hydrogen is produced using the conventional steam reforming process that involves a reformer, a WGS reactor, and a separation unit. On the other hand, membrane reactors that allow both hydrogen production and purification in the single reactor at the same time have recently become the focus of interest for producing hydrogen from biogas as discussed in the following section.

4 Membrane Reactor

Even though conventional steam reforming reactor is widely used to produce hydrogen in industry, it has some disadvantages such as thermodynamic chemical equilibrium limitations, carbon formation, and deactivation of catalyst. Moreover, the material used in reactor wall should be very durable, since this reactor operates at high temperature (650–850 °C) (Lu and Xie 2016).

Membrane reactors where both chemical reactions and purification processes take place in the same reactor have a compact structure. Furthermore, membrane reactors

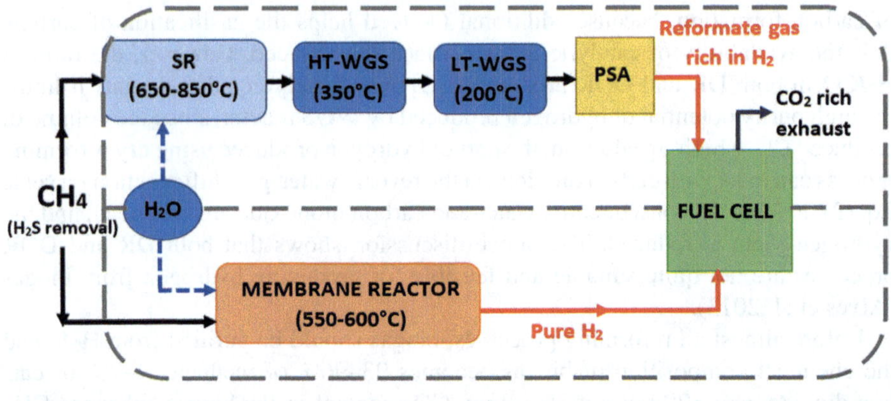

Fig. 7 Comparison of conventional and membrane reactors (Adapted from Marcoberardino et al. (2018))

provide higher hydrogen production with lower energy requirements. Membrane reactors eliminate the thermodynamic chemical equilibrium limitation with using hydrogen selective membrane, and these chemical reactions can be occurred at lower temperature (550–600 °C). Moreover, this reactor allows the use of material at lower temperature and hydrogen yield increases (Lu and Xie 2016). Figure 7 shows the comparison of conventional reactor and membrane reactor that produce hydrogen via steam reforming process for any fuel cell applications.

Hydrogen yield in membrane reactor depends on membrane that have high hydrogen permeation and allows high hydrogen flow rate (Gallucci et al. 2013). A membrane used in membrane reactors is basically defined as a barrier that does not allow any component other than hydrogen in a feed gas mixture to pass through. Membrane reactors consist of reaction side (hydrogen production region) and permeation side (hydrogen separation region), as shown in Fig. 8. In order to separate hydrogen from other gases, membrane should have high selectivity, high chemical and mechanical stability, and low cost (Gallucci et al. 2013).

Membranes can be split into two as biological and synthetic according to their functionality and structure. Biological membranes are easy to manufacture, but have many disadvantages, such as limited operating temperature (below 100 °C) and pH range, difficulty in cleaning and susceptibility to microbial attacks (Xia et al. 2003). Synthetic membranes are subdivided into organic (polymeric) and inorganic (ceramic and metallic). Organic membranes operate between 100–300 °C, whereas inorganic membranes operate above 200 °C. In order to obtain high purity in hydrogen streams, the most suitable membrane material due to its high hydrogen selectivity is the dense metal membrane (mostly palladium (Pd) alloys due to their high solubility and hydrogen permeability) (De Falco et al. 2011). Figure 9 shows the hydrogen transport mechanism through dense Pd-based membrane.

Dense metal membranes are also divided into supported and unsupported membranes. Unsupported membranes have some disadvantages such as low hydrogen permeability, brittleness, and surface poisoning related to mass diffusion resistances.

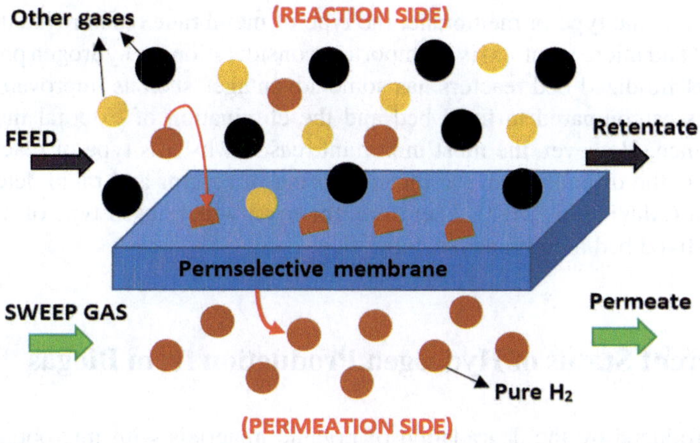

Fig. 8 Schematic diagram of the operating principle of the membrane system

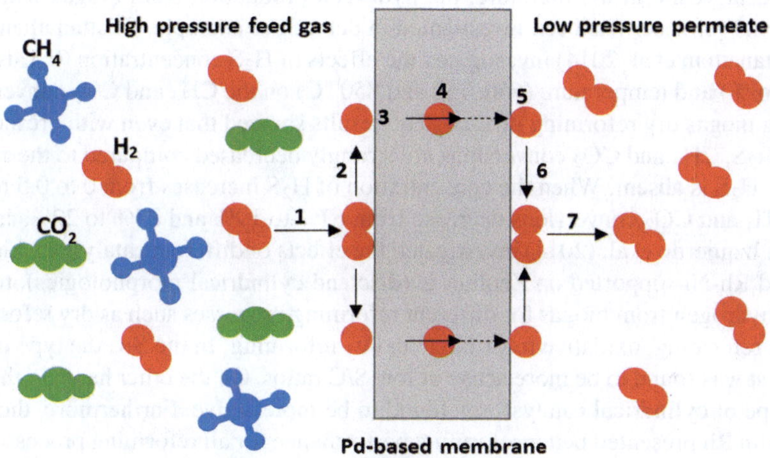

Fig. 9 Hydrogen transport mechanism through dense Pd-based membrane: (1) external diffusion, (2) adsorption, (3) dissociation, (4) diffusion, (5) recombination, (6) desorption, (7) external diffusion (Adapted from Gallucci et al. 2017)

Moreover, the area of membrane should be increased to provide high hydrogen flow. Increasing membrane area increases the thickness of the membrane, and thus the cost of material also increases (Helmi et al. 2014). These reasons clearly show that the composite membrane including a thin selective membrane and porous support should be used to increase mechanical stability and decrease cost. The stability and selectivity of a composite Pd-based membrane depends on the quality of the support material. Porous stainless steel is the most commonly used support material due to its properties such as similar thermal expansion coefficient to Pd-based films, easy to produce, good weldability, and low cost (Li et al. 2007).

As well as the type of membrane, the type of membrane reactor (fixed bed, fluidized bed and micro reactors) is an important consideration for hydrogen production. The use of fluidized bed reactors has some advantages such as improved heat and mass transfer compared to fixed bed and the elimination of external mass transfer resistance. However, the most important reason why this type of reactor is not preferred is the difficulties in the manufacturing of reactor and rapid deterioration of catalyst (Alavi et al. 2017). Therefore, the most widely used type of membrane reactor is fixed bed membrane reactors.

5 Current Status of Hydrogen Production from Biogas

Biogas produced by the degradation of organic materials with microorganisms is a renewable fuel that could be used for the production of clean fuel like hydrogen. The interest in hydrogen production from biogas is gradually increasing. In the recent years, in the literature, the hydrogen production from biogas with conventional reformer has been investigated in detail. For example, Chattanathan et al. (Chattanathan et al. 2014) investigated the effects of H_2S concentration (0.5, 1.0 and 1.5 mol%) and temperature (650, 750 and 850 °C) on the CH_4 and CO_2 conversions during biogas dry reforming process. The results showed that even with presence of little H_2S, CH_4 and CO_2 conversions are strongly decreased compared to the case in which H_2S is absent. When the concentration of H_2S increases from 0 to 0.5 mol%, the CH_4 and CO_2 conversions decrease from 67% to 19% and 87% to 22%, respectively. Izquierdo et al. (2014) investigated the effects of different catalysts, which are Ni and Rh-Ni supported on Zeolites L (disc and cylindrical morphologies), to produce hydrogen from biogas for different reforming processes such as dry reforming, steam reforming, oxidative reforming, and tri-reforming. In the SR, the type of disc catalyst was found to be more active at low S/C ratios. On the other hand, in the DR, the type of cylindrical catalyst was found to be more active. Furthermore, the catalyst with Rh presented better reforming performance for all reforming processes and conditions. Consequently, the most promising catalyst was found as Rh-Ni catalyst based on especially cylindrical Zeolite L. Cipiti et al. (2016) developed a mathematical model to test biogas steam reforming reactor. The effects of inlet temperature and the molar ratio of reactants on the performance of reactor were investigated with parametric analyses, then this model was validated with the experimental data for Ni/CeO_2 catalyst and the temperature range of 700–900 °C. At the $S/C = 3$, when the temperature increases from 700 to 900 °C, CH_4 conversion and H_2/CO change from 65.4% to 99.7% and 4.5 to 2.8, respectively. At 900 °C, when the S/C increases from 1 to 5, CH_4 conversion and H_2/CO change from 99.3% to 98.6% and 1.6 to 3.8, respectively. Pawar et al. (2017) studied the biogas dry reforming and investigated the catalyst deactivation that is formed by poisoning of sulphur and coke formation. This study was examined for different CH_4/CO_2 ratios (1.5 and 2), temperature (700–800 °C), and H_2S content in biogas (0, 5, and 10 ppm). The results showed that the catalyst deactivation is faster in the presence of sulphur. At 700 °C, the deactivation is not reversible, while at 800 °C, the catalyst activation is recovered with removing sulphur from the biogas feed. Moreover, the dry reforming of biogas leads to the carbon formation in the form of multi-walled nanotubes for these conditions. Tuna et al. (2018) evaluated the catalyst used in biogas steam reformer. In this study, two types of catalyst, which are 7.4% $Ni/NiAl_2O_4/g–Al_2O_3$ and 3.1% $Ru/g–Al_2O_3$, are studied. These catalysts showed high H_2 and CO selectivity, high CH_4 conversion, high resistance for carbon formation, zero production of CO or CO_2, and low deactivation. When the CO_2 concentration increases, the ratio of H_2/CO for Ni (at 850 °C) and Ru (at 700 °C) catalysts decrease from 2.9 to 2.4 and from 3 to 2.4, respectively. Calgora and Perez-Lopez (2019) evaluated Ni-M-Al catalysts used in hydrogen production from synthetic biogas (70% of CH_4 and 30% of CO_2) by dry reforming. This catalyst involving 55% of Ni, 33% of Al and 11% of third component (M, Mg, Ca, Cu, Li, La, Zn, Co) was investigated. The catalyst including La provided greater stability in reforming reactions because of its high sintering resistance and stronger acid sides.

On the other hand, the interest in hydrogen production via membrane reactor has increased due to some advantages mentioned above compared to conventional reactors. However, most of the studies in the literature are related to the hydrogen production from methane through membrane reactor. There are very few studies related to hydrogen production from biogas using membrane reactor. For example, Silva et al. (2015) developed the mathematical model of a fixed bed membrane reactor that use Pd-Ag hydrogen selective membrane for hydrogen production from biogas. The results showed that the yields of hydrogen and carbon monoxide are similar below 873 K for conditions both with and without permeation of hydrogen. At 873 K and with permeation of hydrogen, the methane conversion and hydrogen yields were found as 83 and 113% higher than that for condition without permeation of hydrogen. In addition, the hydrogen yields at 823 and 873 K were calculated as 21% and 47%, respectively, under permeation of hydrogen conditions. Iulianelli et al. (2015) investigated biogas steam reforming for clean hydrogen production through membrane reactor that consists of Pd/Al_2O_3 membrane. In this study, at 380 °C, $H_2O:CH_4 = 3:1$, and 2 bars, the permeate purity of hydrogen is approximately found as 96%, although methane conversion and hydrogen recovery are 15% and > 20%, respectively. On the other hand, at 450 °C, $H_2O:CH_4 = 4:1$, and 3.5 bar, methane conversion and hydrogen recovery increase up to more than 30% and 70%, respectively. Nooijer et al. (2018) investigated a fluidized bed membrane reactor that provides pure hydrogen production from biogas with steam reforming and developed a one-dimensional and two-phase model of a fluidized bed membrane reactor. This model was validated with experimental data for 435–535 °C, 2–5 bar, and the ratio of CO_2/CH_4 up to 0.9. Table 4 presents the summary of the studies on the different H_2 production methods using biogas and reaction conditions in the literature.

Table 4 The summary of the studies on the different H_2 production methods using biogas and reaction conditions (CR = Conventional Reactor and MR = Membrane Reactor)

Methods	Reactor type	Biogas composition	T (°C)	Catalyst	H_2/CO	Methane conversion	Ref.
DR	CR (fixed bed)	59% CH_4, 2%N_2, and 39% CO_2	800	Reformax®250	0.98	67%	(Chattanathan et al. 2014)
SR DR POR	CR (fixed bed)	60% CH_4 and 40% CO_2	800	Rh-Ni based on cylindrical Zeolite L	1.6 0.9 1.1	86% 65% 90%	(Izquierdo et al. 2014)
SR	CR (fixed bed)	60% CH_4 and 40% CO_2	900	Ni (7.5wt%)/CeO_2	2.8	99.7	(Cipitì et al. 2016)
DR	CR (fixed bed)	60% CH_4 and 40% CO_2	700 800	Ni/-Al_2O_3	~1.118 ~1.2	~65% ~72%	(Pawar et al. 2017)
SR	CR (fixed bed)	60% CH_4 and 40% CO_2	700 850	3.1% Ru/g-Al_2O_3 7.4% Ni/$NiAl_2O_4$/g-Al_2O_3	3 2.9	60% 60%	(Tuna et al. 2018)
DR	CR (fixed bed)	70% CH_4 + 30% CO_2	500 700	Ni-La-Al	2.6 5.2	35% 75%	(Perez-Lopez 2019)
SR	MR (fixed bed)	74% CH_4 + 26% CO_2	600	Ni (3.31 wt%)/γ-Al_2O_3)	–	47%	(Silva et al. 2015)
SR	MR (fixed bed)	60% CH_4 and 40% CO_2	380 450	Ni (25 wt%)/Al_2O_3	–	15% 34%	(Iulianelli et al. 2015)
SR	MR (fluidized bed)	53% CH_4 and 47% CO_2	435 535	–	–	30% 70%	(de Nooijer et al. 2018)

6 Concluding Remarks

At the present time, the interest on hydrogen production, which has become one of the most important research subjects, increases day-by-day with the increase in global energy demand and progress in fuel cell technology. Especially, clean hydrogen production from renewable energy resources (solar, wind, biomass, etc.) is more on the agenda due to depletion of fossil fuels and its environmental damages. Different hydrogen production methods are available such as electrolysis of water and chemical reforming processes. The production of hydrogen by chemical reforming processes seems to be more preferred in the near future as the electrolysis process is expensive and still under development. These chemical reforming processes are steam reforming, partial oxidation reforming, auto-thermal reforming, dry reforming, and dry oxidation reforming processes. However, the conventional reformer, which consists of a reformer, two separate water-gas shift reactors, and a separation unit, takes significant amount of space, has a complex control, and has low system efficiency. Therefore, innovative membrane reactors that provide the production and separation of hydrogen simultaneously in a single reactor have become more attractive compared to the conventional reformers. According to the literature survey conducted, the following conclusions are derived;

- The biogas purification step is critical and necessary to eliminate the corrosive effects of hydrogen sulfide irrespective of the reforming process.
- In all reforming processes, the catalyst used in hydrogen production reactions must be developed for longer utilization life to prevent carbon formation on the active catalyst surface. Ni-based catalysts are the most preferred catalysts in the reforming process due to their lower cost.
- DR and DOR reforming processes can be appeared as more attractive methods for hydrogen production from biogas; however, these processes are not preferred due to the tendency of carbon formation on the catalyst surface and the need of additional oxidant.
- The auto-thermal reforming process can be attractive due to providing high conversion efficiency and hydrogen yield; however, this process has complex control system.
- In the present time, the most of hydrogen production from biogas is achieved by conventional reformers based on steam reforming process. However, it is also possible to produce hydrogen from biogas using innovative membrane reactors in a single step (i.e. reforming and separations occur at the same place) and at lower temperatures.
- More research should be carried out on the development of materials and design for membrane reactors for hydrogen production from biogas.

References

Ahmed S, Krumpelt M (2001) Hydrogen from hydrocarbon fuels for fuel cells. Int J Hydrog Energy 26(4):291–301

Alavi M, Iulianelli A, Rahimpour MR, Eslamloueyan R, De Falco M, Bagnato G, Basile A (2017) Chapter. 8. Fixed bed membrane reactors for ultrapure hydrogen production: modelling approach. In: Basile A, Dalena F, Tong J Nejat Veziroglu T (eds) hydrogen production, separation and purification for energy, pp 231–257

Alves HJ, Junior CB, Niklevicz RR, Frigo EP, Frigo MS, Coimbra-Araújo CH (2013) Overview of hydrogen production technologies from biogas and the applications in fuel cells. Int J Hydrog Energy 38(13):5215–5225

Avraam DG, Halkides TI, Liguras DK, Bereketidou OA, Goula MA (2010) An experimental and theoretical approach for the biogas steam reforming reaction. Int J Hydrog Energy 35(18):9818–9827

Bereketidou OA, Goula MA (2012) Biogas reforming for syngas production over nickel supported on ceria–alumina catalysts. Catal Today 195(1):93–100

Calgaro CO, Perez-Lopez, OW (2019) Biogas dry reforming for hydrogen production over Ni-M-Al catalysts (M = Mg, Li, Ca, La, Cu, Co, Zn). Int J Hydrog Energy

Chattanathan SA, Adhikari S, McVey M, Fasina O (2014) Hydrogen production from biogas reforming and the effect of H_2S on CH_4 conversion. Int J Hydrog Energy 39(35):19905–19911

Cipitì F, Barbera O, Briguglio N, Giacoppo G, Italiano C, Vita A (2016) Design of a biogas steam reforming reactor: A modelling and experimental approach. Int J Hydrog Energy 41(27):11577–11583

Dantas SC, Resende KA, Rossi RL, Assis AJ, Hori CE (2012) Hydrogen production from oxidative reforming of methane on supported nickel catalysts: an experimental and modeling study. Chem Eng J 197:407–413

De Falco M, Salladini A, Palo E, Iaquaniello G (2011) Reformer and membrane modules (RMM) for methane conversion powered by a nuclear reactor. Nuclear Power-Deployment, Operation and Sustainability, p 467

de Nooijer N, Gallucci F, Pellizzari E, Melendez J, Tanaka DAP, Manzolini G, van Sint Annaland M (2018) On concentration polarisation in a fluidized bed membrane reactor for biogas steam reforming: modelling and experimental validation. Chem Eng J 348:232–243

Di Marcoberardino G, Manzolini G, Guignard C, Magaud V (2018) Optimization of a micro-CHP system based on polymer electrolyte membrane fuel cell and membrane reactor from economic and life cycle assessment point of view. Chem Eng Process-Process Intensif 131:70–83

Gallucci F, Fernandez E, Corengia P, van Sint Annaland M (2013) Recent advances on membranes and membrane reactors for hydrogen production. Chem Eng Sci 92:40–66

Gallucci F, Medrano JA, Fernandez E, Melendez J, van Sint Annaland M, Pacheco-Tanaka DA (2017) Advances on high temperature Pd-based membranes and membrane reactors for hydrogen purification and production. J Membr Sci Res 3:142–156

Galvagno A, Chiodo V, Urbani F, Freni F (2013) Biogas as hydrogen source for fuel cell applications. Int J Hydrog Energy 38(10):3913–3920

Gao Y, Jiang J, Meng Y, Yan F, Aihemaiti A (2018) A review of recent developments in hydrogen production via biogas dry reforming. Energy Convers Manag 171:133–155

Helmi A, Gallucci F, van Sint Annaland M (2014) Resource scarcity in palladium membrane applications for carbon capture in integrated gasification combined cycle units. Int J Hydrog Energy 39(20):10498–10506

Holladay JD, Hu J, King DL, Wang Y (2009) An overview of hydrogen production technologies. Catal Today 139(4):244–260

Iulianelli A, Manzolini G, De Falco M, Campanari S, Longo T, Liguori S, Basile A (2010) H_2 production by low pressure methane steam reforming in a Pd–Ag membrane reactor over a Ni-based catalyst: experimental and modeling. Int J Hydrog Energy 35(20):11514–11524

Iulianelli A, Liguori S, Huang Y, Basile A (2015) Model biogas steam reforming in a thin Pd-supported membrane reactor to generate clean hydrogen for fuel cells. J Power Sources 273:25–32

Izquierdo U, Barrio VL, Bizkarra K, Gutierrez AM, Arraibi JR, Gartzia L, Cambra JF (2014) Ni and RhNi catalysts supported on Zeolites L for hydrogen and syngas production by biogas reforming processes. Chem Eng J 238:178–188

IEA (2019) The future of hydrogen: seizing today's opportunities, report prepared by the IEA for the G20, Japan

Kikuchi E, Tanaka S, Yamazaki Y, Morita Y (1974) Steam reforming of hydrocarbons on noble metal catalysts (part 1). Bull Jpn Pet Inst 16(2):95–98

Lau CS, Tsolakis A, Wyszynski ML (2011) Biogas upgrade to syn-gas (H_2–CO) via dry and oxidative reforming. Int J Hydrog Energy 36(1):397–404

Li A, Grace JR, Lim CJ (2007) Preparation of thin Pd-based composite membrane on planar metallic substrate: Part I: Pre-treatment of porous stainless steel substrate. J Membr Sci 298(1–2):175–181

Lu N, Xie D (2016) Novel membrane reactor concepts for hydrogen production from hydrocarbons: a review. Int J Chem React Eng 14(1):1–31

Minh DP, Siang TJ, Vo DVN, Phan TS, Ridart C, Nzihou A, Grouset D (2018) Hydrogen production from biogas reforming: an overview of steam reforming, dry reforming, dual reforming, and tri-reforming of methane. In: Hydrogen supply chains. Academic Press, pp 111–166

Mosayebi Z, Rezaei M, Ravandi AB, Hadian N (2012) Autothermal reforming of methane over nickel catalysts supported on nanocrystalline MgAl2O4 with high surface area. Int J Hydrog Energy 37(2):1236–1242

Pawar V, Appari S, Monder DS, Janardhanan VM (2017) Study of the combined deactivation due to sulfur poisoning and carbon deposition during biogas dry reforming on supported Ni catalyst. Ind Eng Chem Res 56(30):8448–8455

Rohendi D, Rahmah DR, Yulianti DH, Amelia I, Sya'baniah I, Syarif N, Rachmat N (2018) A review on production of hydrogen from renewable sources and applications for fuel cell vehicles. Int J Sustain Transp 1(2):63–68

Scarlat N, Dallemand JF, Fahl F (2018) Biogas: Developments and perspectives in Europe. Renew Energy 129:457–472

Silva FSA, Benachour M, Abreu CAM (2015) Evaluating hydrogen production in biogas reforming in a membrane reactor. Braz J Chem Eng 32(1):201–210

Tuna CE, Silveira JL, da Silva ME, Boloy RM, Braga LB, Pérez NP (2018) Biogas steam reformer for hydrogen production: evaluation of the reformer prototype and catalysts. Int J Hydrog Energy 43(4):2108–2120

Verma P, Samanta SK (2016) Overview of biogas reforming technologies for hydrogen production: advantages and challenges. In: Proceedings of the first international conference on recent advances in bioenergy research. Springer, New Delhi, pp 227–243

Xia Y, Lu Y, Kamata K, Gates B, Yin Y (2003) Macroporous materials containing three-dimensionally periodic structures. In: Yong P (ed) The chemistry of nanostructured materials. World Scientific Publishing Co., Singapore, pp 69–100

Underground Large-Scale Hydrogen Storage

Hatice Karakilcik and Mehmet Karakilcik

Abstract Hydrogen can be produced and stored by electrolysis of water using 100% renewable and clean energy sources (such as solar and wind energy). It can then be converted back into electricity with fuel cells to meet the energy demand. Today, by physical methods, hydrogen is stored as liquid and gas in tanks and underground geological areas (depleted oil and gas, aquifers and salt caverns). One of the important underground gas storage areas are salt caverns. Salt caverns are built in very tight and sealed underground salt rock formation by solution mining method. In this study, the storage of hydrogen in the underground at very large scales was investigated. As a result, salt caverns built in underground salt domes have very high hydrogen storage capacity. Therefore, in the future, we believe that large-scale underground hydrogen storage areas will play an important role in the hydrogen economy as integrated power plants.

Keywords Underground large-scale hydrogen storage · Salt dome · Salt cavern

1 Introduction

Today, one of the most important energy problems is the efficient storage of renewable energy sources, especially in large quantities, and its cost and safety. In particular, there are some problems in the storage of heat and electricity generated by renewable energy sources. But, by electrolysis of water with these energy sources, water can be stored by separating into hydrogen and oxygen. In case of need, electricity can be regenerated with hydrogen and oxygen by fuel cells. Because, hydrogen is a very important energy carrier and can be stored at very large scales as liquid and gas. Storage of hydrogen in underground caverns at very large scales is very important for the

H. Karakilcik (✉)
Department of Geology Engineering, Faculty of Engineering, Cukurova University, Saricam 01250, Türkiye
e-mail: hkilcik@cu.edu.tr

M. Karakilcik
Department of Physics, Faculty of Sciences and Letters, Cukurova University, Saricam 01250 Adana, Türkiye

T. S. Uyar (ed.), *Accelerating the Transition to a 100% Renewable Energy Era*, Lecture Notes in Energy 74, https://doi.org/10.1007/978-3-030-40738-4_17

solution of the energy problem in the future and therefore for the economy of hydrogen. Pure hydrogen can be stored as a liquid and gas in many ways. These are physical, chemical and adsorption methods (Andersson and Grönkvist 2019). Physical methods; pressure can be stored in large steel tanks and underground geological structures. These structures include depleted oil and gas aquifers and salt caverns. However, it is important that hydrogen can be recycled as much as it is efficient and cost-effective to store. It also plays environmentally key role in the storage processes as well as in the production of hydrogen and storage of 100 percent renewable and clean resources.

One of the most important advantages of underground salt caverns is the large scale, cheaply and safely storage of natural gas, crude oil, heating oil, diesel, gasoline, kerosene, ethylene, propylene, LPG (butan/propane) hydrocarbons and the compressed air and hydrogen. In particular, underground salt domes are important due to the natural sealing properties against gaseous environment and non-aqueous liquids. Thanks to this feature of the domes, the lightest element can be stored in huge amounts in the salt caverns as hydrogen gas, which will be an important place in the hydrogen economy in the future. However, although rock salt sediments are seen all over the world, they are not evenly distributed throughout the world. Moreover, these sediments must have a certain composition and interior structure, thickness and depth range, in order to be suitable for cavern construction and storage.

At the present time, energy resources for example fossil fuels are economically stored underground in some developed countries. So, the demands to explore the sub-surface of the geological areas on the earth are increased. One of the main aim is underground storage of the energy resources. The storage can probably also be used to apply the climate and energy policy targets in the context of the energy transition to renewables (Martens and Kühn 2015). To these are contained the storage of energy from renewable sources (e.g. hydrogen (H_2) and synthetic methane (CH_4)), the use of geothermal energy and likely the long periods storage of carbon dioxide to decrease the let loose of greenhouse gases into the atmosphere.

Because of the growing importance of renewable energy, these green energy sources are becoming more dominant and contributing to the general energy supply. The energy industry is faced with providing energy balance that requires to store more than enough energy quantities for longterm to correct the fluctuation in the seasonal term. But this is not anymore possible by making use of traditional technologies (Crotogino et al. 2010).

Hydrogen is a clean energy source that can be generated and stored in enormous amounts using renewable energy sources. Hydrogen as energy carrier is efficient, and which in common with electricity, may provide overall the energy requirements (Sherif et al. 2003). Many scientists estimate that several different energy sources could be used by society in the 21st century. They consider hydrogen as major carrier of energy according to other energy sources. Hydrogen can take over a substantial mission in the twenty-first century energy industry. Because hydrogen is present nearly everywhere on the surface of the earth as a component of water. It has many trading usage areas for example, containing ammonia (NH_3) manufacture for utilization in fertilizers, methanol (CH_3OH) manufacture, hydrochloric acid (HCl) manufacture and as a rocket fuel. Liquid hydrogen is also made use of in cryogenics and superonductivity. Additionally hydrogen can be used as a fuel (Matos et al. 2016).

The hydrogen is the lightest element in the universe. It is a very poorly soluble in water and a good conductivity with low permeability. Due to its diffusing properties, it passes easily through perforated or porous walls, it is stored in steel vessel protectors for these reasons. Hydrogen forms water if it is exposed to air or oxygen. In the air, under the atmospheric pressure it exist freely distributed. However, hydrogen gas when exposed to pressure in a steel protective container or underground geological structures, it can be compressed to a significant extent. It is extremely sensitive to pressure and temperature. Therefore, one of the most important alternatives to the containers for storage of hydrogen gas in large quantities is underground storage areas where are geological areas (e.g. salt caverns, aquifers, depleted oil and gas areas). Underground salt caverns, aquifers, depleted oil and natural gas reservoirs can be considered as potential geological structures for large scale storage of hydrogen. Because these structures are used to successfully store hydrogen gas under high pressure for long periods of time. Salt caverns are in the form of salt dome up to a depth of 2,000 m and the height of the caverns is 300 m, the width is 60 m and the storage volume is generally between 500,000 and 800,000 m^3. The working pressure is a maximum working pressure of 20 MPa for a salt cavern with a volume of 700,000 m^3. The most important reason for preferring salt caverns for underground storage is that the salt domes have a natural sealing property against gaseous media and non-aqueous liquids. For this reason, hydrogen is safely stored as a gas. In addition, crude oil, natural gas and compressed air can be safely stored in salt caverns. Although rock salt sediments are seen all over the world, they are not evenly distributed to the whole world. In order to be able to design rock salt beds for cavern construction and storage, the interior must have certain characteristics such as a certain depth range, thickness, composition and quality. Hydrogen storage in salt caverns should be assessed taking into account the feasibility analysis results. Because underground energy storage will have a great importance in the future as it is both a raw material in the industry sector and a renewable energy carrier in the natural gas industry. Early market will take place by 2025, but the market will be established by 2050 (Matos et al. 2016).

In this study, the most suitable underground geological structures where hydrogen gas can be stored safely and economically on a very large scale are examined. From these structures, the formation of salt caverns in very large volumes by solution mining method in underground salt domes, sealing properties under pressure and storage capacity are determined. Furthermore, the importance of salt caverns in the hydrogen economy has been tried to be determined by the use of renewable energy resources in the storage and withdrawal of very large scale hydrogen in underground storage caverns through the integration of renewable energy sources.

2 Hydrogen Storage Technologies

Physical properties of hydrogen: colorless, odorless, in the simplest atomic structure in nature, with a proton in its core and a single electron around it. Hydrogen gas is 14 times lighter than air, 9 times lighter than natural gas and 10 times lighter than liquid gas. The volume of liquid hydrogen is only 1/700 of the volume of hydrogen gas. Hydrogen turns into liquid hydrogen at –252.8 °C. In all fuels, it contains the highest energy per unit mass. Energy equivalent to 1 kg mass of hydrogen is equal to the energy of 2.1 kg mass of natural gas and 2.8 kg mass of petroleum, and also on average 1.33 times more efficient than petroleum fuels. Chemical properties of hydrogen: Hydrogen is not present in the free state in nature, it exists as compounds. When used as a fuel, the product is only water and/or water vapor. Due to this property, hydrogen can be stored in high pressure steel tanks as liquid or gas under high pressure and low temperatures. However, this method is not economic in terms of energy efficiency.

As seen in Fig. 1, the hydrogen storage technologies divises into three main categories as physical, chemical and adsorption storages.

Physical Storage: Hydrogen may be stored as gas or a liquid in pure, molecular form without any significant physical or chemical bonding to other materials given as:

(a) **Compressed Gas Storage**: Hydrogen is conveniently stored for many applications in higher pressure cylinders. But, the method is rather expensive and bulky.

(b) **Liquid Storage**: It is as liquid storage as cryogenic storage in vacuum insulated or super insulated storage tanks. The liquid hydrogen fuel used as rocket propellant in the space programme is stored in the large tanks (stored at about –253 °C).

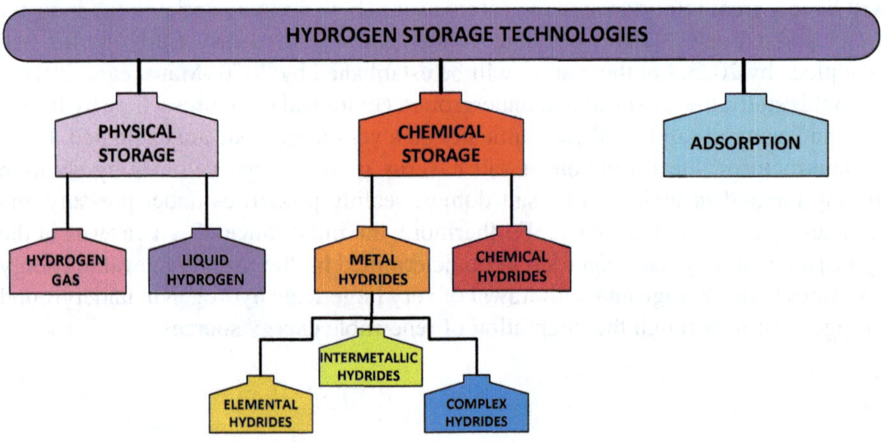

Fig. 1 Hydrogen storage technologies

(c) **Line Pack System**: It is of the line pack system where it is allowed to vary the pressure in the transmission and distribution system.
(d) **Underground Storage**: Suitable geological areas for underground storage include caverns in the salt dome, depleted oil, gas fields and aquifer systems. This is the cheapest way to store large amounts of hydrogen for subsequent distribution. As it is known for too long that salt caverns are preferred for storage of hydrocarbon products. It is very clear that renewable energy sources can be stored in large quantities as an energy source underground. For example, hydrogen produced by renewable energy sources (100 GWh electricity) stored in very large quantities in artificial underground salt caverns of up to 500,000 m^3 at 200 bar can be supplied to the grid by being re-powered with hydrogen fuel cells.

Chemical Storage: Atomic hydrogen may be chemically bonded (absorbed). Further, it is useful to divide the storage technologies based on chemical bonding into two subcategories: the metal hydrides and the chemical hydrides. This division makes sense due to the fundamentally different hydrogen storage properties of these materials. As implied by their designation, metal hydrides contain metal atoms. Hydrogen may either be directly bonded to the metal atom (elemental metal hydrides and intermetallic hydrides) or be part of some complex ion that is bonded to a metal atom (complex metal hydrides). In contrast, the chemical hydrides only consist of non-metallic elements, generally some combination of carbon, nitrogen, boron, oxygen, and hydrogen (Andersson and Grönkvist 2019).

(a) **Metal Hydrides Storage**: As metal hydrides in chemically bound form is another way of storing hydrogen. A number of metals and alloys form solid compounds by direct reaction with hydrogen gas. The metal hydrides can be transported in solid form. When the hydrides is heated, hydrogen is released for use (ENeRG Position 2017).
(b) **Chemical Hydrides Storage**: Like the metal hydrides, the chemical hydrides chemically bond hydrogen. However, since the chemical hydrides consist of lighter elements, their properties are radically different from those of the metal hydrides. The most significant difference is, perhaps, that the chemical hydrides are generally liquids at normal conditions, thereby simplifying their transport and storage as well as heat and mass transfer during dehydrogenation and hydrogenation processes. Several of the chemical hydrides suggested for hydrogen storage, such as methanol, ammonia, and formic acid, are currently bulk chemicals commonly synthesized from natural gas. In other words, the utility of these chemicals goes beyond the storage of hydrogen (Andersson and Grönkvist 2019).

Adsorption Storage: Molecular hydrogen may be adsorbed onto or into a material, held by relatively weak physical Van der Waals bonds. The storage of hydrogen via adsorption exploits physical Van der Waals bonding between molecular hydrogen and a material with a large specific surface area. Due to the weakness of the Van der Waals bonding, low temperatures and elevated pressures must typically be

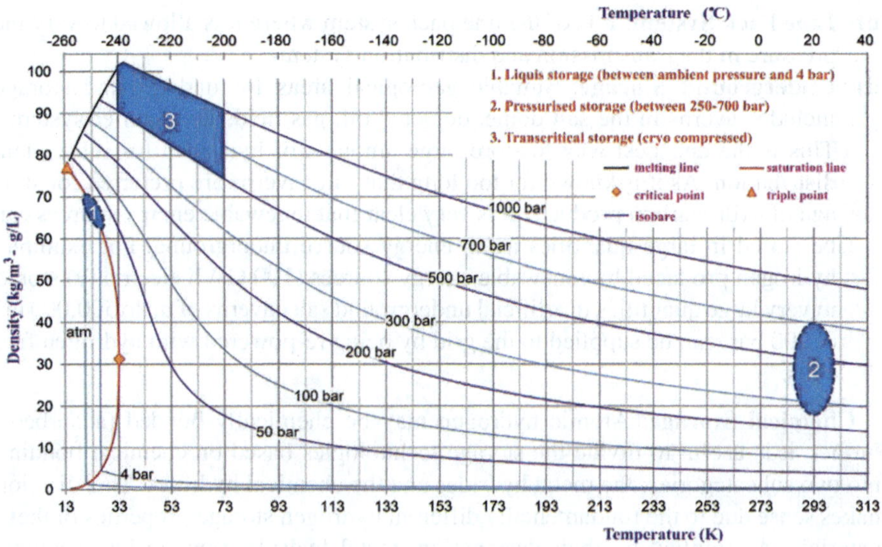

Fig. 2 Storage density of hydrogen under certain pressure and temperature conditions (http://www.ilkdresden.de/en/service/research-and-development-rd/detail/hydrogen-test-area-at-ilk-dresden/)

applied to achieve significant hydrogen storage densities using adsorption. The most common refrigerant used for hydrogen adsorption is liquid nitrogen (boiling point: –196 °C). The applied hydrogen pressure is typically 10–100 bar but varies depending on the adsorbent and the intended application (Berenguer-Murcia et al. 2018). Using higher pressures is not advantageous beyond a certain threshold as the presence of the adsorbent may then no longer improve the hydrogen storage capacity compared to if pressurized gas was stored in the same vessel, the reason being the space occupied by the adsorbent (Klebanoff et al. 2014).

As seen in Fig. 2, hydrogen is physically stored as a gas or a liquid. Storage density of hydrogen under certain pressure and temperature conditions. (1) liquid storage-LH_2, (2) pressurised storage CGH_2, (3) transcritical storage (cryo compressed) (http://www.ilkdresden.de/en/service).

Strong storage tanks capable of operating under pressure of 350–700 bar are required in order to store hydrogen gas. Thus, very large volumes are required to store hydrogen as a gas at atmospheric pressure in the tanks. This type of storage is not practical and economical in terms of viability. Because, cryogenic temperatures (–252.8 is boiling point of hydrogen) are required for liquid storage. Otherwise, hydrogen can also be stored by adsorption on the surface of the solids or by absorption in the solid. One of the best solutions to overcome the difficulties of hydrogen storage is the creation of man-made gigantic artificial caverns in the salt domes of the earth's underground units. The hydrogen economy in the hydrogen energy sector has considerable difficulties in storing large quantities of hydrogen. Presently hydrogen can be stored in three forms as gas, liquid or as a solid combined with a metal hybrid. Until recently, different storage notions for hydrogen have been developed. The notions that are already on the way of commercialization are described below. There are five basic principle methods that have been considered for hydrogen storage (ENeRG Position 2017).

The most suitable storage method is dependant on safety aspects, environmental issues, economic criteria and the end-use of hydrogen. Hydrogen storage should be analyzed as a whole in which the effectiveness of each stage is assessed in relation to each other. These stages are included in production, storage, consumption, transportation and distribution. Underground hydrogen storage is necessary for energy consumption and production in industrial areas. The fluctuations that may affect the hydrogen consumption may be the industry, the lifestyle of the population and the climatic conditions in the country (Basniev et al. 2010).

Another aspect plays an important role in the selection of suitable storage formations for the storage of air and hydrogen: after injection to the medium, the oxygen in the hydrogen can react with the minerals and the microorganisms present in natural reservoirs. This can result in a loss (depletion) of hydrogen, as well as the blockage of the fine pores in the reservoirs by the reaction products. To avoid this, salt caverns are exclusively used as energy storages in those regions which have suitable salt formations available (Kepplinger et al. 2011). Nowadays, this technology has been improved to high levels of perfection, capable of constructing caverns with optimal shapes, conforming to precise specifications at depths down to 3000 m and volumes of up to 8×10^5 m^3 and more. The leaching process is implemented by wells to be drilled from the surface down to the depths required by the specific geological and project pre-conditions (Nanda et al. 2015). The caverns thus created are used for several purposes. Some of the most important uses are the stocking of liquid and gas-type energies.

3 Geological Reservoirs for Hydrogen Storage

The several types of suitable geological reservoirs are the depleted hydrocarbon reservoirs, aquifer systems, salt domes, engineered rock caverns and abandoned mines. Suitable structures for underground storage of gases are divided into two main categories. These are:

(a) **Porous medium storage**: Either in partly depleted oil or gas fields or aquifers, in which the gas occupies the innately occurring pore space among mineral grains or crystals in sandstones or porous carbonates.

(b) **Cavern storage**: This is a type of energy storage application in aquifers, depleted oil and gas deposits or caverns placed in a salt dome. .

Two essential considerations must be taken into account when storing in both porous medium and cavern medium. These are providing sufficient storage capacity and the protection of stored gas. They are expensive to construct traditional energy storage facilities on this scale, such as pumping hydroelectric power plants. In addition to, the suitable sites for the construction of these facilities are limited and have negative effects on ecosystems. Where underground reservoirs of oil and gas for underground energy storage areas are not available, salt dome or salt beds for underground gas storage is the reason for preference. Salt as the main rock has a few extra important features. It is known that rock salt is impermeable for the stored gases and displays self-healing behavior for example fractures or cavities in the salt close over time re-creating the pre-damaged condition (Böttcher et al. 2017). Geological formations have a potential to store large volumes of fluids with minimal impact to environment and society. One of the ways to ensure a large scale energy storage is to use the storage capacity in geological reservoir. There are several viable technologies for underground energy storage for example Underground Gas Storage (UGS), Hydrogen Storage (HS), Compressed Air Energy Storage (CAES), Underground Pumped Hydro Storage (UPHS) and Thermal Energy Storage (TES) (Matos et al. 2016). Because the energy in the hydrogen is chemically bound (just like natural gas), hydrogen caverns can be designed much more flexibly than CEAS caverns, and can therefore also be adapted to the local geological conditions. This means that many more salt formations in the world are potentially suitable for the construction of this type of energy storage than is the case with CAES caverns (Kepplinger et al. 2011). Additionally, geophysical monitoring methods, which enable for a good spatial resolution of the storing operation, induced effects or leakages, are interpreted by these scenario simulations. In this way, methods for the appraisement of underground geological storage sites are improved, which account for the spatial extension of the underground operation itself as well as its induced effects and the spatial necessities of adequate monitoring methods (Kabuth et al. 2017).

The geological, physical and chemical properties of the salt formation play a key role in determining the suitability of salt caverns. The structure of the salt formation is influential on parameters such as age of formation, age, salt quality, cavern type, volume, maximum and minimum endurance pressure, cushion volume, maximum injection rate, maximum recycling rate.

In Fig. 3, the vertical extent of the salt deposits can be up to several hundred metres. The underground salt cavern design and application studies of natural gas and hydrogen are very similar with each other. The main difference concerns the choice of materials for the equipment used and the safety precautions on the surface. Most European stratified salt beds Triassic and Tertiary are found at depths of 1,000 m, while salt deposits at Permian age in Germany are found at depths of more than 2000 m (http://www.kbbnet.de/en/, https://energy.gov/eere/fuelcells/hydrogen-storage).

Underground storage is widely used in oil, natural gas and compressed air energy industries in the developed countries of the world. For this reason, regional geological and geophysical studies should be carried out in the determination of underground storage areas. The underground storage options, these studies are necessary for operational need. One of the first successful gas storage projects was completed in 1915 in Ontario, an active gas field. In the following year, studies were carried out in a

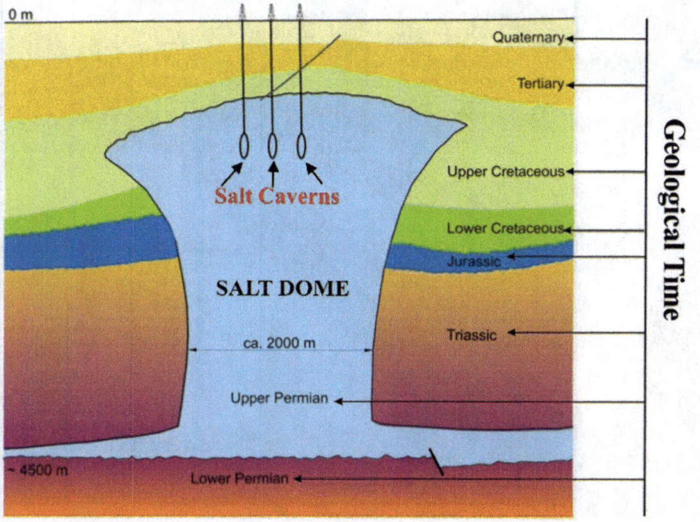

Fig. 3 Geological time of the salt dome (Beckman et al. 1995)

depleted reservoir in the Zoar area near Buffalo (Beckman et al. 1995). To date, pure hydrogen has been stored in three caverns in Teesside (UK) since 1972 and in two caverns near the US Gulf Coast in Texas since 1983, see Table 1 for details. Practical experience in UK and USA hydrogen caverns has shown that hydrogen can also be safely stored in salt caverns for long periods of time (Kruck et al. 2013).

The United States of America (USA) now has much more than many countries with about 400 underground gas storage areas, of which about 800 million barrels of oil and 1000 billion cubic feet of natural gas are geologically stored. The basic driver for geological storage is that the cost per stored volume is 3–5 times less

Table 1 Existing hydrogen storage caverns in USA and UK (Kruck et al. 2013)

	Clemens (USA)	Moss Bluff (USA)	Teesside (UK)
Geology	Domal salt	Domal salt	Bedded salt
Operator	Conoco-phillips	Praxair	Sabic peatroleum
Stored fluid	Hydrogen	Hydrogen	Hydrogen
Commissioned	1983	2007	~1972
Geometrical volume (m^3)	580,000	566,000	3 × 70,000
Reference depth (m)	930	>822	350
Pressure range (bar)	70–135	55–152	~45
Possible working gas capacity H_2 Mio (kg)	2.56	3.72	0.83

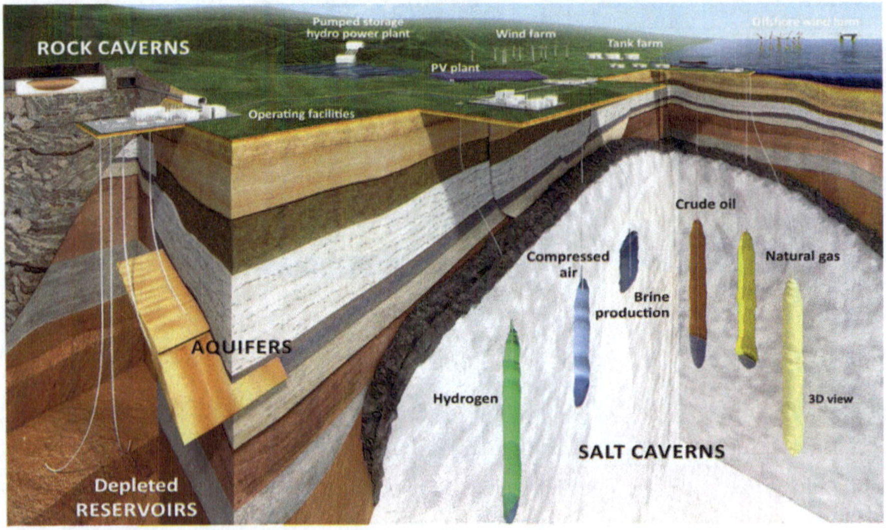

Fig. 4 Variation with depths of liquid and gas storage facilities in the world (http://www.kbbnet.de/en/, http://energy.gov/eere/fuelcells/hydrogen-storage)

than the surface storage area. Very large volumes of energy storage are relatively inexpensive, can be used to meet seasonal demands, provide continuity in supply chain deterioration, and control in the pipeline congestion (Lord et al. 2014).

Until today, UK has a long history of storing natural gas and compressed air in spaces in salt layers. England used salt caverns to store crude oil during the Suez Crisis. Ten years later, Americans and Canadians began using salt caverns for gas storage. Since then, underground gas storage (UGS) has been extensively used and developed in salt layers (Yang et al. 2015).

Figure 4 presents the depths, shapes, and relative volumes of typical salt caverns liquid and gas storage facilities in the world. Such plants are distributed at different depths and widths, different types and large-scale volumes and complex geological formations. The geological underground mainly comprise salt caverns, saline aquifers and depleted gas reservoirs.

Artificial underground caverns are large-volume structures formed as a result of dissolution with salt water by solution mining method. As shown in Fig. 4, the caverns are artificially constructed within the geological salt formations. It is used to store liquid and gas type energy resources in very large quantities. Nowadays, artificial salt caverns are made suitable for storing hydrogen in very large quantities other than compressed air, crude oil and natural gas. Salt caverns are very important in terms of storage capacity, sustainability, economic and strategic features.

Salt is found in solid ground in the form of rock salt (halite) in different layer thicknesses near the surface on the ground in the nature, or in salty water-laced lakes and seas. Rock salt (NaCl) has its own physical, chemical and mechanical properties; which is an ideal place for the development of (large) caverns for the storage of substances that do not react with salt or dissolve themselves. For this reason, salt domes are not only a source of salt, but also an important source for building salt caverns. In addition, the distribution of the salt makes it a source of potential strategic reserves, especially for natural gas and hydrogen storage.

Underground storage is the cheapest cost for storing large quantities of hydrogen gas. In fact, it is estimated that the cost of hydrogen per Nm^3 is twice as cheap as that of storage. Underground geological storage plays an important role in the development of the hydrogen economy.

In the last decade, a few examples have been given for pure hydrogen or synthesis underground storage (Plaat 2009):

- In Britain, Teesside, Yorkshire, the British ICI company stored 1 million Nm^3 of pure hydrogen in three salt caverns at a depth of about 400 m.
- In France, the company Gas de France stored a gas containing 50–60% hydrogen in a 330 million Nm^3 capacity aquifer for about 20 years without loss of gas safely.
- In Russia, pure hydrogen was stored in underground at 90 MPa for the needs of the aviation industry.
- In Germany, 32,000 m^3 of 62% H_2 gas was stored in an underground salt cavern at 8–10 Mbar.
- 50% H_2 syngas was stored in an aquifer in Czechoslovakia.

4 Man-Made Structures

Salts are important in the trapping, sealing and generation of hydrocarbons and various metal deposits. Manufactured salt cavities are used as long term waste repositories, as receptacles for the safe storage and cycling of liquids and gases and even for the manufacture of peak load electricity.

Nowadays, salt caverns are used especially as natural gas storage facilities. If hydrogen is stored in natural gas storage facilities, some new operational difficulties can be added. These are chemical reactions that cause hydrogen loss and hydrogen embrittlement. Globally, there are three facilities that store hydrogen in salt caverns. Two of these are in the United States (Lord et al. 2014). USA companies that do the job of storing hydrogen in salt mines are ConocoPhillips and Praxair. Both companies use caverns created in the Clemens salt dome in Texas to store hydrogen (Leighty 2007). The ConocoPhillips cavern is 2800 ft deep, 1000 ft long and 160 ft in diameter. ConocoPhillips has been storing hydrogen gas for several decades for use as a buffer to store hydrogen gas when their hydrotreating equipment is down or to provide hydrogen when the ethane crackers are not operational. Besides, three small and shallow salt deposits in Teeside in UK store hydrogen. These caverns are 400 m deep and have a storage capacity of approximately 70,000 m^3 hydrogen gas at 50 bar pressure (Crotogino and Huebner 2008).

Salt caverns are artificially created cavities built in salt deposits. They are suitable for the storage of liquid hydrocarbons and in particular for gases under high pressure.

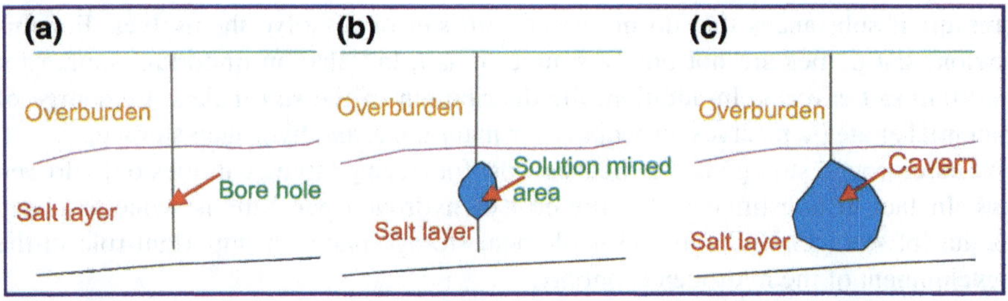

Fig. 5 Construction of a salt cavern storage facility. **a** Initial borehole **b** solution mining and cavern formation in process **c** final solution-mined cavern (Stone et al. 2009)

Large amounts of gas can safely be stored due to large geometrical volumes and high storage pressures. Depending on the cavern depth, pressures of 200 bar and above are common. Geometrical volumes reaching up to more than 1 million m^3 can be constructed.

As seen in Fig. 5 the feasible geometrical volume of salt caverns depends mainly on the thickness of the salt formation. The volume may range typically from 150,000 to 800,000 m^3. Under optimal conditions in a huge salt dome volumes even above 1,000,000 m^3 have already been achieved. It is possible to create small sized salt caverns initially and re-leach them afterwards. However, flooding of the cavern and subsequent gas first fill is required in this case which increases the costs for these operations. Operating pressures depend primarily on the cavern depth. The maximum pressures range from 100 to 270 bar, the minimum pressures from 35 to 90 bar respectively (orders of magnitude only). An increase in cavern volume from 75,000 to 500,000 m^3 reduces the total investment cost by a factor of 1.5 to 2. Even larger caverns, 600,000 m^3, have been mined in salt domes in Germany (Plaat 2009).

It is important to select the gas pressure used in storage in the cages. Because high-pressure gas caverns function with compression and decompression between minimum and maximum pressure. While the maximum pressure corresponds to about 80% of the initial formation pressure at the depth of the cavern ceiling, the minimum pressure is about 30% of the maximum pressure. The net storage capacity of a cavern is known as its working gas. For example, if a salt cavern has a depth of 1,000 m and a volume of 700,000 m^3, the working gas will be approximately 6,000 t. The gas which is caught in the cavern at the minimum pressure is called the cushion gas. In this case, the remaining gas volume is about 3,000 t hydrogen. Therefore, the maximum allowable rates, maximum flow rates in the boreholes and maximum pressure reduction rates for cavern filling and discharge are governed by 1 MPa/day. As a rule of thumb, maximum withdrawal rates correspond to approximately 10% of the storage capacity per day, with a maximum of 10 turnovers per year (Crotogino et al. 2010).

The operational economics of an underground hydrogen storage cavern is further aided by reducing minimum operating pressure and thus the volume of cushion gas, whilst increasing the maximum operating pressure. If the minimum operating pressure is reduced too far, the salt cavern will decrease in volume due to creep of the enclosing salt. By increasing the maximum operating pressure additional hydrogen is stored improving the cycle efficiency. The maximum operating pressure is limited by rock salt permeability, rock strength and depth of cavern. With its smaller molecule size, than natural gas, the diffusion of hydrogen through rock salt is not foreseen as being any different to the larger molecules. Typical maximum operating pressure ranges from 0.019 to 0.021 MPa per metre in depth of overburden (Stone et al. 2009).

In planning and implementation of specific projects for the storage of hydrogen in salt caverns there are the same risks associated as with natural gas storage caverns. The main risk of salt cavern storage projects is the geological uncertainty, which can be reduced by increasing the knowledge base on the specific salt deposit (exploration). For a given cavern location this risk includes e.g. the occurrence of non-halite interbeds, faults, distance to edge of salt structure, etc. All of these issues affect the rock mechanical layout of the cavern and may lead to a reduced cavern size, a reduced operational pressure range or even the requirement to abandon and plug the well. Further risks are associated with the technical processes of drilling and cementing of the well, since failure in cementation will lead to a cavern leakage. During leaching the rock mechanical limitations must not be exceeded, otherwise the leaching strategy must be modified which will also lead to a lower cavern volume.

5 Results and Discussions

Particularly in developed countries, the energy industry is undergoing a major shift in renewable energy sources from dependence on carbon-based and nuclear fuel use. Green electricity can be generated by sources of renewable energy such as solar and wind energy. However, these sources generate electricity and depend entirely on demand, depending on season, meteorological conditions and time (Kepplinger et al. 2011).

Supply-demand balance can be achieved by short and long cycle production and load balancing. Using renewable energy sources, hydrogen is produced in the required quantities, and if it can be stored in salt caverns, the fuel cells re-convert hydrogen to electrical energy for load balancing. This method has a wide application and a strategic potential. Strategic measures to store hydrogen in salt caverns have a key role to play in ensuring energy supply-demand balance in the future. Storage plays an important strategic role as a defence against import or production shortages in periods of interrupted supply or particularly high demand. Subsurface energy storage represents a complex and broadly evolving field of research, as it covers multiple scales of application, a variety of end-user profiles, and different types of energy carriers. Subsurface storage capacities are present in many types of geological formations, each of which has its own criteria for identifying techno-economic viability. Some of the subsurface energy storage technologies (e.g. natural gas storage) have been applied at large scale for decades, while others have thus far been applied in pilot projects or at modest scale only (e.g. compressed air energy storage, hydrogen storage). It is crucial to further increase our level of understanding of subsurface energy storage potential on the basis of new geoscientific data, improved models and common agreed assessment principles (http://www.energnet.eu).

Over the past few decades, an industry based on the construction of salt caverns has developed. The number of hydrogen, hydrocarbon and waste storage caverns formed using solution mining techniques is growing rapidly (Thoms and Gehle 2000). Today's commercial salty water producers use the necessary saltwater feeding methods using controlled solution mining techniques. However, the main purpose of salty water suppliers is not only to produce salty water but also to produce robust and strase-resistant caverns suitable for hydrogen, hydrocarbon or long-term waste storage (Warren 2006).

The key to unravelling the full potential and effective implementation of large-scale subsurface energy storage lies in the integration of geological knowledge, engineering solutions, market-economy information and a comprehensive analysis of the entire energy system. Close cooperation between all actors from science, industry and policy areas is therefore essential to a successful development.

Geological setting is one of the prime considerations for siting underground rock caverns so also has an important contribution towards successful construction of the facilities. Therefore, in the course of designing suitable underground storage caverns, the underground geology of the appropriate salt domes must be well known. From the determination of the salt domes to the completion of the project, the geological prediction is an integral part of any underground project. The results of the fieldwork are based on a seismic survey and an exploration well. The interpretation of seismic data proved the depth and integrity of the proposed storage area. Once hydrogen is produced, it may need to be stored. Options for hydrogen storage include physical storage (liquid tanks and high pressure gas tanks), geologic storage (such as solution-mined salt caverns), and material-based storage (hydrogen carrier materials). The hydrogen atom is the lightest and the smallest element. Hydrogen gas is an ideal gas. Pure hydrogen storage may limit some storage options. Some possible problems must be well known for the hydrogen storage to be achieved. Some peculiar analyzes may be needed for this. The specific characteristics of the salt can make it an ideal bedrock of tight, stress-resistant caverns resistant to the storage of air and hydrogen under pressure.

The reservoirs of salt caverns have a larger gas volume than the aquifers. Salt caverns are filled underground, like a giant gas cylinder, and used to meet high-capacity peak load demands. The construction of caverns in salt-dome caverns for gas storage are between 460 and 1900 m, although they are built close to the ground (Evans et al. 2006).

Salt beds are shallower, thinner formations, the salt generally being interbedded with mudstones and anhydrite beds. Salt beds may range in thickness from a few tens of metres to perhaps 300–500 m thick and are effectively wide, thin formations. Once a salt cavern is introduced they are more prone to deterioration and tend to be more expensive to develop than salt domes.

The technology of cavern development involves the drilling of a well into a salt dome or bedded salt sequence. Cavities are then leached by the controlled circulation of water down the well bore into the salt and back as brine to the surface. The former brine well then serves for gas injection and withdrawal. The withdrawn gas may need to be dried because brine residues in the cavern sumps can saturate the gas with water vapor. No other contamination is expected during storage because rock salt does not react with hydrogen. The volumetric storage capacity of a cavity of several hundred thousand cubic meters is proportional to the maximum working pressure which is dependent on the depth. Some factors affecting the stability of the cavern are being investigated and compared with the results of basal studies. These should be included in parameters such as cavern's pressures and operations, cavern size and geometry, roof thickness, overload hardness and interface properties. Artificially constructed salt caverns in salt domes provide significant advantages over other natural underground cavities due to their ease of loading in large quantities depending on the operating pressure of the gas. Salt caverns are underground gas storage facilities that do not pass the ideal gases due to their impermeable geological structures due to thick layers.

Typical dimensions are volumes of 500,000 m^3 with pressure ranges between 60 and 180 bar, which corresponds to a working storage capacity of around 4,200 t (Kepplinger et al. 2011).

The caverns should be considered as a whole from the structural point of view. Because of the change of gas pressure in the caverns, the stress distribution of the salt rocks also changes. The salt walls between the caverns are affected by the stress, and due to the creep properties of the salt, there is a gradual convergence behavior, which depends on the temperature. For this reason, factors such as static, creep and combination of thermal loads must be carefully examined. The depth, thickness, purity and structural integrity of the salt body are important in hydrogen storage in salt caverns.

The salt dome is an ideal host rock for storing high-pressure hydrogen in caverns for a few reasons. Salt caverns formed in the salt rock are the most economical storage areas because of their multi-cycle capacity and high gas flow. These are typically used to meet the peak load demands that are most needed. They have a key role in terms of energy economy and strategy. Technologically, salt mines seem to have very convenient storage areas for high pressure hydrogen storage, although they have to be adapted to satisfy safety regulations. Underground salt cavern storage has an important place in international energy reserves, and in the future a wide range of salt caverns will be built as strategic energy storage areas.

6 Conclusions

Due to the growing importance of renewable energy sources it becomes more and more necessary to investigate energy storage potentials. It is necessary to overcome various technical difficulties in order to produce, use and safely store the hydrogen energy in a widespread manner. Reliable, compact and low cost hydrogen storage technologies need to be developed and implemented to overcome these challenges. For this reason, the best storage options should be determined to improve the storage techniques, and some important storage parameters have to be taken into consideration such as size of hydrogen storage area, energy density, time-dependent storage period, maintenance, cost and also safety.

Underground hydrogen storage areas can be constructed in different structural types and models. However, underground depots require different geological and hydrogeological conditions associated with different rock engineering problems. Storage of hydrogen and other hydrocarbons in underground salt caverns have many advantages over depleted reservoirs or aquifers, such as high efficiency, lower cushion gas requirement, lower development cost, and gas flow initiation and refill efficiency.

Salt cavern is a strategic facility not only because of its huge capacity but also because of its high injection rate and retreat potential. Most salt caverns have been developed in underground salt dome. Although of the artificial salt cavern construction cost is more expensive than depleted gas and oil areas thousand cubic feet (tcf) per dollar in terms of gas capacity, the cost is reduced due to the multi-operation feature (injection and re-injection) and it is used as an active storage reservoirs areas during the year (owing to the ability to withdraw several times a year and perform an injection cycle). These reduces the cost of the caverns for per unit of cubic feet of hydrogen gas.

Hydrogen gas can be stored economically in huge amounts in underground salt caverns. This is possible with the identification of sustainable and safe geological areas. Geophysical methods have an important role in the discovery of these areas due to the reservoir parameters of such areas, the location, structure, size of the underground salt caverns desired to be formed in salt domes. An important factor for cavern construction is the solubility of the salt rock. The salt rock is affected by various conditions during salt sedimentation and seldom consists of pure halite, but usually a mixture with a number of other minerals. Because some salt domes may contain a certain amount of insoluble rock like anhydrite and claystone, as well as highly soluble potassium salt. Investigation area may be also not free of faulting. There must be carried out substantial structural analysis and structural modelling, followed up by a seismic investigation of the whole site and its vicinity. This investigation carefully identify faults to ensure that they are at a safe distance from the cavern installations.

Underground energy storage represents a complex and widespread field of research in large-scale applications, depending on the geological structure of the site, the nature of the material to be stored and the purpose of storage such as displacement and recovery. Underground storage capacities are available in many geological formations, each of which has its own criteria for determining geo-economic viability. While some of the underground energy storage technologies (such as natural gas storage) have been applied on a large scale for decades, others have only been applied to energy resources up to now only on a modest scale such as compressed air energy storage, hydrogen storage. Underground geological surveys have shown that the hydrogen storage potential of suitable geological structures such as the appropriate salt domes can be easily determined using geophysical methods in order to form suitable hydrogen storage salt caverns. However, it is crucial to further develop the understanding level of underground energy storage potential on the basis of new geological scientific data, improved models and common accepted assessment principles.

References

Andersson J, Grönkvist S (2019) Large-scale storage of hydrogen. Int J Hydrog Energy 44:11901–11919

Basniev KS, Omelchenko RJ, Adzynova FA (2010) Underground hydrogen storage problems in Russia. In: Stolten D, Grube T (eds) Parallel sessions book 4: storage systems/policy perspectives, initiatives and co-operations. 18th World Hydrogen Energy Conference 2010-WHEC 2010, Essen

Beckman KL, Determeyer PL, Mowrey EH (1995) Natural gas storage: historical development and expected evolution: December 1994–February 1995. GRı-95/0214. Gas Research Institute, Houston, TX

Berenguer-Murcia A, Marco-Lozar JP, Cazorla-Amoros D (2018) Hydrogen storage in porous materials: status, milestones, and challenges. Chem Rec 18(7–8):900–912

Böttcher N, Görke UJ, Kolditz O, Nagel T (2017) Thermo-mechanical investigation of salt caverns for short-term hydrogen storage. Environ Earth Sci 76:98

Crotogino F, Huebner S (2008) Energy storage in salt caverns/Developments and concrete projects for adiabatic compressed air for Hydrogen Storage, Spring 2008 Solution Mining Research Institute Technical Conference, Porto, Portugal, April 28–29. https://doi.org/10.1007/s12665-017-6414-2

Crotogino F, Donadei S, Bünger U, Landinger H (2010) Large-scale hydrogen underground storage for securing future energy supplies. In: Stolten D, Grube T (eds) Proceedings of the WHEC2010 18th world hydrogen energy conference 2010, Parallel Sessions Book 4: Storage Systems/Policy Perspectives, Initiatives and Co-operations

ENeRG Position (2017) ENeRG, the European network for research in geo-energy (www.energnet.eu), ESTMAP project (www.estmap.eu)

Evans DJ, Reay DM, Riley NJ, Mitchell WI, Busby J (2006) Appraisal of underground energy storage potential in Northern Ireland. British Geological Survey Internal Report. IR/06/095, p 17171

http://www.ilkdresden.de/en/service/research-and-development-rd/detail/hydrogen-test-area-at-ilk-dresden/. Accessed 10 Jan 2018

http://www.kbbnet.de/en/. Accessed 10 Feb 2017; https://energy.gov/eere/fuelcells/hydrogen-storage. Accessed 1 May 2017

Kabuth A, Dahmke A, Beyer C, Bilke L, Dethlefsen F, Dietrich P, Duttmann R, Ebert M, Feeser V, Görke UJ, Köber R, Rabbel W, Schanz T, Schafer D, Wüdemann H, Bauer S (2017) Energy storage in the geological subsurface: dimensioning, risk analysis and spatial planning: the ANGUS + Project. Environ Earth Sci 76:23. https://doi.org/10.1007/s12665-016-6319-5

Kepplinger J, Crotogino F, Donadei S, Wohlers M (2011) Present trends in compressed air energy and hydrogen storage in Germany. In: SMRI Fall 2011 technical conference 3–4 October, New York

Klebanoff LE, Ott KC, Simpson LJ, O'Malley K, Stetson NT (2014) Accelerating the understanding and development of hydrogen storage materials: a review of the five-year efforts of the three DOE hydrogen storage materials centers of excellence. Metall Mater Trans 1(2):81e117

Kruck O, Crotogino F, Prelicz R, Rudolph T (2013) Assessment of the potential, the actors and relevant business cases for large scale and seasonal storage of renewable electricity by hydrogen underground storage in Europe. HyUnder, Grant agreement no: 303417, Deliverable No: 3.1. Status: D (4). Overview on all Known Underground Storage Technologies for Hydrogen, 93

Leighty WC (2007) Running the world on renewables: hydrogen transmission pipelines with firming geologic storage. In: Spring 2007 Solution mining research institute technical conference, Basel, Switzerland, April 29–May 1, 2007

Lord AS, Kobos PH, Borns DJ (2014) Geologic storage of hydrogen: scaling up to meet city transportation demands. Int J Hydrogen Energy 39(28):15570–15582

Martens S, Kühn M (2015) Geological subsurface will contribute significantly to the implementation of the energy policy towards renewables in Germany. EGU General Assembly 2015. Geophysical Research Abstracts 17, EGU2015-6097

Matos C, Carneiro JF, Pereira da Silva P (2016) Large scale underground energy storage for renewables integration: general criteria for reservoir identification and viable technologies. In: The 11th conference on sustainable development of energy, water and environment systems—SDEWES, Lisbon

Nanda A, Rath R, Usmani A (2015) Underground storage technologies, engineers india ltd., New Delhi, India, ISBN No. 978-93-5254-383-0

Plaat H (2009) Underground gas storage: why and how. In: Evans DJ, Chadwick RA. (eds) Underground gas storage: worldwide experiences and future development in the UK and Europe, Special Publications, vol 313. Geological society, London, pp 25–37

Sherif SA, Barbir F, Veziroglu TN (2003) Principles of hydrogen energy production, storage and utilization. J Sci Ind Res 62:46–63

Stone HBJ, Veldhuis I, Richardson RN (2009) Underground hydrogen storage in the UK. Geological Society, London, Special Publications 3:217–226

Thoms RL, Gehle RM (2000) A brief history of salt cavern use (keynote paper). In: Geertman RM (ed), Proceedings of 8th world salt symposium, part 1. Elsevier B.V., pp 207–214

Warren JK (2006) Evaporites: sediments, resources and hydrocarbons. Springer, P.O. Box 17, 3300 AA Dordrecht, The Netherlands, p 1035. ISBN 3-540-26011-0

Yang C, Wang T, Li Y, Yang H, Li J, Qu D, Xu B, Yang Y, Daemen JJK (2015) Feasibility analysis of using abandoned salt caverns for large-scale underground energy storage in China. Appl Energy 137:467–481

Biomass and Its Thermochemical Conversion: Can It Be a Road Map for Transition to 100% Renewable Energy?

Atakan Ongen, Emine Elmaslar Özbaş, Hüseyin Kurtuluş Ozcan, Serdar Aydın and Nazlıcan Karabağ

Abstract In the globalizing world, the rapidly increasing energy need and environmentally friendly methods have made it obligatory to tend to alternative energy production methods which are independent from fossil fuels. Biomass, among all the energy resources that are considered as alternative to conventional energy production based on fossil fuels, plays a significant role as a local resource that can be used. Pyrolysis and gasification are promising technologies used for the conversion of biomass to valuable hydrocarbons and alternative fuels. The purposes of this study are to examine the status of renewable energy and biomass as an alternative energy carrier or resource and to assess the gasification in thermochemical degradation of biomass. All those topics have been supported by up-to-date scientific literature. In this study, the potential of biomass energy, which is one of the most important renewable energy sources, has been examined and the efficiency of gasification process, which is one of the important methods in biomass conversion, has been demonstrated. In the light of the information obtained from the literature, it was concluded that gasification technologies have an important potential in obtaining renewable synthetic fuels from biomass as transition to 100% renewable energy has been targeted widespread.

A. Ongen (✉) · E. E. Özbaş · H. K. Ozcan · S. Aydın · N. Karabağ
Department of Environmental Engineering, Engineering Faculty, Istanbul University-Cerrahpasa, Avcilar, 34320 Istanbul, Turkey
e-mail: aongen@istanbul.edu.tr

E. E. Özbaş
e-mail: elmaslar@istanbul.edu.tr

H. K. Ozcan
e-mail: hkozcan@istanbul.edu.tr

S. Aydın
e-mail: saydin@istanbul.edu.tr

N. Karabağ
e-mail: nzlcnkrbg@gmail.com

A. Ongen · E. E. Özbaş · H. K. Ozcan · S. Aydın
Environmental and Earth Sciences Research and Application Center (ÇEYBAM), Istanbul University-Cerrahpasa, Istanbul, Turkey

T. S. Uyar (ed.), *Accelerating the Transition to a 100% Renewable Energy Era*, Lecture Notes in Energy 74, https://doi.org/10.1007/978-3-030-40738-4_18

Keywords Biomass · Gasification · Pyrolysis · 100% renewable energy · Thermochemical conversion

1 Introduction

In the globalizing world, the rapidly increasing energy need and environmentally friendly methods have made it obligatory to tend to alternative energy production methods which are independent from fossil fuels. The production processes and transfer of fossil fuels damage the environment to a large extent. These damages cause air, water and soil pollution, and global warming. Considering the effects of air pollution on human health, the use of cleaner alternative energy resources instead of fossil fuel will ensure improvement also in physical health conditions. In addition, it is not being expected for the fossil fuels to meet the energy requirement of the world in the future. Along with the increasing world population, the requirement for energy in order to bring the life standards to better conditions in developing countries is gradually gaining importance. Fossil fuels, which meet a significant part of the world's energy requirement, are energy resources that gradually diminish, and that cause significant environmental problems. Many countries are still supporting the consumption of fossil fuels, and thus the support for fossil fuel consumption had increased by 11% in 2017 (Aydın et al. 2019). Changing the energy systems requires urgent action plans, and in case of maintaining the actual energy systems in order to increase renewable energy, energy efficiency and access to energy, it will not be possible to meet the United Nations' Sustainable Development Goals (SDGs).

The fact that the world is facing a growing crisis of supplying raw material cannot be ignored. Two main difficulties that the current energy sector faces are the increasing environmental effects of fossil fuels, and their variable prices. The increasing air pollution has required the discovery of environmentally friendly fuel alternatives. Environmental problems that arise from excessive use of conventional fossil fuels is pushing the countries for the integration of effective policies and renewable energy technologies (Aydın et al. 2019). The increase of environmental concerns, and the increase of political dependency, the researches in the field of alternative energy resources are increasing. United Nations climate panel has intended a decrease of 50–80% until year 2050 in greenhouse gas emissions. In order to attain this objective, it is required to transit to renewable energy resources from fossil fuels. The combination of renewable energy requirements, and diversity of materials at landfills have caused the waste materials to be used as renewable energy resources.

Despite making progress in effective use of renewable energy and of energy efficiency, it has not yet been reached to international climate objectives formed within the scope of Paris Agreement, and to the international objectives required for sustainable development. Renewable energy, causing very low or no emission, is very important in attaining these international objectives. The use of renewable energy resources provides many advantages and opportunities from environmental factors to socio-economic factors and politics against the fossil fuels. Renewable energy resources have great effect in struggle with climate change. The combustion of fossil fuels for energy causes significant amount of greenhouse gas emission that negatively contribute to global warming.

Solar energy, wind energy, hydraulic (hydroelectric) energy, geothermal energy, biomass energy, hydrogen energy, wave energy and tide energy may be referred as renewable energy resources which will lead to achieve "100% transition to renewable energy" target of the countries. Biomass, being among the renewable energy resources, is a promising source due to being found in plenty amount in nature. The increasing energy consumption along with the increasing energy resources, and the negative contribution of fuels to pollution and greenhouse gas emissions is changing the energy balance of the world. Biomass, among all the energy resources that are considered as alternative to conventional energy production based on fossil fuels, plays a significant role as a local resource that can be used (Tańczuk et al. 2017). Biomass energy is a type of energy that is being stored in plants, animals, or in the wastes they generate. Most of the biomass energy is being generated from forest waste/residues (64%), domestic/industrial solid wastes (24%), agricultural wastes (5%), and landfill gas (5%). Experience in production technologies, conformity of energy for storage, and its harmlessness in environmental terms are bringing this energy type to the forefront compared to other alternative energy resources. This energy may be recovered along with the processing of biomass as fuel through thermochemical, biological or chemical processes.

Pyrolysis and gasification are promising technologies used for the conversion of biomass to valuable hydrocarbons and alternative fuels. Thermochemical methods such as combustion, pyrolysis and gasification are extensively being used in energy production from biomass. Biomass gasification process transforms the biomass to green energy, or to fuels in the form of clean gas which are known as synthesis gas. This promising technology has faced an increasing interest over the last two decades due to increasing demand for clean fuels and chemical raw materials, and due to the requirements of decreasing the dependency on fossil fuels, decreasing the greenhouse gas emission and disposal of current fuels. Considering the problems of environment and energy, thermal technology is progressing towards gasification systems in order to decrease the greenhouse gas emissions and also for the by-products of production (Kalıncı and Dinçer 2018). According to International Energy Agency information for the year 2019, there are 114 operational biomass gasification facilities, 14 "idle" and or "on hold" biomass gasification facilities, and 13 biomass gasification facilities being under construction/being planned in the whole world. This corresponds to 141 facilities in total, and for biomass based global liquid fuel ($\cong$750.000 tons/year) is being produced at 24 facilities (methanol, ethanol, DME, FTS, diesel, gasoline), and synthesis gas derived from biomass is 3.2×10^8 Nm^3/year.

The purposes of this study are to examine the status of renewable energy and biomass as an alternative energy carrier or resource and to assess the gasification in thermochemical degradation of biomass with the support of up-to-date scientific literature.

2 Economical Effect of Renewable Energy, and Its Current and Future Status

Renewable energy is at the center of transition to a more sustainable energy system. Renewable energy resources have rapidly increased in recent years, and especially solar energy and wind energy have gained momentum, and have become alternative energy resources. Today, renewable energy resources have started to experience their golden age in terms of electricity, but electricity is forming only one fifth of global energy consumption, and the role of renewable energy in transportation and heating sectors is continuing to be critical for energy transition. According to data of IEA (International Energy Agency) for the year 2018, it is being expected for the share of renewable energy in meeting the global energy demand to increase by one fifth percent in the following five years, and to reach to 12.4% by 2023. Global economy requires increasing amount of energy in order to sustain its growth. Energy is not just a means of economical development, it is also a vital factor for increasing the life standards and poverty, because 1.1 billion people are still unable to access electricity (World Bank 2017).

As the world population and economy increase, the pressures on electricity demand and world's ecosystems and societies are increasing. Until 2040, it is being expected for the renewable energy resources to reach a share of over 50% in European Union, 30% in China and Japan, and 25% in USA and India (Büyüközkan and Karabulut 2017; IEA 2015). Renewable energy resources are continuing to grow in electricity sector, and they have met 24% of energy demand in 2017, and it is being expected for them to meet about 30% of energy demand in 2023. In this period, it is being estimated for the renewable energy resources such as solar, wind, hydro and bioenergy to meet more than 70% of global electricity production (https://www.iea.org/topics/renewables/ 2019).

Renewable energy resources are providing lower costs as they increase the reliability and flexibility of energy system and energy supply. Through renewable energy rapid access to energy with the lowest cost is ensured. At many locations of the world, renewable energy resources represent new energy production technology with the lowest cost that ensure 100% transition to renewable energy. Renewable energy resources are also increasing the energy safety. Using the renewable energy in heating, cooling and transportation has many opportunities. Renewable energy resources may meet the heating demands of the buildings including the use of thermal water heaters with solar energy, biomass boilers and heat pumps. In industry, renewable products may also be used for heating and cooling operations in industrial operations such as food processing, and production of pulp and paper. Renewable energy has critical importance in decreasing the total fuel demand in transportation sector, and it is able to be actualized through policies encouraging energy efficiency and preservation of energy. For this reason, a policy approach integrated to renewable energy and energy efficiency is the basis. Progress in renewables remains concentrated in the power sector, while far less growth has occurred in heating, cooling and

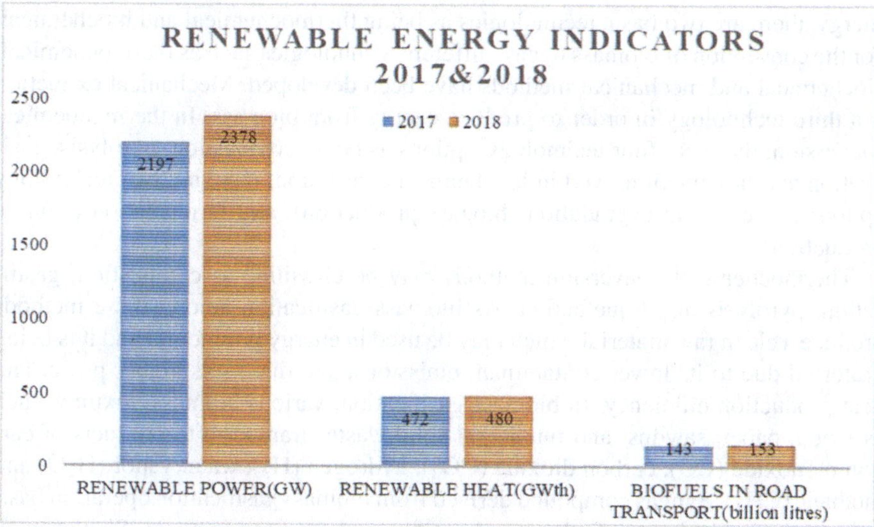

Fig. 1 Indicators of renewable energy in power, heating and transportation in 2017 and 2018 (Murdock et al. 2019)

transport. The year 2018 saw a relatively stable market for renewable energy technologies. A total of 181 gigawatts (GW) of renewable power was added, a consistent pace compared to 2017, and the number of countries integrating high shares of variable renewable energy keeps rising. The renewable energy sector overall employed (directly and indirectly) around 11 million people worldwide in 2018. Renewable energy is expanding in the power sector, with 181 GW newly installed in 2018. Global renewable power capacity grew to around 2,378 GW in 2018. In 2018, more than 90 countries had installed at least 1 GW of generating capacity, while at least 30 countries exceeded 10 GW of capacity. Figure 1 shows the indicators of renewable energy in power, heating and transportation in 2017 and 2018 (Murdock et al. 2019).

3 Biomass Energy Conversion Technologies

The conversion of biomass to energy is able to be actualized from highly various different types and resources, and through highly various technologies. Biomass may be obtained from perennial plants, forest and plant wastes, organic industrial wastes, and organic fraction of municipal solid wastes. The use of biomass in energy production is renewable energy form which may used to decrease the effect of energy production and use on global environment. As in any energy resource, there are limitations regarding the use and applicability of biomass, and it is an energy type that competes not only with fossil fuels, but also with other renewable energy resources such as wind, solar and wave force (Adams et al. 2018). In the conversion of biomass to energy, there are two basic technologies as being thermochemical and biochemical. For the conversion of biomass to gas, different technologies such as thermochemical, biochemical and mechanical methods have been developed. Mechanical extraction is a third technology in order to produce energy from biomass. In thermochemical conversion, there are four technology options as being combustion, pyrolysis, gasification and liquefaction. And in biochemical conversion, there are two technology options as being bio-degradation (biogass production), and fermentation (ethanol production).

Thermochemical conversion methods may be classified as combustion, gasification, pyrolysis and liquefaction. As biomass gasification among these methods produce a clean raw material which may be used in energy production and it is being preferred due to its lower contaminant emissions, and due to its higher power and heat production efficiency. In biomass gasification, various biomass mixtures such as wood, paper, sawdust and municipal solid wastes transform to products of carbon monoxide (CO), carbon dioxide (CO_2), hydrogen (H_2), water vapor (H_2O) and methane (CH_4). Syngas compound derived from biomass gasification operation basically depends on biomass fuel, type of gasifier, and gasification agent (Shayan et al. 2018). The biomass being rich in organic substance may both be used as industrial raw material by the use of thermochemical methods such as pyrolysis and gasification as alternative to combustion, and it may also be re-integrated to economy through energy recycling. Gasification, which is a clean and environmental friendly process, is attracting the attention of researchers in recent years. This process has limitations and economic disadvantages despite being a good alternative to the combustion of fossil fuels. These economical disadvantages are delaying the commercialization of the process (Li et al. 2018; Öngen et al. 2017). Thermochemical conversion methods such as gasification and pyrolysis are options which may be industrially applied with high efficiency compared to biochemical conversion methods depending on high thermochemical reaction rates.

3.1 Thermochemical Technologies

Pyrolysis, from among thermal energy production methods, is being performed for the recovery of biofuel having medium-low calorific value. Gasification is the controlled partial oxidation of a carbonous material, and it is being performed by providing less oxygen compared to stoichiometric requirement for complete combustion. Gasification, being a central operation in between combustion (thermochemical decomposition in the presence of excessive oxygen) and pyrolysis (thermochemical decomposition in the absence of oxygen), proceed by the temperatures varying in between 600 and 1500 °C (Ongen and Arayıcı 2014). Biomass/organic waste gasification attracts more interest due to provision of higher efficiency compared to pyrolysis. Thermochemical processes form a wider range by the concept of formation of biomass energy (Brunner and Rechberger 2015; Arena and Di Gregorio 2013; Brems et al. 2013). Thermochemical operations such as pyrolysis and gasification are used to transform biomass to a more beneficial energy or energy carriers, such as hydrogen. At high temperatures, the rate and degree of pyrolysis arising from temperature, pressure, heating rate, waiting period and gasification operation have great effect on these thermochemical operations (Ongen and Arayıcı 2014).

3.1.1 Pyrolysis

Pyrolysis means thermochemical decomposition of organic (biomass/waste) materials under inert environment. Very complex reactions occur during the pyrolysis of biomass/waste. The products obtained from pyrolysis operation are being divided to two as being primary products and by-products. The obtained products cover high quality and very valuable fuels and chemical compounds. The primary products may be in the form of solid, or liquid or gas rich in carbon. The primary product may used to being transformed to by-products through direct and chemical operations. The by-products are high quality fuels and valuable chemical compounds (Sadikoglu and Ongen 2016). In the studies where the formed products are being examined (Özcan et al. 2016; Öngen et al. 2019), it has been determined that the quality and efficiency of products generated by pyrolysis are high, and that the calorific value of synthesis gas may reach up to 4400 kcal/m^3.

3.1.2 Gasification

The gasification technologies cause the carbon based fuel to be completely transformed into the form of combustible gas, volatile substance, coal, dust and ash. In other words, gasification is the operation of conversion of any carbon fuel (solid or liquid) to fuel in gas form through partial oxidation by an oxidizing agent (air, O_2, H_2O vapor, CO_2, or their mixture). It may also be defined as a thermochemical operation limited with partial combustion and pyrolysis (Násner et al. 2017). It is four phased endothermic operation as starting from drying phase which include a series of complex chemical reactions occurring without oxygen and vapor. And real gasification reaction phase follows which breaks up the pyrolysis products more through partial oxidation in the presence of an oxidizing agent (vapor and/or a sub-stoichiometric amount of oxygen) (Assima et al. 2018).

Pyrolysis and gasification processes, among the thermochemical conversion methods, seems promising for the production of hydrogen rich gas (H_2 and CO) by the purpose of gas or synthesis of medium calorific value for electric energy production. It is being known that the distribution of products and gas efficiency rich in H_2 basically depends on fuel type, reactor's configuration, reaction period, reaction temperature, pressure, gasification agent and catalyst of gasification operation. In gasification process, complex chemical reactions occur, and mainly H_2 and CO mixture is produced along with H_2O, CH_4 and CO_2. The synthesis gas, which is being obtained as a result, may have many potential applications such as raw material for producing energy and pure hydrogen to motors and generators, and raw material for producing fuel and methanol and ammoniac for fuel cells. Tar (formed of heavy hydrocarbon wastes) arises as an undesirable by-product. The tar content, formed in gasification systems, greatly depends on operational parameters such as heating speed, gasification environment and reactor's temperature.

Managing hazardous wastes is an extraordinary advantage of thermochemical processes compared to biochemical processes (Öngen et al. 2017; Ongen and Arayici 2014, 2015). Öngen et al. (2013), had succeeded to produce synthetic gas with a heating value between 1000 and 1500 kcal/m^3 while the air flow rate was 0.05–0.1 l/min via gasification of treatment sludge from a wastewater treatment plant of a tannery industry.

When the gasification process, forming the basis of the study, is considered, it is possible to find in literature many scientific studies characterizing the biomass gasification operation by the use of pilot and laboratory scaled reactors. Highly various biomasses (nutshell, olive shell, grape remains, beech tree, wood pellet), and waste derivatives had been analyzed in order to compare the gasification features. Kalıncı and Dinçer (2018) had determined in their study that in a gasification system in which 8.5 kg/s fuel was fed -with a cost of 3.33 $/kg, 1.17 MW power and 0.521 kg/s hydrogen could be produced. Aydın et al. (2019) had intended in their study to examine the gasification of pine cone pieces and wood lumps by the use of dry air as gasification agent at a 10 kWh downdraft fixed bed gasifier at a pilot scale. In the study, it had been determined that the synthesis gas produced by the gasification of pine cone pieces is a clean alternative energy carrier for environmentally friendly clean energy production rich in hydrogen. During the experimental studies, the gasification had been actualized for various air rates (0.14–0.37), and under atmospheric pressure, and through air at the temperature range of 701–1046 °C. It had been determined that production of H_2 and CO is increasing along with the increase of reactor's temperature. Another finding was that average syngas yield was 65% by the gasification of pine cone pieces, and was 80% by the gasification of wood lumps.

The waste gasification process, which is one of the energy technologies from waste and which is a promising one, transforms the solid wastes to green energy and to fuels in clean gas form which are being known as synthesis gas, and it assists decreasing of dependency on fossil fuels, and decreasing of greenhouse gas emissions. Gasification is a significant synthesis gas production method for energy production and for chemical synthesis application by its high H_2 content, specific H_2/CO rate, and low tar rate (Šuhaj et al. 2019).

Significant parameters affecting gasification are the features of biomass (fuel), operating temperature, type of gasification agent and pressure. High moisture content is among the significant parameters affecting the yield of gasification. High moisture content decreases the yield of gasification, it affects the performance of the gasifier, and affects the quality of product. It is required for the moisture content of fuel not to exceed 35%. Along with the rate of moisture, the calorific value of gas decreases. High ash content is a parameter affecting and decreasing the calorific value of organics. And it is being preferred for the ash content not to exceed 10% (Sansaniwal et al. 2017). High operating temperature is a parameter which decreases tar content in gasification operation, which forms high carbon conversion efficiency, and which causes more gas production (Sansaniwal et al. 2017). During gasification process, air, vapor or oxygen may be used as agent gas. Type of agent has effect on the yield of gasification. While a synthesis gas with low calorific value is being formed when air is used as gasification agent due to dilution of nitrogen, the combination of vapor and oxygen provides a synthesis having medium heating value. During gasification, operation under pressure has great effect of the yield of gasification. Operation under pressure has many advantages such as saving from compression energy, and decreasing the size of equipments.

The design of a gasifier depends on the availability of fuel, its form and size, moisture content, ash content and its recent applications. There are different and various gasifiers of various types and sizes, and the gasifiers are being classified as per the conditions of gas flow and fuel flow compared to each other, and as per the method of removal of ash from the gasifier. Gasifiers may be classified as upward draft or counter-current gasifier, downdraft or co-current gasifier, cross draft gasifier and fluidized bed gasifier. The recent researches made on the high efficient gas production with less tar have become a significant problem in gasification of biomass. Downstream gasifiers is more efficient in terms of tar removal, and due to simple design, construct and synthesis gas production with low tar content (1–2 g/Nm^3), it is more efficient for small-scaled applications (<10 MWth) compared to other mostly preferred gasification systems (Aydın et al. 2019). Fixed bed type gasifiers have a bed in which the gasification agents (air, oxygen and vapor) and gas flow up or down. It is the most simple gasifier type which consists of fuel and gasification agents, fuel feeding unit, ash collection unit, and a cylindrical container and seal material for outgassing. These type of gasifiers have been designed in order to operate under medium pressure conditions in between 25–30 atm. The production of these gasifiers is simple, they are manufactured of concrete or steel, and they generally operate under low gas speed, high carbon conversion and long period. Fixed bed gasifiers, which are significantly being affected from the formation of tar contents, are suitable for small-scaled heat and energy production applications. Fixed bed gasifiers divide to three as being downstream, upstream and cross stream fixed bed gasifiers (Sansaniwal et al. 2017). Fixed bed gasifiers are the most used technology in small-scaled applications. These systems are both non-complex, and they provide high synthesis gas yield in order to obtain energy from biomass. Producing synthesis gas, which is rich in CO and H_2, and which include low amounts of CH_4, is possible with fixed bed gasifiers (Öngen et al. 2016). The basis of fluidized bed gasifiers is based on a principle in which both the fuel and bed material act as a fluid. In these gasifiers, back mixing is used for effective mixing of substances being subjected to gasification operation. Silica is the most extensively used inert bed material for this type of gasifiers. The operating temperature of a fluidized bed reactor depends on the melting point of the bed material, and it generally varies in between 800–900 °C, and this is relatively low, and thus gasification reactions don't reach to chemical balance under such low temperature conditions unless any catalyst is used. Carbon conversion efficiency reaches to 95% at fluidized bed gasifiers. This gasification type is being categorized under two types as being blistery fluidized bed gasifiers, and circulating fluidized bed gasifiers.

Technologies of catalytic gasification, plasma gasification and hydrothermal gasification are being developed day by day for high efficiencies with lower costs. The valuable products that formed by catalytic method is not just being limited with liquids, and synthesis gas with high hydrogen content and high quality coal are able to be obtained (Dincer and Acar 2015). The reactions of catalysts play a significant role in the procurement of primary products providing high quality fuels or chemical substances (Kudo et al. 2012; Mohanty et al. 2014). Clay with sulphur, calcium hydroxide, zirconium phosphate, dolomite, cement kiln dust, potassium, iron, rubidium and nickel based compounds are catalyst types use to increase the efficiency of gasification. Biomass is extensively used food stock for catalytic gasification (Hamad et al. 2016; Ding et al. 2015; Ganesh et al. 2015). In the study (Šuhaj et al. 2019), catalytic gasification of waste materials was performed at two phased laboratory scaled pyrolysis/gasification reactor. A catalyst prepared from natural clay had been tested in the decomposition of p-xylene, and in the breaking up of tar generated at Refuse Derived Fuel (RDF) gasification experiments. As the result of the study, high catalytic activity of the prepared catalyst in the decomposition of p-xylene had been observed. At catalytic gasification experiments, different amounts of catalysts (0.75, 1, 1.25, 1.5, 2 and 4 g) per 1 g RDF had been applied under reactor temperatures of 700–850 °C. In the study (Kraisornkachit et al. 2018), in which simultaneous O_2-vapor-CO_2 fed double fixed bed catalytic gasification experimental studies had been actualized for the production of synthesis gas or hydrogen, the catalytic gasification of biochar by the presence of Ni/SiO_2 catalyst in a O_2-vapor-CO_2 gas fed fixed bed reactor had been searched. While the results of the study had shown that the highest biochar-carbon conversion is able to be obtained under about 800 °C, it had been determined that the highest synthesis gas yields are being produced by 10% Ni/SiO_2 catalyst. By changing the O_2/C mol rate in between 0 and 0.5, and the S/C mol rate in between 0 and 2, the obtained H_2/CO mol rate had changed at a small range in between 0.2114 and 0.5648.

And in plasma technology, an ionized gas stream or plasma phase which separate the electrons from gas molecules forms due to the electrical resistance in the system, and the heat is released at a very high temperature. At 2000 °C, the gas molecules split to its atoms, and when the temperature increases to 3000 °C, the gas molecules lose electron, and become ionized. The gas being formed under this condition has a viscosity as in the fluids, and free electric loads ensure an electrical conductivity higher than zeolite and metals (Dincer and Acar 2015). Hydrothermal gasification applications for producing hydrogen—which is a renewable energy source- from biomass and wastes with high moisture content are relatively new approaches on this subject. As supercritical water conditions are being ensured at $T > 374$ °C and $P > 22$ MPa, hydrothermal gasification is generally ensured at 250–374 °C under a pressure of 4–22 MPa. Disintegration of raw materials having high moisture content reveals the importance of hydrothermal gasification. Hydrothermal applications are suitable for biomasses/wastes having high water content (>80% moisture) such as food wastes and animal manure. High hydrogen content (50–60% volume), low contaminant level and syngas production which do not cover tar are the main advantages of the operation. Calorific value of the produced gas generally varies in between 12 and 18 MJ/N m^3. But the process also has some difficulties required to be solved, and the studies on the subject are under development yet.

Biomass, and Gasification of Biomass

Gasification is the thermochemical conversion process of a solid fuel to a mixture of combustible gases through partial oxidation at high temperatures over 700 °C. Moreover, a solid residual forms after the biomass decomposition which is called as char. Syngas forms from the mixture of CO, H_2, CO_2, CH_4 (primary components), and H_2O, H_2S, NH_3, tar and other trace types (secondary components) depending on the type and features of raw material, and operating conditions. One of the main problems in gasification of biomass is the issue relevant to the removal of tar being formed in the pyrolysis (Rios et al. 2018).

In the thermochemical conversion of biomass, carbon monoxide (CO), carbon dioxide (CO_2), methane (CH_4) and hydrogen (H_2), volatile organic compounds (VOC), tar, water vapor, H_2S and solid residue forms (Öngen and Arayici 2015; Burra et al. 2016). Specific fractions of obtained different types depend on gasification type (fixed bed, and fluidized bed), operation conditions and gasification agents used during the gasification operation (air, vapor, oxygen and/or their mixtures). Biomass gasification covers many complex and interdependent thermochemical processes. Drying starts in biomass until reaching 150 °C where the moisture turns to vapor. Then the thermal disintegration of solid biomass occurs, and volatile substances evaporate as the temperature increases, and they form a gas mixture generally being formed of H_2, CO, CO_2, CH_4, hydrocarbons in gas phase and vapor. As the result of release of these gases, the biomass degrades into char and hydrocarbons in gas phase concentrate at low temperatures in order to produce tar. The oxygen in the air reacts with the available combustibles (pyrolysis gases, the produced tar and coal), and forms and vapor by the completely or partially oxidizing. Due to the oxidation reactions (exothermic reactions), the temperature of operation increases, and it ensures the procurement of a gas rich in H_2, CO, CH_4 (Rios et al. 2018).

Gasification consists of drying, oxidation, pyrolysis (distillation), reaction (carbonization), gasification (reduction) phases respectively. During drying operation, the moisture in biomass is removed by the temperature arising from oxidation operation. Besides the drying operation, the heat arising during oxidation operation is also used for pyrolysis and reduction operations (Öngen et al. 2016; Susastriawan and Saptoadi 2017).

One of the most significant environmental benefits of gasification is about the air emissions. By the gasification technologies, it is possible to decrease the total sulphur gases, and to decrease the odor. Gasification ensures the reduction of solid residue amount to be sent to landfill, and it ensures the joint gasification of different types of wastes (Öngen and Arayici 2015). The use of biomass in energy conversion is an efficient, low-priced and effective method, the calorific value of synthesis gas obtained as the result of gasification of biomass starts from the values of 2000 kcal/m^3, and increases as per the operation performed and the biomass content being used. Shayan et al. (2018) achieved hydrogen production by actualizing biomass gasification in the presence of various agents. In the study, the results had been assessed based on the first and second principles of thermodynamics, wood and paper had been selected as biomass, air/air enriched with oxygen/oxygen/vapor had been used as gasification agent. According to variable parameters, concentration of hydrogen, calorific value of gas, and effects of energy and exergy had been examined. When the results are considered, higher hydrogen production is achieved by the use of vapor, oxygen, air enriched with oxygen and air respectively. Moreover, the highest energy efficiency had been obtained when air agent was used in gasification, and it had been determined that the produced gas had a calorific value higher than 11 MJ/Nm^3 when vapor was used. Öngen et al. (2016), had gasified woody biomasses in the study, and they had compared the yield of synthesis gas by the synthesis gas yield of textile industry's treatment sludge. Synthesis gas calorific value of woody biomasses had been 2500–3000 kcal/m^3, and synthesis gas calorific value of treatment sludge had been 2500–2650 kcal/m^3. And in another study in which different biomass wastes had been used (Ongen and Arayıcı 2014; Öngen and Arayici 2015), it had been observed that the calorific value of produced synthesis gas may reach to 2800–3000 kcal/m^3. There are many other studies in literature which indicate the efficiency of gasification of biomass, and the suitability of use of biomass in energy generation (Li et al. 2018; Rios et al. 2018; Molino et al. 2018; Öngen et al. 2018).

4 100% Transition to Renewable Energy

Renewable energy resources have started to expand to nearly all the locations of the world, and they have become one of the main options in energy sector. In 2018, renewable power capacity of more than 1 GW has been established at more than ninety countries, and at thirty countries the renewable power capacity is more than 10 GW. Renewable energy resources currently form more than one third of the global power capacity. In 2018, at least nine countries have produced more than 20% of their electricity through variable renewable energy resources.

Renewable energy has become an essential part of the solution to a number of issues, from climate change and air pollution to geopolitical risks, price volatility and local and regional economic development. The sustained decrease in production costs means that renewable energy is already a competitive source of energy in many regions of the world and is set to become so in most very soon (https://energy-cities.eu/publication/cities-heading-towards-100-renewable-energy-by-controlling-their-consumption/ 2019).

Communities, cities, regions, islands and even countries across the world have embarked on a 100% renewable energy path. An increasing number of local authorities are proving that a 100% renewable energy target–of course combined with energy efficiency and low energy use policies—is not only technically feasible but also economically and socially beneficial (https://energy-cities.eu/publication/cities-heading-towards-100-renewable-energy-by-controlling-their-consumption/ 2019).

In recent years, the cities and countries have an increasing movement in covering 100% of the energy from renewable resources. The countries such as Costa Rica, Djibouti and Sweden have determined their objectives of 100% of their energy sectors to be renewable. And Iceland and Norway are already producing all the electricity from renewable energy resources. Currently, renewable energy resources in energy sector are playing a smaller role in heating, cooling and transportation, and unfortunately this is a great problem, because these sectors constitute 80% of global energy demand. Today, the renewable energy resources supply more than one fourth of electricity demand, but renewable energy options provide only 10 and 3% of the energy being required in heating and transportation respectively. In order to improve the integration among energy, heating, cooling and transportation sectors, increasing the renewable technologies and opportunities will ensure transition to a world completely based on renewable energy (https://www.iea.org/topics/renewables/ 2019).

The Table 1 shows a summary of the recent trends and required levels of ambition for each indicator according to the renewable energy roadmap analysis by IRENA (REmap) analysis, with respective compound annual growth rates (CAGR) (Global Energy Transformation 2019).

Electrification, energy efficiency improvements and the deployment of renewables must all accelerate throughout the years up to 2050 to achieve the aims of the Paris Agreement. Renewable energy and energy efficiency, in combination with electrification, are the key ingredients to ensure a sustainable energy future. By 2050, renewables could dominate the transport and buildings sectors reaching 57 and 81% of the sectors' total final energy consumption, respectively. Renewables would cover one-quarter of final energy use for industry. In all sectors, electricity would account for the largest share of renewable energy use, complemented by direct uses of biomass, geothermal and solar thermal (Global Energy Transformation 2019).

The installed capacity of biomass energy, which was 66.929 MW in 2010, showed a great upward trend and reached 115.731 MW in year 2018 (IRENA 2019). This increasing trend of biomass energy will play an important role to reach 100% renewable energy targets of the developed countries. Biopower or electricity from biomass is the 4th largest renewable electricity generation source. In the future, solar and wind technologies will play the most prominent role followed by significant contributions from hydro, geothermal and biomass sources (WBA 2018). With the need for biomass, interest in biomass conversion technologies has also increased. In this context, thermochemical technologies used for the production of synthetic fuels which have an effective place in electricity generation are being developed rapidly (Mathiesen et al. 2015; Brown et al. 2019).

Table 1 Summary of recent trends and required levels of selected indicators (Global Energy Transformation 2019)

	Historical		Estimated	REmap case		CAGR (%)	
	2010	2016	2018	2030	2050	2010–2016	2016–2050
Share of renewable energy in electricity generation (%)	20	24	26	57	86	+3.1	+3.8
Energy intensity of GDP based on primary supply (MJ/USD-PPP, 2011)	5.9	5.1	5.0	4.6	2.4	−2.3	−3.2
Share of modern renewable energy in total final energy consumption (%)	10.1	9.8	10.5	28	66	−0.4	+5.8
Electricity share of total final energy consumption (%)	18	19	19	29	49	+1.1	+2.9
Annual energy-related emissions (Gt CO_2/year)	29.7	33.5	34.3	24.9	9.7	+2.0	−3.5

5 Conclusion

Energy, which is the main input of social and economic development in the world, is increasingly needed. Global energy use shows an increase of about 2% per year. The main reason for this increase is the increasing population and efforts to reach high standards of living. Concurrently with increasing energy demand, there is an increase in environmental problems such as acid rain, ozone depletion and global climate change. Energy, economy and the environment are three key elements of sustainable development, and these three components must be in balance to achieve truly sustainable development. A negative change in one of these three components affects the others and may cause difficulties in the implementation of the concept of sustainability.

Although energy is the main element of development, it also brings environmental risks and problems, and the interaction between energy and environment is an important issue to be considered in energy related activities. Energy sources, whether fossil fuels such as coal, oil, natural gas, or renewable energy sources such as hydraulic energy, geothermal, wind and biomass, are obtained from nature. On the other hand, the use of these resources creates negative effects on the environment, in other words, it turns to nature as a negative effect.

In order for the energy resources to be environmentally benign, energy production from all primary energy sources should be realized with high efficiency and clean technologies. In addition, the use of renewable energy sources as much as possible instead of fossil resources is important in terms of minimizing environmental damages. Nowadays, research and investments in clean energy obtained by thermal decomposition of organic materials by pyrolysis and gasification are of increasing interest. In particular, the use of biomass for this purpose appears as an efficient and clean energy source. Biomass has an important place among renewable energy sources. In developing countries, wood and traditional biomass materials are already used for energy recovery and heating.

In this study, the potential of biomass energy, which is one of the most important renewable energy sources, has been examined and the efficiency of gasification process, which is one of the important methods in biomass conversion, has been investigated. In the light of the information obtained from the literature, it was concluded that biomass and its thermochemical conversion have an important potential for the transition to 100% renewable energy.

References

Aydın ES, Yucel O, Sadikoglu H (2019) Experimental study on hydrogen-rich syngas production via gasification of pine cone particles and wood pellets in a fixed bed downdraft gasifier. Int J Hydrogen Energy 44:17389–17396

Tańczuk M, Skorek J, Bargiel P (2017) Energy and economic optimization of the repowering of coal-fired municipal district heating source by a gas turbine. Energy Convers Manag 149:885–895

Kalıncı Y, Dinçer I (2018) 5.3 waste energy management, pp 91–133

World Bank (2017)

Büyüközkan G, Karabulut Y (2017) Energy project performance evaluation with sustainability perspective. Energy 119:549–560

IEA (2015)

https://www.iea.org/topics/renewables/. 07 Nov 2019

Murdock HE, Gibb D, André T, Appavou F, Brown A, Epp B, Sawin JL (2019) Renewables 2019 global status report. IEA, Renewables

Adams P, Bridgwater T, Lea-Langton A, Ross A, Watson I (2018) Biomass conversion technologies. In: Greenhouse gas balances of bioenergy systems. Academic Press, pp 107–139

Shayan E, Zare V, Mirzaee IJEC (2018) Hydrogen production from biomass gasification; a theoretical comparison of using different gasification agents. Energy Convers Manag 159:30–41

Li Z, Han C, Gu T (2018) Economics of biomass gasification: a review of the current status. Energy Sources Part B 13(2):137–140

Öngen A, Aydın S, Arayıcı S (2017) Composition and energy potential of industrial sludge derived synthetic gas. Int J Glob Warm 11(3):285–293

Ongen A, Arayıcı S (2014) Thermal treatment of fleshing residue for producing syngas. Int J Glob Warm 6(2–3):149–159

Brunner PH, Rechberger H (2015) Waste to energy–key element for sustainable waste management. Waste Manag 37:3–12

Arena U, Di Gregorio F (2013) Element partitioning in combustion-and gasification-based waste-to-energy units. Waste Manag 33(5):1142–1150

Brems A, Dewil R, Baeyens J, Zhang R (2013) Gasification of plastic waste as waste-to-energy or waste-to-syngas recovery route

Sadikoglu HO, Ongen A (2016) Stabilization of galvanic sludge by microwave pre-treated pyrolysis. Int J Environ Sci Technol 13(2):691–698

Özcan HK, Öngen A, Özbaş EE (2016) Süt endüstrisi arıtma çamuru pirolizinde sıcaklığın ve parçacık boyutunun sentez gaz oluşumuna etkisi. Çukurova Üniversitesi Mühendislik-Mimarlık Fakültesi Dergisi 31(1):301–312

Öngen A, Karabağ N, Hunce SY, Özcan HK, Özbaş EE, Aydın S (2019) An assessment of pyrolysis process for the treatment of agricultural and forest wastes. In: Recycling and reuse approaches for better sustainability. Springer, Cham, pp. 97–110

Násner AML, Lora EES, Palacio JCE, Rocha MH, Restrepo JC, Venturini OJ, Ratner A (2017) Refuse derived fuel (RDF) production and gasification in a pilot plant integrated with an Otto cycle ICE through Aspen plus™ modelling: thermodynamic and economic viability. Waste Manag 69:187–201

Assima GP, Marie-Rose S, Lavoie JM (2018) Role of fixed carbon and metal oxides in char during the catalytic conversion of tar from RDF gasification. Fuel 218:406–416

Öngen A, Arayici S (2015) Monitoring of waste gasification products: solid, liquid and syngas. Hittite J Sci Eng 2(1)

Öngen A, Ozcan HK, Arayıcı S (2013) An evaluation of tannery industry wastewater treatment sludge gasification by artificial neural network modeling. J Hazard Mater 263:361–366

Šuhaj P, Haydary J, Husár J, Steltenpohl P, Šupa I (2019) Catalytic gasification of refuse-derived fuel in a two-stage laboratory scale pyrolysis/gasification unit with catalyst based on clay minerals. Waste Manag 85:1–10

Sansaniwal SK, Pal K, Rosen MA, Tyagi SK (2017) Recent advances in the development of biomass gasification technology: a comprehensive review. Renew Sustain Energy Rev 72:363–384

Öngen A, Ozcan HK, Ozbas EE (2016) Gasification of biomass and treatment sludge in a fixed bed gasifier. Int J Hydrogen Energy 41(19):8146–8153

Dincer I, Acar C (2015) Review and evaluation of hydrogen production methods for better sustainability. Int J Hydrogen Energy 40(34):11094–11111

Kudo S, Norinaga K, Hayashi JI (2012) Application of catalysis in the selective conversion of lignocellulosic biomass by pyrolysis. J Novel Carbon Resour Sci 6:1–8

Mohanty P, Pant KK, Naik SN, Parikh J, Hornung A, Sahu JN (2014) Synthesis of green fuels from biogenic waste through thermochemical route–the role of heterogeneous catalyst: a review. Renew Sustain Energy Rev 38:131–153

Hamad MA, Radwan AM, Heggo DA, Moustafa T (2016) Hydrogen rich gas production from catalytic gasification of biomass. Renew Energy 85:1290–1300

Ding J, Wang H, Li Z, Cui K, Karpuzov D, Tan X, Mitlin D (2015) Peanut shell hybrid sodium ion capacitor with extreme energy–power rivals lithium ion capacitors. Energy Environ Sci 8(3):941–955

Ganesh KP, Maria SS, Thi PTP, Russell M, Rajasekhar B (2015) Heterogeneous catalyst-assisted thermochemical conversion of food waste biomass into 5-hydroxymethylfurfural. Biores Technol 178:19–27

Kraisornkachit P, Vivanpatarakij S, Powell J, Assabumrungrat S (2018) Experimental study of dual fixed bed biochar-catalytic gasification with simultaneous feed of O_2-steam-CO_2 for synthesis gas or hydrogen production. Int J Hydrogen Energy 43(32):14974–14986

Rios MLV, González AM, Lora EES, del Olmo OAA (2018) Reduction of tar generated during biomass gasification: a review. Biomass Bioenerg 108:345–370

Burra KG, Hussein MS, Amano RS, Gupta AK (2016) Syngas evolutionary behavior during chicken manure pyrolysis and air gasification. Appl Energy 181:408–415

Susastriawan AAP, Saptoadi H (2017) Small-scale downdraft gasifiers for biomass gasification: a review. Renew Sustain Energy Rev 76:989–1003

Molino A, Larocca V, Chianese S, Musmarra D (2018) Biofuels production by biomass gasification: a review. Energies 11(4):811

Öngen A, Özbaş EE, Karabağ N, Özcan HK, Aydın S, Kaya E (2018) Thermal behavior of waste-derived fuels and determination of optimum mixture ratio for gasification. Eurasian J Environ Res 2(2):1–7

https://energy-cities.eu/publication/cities-heading-towards-100-renewable-energy-by-controlling-their-consumption/. 27 Nov 2019

Global Energy Transformation (2019) The REmap transition pathway, a ROADMAP TO 2050. 2019 Edition-Background Report. https://www.irena.org/publications/2019/Apr/Global-energy-transformation-The-REmap-transition-pathway. 27 Nov 2019

IRENA (2019) Renewable capacity statistics 2019. Int Renew Energy Agency (IRENA), Abu Dhabi. https://irena.org/, 27 Nov 2019

WBA (2018) Global bioenergy statistics 2018. https://worldbioenergy.org, 27 Nov 2019

Mathiesen BV, Lund H, Connolly D, Wenzel H, Østergaard PA, Möller B, Hvelplund FK (2015) Smart energy systems for coherent 100% renewable energy and transport solutions. Appl Energy 145:139–154

Brown T, Schäfer M, Greiner M (2019) Sectoral interactions as carbon dioxide emissions approach zero in a highly-renewable european energy system. Energies 12(6):1032

Role of Energy Storage in 100% Renewable Urban Areas

Halime Paksoy, Nurten Şahan and Burcu Koçak

Abstract Transition from hydrocarbon to renewable era has already started, but not everywhere around the world and not at the same pace. The gap between supply and demand of renewables has to be closed for providing uninterrupted and secure energy. Reaching 100% renewable target in urban areas can only be attainable with energy storage. More cost-effective energy storage options will increase exploitation of renewable potential. Energy storage can be used in several ways and points in the energy value chain. Energy end-users can benefit from storage technologies to meet their electrical, heat and cold demands. Energy storage technologies vary in terms of maturity and services they can provide. Using energy storage with different concepts such as sector coupling may increase its value and widen its users and benefits. Buildings are the focus of 100% renewable energy urban areas. Increasing net-zero energy buildings will accelerate transition to ultimate goal of 100% renewables target. There are several ways of using energy storage in buildings for using renewables and also preventing urban heat island effects. This chapter will give an overview on energy storage and its current applications in urban areas. The roles of energy storage with a special focus on 100% renewable urban areas are discussed.

1 Introduction

Urbanization trend around the world increases population in the cities. According to UNESCO (www.unesco.org), population in urban areas is growing by 60 million persons per year. The growth is larger in developing countries. It is expected that by 2030, about 5 billion (61%) of the world's 8.1 billion people will live in cities (www.unesco.org). This trend increases the energy demand in urban areas even more. The services and comfort requirement of the citizens is increasing energy demand even more. Today's energy use is highly dependent on fossil fuels, which are not only depleting, they also threaten the future of the world by global warming they cause. In addition, fossil fuels not being found everywhere around the world

H. Paksoy (✉) · N. Şahan · B. Koçak
Chemistry Department, Faculty of Arts and Sciences, Çukurova University, Adana, Turkey
e-mail: hopaksoy@cu.edu.tr

T. S. Uyar (ed.), *Accelerating the Transition to a 100% Renewable Energy Era*,
Lecture Notes in Energy 74, https://doi.org/10.1007/978-3-030-40738-4_19

have to be imported, which may cause economic and political problems. Providing clean, secure, local and cost-effective energy to consumers is more urgent than ever. Renewable energy resources are the alternatives to fossil fuels. Intermittent characteristic of renewables hinders their capability to replace fossil fuels entirely. That is why energy storage technologies should be included in renewable energy systems for more efficient and continuous utilization. Energy storage technologies can provide power, heat, cold, domestic and industrial hot water and thermal management for more efficient energy use (Dincer and Rosen, 2010).

Targets to increase the use of renewables and decrease CO_2 emissions have been set by several organizations, governments and communities. Intergovernmental Panel on Climate Change (IPCC's) recommends that global carbon emissions should reach zero by mid-century in order to limit warming below 1.5 °C (www.ipcc.ch). Danish government aims to go 100% renewable in power and heat by 2035 and 100% renewable energy in all sectors by 2050 (http://www.go100percent.org). Many cities have committed to IPCC's recommendations even though their central governments are not taking any actions. One of these is San Diego, who has an aim of reaching 100% carbon-free electricity by 2035 (www.greentechmedia.com).

The building sector has huge share in global energy consumption and is expected to increase by about 48% from 2012 to 2040 (Drissi et al. 2019). Different categories of energy efficient buildings according to recent standards and guidelines on improving energy performance in buildings are given below (Kuznik et al. 2015):

- Low energy building
- Ultra-low energy building
- Zero energy house
- Autonomous building
- Energy plus house.

Energy Performance of Buildings Directive of European Union imposes that all new buildings to be nearly zero-energy by the end of 2020 (Directive 2010/31/EC). Achieving net-zero CO_2 emissions is an ambitious target for buildings sector and needs drastic changes in energy use patterns. The current interest and research are in tendency to set up zero energy buildings by combining with renewable energy sources, energy storage systems, saving and insulations technologies and smart technologies (Sun et al. 2018). Thermal energy storage plays a key role in providing thermal comfort to occupants and also meets these demanding targets (Kuznik et al. 2015).

Energy storage can be used in a wide range of renewable energy applications especially for buildings. This chapter aims to show how this can be achieved, current situation of applications in urban areas and roles and different models of operation of energy storage in 100% renewable era.

2 Energy Storage

Energy storage can be realized by transformations between different energy forms according to the laws of thermodynamics. Energy storage methods can be classified according to these transformations a shown in Fig. 1 (Konuklu et al. 2018).

2.1 *Sensible Heat*

Sensible heat is stored in sensible thermal energy storage materials (STESMs) through changing temperature of material without changing phase (Alva et al. 2017). Sensible thermal energy can be expressed as a function of materials' mass (m, kg), specific heat capacity (C_p, kJ kg^{-1} K^{-1}) and temperature difference between lower and upper temperature (ΔT, K), as seen in Eq. 1.

$$Q = m * C_p * \Delta T \tag{1}$$

STES systems provide superiority over latent and thermochemical storage systems in terms of simple system design. However, STES systems need more storage volume because of their lower energy capacity per unit volume. Selection of STESMs is crucial for storage volume and performance (Ataer, 2006). The most common STESMs

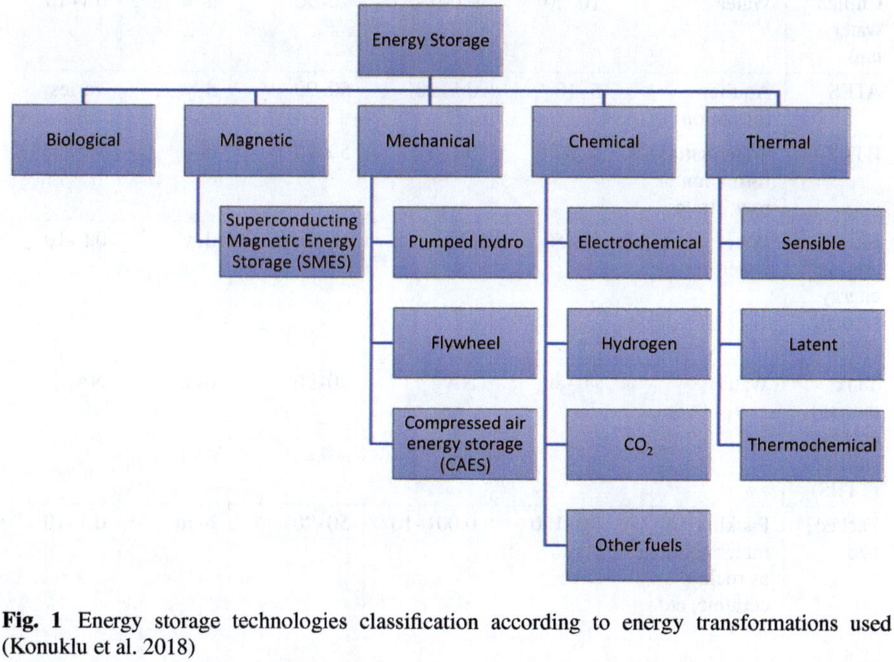

Fig. 1 Energy storage technologies classification according to energy transformations used (Konuklu et al. 2018)

are water, rocks like silica, quartz, basalt and granite (Dincer and Rosen 2010; Bruch et al. 2014; Schlipf et al. 2015; Allen 2010), alumina beads (Cascetta 2015), mineral oil (Alva et al. 2017), molten salt (Alva et al. 2017). In recent years, the usability of waste materials as storage material has been investigated to prevent the depletion of natural resources. Demolition wastes (Koçak and Paksoy 2019), asbestos containing wastes (Py et al. 2009, 2011), industrial by-products (Grosu et al. 2018a, b; Navarro et al. 2012; Miró et al. 2014) and industrial furnace slags (Motte et al. 2015; Fernández et al. 2015; Agalit et al. 2017) are good candidates for STES systems.

STES technologies can be divided in 3 groups as underground thermal energy storage (UTES), water tanks and packed beds. Comparison of typical parameters of STES technologies are given in Table 1. Storage period, installation cost, operation temperature, heat loss, storage capacity, charge-discharge time and efficiency are the decisive factors for selection for sustainable TES technologies (Dincer and Rosen 2010; González Roubaud et al. 2017). Storage period in TES systems can be divided

Table 1 Typical parameters of thermal energy storage systems

Testes system	Storage material	Capacity (kWh/t)	Power (MW)	Efficiency (%)	Storage period (h, w, d, m)	Cost (€/kWh)
Hot water tank	Water	20–80	0.001–10	50–90	d–m	0.1–10
Chilled water tank	Water	10–20	0.001–2	70–90	h–w	0.1–10
ATES	Aquifer formation	5–10	0.5–10	50–90	d–y	Varies
BTES	Water-saturated formation or rock strata	5–30	0.1–5	50–90	d–y	Varies
Pit thermal energy storage (PTES)	Water, water/gravel	30–50	NA	NA	d–y	0.5–10
Tank thermal energy storage (TTES)	Water	40–80	NA	70–90	d–y	NA
Packed bed	Packing materials such as rock, ceramic, etc.	20–100	0.001–10	50–90	h–m	0.1–10

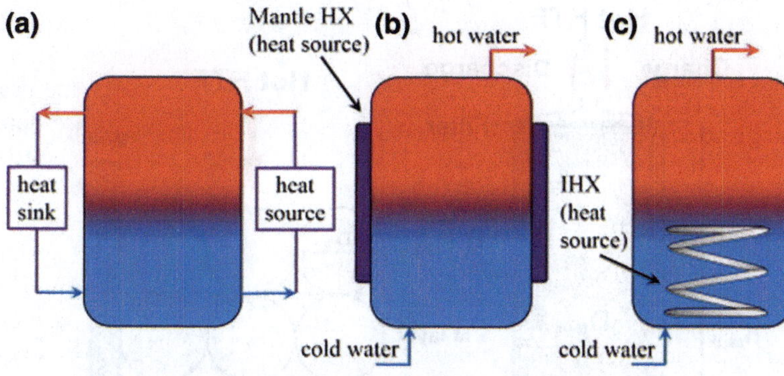

Fig. 2 Water tank operation modes; **a** External heat exchanger, **b** Mantle heat exchanger, **c** Immersed coil heat exchanger (Nash et al. 2017)

as short-tern (daily) or long-term (seasonal) (Guelpa and Verda 2019). Two systems have distinct differences in installation cost, heat loss and installation area.

Water tank storage systems are most common storage technologies that can be used for both cold and hot storage. Water tanks can be installed on the ground or underground. Tanks are made from stainless steel, concrete or fiberglass (Ataer 2006). An insulation layer should be used to reduce heat losses. Storage tanks are full of water. Water is cheap and abundant storage material. Its heat capacity (4.2 kJ kg^{-1} K^{-1}) is higher than rocks (app. 0.82 kJ kg^{-1} K^{-1}). Only drawback is limited storage temperature range (4–90 °C) (Dincer and Rosen 2010; Guelpa and Verda 2019). A pressurization system should be used to avoid evaporation at higher temperature (Guelpa and Verda 2019). Water tank technologies are most preferable for cold storage because of high density of water at 4 °C, this is the minimum temperature for water tank storage technologies (Guelpa and Verda 2019) As seen in Fig. 2, there are 3 alternative modes for water tank operations. In the first mode, heat is transferred from heat sink to water tank by an external heat exchanger. External heat exchanger maintains flexible and cheap solutions for indirect systems. In the mantle heat exchanger mode, HTF comes from heat sink released heat by contacting with tanks walls. The system brings special tank design and high cost. In the third mode, coil heat exchanger is placed inside the tank. HTF flows inside this coil to transfer heat between HTF and water (Nash et al. 2017).

Packed-bed storage systems store heat in packing material. During charging, packing material absorbs heat from HTF then releases heat to HTF during discharging (Bruch et al. 2014; Koçak and Paksoy 2018). Figure 3 shows a single tank packed-bed thermocline system. In single-tank packed-bed system, packing materials can be cheap, high density and abundant storage materials like rock, ceramic, waste materials etc. used for high temperature storage application.

2-tank storage system used in a solar field consists of cold tank filled with molten salt and empty hot tank. When the system starts, molten salt in cold tank starts to heat

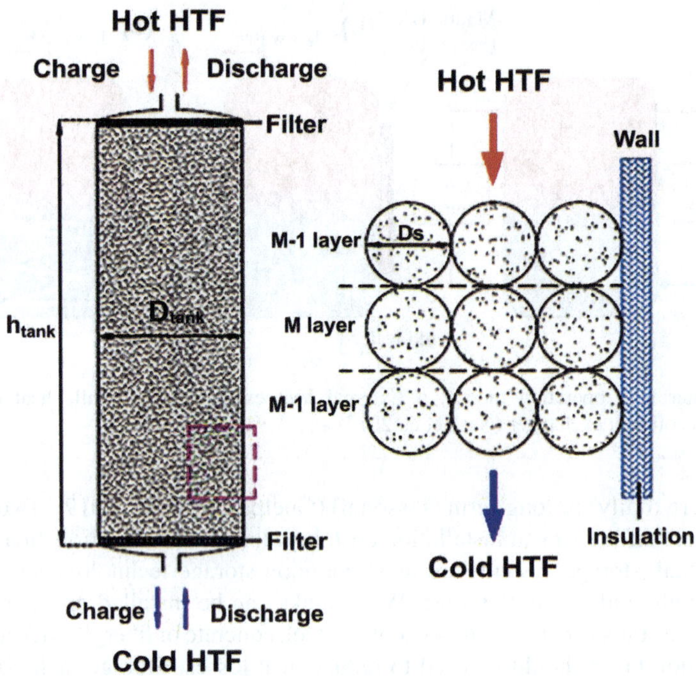

Fig. 3 Packed-bed storage tank with packing material (Koçak and Paksoy 2018)

up and transfers to hot tank (Dincer and Rosen 2010). Both single-tank and 2-tank storage systems can be used for high temperature storage application.

2.1.1 Underground Sensible Thermal Energy Storage (UTES)

UTES systems are generally preferred for long-term (seasonal) storage. There are four main systems shown in Fig. 4 (Ochs et al. 2008).

- Tank thermal energy storage (TTES)
- Borehole thermal energy storage (BTES)
- Pit thermal energy storage (PTES)
- Aquifer thermal energy storage (ATES).

Tank thermal energy storage (TTES): TTES systems is installed underground independent of geological formations. The system consists of concrete or steel tank filled with storage medium, generally water (Guelpa and Verda 2019; Schmidt et al. 2004). Tank geometry may be cuboids, cylinders, inversed or truncated pyramids. Optimum aspect-ratio of h/d is 1 to reduce heat losses (Ochs et al. 2008). For long term storage systems, minimum 1 m insulation layer from glass wool or polyurethane should be used (Guelpa and Verda 2019).

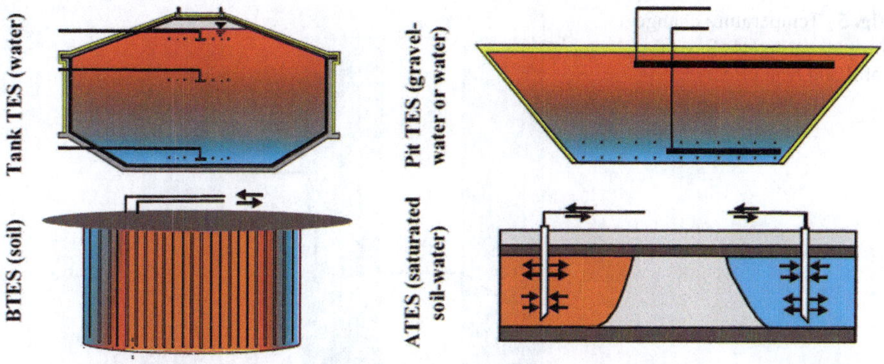

Fig. 4 UTES systems (Ochs et al. 2008)

Borehole thermal energy storage (*BTES*): In BTES systems, heat is stored completely in underground by several U pipes, which are called ducts. U pipes from plastic like HDPE are lined in underground and filled with gravel and water. Insulation layer is used at the top of the system to reduce heat loss. Storage capacity per unit volume of BTES systems are less than 40% of TTES systems (Mangold and Deschaintre 2019).

Pit thermal energy storage (*PTES*): The storage medium is water or water-gravel mixture. Gravel's specific heat capacity is lower than water. Hence, water-gravel mixture system needs more storage volume than TTES system for unit storage capacity. PTES system is more economical than other UTES systems (Alva et al. 2018; Guelpa and Verda 2019; Ochs et al. 2008).

Aquifer thermal energy storage (*ATES*): ATES systems need suitable geological and design conditions. ATES utilizes groundwater to transfer heat to aquifer during summer. During winter while recovering heat, cold is stored to be used in summer (Dincer 2002; Paksoy et al. 2009). ATES systems do not pollute groundwater (Dincer 2002; Guelpa and Verda 2019).

2.2 Latent Heat Storage

The enthalpy change accompanying phase change so called latent heat can be used in thermal energy storage. The phase change that is preferred in latent heat storage is solid to liquid. The latent heat of a solid-solid phase change is generally small. Phase changes that involve gas phase require adaptability to volume changes making operation more complex. Figure 5 shows enthalpy change with respect to temperature during heating/cooling of a pure solid material. The graph is divided into latent heat (white) and sensible heat (blue) zones. The constant temperature regions on the heating/cooling curve correspond to melting/freezing and the second one corresponds to evaporation/condensation. At these plateaus phase change occurs isothermally.

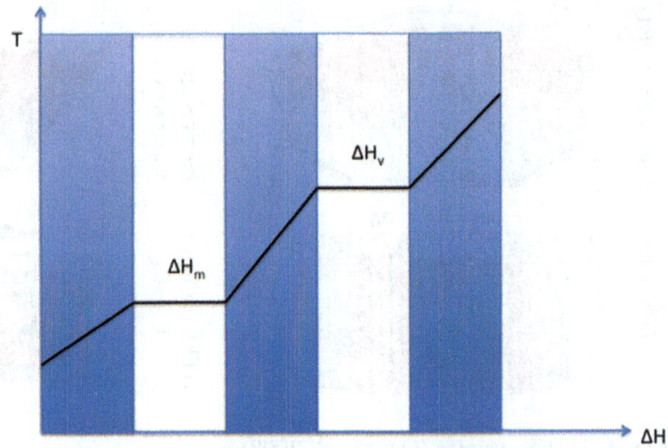

Fig. 5 Temperature change during heating/cooling of a solid and heat effects

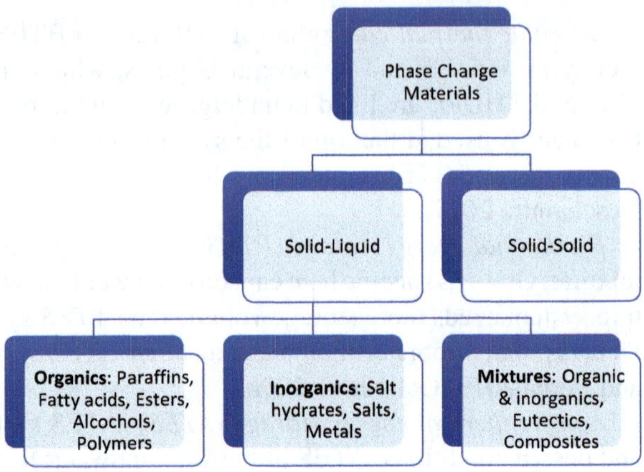

Fig. 6 PCM categories

Materials with behavior shown in Fig. 6 so called Phase change materials (PCM) are used in latent heat storage. Any material that has a reproducible phase change at a temperature suitable for the heating or cooling application can be used as PCM, but there are some more anticipated properties based on type of application. The general criteria of a PCM are listed in Table 2 (Konuklu et al. 2018).

In latent heat storage systems using solid-liquid PCM, the significant part of storage capacity comes from enthalpy of melting. Latent heat storage capacity can be about 10 times higher than the sensible heat storage capacity depending on the PCM. Depending on the application temperature sensible heat below and above the phase change temperature becomes important. For a PCM made up of pure material and has mass of m (g), total heat stored, Q (J) between temperatures T_1 and T_2 can be calculated from Eq. (2) (Koçak and Paksoy, 2019):

$$Q = m\Delta H_m + \int_{T_1}^{T_m} mC_{p,s}dT + \int_{T_m}^{T_2} mC_{p,l}dT \tag{2}$$

Table 2 General criteria of PCM

Properties	Requirements
Thermal	Suitable phase change temperature High latent heat of phase change Good heat transfer
Physical	Favorable phase equilibrium Low vapor pressure Small volume change High density
Kinetic	No supercooling Sufficient crystallization rate
Chemical	Long term chemical stability Compatibility with materials of construction Nontoxic Nonflammable No nuisance factor
Economics	Abundant Available Cost-effective
Sustainability	Recyclability Embodied energy

where ΔH_m (J/g) is the enthalpy of melting per unit mass, T_m (°C) is the melting temperature, $C_{p,s}$ (J/g °C) and $C_{p,l}$ (J/g °C) are heat capacity of PCM in solid and liquid phases, respectively.

PCMs can be categorized as given in Fig. 6. There are many PCMs that are being investigated by several researchers. These can be grouped according to application areas as follows:

- Buildings (Cabeza et al. 2011; Rathore et al. 2019)
- Free cooling in buildings (Soares et al. 2013; Stritih et al. 2018)
- Solar applications (Kenisarin and Mahkamov 2007; Pandey et al. 2018)
- Concentrated solar power (Medrano et al. 2010; Xu et al. 2015)
- Cold storage (Souayfane et al. 2016)
- Vehicles (Jankowski and McCluskey 2014; Jaguemont et al. 2018)

There are also commercial PCMs available on the market. The list of manufacturers and suppliers can be found in Bruno et al. (2015).

3 Themochemical Heat Storage

3.1 Chemical Reactions

Thermal energy storage in chemical reactions can be summarized simply as follows:

Table 3 TCM chemical reaction systems

Solid–gas reactions	Solid–liquid reactions	Gas–gas reactions
Carbonates Hydroxides Metal hydrides Metal oxides	Ammonium hydrogen sulphate Isopropanol/acetone/hydrogen system	$NH_3/N_2/H^2$ system $SO_3/O_2/SO_2$ system Methane reforming

$$\text{Charging: } AB + \Delta H \rightarrow A + B \tag{3}$$

$$\text{Recovery: } A + B \rightarrow AB + \Delta H \tag{4}$$

Here the reactant AB produces products A and B using reaction enthalpy ΔH, which is provided by a renewable or waste heat source. During this endothermic forward reaction (Eq. 3) the heat is stored in products A and B. When stored heat is needed, it is recovered with the exothermic reverse reaction (Eq. 4) when A and B reacts to produce AB and give the reaction heat, ΔH as output. These systems are based on storing heat as chemical product; therefore, there is almost no heat loss from the storage. Thermochemical heat storage materials (TCM) should be abundant, economic, have high reaction enthalpy with no side reactions. The reactions that can be used TCM applications are categorized as shown in Table 3 (Prasad et al. 2019). Wu et al. states that a general rule-of-thumb for energy density is highly dependent on the more reactive materials in the components.

3.1.1 Sorption Storage Systems

The sorption storage systems involve a sorbent and a second material. When the second material accumulates on the solid sorbent's surface, the process is called adsorption. In this case solid material is called adsorbent (e.g. zeolite) and second material is the adsorbate (e.g. water vapor). In the case, when it accumulates in the liquid sorbent, it is called absorption. These systems are classified as open and closed (Fig. 7). The open system is open to atmospheric conditions as seen on the right in Fig. 7.

The design of open systems may be simpler since they do not involve evaporator and condenser. With closed systems, it may be possible to reach higher temperatures. For open systems, mass and heat transfer conditions may be better (Krese et al. 2018).

3.2 Energy Storage for Electricity

Power generation from intermittent sources like renewables necessitates use of energy storage for reliable and qualified energy supply to users. Based on discharge

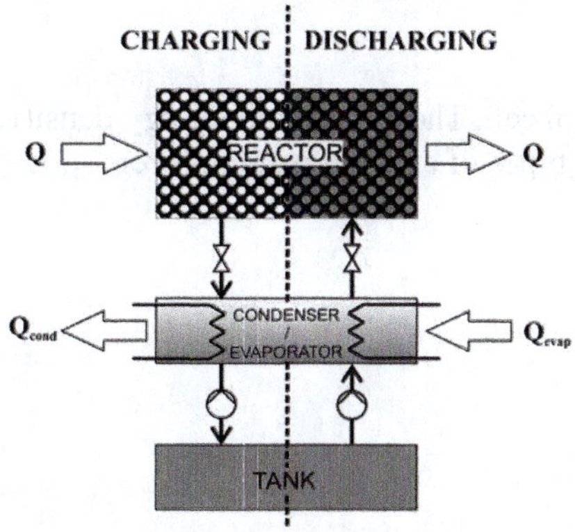

Fig. 7 Closed (left) and open (right) sorption storage systems (Krese et al. 2018)

time and power requirements there are various energy storage technologies as shown in Fig. 8. This Figure is divided into the following three regions depending on how energy storage for power systems and electricity can serve following services:

- Frequency and power quality and uninterrupted power supply (UPS)
- Transportation and grid support for load shifting and bridging power
- Energy management for bulk power management.

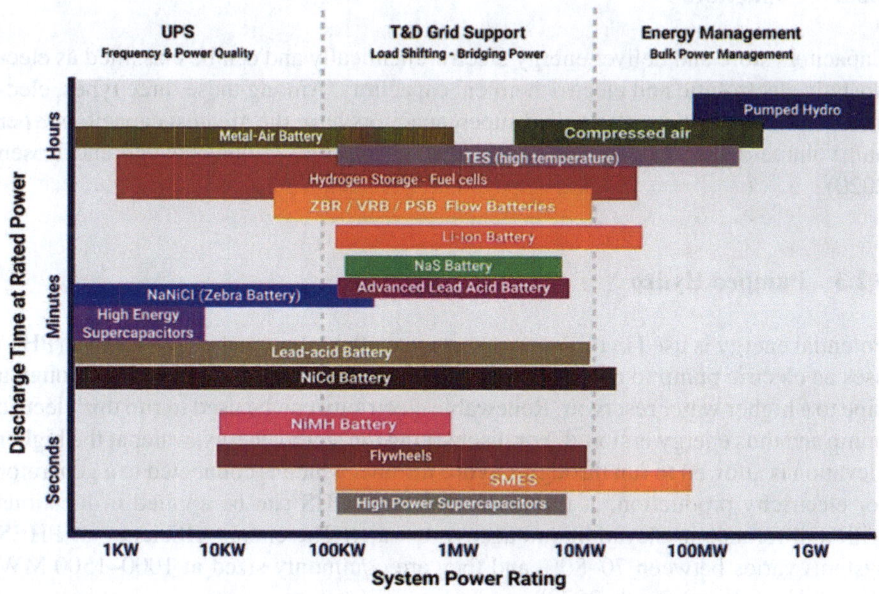

Fig. 8 Energy storage for power applications (Argyrou et al. 2018)

3.2.1 Batteries

Batteries store energy using an electrochemical cell. They offer high-energy-density, low-power-density technologies. Among the types of existing batteries are:

- Lithium- ion (Li-ion)
- Sodium-sulphur (NaS)
- Nickel-cadmium (NiCd)
- Lead acid (Pb-acid)
- Lead-carbon batteries
- Zebra batteries (Na-NiCl2)
- Flow batteries.

Depending on the power need and duration of usage, appropriate battery storage can be chosen. Among the various battery types, lithium batteries are becoming increasingly important in energy storage due to their high specific energy (energy per unit weight) and energy density (energy per unit volume) (Koohi-Fayegh and Rosen 2020). It is estimated that the world production of Li-ion batteries in 2017 was about 148 GWh/year with about 60% made in China (Ferg et al. 2019). They are mostly used for portable devices, electric vehicles, scooters and bikes. Thermal management of Li-ion batteries is extremely important due to the risk of fire, which has been seen in some consumer products (Ferg et al. 2019).

3.2.2 Capacitors

Capacitors store and deliver energy electro-chemically and can be classified as electrostatic electrolytic and electrochemical capacitors. Among these three types, electrochemical capacitors, also called supercapacitors have the greatest capacitance per unit volume due to having a porous electrode structure (Koohi-Fayegh and Rosen 2020).

3.2.3 Pumped Hydro

Potential energy is used in this energy storage method. Pumped hydro storage (PHS) uses an electric pump to move water from a water body at a low elevation through a pipe to a higher water reservoir. Renewable electricity can be used to run this electric pump and thus energy is stored. For discharging the stored energy, water at the higher elevation is allowed to run through a hydro turbine, which is connected to a generator for electricity production, at the lower elevation. PHS can be applied in locations with a difference in elevation and access to water. The energy efficiency of PHES systems varies between 70–80% and they are commonly sized at 1000–1500 MW (Koohi-Fayegh and Rosen 2020).

3.2.4 Compressed Air Energy Storage

This energy storage method uses mechanical energy to compress a gas—usually air. In the compressed air energy storage (CAES), renewable electricity can be used to compress the air and charge it to a tank or underground cavern. In order to discharge the storage system, the stored air is expanded in turbines again. Compared to other energy storage for electricity methods, CAES has higher energy storage efficiency, longer life cycle, large storage/release capacity and relatively low investment cost (Zhou et al. 2019).

3.2.5 Flywheel

This energy storage method uses kinetic energy. Electric motor or generator is used to drive the flywheel to rotate at a high speed so that the electrical power is transformed into mechanical power. Charging energy is input to the rotating mass of the flywheel and stored as kinetic energy. This stored energy can be released as electric energy on demand (Koohi-Fayegh and Rosen 2020). Flywheels are commonly used in vehicles for transportation.

3.2.6 Magnetic Storage

Energy can be stored in a magnetic field. Large superconducting magnets are capable of storing 1,000–10,000 MWh of electricity could be attractive as load-leveling devices for central power stations: Superconducting Magnetic Energy Storage (SMES). Smaller magnets with storage capacities in the 10-kWh range may be cost-effective in smoothing out transmission line loads. The potential for highly efficient storage is especially attractive for utilities, particularly when energy costs increase (Koohi-Fayegh and Rosen 2020).

3.3 Hydrogen as Energy Carrier

Renewable energy can also be used for production of hydrogen (through electrolysis of water) to use it as an energy carrier. Using hydrogen produced from renewable source in a fuel cell to generate power. The emission from this operation is only water without any adverse effects to the environment. This makes hydrogen a clean alternative that can be used in almost any sector where fuel cells can be integrated. The following sectors are already using hydrogen:

- Vehicles and aircrafts
- Buildings—electricity and heating

- Space and military applications
- Stationary and portable power systems.

Hydrogen can also be produced by photochemical catalytic reaction. This process is also independent of fossil fuels and can make use of solar energy during the photoreaction.

4 Roles of Energy Storage

Transition from traditional energy system to 100% renewable energy society requires new approaches to manage the new energy system. In traditional system energy is generated by central power plants and transmitted through the grid to consumers. In this system, there may also be a second grid (pipelines) for distributing natural gas from the source to the consumers. District heating and cooling that are run by central power plants may also be included in the traditional system in some countries. The new system, in addition to existing power plants has renewable power generation from several producers—large and small. Some of these may be located on the consumer side, who can produce their own renewable electricity (e.g. through PVs on their roofs), consume part of the production and sell the extra to the grid. These are called "prosumers". There may be larger renewable electricity producers that are connected directly to the grid. With the percentage of renewable energies reaching 100, new challenges and also new functions for the grids are expected. The amount of fluctuating energy leads to a requirement of more flexibility and storage capacity. In addition, with the increase in urbanization trend the demand itself may vary extremely. Energy storage is one of the key technologies to meet these challenges.

4.1 *Optimizing Renewable Energy Use*

The mismatch between supply and demand causes renewables whenever it is available and if there is a need for it at that time. Considering sun not being available 24 h and 365 days or wind blowing at the required speed all the time, there are a lot of worries about continuity of renewable energy production. Energy storage—electrical and thermal—can close this mismatch. Energy storage has the role of making the excess renewable energy production, which would otherwise be wasted, accessible for later use when renewables are not available. This optimizes the use of renewables and increases amount of renewable energy produced.

4.2 Grid Stability

Renewable energy entering the grid at various locations by diverse producers at varying times and capacities leads to instabilities in the grid. Fluctuating nature of renewables leads to steep downward or upward ramps, which can alter directions quickly. The grid should be able to react quickly to these changes and start with short notice from zero power to demand level. In addition, power generation may start and stop several times per day. The role of energy storage in this case is to manage the grid by storing during peak hours (load leveling) and provide fast response to demand when renewables are not available. Currently, in countries like Germany, where share of renewables in energy production is higher, grid instability problems are seen. One example is the negative renewable electricity prices that occurred due to a combination of high wind and solar production and low energy consumption (due to time being a Sunday).

4.3 Flexibility

The services provided to end-users through electricity, heat and cold networks are operated separately. If demand for services like power, heating, cooling, hot water etc. are judged together, the supply side can be more controllable. This makes energy system more flexible and increases the value of renewables. All types of storages have to be taken into account to find the optimum in a given supply and demand situation. This means that even thermal energy storages are suitable for balancing the net. One example is the use of cold storages for decreasing the installed cooling capacity in buildings in summer. This helps to avoid blackouts as the electricity peak demand decreases. But even transforming surplus electrical energy and storing it e.g. in decentralized latent storages for refrigeration applications in a way that they have no electricity demand when the total demand exceeds the supply, may be energetically and economically efficient solutions. There are also cross-sectoral applications of energy storage. Some of the examples are:

- Power ←→ Buildings ←→ Industry
- Power ←→ Buildings ←→ Transportation
- Power ←→ Buildings ←→ Agriculture
- Agriculture ←→ Industry.

In such applications, surplus energy produced in one sector can be used as supply in another sector. Energy storage closes the gap between these sectors in terms of distance and time. Surplus energy from one sector can also be transformed into other energy forms before it is used in another sector. Some of these flexible concepts are called:

- Power-to-Heat
- Power-to-X
- Vehicle-to Grid (V2G).

Power-to-Heat concept is used for power generated from renewables to be used to produce heat before it is provided to the customers. Power-to-X is used for concepts where X stands for any chemical (gas or liquid) that can be produced using renewable power. This can be H_2 that is produced by photocatalysis or by solar thermal. X can also be another chemical such as CO_2, which can be converted to methanol by using renewable power. In V2G concept plug in hybrid electric vehicles can provide storage to match supply with demand.

4.4 Increasing Energy Efficiency

Energy storage can be integrated in different points in the energy value chain to increase energy efficiency. These are:

- Generation: Power, Heat and cold production
- Distribution: Power grids, district heating and cooling networks
- End-users: Cost-effectiveness, security, controlled energy use, independence.

Energy storage methods that can be used for the options given in Fig. 1 are shown in the matrix in Table 4. On the generation side of energy value chain bulk energy storage methods are needed. For power generation from solar and wind, PHS, CAES and SMES shown in the leftmost region of Fig. 8 are recommended. For solar thermal power plant—Concentrated Solar Power (CSP)—thermal energy storage at high temperature is needed. For this purpose, sensible heat storage materials are used and PCM and TCM are under development. PV-T is a new concept developed where both power and hot water can be produced from PV cells equipped with a cooling loop. This provides more efficient operation of PV by lowering its operation

Table 4 Matrix for energy storage technologies for different option in renewable energy value chain

		Energy storage	
		Electrical	Thermal
Generation			
Power	Wind, Solar-PV	PHS, CAES, SMES	PCM (for PV-T)
	Solar CSP		Sensible (e.g.molten salt), PCM, TCM
Heat and cold			Sensible, PCM (e.g.ice sorage)
Distribution			
Power grid		Flywheels, supercapacitors, SMEs	
Heating/cooling network			UTES, PCM, TCM
End-users		Batteries	Sensible, PCM, TCM

temperature. In addition, heat removed is transferred to hot water cycle to support hot water demand of the building.

On the distribution side, for the electricity grid flywheels, supercapacitors, SMEs, and batteries can be used to shift load. For heating and cooling networks, load shifting can be realized with thermal energy storage methods. Chilled water tanks or different ice storage methods are used for cooling networks. For heating networks, UTES systems and water tanks are preferred. For the end-users, there are several ways of using energy storage to increase energy efficiency. These include incorporation of energy storage in heating and cooling systems, appliances, building materials and elements for more efficient renewable energy use.

4.5 Urban Heat Island Effect

The number of metropolitan areas is increasing with increasing urbanization trend. The high-rise buildings dominating the skylight of the metropolitan areas are providing shelters to increasing population and business activities. These buildings radiate heat that they have absorbed and cause urban heat island effect. This effect increases temperature level of the metropolitan areas in summer. This causes cooling demands to increase even higher than expected, bringing additional burden to the electricity grid and to the consumers.

The urban heat island effect can be controlled by several methods. Using latent heat storage in PCM roof technology showed 40% better performance than the alternative cool roof technology (Roman et al. 2016).

5 Applications of Energy Storage in Transition to 100% Renewables

According to Renewables 2019 Global Status Report (REN21 2019), at the end of 2018, 100 cities in worldwide have used at least 70% of their energy demands from renewable sources. More than 40 of them are 100% renewable cities. TES systems are crucial for heating and cooling systems in renewable cities to ensure balance between energy supply and demand (Guelpa and Verda 2019). When the demand exceeds the consumption, thermal energy storage systems provide uninterrupted energy to buildings and industry by charging heat and cold. USA, Spain, Denmark and Germany are leading the way in the transition to %100 renewable energy (REN21 2019). There are many applications of 100% Renewable Energy cities for both short-term and long-term storage systems. Possible integration of thermal energy storage technologies in urban areas can be done according to following categories:

- Passive short-term storage: Using the building's components for thermal energy storage in form of sensible (Thieblemont et al. 2016) or latent (Bastani et al. 2015)

- Active short-term storage: Water tanks with or without PCMs (thermal sensible/latent), ice storages (thermal latent), batteries (electrochemical), flywheels (mechanical), super-capacitors (electrochemical), compressed air energy storage (mechanical), hydrogen (chemical)
- Active seasonal storage: Typically using underground thermal energy storage (UTES) (thermal sensible) or thermochemical.

5.1 Sensible Heat Storage Applications

Denmark is the only country with a target of 100% renewable in total final energy consumption by 2050 (REN21 2019) The PTES system with 75,000 m^3 storage volume in New Marstal in Denmark is the largest UTES system in Europe. The system provides 7500 MWh of solar heat for space heating (Bespalko et al. 2018).

Cities also apply district heating and cooling systems to achieve their future renewable goals. According to IRENA 2016 report (REN21 2016), around 2000 ATES systems have been installed in cities in the Netherlands. It is expected that ATES applications will increase 10 times in next ten year.

Sweden is also leading country in UTES applications for buildings and districts. Total energy delivered as heat and/or cold with these storage units was 23,000 GWh of which 700 GWh is installed in 2016 (Gehlin et al. 2018a, b). One of the first 100% renewable energy applications for urban area was realized in Bo01 project in Malmö, Sweden (Andersson 2003).

District heating and cooling systems are widespread in Germany. Figure 9 shows a central solar heating plant with seasonal TES (CSHPSS), which is funded by Solarthermie-2000 program in Hamburg, Germany. CSHP plant with 3000 m^2 collector includes 4500 m^3 water-filled concrete tank. The system supplies 50% of heat demand for space heating and domestic hot water for 124 terraced single-family houses (Schmidt et al. 2004).

Another CSHPSS system was installed in Friedrichshafen, Germany in 1996. 12,000 m^3 water storage tank supply heat for 570 apartments (Schmidt et al. 2004). Design configuration of both seasonal storage systems in Hamburg and Friedrichshafen is given in Fig. 10.

In another solar heating project, 2750 m^3 of storage tank filled with water was built in Hannover, Germany. The hot water storage tank provides heat for space heating of 106 apartments (Bespalko et al. 2018). Julih Solar Tower project was built in Julih, Germany, with a design power of 1, 5 MWe. The system has ceramic heat sink for short-term (1, 5 h) STES (solarpaces.nrel.gov).

In Am Ackerman-Bogen solar district heating project 50% of annual heat demand of 320 apartments in Munih is covered by solar energy (See Fig. 11). Solar heat is provided from flat plate collectors with 2700 m^2 surface area and stored in a 6000 m^3 underground seasonal hot storage (IEA-ETSAP and IRENA 2013).

Fig. 9 Central solar heating plant with seasonal heat storage in Hamburg, Germany (Schmidt et al. 2004)

In Eggenstein-Leopoldshafen, 1600 m^2 of flat plate collectors provide solar heat for 1200 m^2 area consisting of gym, public swimming pool, school and buildings. The system shown in Fig. 12 consists of 4500 m^3 PTES tank filled with water/gravel mixture (Ochs et al. 2008).

In Saint Paul Minnesota, short-term storage system was built to cool downtown office buildings. Cold water is stored in 15,000 m^3 water storage tank during nights and distributed during day (www.bwbr.com).

Concentrated solar power plants (CSP) to produce solar thermal power providing electricity to districts can have significant share in transition to 100% renewable era. The world's first commercial CSP tower PS10 project is in Seville, Spain. PS10 generates 24.3 GWh/yr of solar power. Four pressurized water tanks (called Ruth system) filled with ceramic alumina bed condensate steam from the solar field are used for short term storage (50 min). 600 m^3 of Ruth system has 20 MWh storage capacity with 92.4% efficiency (www.peacelink.it, STS-med 2018). PS20 is another CSP plant in Seville, near PS10. The system is designed for a capacity of 20 MW. Excess heat from PS20 is stored in a short term (1 h) storage system. By this way, PS20 supplies energy for 10000 houses (STS-med 2018).

Andasol 50 is a 50 MW solar power plant with 9 full-load hours of storage system in Granada, Spain. Synthetic oil is heated up to 400 °C trough collectors. Heat is stored in 2-tank molten salt storage system. Storage capacity is 1,350 MWh. Andasol 50 Project decreases 72,000 tons/year of CO_2 emissions in Spain (Aringhoff et al. 2002). STES system integration in Andasol 50 project is given by Fig. 13.

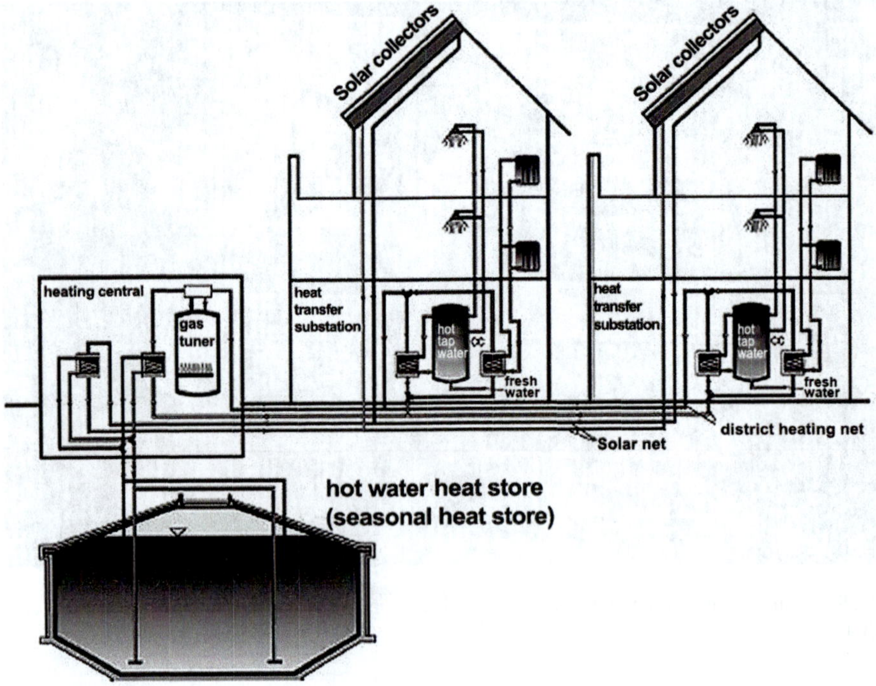

Fig. 10 Design configuration of a CSHPSS in Hamburg and Friedrichshafen, Germany (Schmidt et al. 2004)

Fig. 11 Large hot water storage (construction and final state) combined with solar thermal district heating "Am Ackerman-Bogen" in Munih Germany (IEA-ETSAP and IRENA 2013)

Crescent Dunes Solar Energy Project is a CSP technology in Nevada desert. Since 2015, the Project generates app. 500,000 MWh/yr electricity. 2-tank direct molten salt storage system is running for uninterrupted solar power with 1,100 MWh storage capacity. 10 h storage system has 99% efficiency.

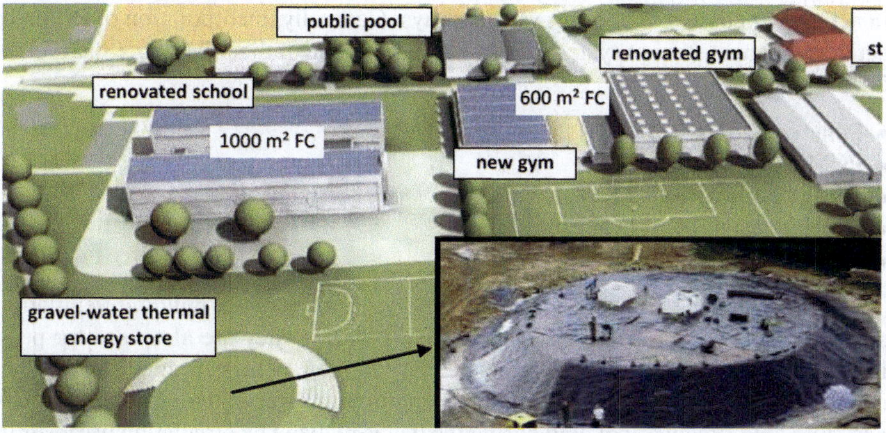

Fig. 12 Seasonal storage in Eggenstein-Leopoldshafen, Germany (Ochs et al. 2008)

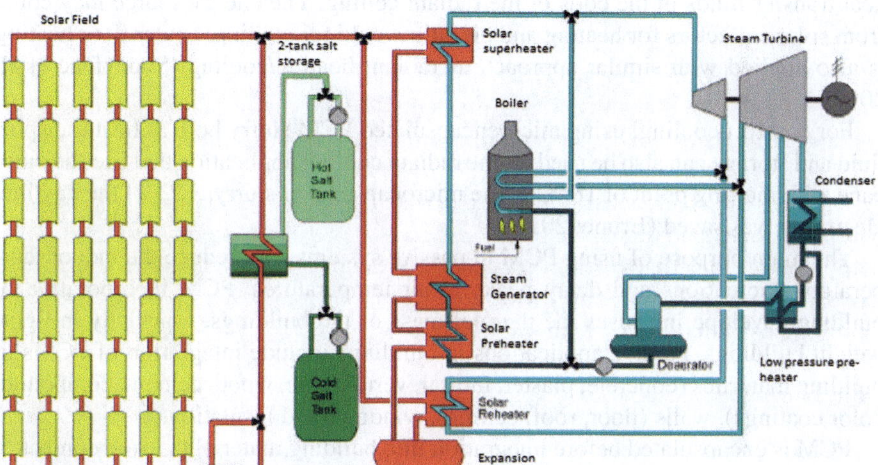

Fig. 13 System diagram of Andasol power plant with 2-tank molten salt storage system (Aringhoff et al. 2002)

5.2 Latent Heat Storage Applications

Significant increase in number of articles on latent heat building applications in literature (324% in the last decade) shows the growing interest and applications in these systems. The main reason for this is using phase change material (PCM) can decrease the annual energy demand of cooling and heating of the buildings by 50% in a more compact volume (Drissi 2019). This is especially important in buildings with limited space available for storage. Latent heat storage in PCMs can be used in active and passive systems in various ways. Generally, incorporation of PCMs in active latent heat storage systems can be done with one of the following (Sharif et al. 2015):

- Solar collectors with PCMs
- PCM storage tanks
- Packed beds with PCM as packing
- Special design PCM heat exchangers
- Ventilation duct networks.

PCM to air heat exchangers can be integrated directly in building ventilation system for heating process (Tyagi and Buddhi 2007). They are also used for peak load shaving by placing PCM composite slabs in the radiant ceiling or floor. "Free cooling" systems use cold ambient air during night to freeze the PCM in the slabs and store cold. Recovery of cold during the day is realized by circulating warm air in the building over the slabs to melt the PCM. These systems can only operate when there is a temperature difference between day and night. Water or air can be used as heat transfer fluids in the coils of the radiant ceiling. The energy source may come from solar collectors for heating and nighttime cold for cooling. Under floor heating is also applied with similar approach as radiant heating/cooling (Souayfane et al. 2016).

For comfort cooling, using microencapsulated PCM slurry both as heat transport fluid and storage can also be used. In the radiant cooling application that uses hexadecane with melting point of 18 °C in the microcapsules of slurry, 33% of the daytime electricity was saved (Brunet 2013).

The main purpose of using PCM in passive systems is to reduce the indoor temperature fluctuations and delay peaks in air temperatures. PCM incorporation in building envelope improves the thermal mass of the buildings, especially in lightweight buildings. Passive applications in buildings include integration of PCMs in building materials (concrete, plaster, mortar, vermiculite, wood, cement, compound, color coatings), walls (floor, roof, ceiling), windows and insulation.

PCM is encapsulated before integration into building material to avoid contact of PCM with the components of the material and to avoid changing its properties, to prevent leakage of PCM through the porous structure of the building material and to enhance heat transfer by increasing surface area. The required properties and types of PCM can vary based on climate conditions and building component where it will be incorporated i.e. roof, wall, floor etc. For example, the energy saving in a typical Mediterranean climate can be approximately 23% for heating and 37% for cooling (Panayiotou et al. 2016) using PCM. In cold climate, just heating demand can be decreased by 15% (Athienitis et al. 1997).

Cellat et al. (2019), Ravikumar and Srinivasa (2012), Zhu et al. (2018) and Song et al. (2018) added PCM in the building envelope (wall, roof, floors) and building systems (heat pump, ventilation systems etc.). Cellat et al., investigated the energy saving effect encapsulated PCM mixtures incorporated in the mortar of their prefabricated walls of test cabins. Their results showed that energy saving can reach up to 13% (Cellat et al. 2019).

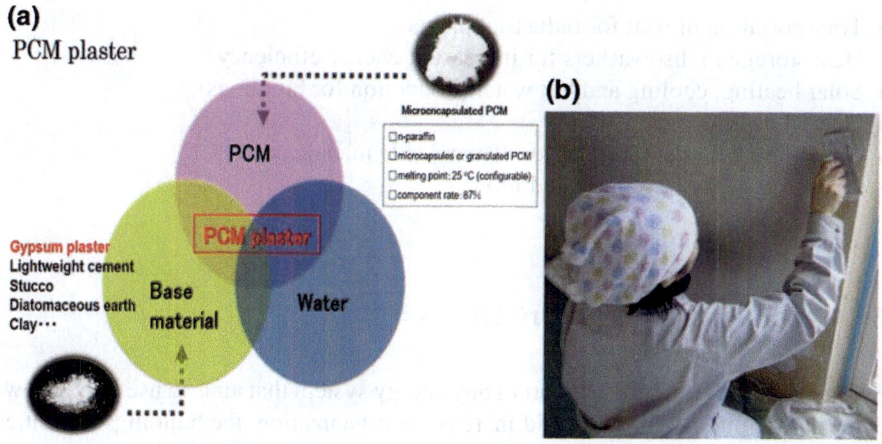

Fig. 14 **a** Mixture of PCM plaster, **b** Applying of PCM plaster to the wall (Kusama and Ishidoya 2017)

In another study, Kusama and Ishidoya developed PCM plaster and applied it in ceiling and walls at their test cell (Fig. 14). Heating demand could be decreased by 35% with this application (Kusama and Ishidoya 2017).

Ice is the best PCM to use near 0 °C. It is also one of the first PCMs used in cooling applications. Low cost and high-energy storage are among the advantages of ice storage in cooling applications. However, ice storage systems have disadvantages such as requiring chillers capable of energy production below freezing point of water and large volume expansion when solidified (Ryu et al. 1991). Ice storage is mainly used for cooling purposes in places where electric energy is more plentiful and less expensive. In countries where multiple tariff electricity pricing is used, nighttime cheaper electricity is used to produce ice with chillers. Technologies used in ice storage are ice on coil (external or internal melt), ice slurry (harvested ice) and encapsulated ice.

One of the most important benefits of ice storage system is the reduction of electricity demand at times of heavy electricity use and the shift of energy usage from non-peak hours. In the United States, it is estimated that compared to the current use, 40% less energy production and transmission lines will be needed, if ice storage is implemented in all the buildings in the US that require air conditioning (McCracken 2006). In Japan ice storage applications increase energy efficiency by 15%, and reduce CO_2 emissions by 20% (Sakai 2000).

5.3 Thermochemical Storage Applications

Thermochemical storage can be used for short term and seasonal storage purposes. Current applications include:

- Transportation of heat for industrial plants
- Heat storage in dishwashers for increasing energy efficiency
- Solar heating, cooling and hot water production for buildings
- District heating
- CHP plant with heat storage for demand side management
- Cold storage for beverages and food.

6 Conclusions and Future Outlook

Energy storage is a significant part of any energy system that aims to use only renewables in the future. With the rapid increase in urbanization, the buildings and cities have an increasing share in global energy consumption. For more ubiquitous applications of energy storage technologies, different stakeholders should work together to realize transition to 100% renewable energy era. New systems thinking is needed to manage future energy systems that are transforming into more complex structures. Energy storage can play the roles of optimizing renewable energy, increasing energy efficiency and providing diverse flexible options that also stabilize the grid. For rapid introduction of energy storage efforts to increase public awareness and acceptance are indispensable. Well-organized demonstration projects that show the various benefits of energy storage can draw the attention of stakeholders. Training activities to increase skilled technical staff and education of especially young researchers are needed. Standards and guidelines for proper installation of these technologies have to be developed to avoid any failures that will prevent further market development. Clear policy measures are needed for energy storage to take position in the 100% renewable energy era. For more economic viability of energy storage applications can be realized with new planning principles for energy investments and suitable political framework.

References

Agalit H, Zari N, Maaroufi M (2017) Thermophysical and chemical characterization of induction furnace slags for high temperature thermal energy storage in solar power plants. Sol Energy Mater Sol Cells 172:168–176

Allen KG (2010) Performence characteristics of packed bed thermal energy storage for solar thermal power plants, University of Stellenbosch Department of Mechanical and Mechatronic Engineering, Master of Science Thesis

Alva G, Liu L, Huang X, Fang G (2017) Thermal energy storage materials and systems for solar energy applications. Renew Sustain Energ Rev 68:693–706

Alva G, Lin Y, Fang G (2018) An overview of thermal energy storage systems. Energy 144:341–378

Argyrou MC, Christodoulides P, Kalogirou SA (2018) Energy storage for electricity generation and related processes: technologies appraisal and grid scale applications. Renew Sustain Energy Rev 94:804–821

Aringhoff R, Geyer M, Herrmann U, Kistner R, Nava P, Osuna R (2002) AndaSol-50 MW solar plants with 9 hour storage for Southern spain. In: Steinfeld A (ed) Proceedings of 11th solar PACES international symposium on concentrated solar power and chemical energy technologies. 4–6 Sept 2002, Zurich, Switzerland, pp 37–42. ISBN: 39521409-3-7
Ataer OE (2006) Storage of thermal energy, in energy storage systems in Encyclopedia of Life Support Systems (EOLSS), Developed under the Auspices of the UNESCO. Eolss Publishers, Oxford
Athienitis AK, Liu C, Hawes D, Banu D, Feldman D (1997) Investigation of the thermal performance of a passive solar test-room with wall latent heat storage. Build Environ 32(5):405–410
Bruch A, Fourmigue JF, Coutirier R (2014) Experimental and numerical investigation of a pilot-scale thermal oil packed bed thermal storage system for CSP power plant. Solar Enery 105:116–125
Brunet Y (2013) Energy storage: applications to the electricity vector. Low Emission Power Gener Technol Energy Manage, 1–49
Bruno F, Belusko M, Liu M, Tay NHS (2015) Using solid-liquid phase change materials (PCMs) in thermal energy storage systems. In: Advances in thermal energy storage systems. Woodhead Publishing, pp 201–246
Cabeza LF, Castell A, Barreneche C, de Gracia A, Fernández AI (2011) Materials used as PCM in thermal energy storage in buildings: a review. Renew Sustain Energy Rev 15(3):1675–1695
Cascetta M, Cau G, Puddu P, Serra F (2015) Experimental investigation of a packed bed thermal energy storage system. In: Journal of Physics: Conference Series, vol. 655, no. 1. IOP Publishing, p 012018
Cellat K, Beyhan B, Konuklu Y, Dündar C, Karahan O, Güngör C, Paksoy H (2019) 2 years of monitoring results from passive solar energy storage in test cabins with phase change materials. Solar Energy
Dincer I (2002) Thermal energy storage systems as a key technology in energy conservation. Int J Energy Res 26(7):567–588
Dincer I, Rosen MA (2010) Thermal energy storage: systems and applications, 2nd edn. Wiley, UK. ISBN: 978-0-470-74706-3
Drissi S, Ling TC, Mo KH, Eddhahak A (2019) A review of microencapsulated and composite phase change materials: alteration of strength and thermal properties of cement-based materials. Renew Sustain Energy Rev 110:467–484
Ferg EE, Schuldt F, Schmidt J (2019) The challenges of a Li-ion starter lighting and ignition battery: a review from cradle to grave. J Power Sources 423:380–403
Fernández AI, Barreneche C, Miró L, Brückner S, Cabeza LF (2015) Thermal energy storage (TES) systems using heat from waste. In: Advances in thermal energy storage systems. Woodhead Publishing, pp 479–492
González Roubaud E, Pérez Osorio D, Prieto C (2017) Review of commercial thermal energy storage in concentrated solar power plants: steam versus molten salts. Renew Sustain Energy Rev 80:133–148
Grosu Y, Fernández I, Fernández L, Nithiyanantham U, Baba Y, Mers A, Faik A (2018a) Natural and by-product materials for thermocline-based thermal energy storage system at CSP plant: structural and thermophysical properties. Appl Therm Eng 136:185–193
Grosu Y, Ortega-Fernández I, López del Amo JM, Faik A (2018b) Natural and by-product materials for thermocline-based thermal energy storage system at CSP plant: compatibility with mineral oil and molten nitrate salt. Appl Therm Eng 136:657–665
Guelpa E, Verda V (2019) Thermal energy storage in district heating and cooling systems: a review. Appl Energy 252:113474
IEA-ETSAP and IRENA (2013) https://www.irena.org/publications/2013/Jan/IRENA-IEA-ETSAP-Technology-Briefs, https://iea-etsap.org/
Jankowski NR, McCluskey FP (2014) A review of phase change materials for vehicle component thermal buffering. Appl Energy 113:1525–1561
Jaguemont J, Omar N, Van den Bossche P, Mierlo J (2018) Phase-change materials (PCM) for automotive applications: a review. Appl Therm Eng 132:308–320

Kenisarin M, Mahkamov K (2007) Solar energy storage using phase change materials. Renew Sustain Energy Rev 11(9):1913–1965

Koçak B, Paksoy H (2018) Sensible thermal energy storage in packed bed for industrial solar applications. In: Häberle A (ed) Proceedings of the ISES EuroSun 2018 conference—12th international conference on solar energy for buildings and industry, 10–13 Sept 2018

Koçak B, Paksoy H (2019) Using demolition wastes from urban regeneration as sensible thermal energy storage material. Int J Energy Res 1–7. https://doi.org/10.1002/er.4471

Konuklu Y, Sahan N, Paksoy H (2018) Comprehensive energy systems, Dicer I (ed) Elsevier, Jan 2018

Koohi-Fayegh S, Rosen MA (2020) A review of energy storage types, applications and recent developments. J Energy Storage 27:101047

Krese G, Butala V, Stritih U (2018) Thermochemical seasonal solar energy storage for heating and cooling of buildings. Energy Buildings 164:239–253

Kusama Y, Ishidoya Y (2017) Thermal effects of a novel phase change material (PCM) plaster under different insulation and heating scenarios. Energy Build 141:226–237

Kuznik F, Johannes K, David D (2015) Integrating phase change materials (PCMs) in thermal energy storage systems for buildings. In: Cabeza LF (ed) Advances in thermal energy storage systems. Woodhead Publishing, Methods and Applications, pp 325–347

MacCracken M (2006) Ice thermal storage and LEED Gold. In: ECOSTOCK, 10th international thermal energy storage conference, New Jersey, USA, 31 May–2 June

Mangold D, Deschaintre L (2019) Task 45 large systems, seasonal thermal energy storage, SHC. http://www.solarthermalworld.org/sites/gstec/files/news/file/2016-07-27/task45_b_saisenal_storages.pdf

Medrano M, Gil A, Martorell I, Potau X, Cabeza LF (2010) State of the art on high-temperature thermal energy storage for power generation. Part 2—case studies. Renew Sustain Energy Rev 14(1):56–72

Miró L, Navarro ME, Suresh P, Gil A, Fernández AI, Cabeza LF (2014) Experimental characterization of a solid industrial by-product as material for high temperature sensible thermal energy storage (TES). Appl Energy 113:1261–1268

Motte F, Falcoz Q, Veron E, Py X (2015) Compatibility tests between solar salt and thermal storage ceramics from inorganic industrial wastes. Appl Energy 155:14–22

Nash AL, Badithela A, Jain N (2017) Dynamic modeling of a sensible thermal energy storage tank with an immersed coil heat exchanger under three operation modes. Appl Energy 195:877–889

Navarro ME, Martı´nez M, Gil A, Fernandez AI, Cabeza LF, Olives R, PY X (2012) Selection and characterization of recycled materials for sensible thermal energy storage. Solar Energy Mater Solar Cells 107:131–135

Ochs F, Nußbicker-Lux J, Marx R, Koch H, Heidemann W, Muller-Steinhagen H (2008) Solar assisted district heating system with seasonal thermal energy in Eggenstein-Leopoldshafen, Eurosun 2008, Lisboa

Paksoy H, Evliya H, Bozdağ S, Mazman M, Konuklu Y, Turgut B, Gök Ö, Yılmaz M, Yılmaz S, Beyhan B (2009) CO2 mitigation with thermal energy storage. Int J Glob Warm 1(1/2/3)

Panayiotou GP, Kalogirou SA, Tassou SA (2016) Evaluation of the application of Phase Change Materials (PCM) on the envelope of a typical dwelling in the Mediterranean region. Renew Energy 97:24–32

Pandey AK, Tyagi HVV, Rahim NA, Selvaraj JAL, Sari A (2018) Novel approaches and recent developments on potential applications of phase change materials in solar energy. Renew Sustain Energy Rev 82, 281–323

Prasad JS, Muthukumar P, Desai F, Basu DN, Rahman MM (2019) A critical review of high-temperature reversible thermochemical energy storage systems. Appl Energy 254:113733

Py X, Calvet N, Olivès R, Echegut P, Bessada C, Jay F (2009) Thermal storage for solar power plants based on low cost recycled material. In: EFFSTOCK, the 11th international conference on thermal energy storage, 14–17 June 2009, Stockholm, Sweden

Py X, Calvet N, Olives R, Meffre A, Echegut P, Bessada C, Veron E, Ory S (2011) Recycled material for sensible heat based thermal energy storage to be used in concentrated solar thermal power plants. J Sol Energy Eng 133:031008
Rathore PKS, Shukla SK, Gupta NK (2019) Potential of microencapsulated PCM for energy savings in buildings: a critical review. Sustain Cities Soc 225:723–744
Ravikumar M, Srinivasan PSS (2012) Analysis of heat transfer across phase change material filled reinforced cement concrete roof for thermal management. Proc Inst Mech Eng, Part C: J Mech Eng Sci 226(12):2933–2940
Ryu HW, Hong SA, Shin BC, Kim SD (1991) Heat transfer characteristics of cool-thermal storage systems. Energy 16:727–737
Sakai H (2000) Application of ice storage system to cold air distribution system. In: IEA ECES Annex 14 cooling in all climates with thermal energy storage workshop, Tokyo, Japan, 10 Nov
Schlipf D, Schicktanz P, Maier H, Schneider G (2015) Using sand and other small grained materials as heat storage medium in a packed bed HTTESS. Energy Procedia 69:1029–1038
Schmidt T, Mangold D, Muller-Steinhage H (2004) Central solar heating plants with seasonal storage in Germany. Sol Energy 76:165–174
Sharif MA, Al-Abidi AA, Mat S, Sopian K, Ruslan MH, Sulaiman MY, Rosli MAM (2015) Review of the application of phase change material for heating and domestic hot water systems. Renew Sustain Energy Rev 42:557–568
Soares N, Costa JJ, Gaspar AR, Santos P (2013) Review of passive PCM latent heat thermal energy storage systems towards buildings' energy efficiency. Energy Buildings 59:82–103
Song M, Niu F, Mao N, Hu Y, Deng S (2018) Review on building energy performance improvement using phase change materials. Energy Build 158:776–793
Souayfane F, Fardoun F, Biwole PH (2016) Phase change materials (PCM) for cooling applications in buildings: a review. Energy Buildings 129:396–431
Stritih U, Charvat P, Klimes L, Osterman E, Ostry M, Butala V (2018) PCM thermal energy storage in solar heating of ventilation air—experimental and numerical investigations. Sustain Cities Soc 37:104–115
STS-med (2018) http://www.stsmed.eu/
Sun X, Gou Z, Lau SSY (2018) Cost-effectiveness of active and passive design strategies for existing building retrofits in tropical climate: case study of a zero-energy building. J Clean Prod 183:35–45
Tyagi VV, Buddhi D (2007) PCM thermal storage in buildings: A state of art. Renew Sustain Energy Rev 11(6):1146–1166
Xu B, Li P, Chan C (2015) Application of phase change materials for thermal energy storage in concentrated solar thermal power plants: a review to recent developments. Appl Energy 160:286–307
Zhou Q, Du D, Lu C, He Q, Liu W (2019) A review of thermal energy storage in compressed air energy storage system. Energy 188:115993
Zhu N, Li S, Hu P, Wei S, Deng R, Lei F (2018) A review on applications of shape-stabilized phase change materials embedded in building enclosure in recent ten years. Sustain Cities Soc 43:251–264

Efficient Use of Energy in Buildings—New Smart Trends

Hasan Heperkan, Büşra Selenay Önal and Tanay Sıdkı Uyar

Abstract Depletion of the fossil fuels which provide most of the world's energy demand today, increase in the consumption in industrial processes and the rise in the concentration of greenhouse gases in the atmosphere to dangerous levels to human health as a result of the destruction of the ozone layer, have put energy supply and efficient use of energy among the most challenging topics today. Energy saving and the efficient use of energy are becoming more and more important. World states and international organizations are competing with each other to access energy resources (oil, natural gas, coal…). Energy is the infrastructure of industrialization and an indispensable element of daily life. Therefore, the need for energy occupies an important place in the national and international agenda. Due to the depletion of energy resources, the existence of external dependence and environmental impacts; producing safe, sufficient amounts of cheap and clean energy for countries is among the main problems of economic and social life today. Our country is growing rapidly with its industry, economy and population. For this reason, it is of great importance to use the generated energy with high efficiency and to evaluate the potential of alternative and renewable energy sources as well as the existing energy sources. The key issue to a sustainable development is the balance between the supply and demand of energy, keeping the environment clean, healthy and pollutant free. Energy efficiency is the reduction of energy consumption per unit or product amount without causing a decrease in the quality of life and quality of services in buildings and production quality and quantity in industrial enterprises. Buildings are responsible for about a third of the global energy consumption. Especially in developing countries like Turkey, during the rapid urban transformation in renovated buildings, energy performance

H. Heperkan (✉) · B. S. Önal
Engineering Faculty, Istanbul Aydın University, Beşyol Mevkii, K.Çekmece, 34349 Istanbul, Turkey
e-mail: hasanheperkan@aydin.edu.tr

B. S. Önal
e-mail: bselenayonal@aydin.edu.tr

T. S. Uyar (✉)
Engineering Faculty, Marmara University, Istanbul, Turkey
e-mail: tanayuyar@marmara.edu.tr

T. S. Uyar (ed.), *Accelerating the Transition to a 100% Renewable Energy Era*, Lecture Notes in Energy 74, https://doi.org/10.1007/978-3-030-40738-4_20

should also be given importance as well as earthquake-resistance. Energy performance of buildings should be sustainable. In this context, smart buildings represent an important stage of energy efficiency and constitute the basic element of the smart micro grid. Large amounts of data must be stored and processed during the planning, implementation, control and correction cycle of air conditioning and mechanical systems. To ensure energy efficiency in buildings, a network structure consisting of wireless sensors and micro control elements is used. This chapter focuses on the new trends in smart buildings.

Keywords Energy · Efficiency · Energy performance of buildings and industry · Energy performance software

1 Introduction

Energy is required to meet the needs of people and to maintain a healthy development. It is used especially in the industry, building and transportation sectors. Today, producing safe, sufficient amounts of cheap and clean energy for countries is among the main problems of economic and social life. Our country is growing rapidly with its industry, economy and population. Therefore, it is of utmost importance to utilize the generated energy with high efficiency, to control the existing energy sources and to evaluate the potential of alternative and renewable energy sources.

Our atmosphere includes carbon dioxide, methane, water vapor, ozone, nitrogen oxide and so on, besides oxygen and nitrogen. Thanks to these gases, it sends some of the sun rays reflected from the earth surface back to the earth. With the help of the greenhouse gases that function as a blanket, the average temperature on earth captures a value of 15 °C, which allows humans, animals and plants to survive. Without greenhouse gases, the average temperature of the earth would be around −18 °C. This natural effect of the greenhouse gases is called the "greenhouse gas effect".

The concentration of greenhouse gases in the atmosphere started to increase after the industrial revolution that started in the 1750s, the carbon dioxide concentration increased by 40% and reached from 280 ppm to 394 ppm. According to the Intergovernmental Panel on Climate Change (IPCC), the increase in carbon dioxide is primarily due to the use of fossil fuels. The second notable factor is the change in land use, particularly deforestation. The Intergovernmental Panel on Climate Change has shown that global average temperatures have increased as a result of the impact of human activities in the atmosphere (https://www.wwf.org.tr/ne_yapiyoruz/iklim_degisikligiveenerji/iklim_degisikligi/).

The results of a scientific research conducted by IPCC, indicates that global average warming should be kept below 1.5 °C. It can reduce the destructive effects of climate change on our planet and humanity, prevent the destruction of the ecosystem in the Arctic, drought and lack of potable water that will affect the lives of more than 300 million people. The destruction of coral reefs that have a very important role in marine life, rise of sea levels which significantly affect island countries, can be prevented. Although further efforts are needed to become zero carbon or carbon neutral, it is planned to prevent possible disasters by keeping global warming below 1.5 °C (https://www.semtrio.com/cop24-birlesmis-milletler-iklim-degisikligi-konferansi).

2 Road Maps and Legislation

There has been a great increase in the number of laws issued in Europe related to energy and the environment. This number has tripled in the last twenty years. An important portion of them is related to building services and HVAC systems. HVAC (Heating Ventilating and Air Conditioning) is a system that includes heating, ventilation and air conditioning and finds wide application areas such as factories, hospitals, shopping centers and areas where energy consumption is intense. Depending on the structure of the buildings, the energy consumed for HVAC constitutes between 15 and 60% of the total energy. Therefore, optimization of HVAC systems is very important in terms of energy efficiency.

HVAC is indispensable for energy efficiency and control; it minimizes operating costs. The best comfort conditions can only be created by HVAC systems. The desired clean, antibacterial environments and humidity conditions can be created and maintained by HVAC systems for all sectors. It reduces personnel costs and provides management from a single center. The selection and design of an HVAC system depends on several parameters, such as, how the building is used (residence, commercial, office, industrial, etc.), the budget, primary energy source and price, duration of the utilization of the spaces, hygienic properties, etc. The relation between the different factors should be studied carefully to find the ideal solution. Usually there are more than one alternative (Heperkan et al. 1994a, b).

2.1 Energy Balances

Turkey imports more than half of her energy although she has a variety of primary sources. According to the OME estimates in our country, where only one quarter of the energy demand is met by domestic production, this share will increase to 43% in the proactive scenario in 2030. Turkey, with 126.9 million TOE, was responsible for 1% of the world total energy consumption in 2015. In terms of primary energy sources, the amount of energy supply in Turkey was 129.2 million TOE, which was calculated to reach a total of 136.2 million TOE in 2016. (MMO 2018; www.enerji.gov.tr).

When the distribution of primary energy consumption in Turkey is examined for the sectors, the industry has the largest share by 25%. This corresponds to 33,264 MTOEs (million tons of oil equivalent). Following this primary energy consumption, the building sector ranks second with a rate of 24.80% and the energy sector ranks third with a rate of 24.60% (Koç et al. 2018).

According to the highly probable scenario of an increase of 4.8% to 375.8 TWh in the base scenario, electricity consumption in the year 2023 is expected to rise by 5.5% to 357.4 TWh. By the end of the first half of 2019, power plants containing a total of 1.8202 MW additional capacities were added to the system, and as of the end of the first half of 2019 our capacity has risen to around 90.421 MW (Electricity 2019).

Figure 1 shows the distribution of resources of the primary energy production, Fig. 2 the energy distribution for the sectors and Fig. 3 the distribution of the resources utilized for electric power generation in Turkey as of the end of 2017. Natural gas is

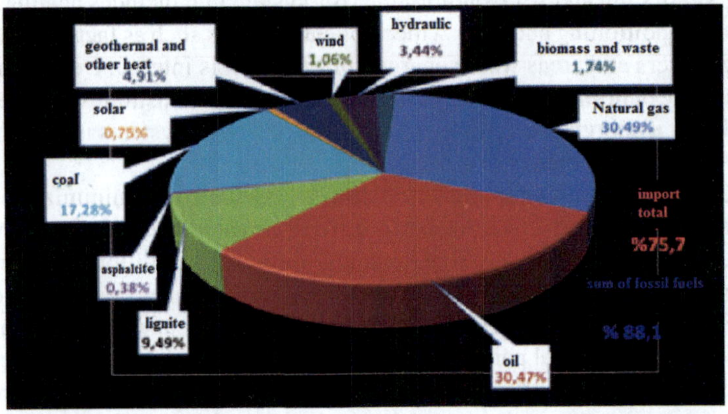

Fig. 1 Distribution of primary energy production by resources, 145,3 million of TOE (https://www.enerji.gov.tr/tr-TR/EIGM-Raporları)

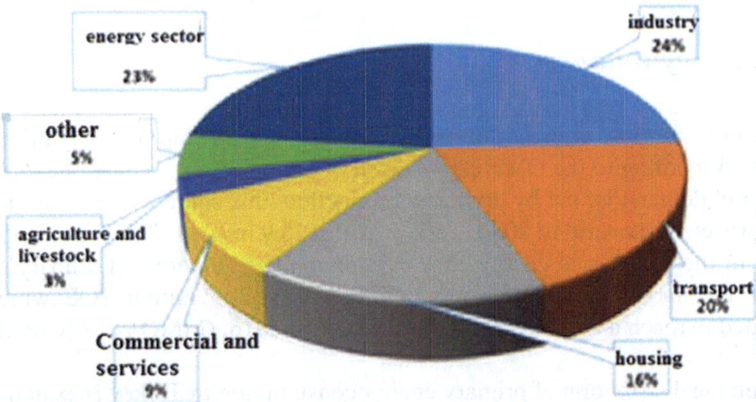

Fig. 2 Breakdown by sector energy consumption of Turkey, 145,3 million of TOE. (https://www.enerji.gov.tr/tr-TR/EIGM-Raporları)

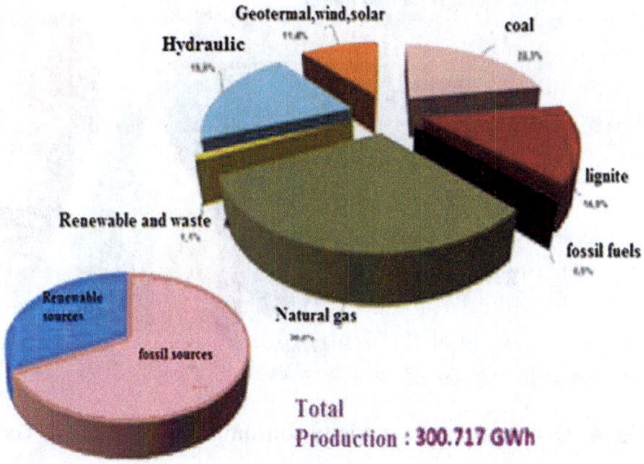

Fig. 3 Distribution of Turkey electricity generation by source (http://www.teias.gov.tr)

the dominant fossil fuel used in the generation of electricity. This picture contradicts the general tendancy seen in the World. Moreover natural gas is an import item for the economy, which makes the situation even harder to understand.

As of the end of the first half of 2019, 15.2% of our electricity production is being produced from natural gas, 33.2% from coal, 36.6% from hydropower, 7.2% from wind, 3.0% from geothermal, 3.0% from solar energy and 1.7% from other sources. As of the end of the first half of 2019, EUAS (State Electricity Generating Company) had a share of 18.6% in the installed capacity of Turkey, the private sector 73.0%, build-operate plants 4.8%, build-operate-transfer plants 0.4% and unlicensed power plants 3.1%. By the end of the first half of 2019, the distribution of our installed power by resources was 31.4% hydraulic, 29.0% naturel gas, 22.4% coal, 8.0% wind, 1.5% geothermal, 6.0% solar and 1.7% other sources. In addition, as of the end of the first half of 2019, the number of electricity energy production plants in our country was 7.957 (Electricity 2019).

Residential buildings account for 21.8% of the total electricity consumption, while the industry 46.8% and commercial buildings 26.9% (Fig. 4).

However, Turkey has various renewable energy resources and has a great potential. For example, the geothermal energy potential accounts for about 8% of the world's total. Due to its geographical location, the solar potential is also quite high. The measurements show that it receives an average of 3.6 kWh/m^2 of solar radiation per day. It also has many water sources that can produce hydraulic energy. The wind energy potential is estimated to be 160 TWh.

2.2 *European Union*

Commercial energy consumption in the world depends approximately 81% on fossil fuels, 5% on nuclear, 2% on hydraulic, 5% on biomass and 7% on renewable energy sources (Fig. 5). Electricity is produced from non-renewable sources (coal,

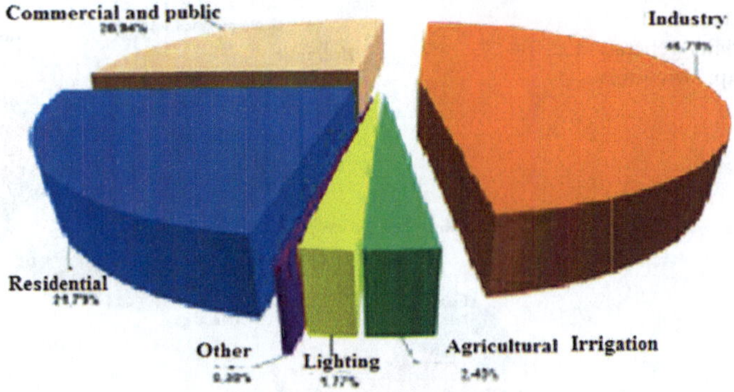

Fig. 4 Distribution of electricity consumption by sector in Turkey (http://www.teias.gov.tr)

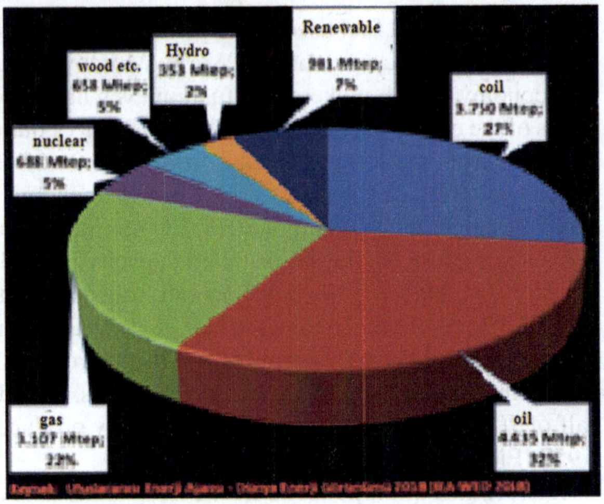

Fig. 5 World primary energy production 2018 (International Energy Agency, World Energy Outlook 2018)

natural gas, oil and uranium) with a share of 80%. The rest uses renewable resources. Hydraulic energy has 19% share among them. Solar, wind, biomass and geothermal resources have only 1% (International Energy Agency, World Energy Outlook 2018). The building sector is responsible for approximately 50% of the global electricity demand and 25% of the global greenhouse gas emissions; about 30% of this are direct emissions (e.g., space heating and hot water production) and 70% indirect emissions (e.g. electrical appliances and lighting).

The EU countries have acted with responsibility although their share in the global greenhouse gas emission is only 10%, and have accepted as a policy to bring the levels to 85–90% of the 1990 level by 2050. We can summarize this concept with the famous 20–20–20 target of EU in those days. They have achieved the 2012 target and some of the member states have beaten the 2020 goal today. They prepared a road map (March 8, 2011) to establish a carbon economy by 2050 and determined the key elements.

Member states have been promoting renewable energies and energy efficiency in their individual state policies since a long time. They have also been playing a pioneer position in the international arena. The reflection of these policies has led to the establishment of zero energy buildings. These buildings have no net energy consumption and do not produce carbon dioxide. According to EU targets, public buildings by 2018/2019 and all other buildings by 2020/2021 should be zero or near zero buildings.

The EU has issued the Energy Performance for Buildings Directive, EPBD, (2002/91/EC), to serve the purpose. This directive can be considered as a progression of the previous directives, Hot Water Boiler Directive (92/42/EEC), Construction Materials Directive (89/106/EEC) and SAVE Directive (93/76/EEC) which aims on energy efficiency to reduce carbon dioxide emissions.

2.3 Turkey

The Ministry of Environment and Urbanization has been working on the subject for several years. The legislation is summarized chronologically.

- May 2, 2007 Energy Efficiency Law, No. 5627
- October 9, 2008 Heat Insulation Regulation, No. 27019
- October 27, 2011 Regulation for the Efficient Use of Energy Sources and Energy, No.28097
- September 18, 2012 Communiqué on Energy Efficiency Training and Certification Activities, No. 28415
- April 28, 2017 Energy Performance of Buildings Regulation, No. 30051.

Turkey has performed better in this subject and has passed the Energy Efficiency Law in 2007 and the related by – laws and regulations in the following couple of years. One of the milestones was the revision of the Thermal Insulation Regulation, TS 825 in 2008. This regulation is now under development to include cooling as well as heating.

3 Energy Efficiency in Buildings

The purpose of the Energy Performance of Buildings Directive is to regulate the principles and procedures regarding the efficient use of energy and energy resources in buildings, prevention of wasting energy and protection of the environment.

This regulation covers, the calculation methods, standards, minimum performance criteria and methods related to architectural design, mechanical installation, lighting and electrical installations in new and existing buildings and implementation of building projects.

It also entails the issuance of energy identity documents, authorization for building controls and audit activities, meeting the energy need from cogeneration systems and renewable energy sources, creating and keeping an inventory of buildings and their energy systems throughout the country, training the technical staff and managers as well as organizing awareness raising activities in the society. This standard covers the work and procedures related to the measures and practices to increase energy efficiency in buildings registered as cultural assets that need to be protected, and to carry out applications that increase the energy efficiency (www.mevzuat.gov.tr).

Energy performance criteria for buildings have existed since the 1970 oil crisis. These initiatives have started as pilot projects; high performance buildings have been designed, built and used. Successful applications have been adopted by construction companies due to the interest they have on the attraction they create on the public. Within the years they were converted into standard applications. The process has ended up in the release of national standards and regulations. Figure 6 illustrates the course of the process for Germany. Other countries have shown similar developments. The top line demonstrates the minimum requirement for the "Wärmeschutzverordnung", insulation regulation (TSE 825 in Turkey) and the "Energieeinsparverordnung", Energy Efficiency Regulation (BEP in Turkey) for Germany. The bottom line indicates the level for pilot projects. The intermediate area accounts for the variation of the average level of performance in construction practice. The bottom right corner represents the energy plus buildings that are becoming more and more popular recently.

The European Union took the first step by issuing the Energy Performance of Buildings Directive (2002/91/EC), EPBD. A revision became necessary within the years to account for the encountered difficulties. The directive which was reorganized

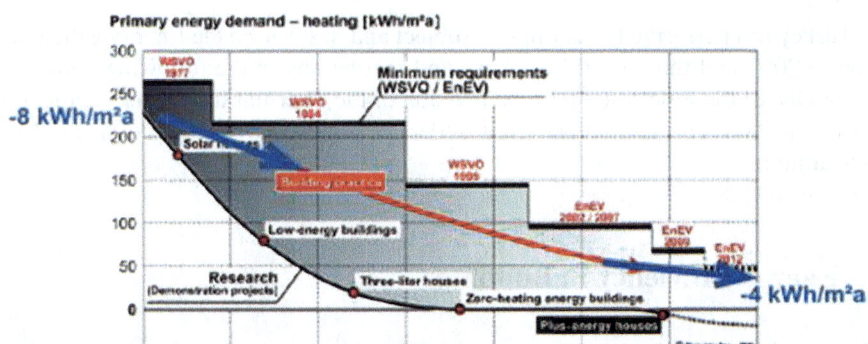

Fig. 6 Development of energy efficiency of buildings in Germany (Erhorn and Erhorn-Kluttig 2012)

in 2010 (2010/31/EU), introduced concepts such as "reference building", "optimum cost" and "buildings with nearly zero energy".

The directive also required the states to ensure that all new buildings built after 2020 to be nearly net zero energy buildings, nnZEB and public buildings after 2018. The member states were to issue national plans for;

- A definition to give an approximate annual kWh/m^2 primary energy consumption value that takes into account local, national and regional elements. Factors for the calculation of primary energy can be based on national or European standards.
- Financial and political benefits for nearly net zero energy buildings.

The 2010 revision of the EPBD requested the states to determine the energy demand of buildings using the cost – optimal concept. The report set out the information the member states must provide annually to the commission under Article 24(1) of the Energy Efficiency Directive and the progress achieved towards national energy efficiency targets in accordance with Annex XIV Part 1. It also included the information as a requirement of the directive under the Article 24(2), that member states must submit a national energy plan to the commission every three years (UK 2017).

Moreover, the economic life of the building may vary and should be defined by each state. Comparable methodologies can be guided by;

- Defining reference buildings
- Defining energy efficiency criteria
- Detecting the actual and primary energy requirements of the reference building
- Calculating the implementation cost of these measures.

We have to be careful with the cost effective (economic) and cost—optimal concepts. An energy efficiency project is economic if the cost of the application is lower than the profit during the life cycle (net present value is positive). The cost—optimal solution is the one that maximizes the net present value. In the case of heat losses from the outer periphery of a building, the calculation becomes complicated compared to a simple insulation thickness problem when the whole building is considered. The optimum should be selected among several points, all representing economic points as depicted in Fig. 7.

The definition of the reference building is critical. Differences with respect to their function, location, indoor and outdoor temperatures should be considered for buildings. At least nine different reference buildings should be defined (one for new and one for existing family houses, complexes and office buildings). In addition reference buildings should be considered for the following cases;

- Different private houses
- Apartments and block complexes
- Office buildings
- Education complexes
- Hospitals
- Hotels and restaurants

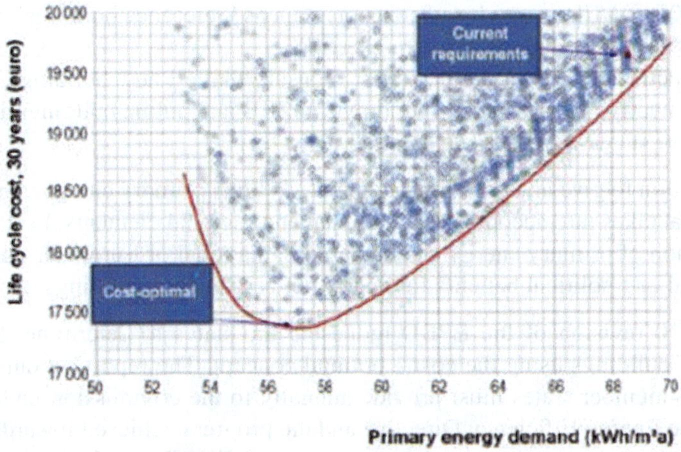

Fig. 7 Cost—optimal point for a specified country (Wittchen and Thomsen 2012)

- Sport complexes
- Shopping centers
- Other.

3.1 Energy Certification

An energy identity certificate includes, according to the Energy Efficiency Law No. 5627 and the Energy Performance Regulation in Buildings, the minimum energy consumption and energy consumption classification, greenhouse gas emission level, insulation characteristics of the building used to ensure the efficient use of energy and information on the efficiency of the heating and/or cooling systems.

The energy performance ratings in white goods or air conditioners, such as refrigerators, washing machines, now also apply to buildings. These classifications are made from A to G. Class A indicates the most efficient level, while Class G indicates the least efficient level. The certificate shows the following information and a sample for Turkey is depicted in Fig. 8.

- Building overview
- Energy consumption class
- Carbon dioxide emission class
- Sanitary hot water release class
- Ventilation energy consumption class
- Building image or model
- Renewable energy rate

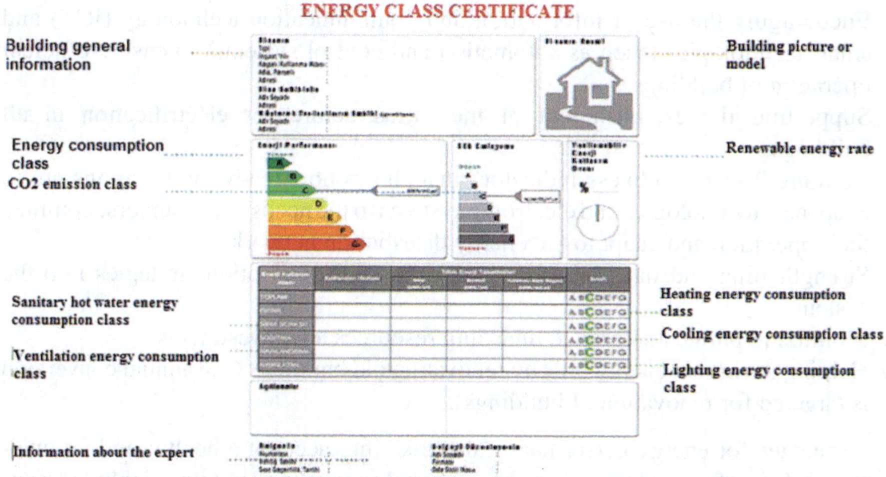

Fig. 8 Energy class certificate (http://www.enerjikimlikbelgesi.com/)

- Heating energy consumption class
- Cooling energy consumption class
- Lighting energy consumption class.

Certificates are being issued since January 1, 2011 in Turkey.

Approximately a year after the "Clean Energy for All Europeans" package announced in November 2016, member countries agreed to revise the energy performance directive, EPBD, in buildings. The European Parliament approved the revision on April 17, 2018. Members were given 20 months to adapt themselves to the new measures. The new revision includes issues such as strengthening the indoor environment quality, proper maintenance and effective supervision and setting more ambitious energy efficiency targets in line with the views of stakeholders and Rehva. In order to ensure efficient operation of buildings, the promotion of information and communication technologies (ICT) and the use of smart technologies (smart meters, building automation and control systems), energy storage, the definition of the "smart readiness indicator" showing how ready the buildings are to comply with the energy distribution network and to start an initiative for the renovation of old buildings within the state.

The main developments can be summarized as;

- Creating a low and zero-emission building stock by 2050, supported by national road maps to ensure that buildings are de-carbonised.
- To reduce greenhouse gas emissions by 80–95% compared to 1990 levels by 2050. Determination of measurable milestones for 2030 and 2040.

- Encouraging the use of information and communication technology (ICT) and smart technologies (such as automation and control systems) to ensure efficient operation of buildings.
- Supporting the establishment of the infrastructure for electrification in all buildings.
- Defining "smart readiness indicator" that shows how ready the buildings are to adapt new technologies and electronic systems to the needs of consumers, optimize their operation and adapt to the energy distribution network.
- Strengthening and integrating long-term building renovation strategies into the system.
- Mobilizing public and private financing resources and investments.
- Helping to combat energy bills by renovating old buildings (3% annual conversion is targeted for renovation of buildings).

Strategies for energy performance also take into account a healthy indoor environment, fire safety and seismic risks. National renovation strategies include issues such as;

- National building stock inventory and renewal rate for 2020
- Defining cost-optimal renovation approaches according to building type and climate and considering the life of the building (life cycle)
- Creating policies to activate cost-optimal major renovations
- Establishing policies to solve the incentive dilemma, market errors and energy waste by targeting the worst performing part of the national building stock
- Establishing policies to support smart technologies, skills and education in the construction and energy efficiency sectors, especially targeting public buildings
- Adding importance to health, safety and air quality in addition to energy saving.

The regulation proposes to pay attention not only to the building envelope, but also to all technical equipment. As long as it is economically and technically realistic, high efficiency devices should be preferred. The technical systems of the buildings not only contribute to energy saving but also help establish a good indoor quality. Financial measures related to energy efficiency should be quality oriented, this should be determined by energy audits and should be checked periodically; it should even be mandatory. Audits must be documented.

In commercial buildings and sites, a cost effective alternative to energy audits is the building automation and control (BAC) systems. In commercial buildings with a capacity of more than 290 kW (heating and ventilation together) according to the directive, building automation and control systems should be installed until 2025 (if technically and economically feasible).

3.2 *Energy Simulation Software*

There are quite a few software available on the market related to the evaluation of the energy performance of a building; only a couple have been given attention here, in this article.

3.2.1 BEP—TR Software

This is the energy performance evaluation software to assess the energy consumption of a building and issue the Energy Class Certificate shown in Fig. 4. National building energy performance calculation method for Turkey (BEP-TR) has been developed to determine the energy performance class of the building, taking into account the impact of all parameters of the building's energy consumption on the energy efficiency of the building. BEP-TR is an internet-based software. Calculates the amount of energy consumed per m^2 per year of the buildings covered by the BEP regulation and their CO_2 emissions (Korkmaz 2019).

In the BEP-TR software, the method for calculating the net amount of energy for heating and cooling of buildings, is a simple hourly based calculation (EN 13790). This method requires the determination of the heating and cooling seasons. It also allows the calculation of the net energy amount during the transition seasons (Korkmaz 2019).

The energy identity certificate is valid for 10 years from the date of issue. At the end of this period, it can be renewed in accordance with a report to be prepared. The Energy Identity Certificate is prepared by an expert authority and approved by the relevant administration (Korkmaz 2019). BEP—TR is a modular software composed of the following components referring to the standards mentioned on the side.

- Geometry and Material Properties
- Net Energy: ISO 13790
- Lighting: EN 15193 (Pre-standard)
- Solar Gains: EN ISO 13790:2004, EN ISO 13790:2008, EN ISO 15255:2007, EN ISO 13792:2005, Ashrae Fundamentals:2009
- Internal Loads: EN 15316, EN15241, EN 15243, EN 15193, Ashrae:2009
- Mechanical Installations: DIN V 18599:2007.

It has four main modules. The first one establishes the geometry of the building, material properties, zones and the heat loss and gain of the structure on an hourly basis. The geometry is defined in a hierarchic flow associated with the component properties. It also provides the opportunity to read data from CAD software.

Properties of the building elements can also be introduced in a hierarchic pattern. Objects holding the geometry together (building, floors, zones, rooms, walls, windows, etc.) have been defined relative to each other. The drawing can be performed using a Professional CAD program or an existing CAD software output can be adopted. Curved contours are taken as polygons having 15° increments. Properties of the materials are given at every related step.

The second module, net energy demand of the building involves several parameters both active and passive. It is based on ISO 13790 standard. The building is broken up into zones and typical zone relations are defined as depicted in Fig. 9. Heat storage characteristics of the materials are also considered during the calculations. The components of the structure are interrelated on an unsteady basis using an electrical network analogy.

Heat storage characteristics of the materials are also considered during the calculations. The components of the structure are interrelated on an unsteady basis using an electrical network analogy as illustrated in Fig. 10.

The third module handles the lighting and the last one the mechanical system serving the building. The system has heating, cooling, ventilation, domestic hot water, cogeneration and photovoltaic elements.

The lighting module uses a new methodology based on measurements and equations taken from the literature. Several new lighting alternatives have been included. Moreover, the height of the lighting fixtures and the color of the walls and ceiling have also been included to the calculations.

The net energy demand for the heating and cooling equipment to cover the overall building is transferred from the net energy module. The energy consumption of the auxiliary elements, like pumps, fans, etc. are calculated and added. Control system losses and consumption are considered. The components of the mechanical system are identical to the ones in DIN V 18599:2007 as referenced in Fig. 11.

The net energy flow of the building is assessed by including renewables like heat pumps, solar collectors, photovoltaic panels, cogeneration, etc. This consumption is

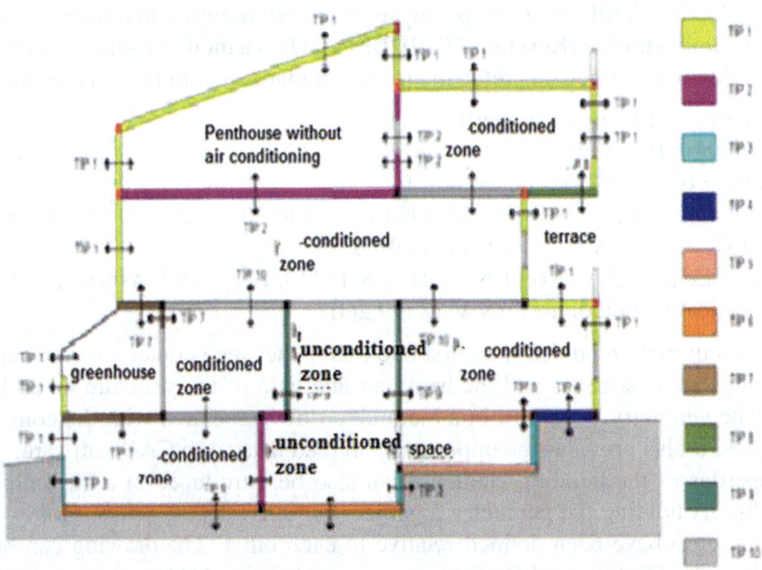

Fig. 9 Characteristic relations between zones

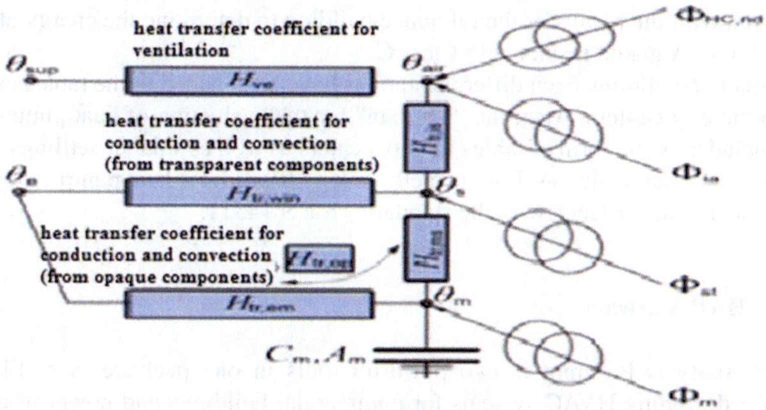

Fig. 10 Electrical network analogy for the calculations (Voss and Musall 2012)

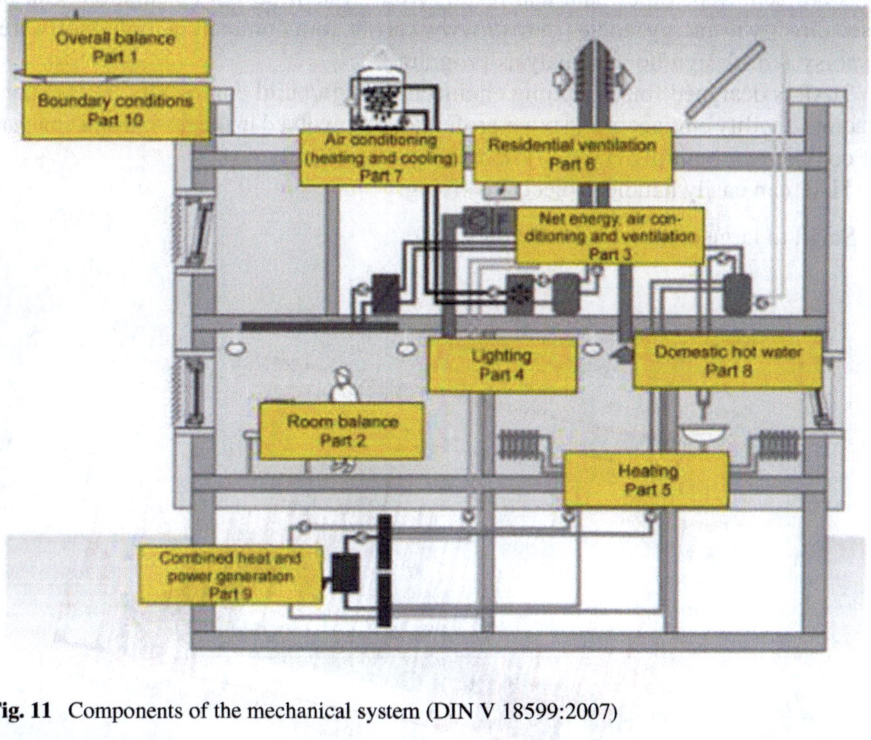

Fig. 11 Components of the mechanical system (DIN V 18599:2007)

compared with the results of the reference building to determine the energy class of the building. A match results with Class C.

Climatic conditions from different stations have been added to the tables in addition to the city centers. Also, the "bin data" for the evaluation of heat pumps have been included in the form of tables for city centers as well as smaller settlings, based on 1 °C temperature classes. For the methodology to work for heat pumps, they have to be tested in accordance with the standard TS EN 14511.

3.2.2 HAP Software

Hourly Analysis Program is two powerful tools in one package- versatile features for designing HVAC systems for commercial buildings and powerful energy analysis capabilities for comparing energy consumption and operating costs of design alternatives. By combining both tools in one package, significant time savings are achieved. Input data and results from system design calculations can be used directly in energy studies (https://www.carrier.com/commercial/en/us/software/hvac-system-design/hourly-analysis-program/).

HAP is designed for consulting engineers, design/build contractors, HVAC contractors, facility engineers and other professionals involved in the design and analysis of commercial building HVAC systems (Fig. 12).

HAP can easily handle projects involving:

- Small to large commercial buildings;

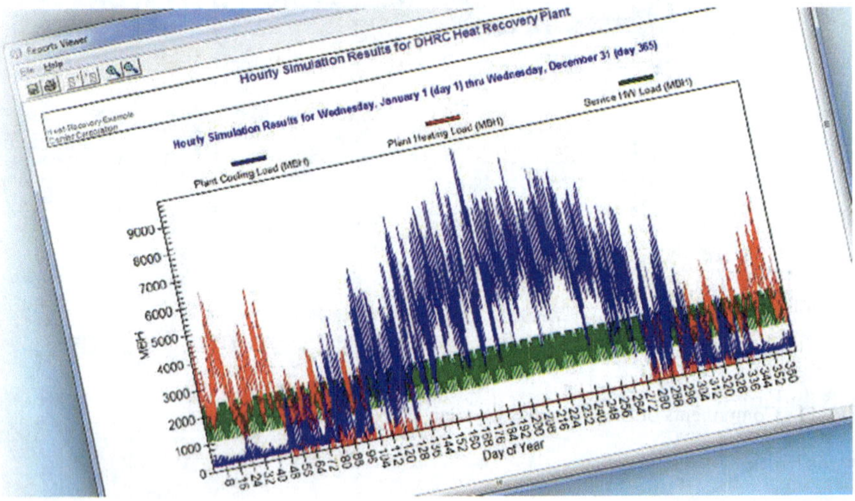

Fig. 12 HAP software (https://www.carrier.com/commercial/en/us/software/hvac-system-design/hourly-analysis-program/)

- Systems including rooftops, central air handlers, WSHPs, GSHPs, fan coils, VRF, chilled water and hot water plants and more;
- Many types of constant volume and VAV system controls;
- Small office buildings, retail stores, shopping centers, schools, churches, restaurants, large office buildings, hotels, malls, hospitals, factories, and multi-use buildings;
- New design, retrofit or energy conservation work. (https://www.carrier.com/commercial/en/us/software/hvac-system-design/hourly-analysis-program/).

HAP performs a true hour-by-hour energy analysis, using measured weather data for all 8,760 h of the year to calculate building heat transfer and loads, air system operation and plant equipment operation. Performs detailed hour-by-hour simulations of the thermal and mechanical behavior of air handling systems for both system design and energy analysis. (https://www.carrier.com/commercial/en/us/software/hvac-system-design/hourly-analysis-program/).

3.2.3 Energy Plus Software

EnergyPlus is a whole building energy simulation program that engineers, architects, and researchers use to model both energy consumption—for heating, cooling, ventilation, lighting and plug and process loads—and water use in buildings. Some of the features of EnergyPlus include (https://energyplus.net/);

- Integrated, simultaneous solution of thermal zone conditions and HVAC systems,
- Heat balance-based solution of radiant and convective effects that produce surface temperatures, thermal comfort and condensation calculations,
- Sub-hourly, user-definable time steps for interaction between thermal zones and the environment,
- Combined heat and mass transfer model that accounts for air movement between zones,
- Advanced fenestration models including controllable window blinds, electrochromic glazings, and layer-by-layer heat balances that calculate solar energy absorbed by window panes,
- Illuminance and glare calculations for reporting visual comfort and driving lighting controls,
- Component-based HVAC that supports both standard and novel system configurations,
- Standard summary and detailed output reports as well as user definable reports with selectable time-resolution from annual to sub-hourly, all with energy source multipliers.

A typical screen from the working page of the software can be seen in Fig. 13, together with the available menus.

4 Energy Efficiency Technology in Buildings

Energy efficiency is the reduction of energy consumption without changing the standard of living and quality of service. Efficiency is ensured in every operation where the same product output is achieved with less energy input. Improvements are made in the areas of heating, cooling, ventilation, lighting, CO_2 emission and hot water, depending on climate conditions, architectural design and insulation standards for buildings. In 2020, according to the Energy Performance in Buildings Directive, all new buildings are required to fall into approximately zero energy class.

During the energy efficiency improvement applications in buildings, the insulation of the outer wall designed and implemented according to the climate zone of the building, directly affects the efficient use of energy. For example; insulation applied to two buildings at the same location, with the same architectural and mechanical system properties but with two different thicknesses, leads to a significant change in the energy performance of the building. When the insulation thickness is increased at appropriate levels, energy losses decrease.

In the order of importance in terms of insulation, roof structures come first. On the basis of the rising principle of heated air, keeping the indoor air heated in the winter is dependent on the insulation properties of the roof. It is important that windows and doors are tight against leaks from the ambient air. Therefore, the use of double glazing and insulating glass is widespread today and heat losses are reduced.

The compliance of the mechanical systems to the project to be implemented should also be considered. For example, the use of condensing combi boilers instead

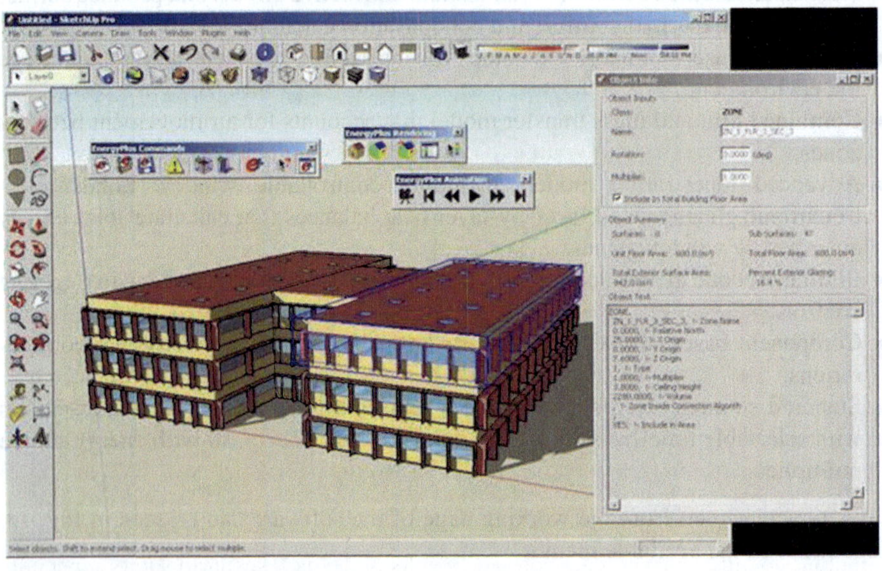

Fig. 13 Energy plus (https://energyplus.net/)

of standard combi boilers reduces the fuel consumption in heating season and reduces the energy consumed for heating purposes. Utilization of natural gas instead of coal as the fuel used in central heating systems, increasing the use of renewable energy sources such as solar, wind and thermal energy are among alternatives for increasing energy efficiency.

Apart from these, in order to meet the lighting needs of buildings with minimum consumption, suitable low energy lighting systems should be selected. The compliance of the architectural design of the buildings with the energy efficiency standards is also among important measures. The exposure time and angle of the sun rays incident on the building during the day are among issues to be considered in this respect.

Examples of energy efficiency technologies in buildings include heat pumps, cogeneration, renewable energy, condensing combi boilers and boiler systems.

4.1 Heat Pumps

Heat pumps are devices that transfer heat between different heat sources; they extract energy from the environment and convert it into usable energy at a higher temperature for space and water heating. Experts estimated that the worldwide heat pump market increased at a compound annual growth rate of ~10.3% between 2016 and 2020. Heat pump technology is a well known technology. According to the data of the European Heat Pump Union, 2,640,000 units were sold between 2005 and 2010. The market has grown at a rate of 30% since 2003. IEA estimates that the market volume will reach 3.6 billion by the year 2050.

As known from the thermodynamic rules, an external energy source (in the form of work) must be used to transfer energy from a low temperature environment to a higher temperature environment. Otherwise would violate the second law of thermodynamics. The majority of the heat pumps used in conventional systems include compressors operating with electrical energy. Heat pumps are named according to the environments in which they draw and transfer heat. Air, water or soil may be present where heat is drawn, while air or water may be transmitted (https://midori.com.tr/isi-pompasi/isi-pompasi-nedir.html).

Air conditioners are essentially 'air to air' heat pumps (Fig. 14). In industrial applications involving larger systems, the preferred devices are soil-to-air and water-to-air heat pumps, which are different types of air source heat pumps. Commonly used heat pumps are designed on the basis of the heating function and can produce domestic hot water and cooling effect if necessary. The characteristics and functions of the heat pump used may vary according to the requirements of the system (https://midori.com.tr/isi-pompasi/isi-pompasi-nedir.html).

Heat pumps offer a high potential to rapidly increase the proportion of renewable energy share within the total heating energy consumption. Due to their built in multiplication effect, by using environmental heat sources. The supplied energy

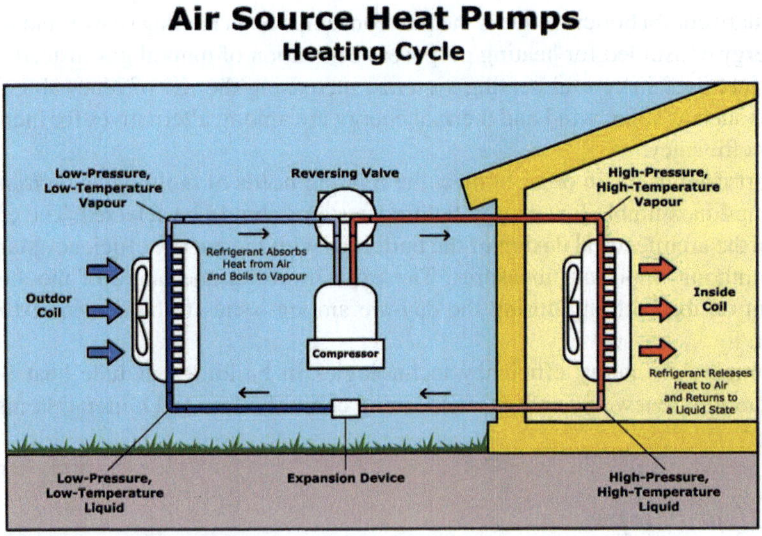

Fig. 14 Heat pump (https://midori.com.tr/isi-pompasi/isi-pompasi-nedir.html)

is transformed into more thermal energy through a factor called the Coefficient of Performance or COP.

The advantages of a heat pump include (https://midori.com.tr/isi-pompasi/isi-pompasi-nedir.html);

- Since it can meet the cooling, heating and hot water needs with a single device, it requires lower investment cost compared to other systems.
- Energy efficiency is high. Compared to other systems, monthly operating costs are considerably lower and savings up to 75% are possible.
- Renewable energy systems can be combined with fan-coil units to cool the space.
- It is safe. There is no risk of gas or liquid fuel leakage and fire in the system.
- There is no danger in the system during an earthquake.
- It is easy to install. No need for ventilation, flue and fuel tank in the installation room.
- It can be easily adapted to existing buildings.
- Annual maintenance costs are very low.

A large part of the use of heat pumps in our country consists of air-to-air operating air conditioners. The use of air-to-water heat pumps is becoming widespread with the promotional activities of the companies, the awareness of the users and the emergence of sample applications. Turkey's geographical position is suitable in particular for water-water heat pumps and air-air heat pumps in the coming years and predictions indicate an increase in the number of users rapidly.

The determination of performance rates for heat pumps is largely standardized throughout Europe on the basis of EN 14511. It defines the standardized conditions for testing electric driven heat pumps and determining their COP using a test rig. This involves using steady state and virtual steady state measuring points, which form the basis for the comparison of individual devices, as well as the dimensioning and design of whole systems. EN 14511 was revised in 2011 and has, for a number of years, replaced EN 255-1 + 2. Environmental heat sources which can be exploited include outside air, exhaust air, ground or surface water and ground heat (brine). The heat source side is described in the standard as the external heat exchanger, which corresponds to the evaporator of the refrigeration circuit in the heating case. The transfer of usable energy can be via air or water. The usable heat side is described as the internal heat exchanger. In the heating case this corresponds to the condenser. The standard defines so called nominal standard and nominal working conditions for each category of device. A device using brine as heat source and water as heat transfer for example is called a brine/water heat pump. In the case of brine/water heat pumps, the nominal standard conditions at low temperature are for brine at 0 °C (B0, B = brine) and a heating water temperature of 35 °C (W35, W = water). This boundary condition is abbreviated to B0W35. If the heat transfer is via water, the temperature of the heating circuit influences the efficiency of overall heat generation. The lower this temperature level, the higher the efficiency of the heat pump.

EN 14511 specifies four temperature levels at which devices can be tested. These are distinguished according to their outlet temperatures and are defined as follows: low temperature (35 °C, W35), medium temperature (45 °C, W45), high temperature (55 °C, W55) and very high temperature (65 °C, W65). The efficiency of the heat pump however, depends on the mean condensation temperature. Therefore the input temperature which depends on the mass flow rate of the water must also be specified. The mass flow rate is set to produce a 5 K temperature difference on the heating side. If the heat transfer is via air, a uniform input air temperature of 20 °C (A20, A = air) in the condenser is used. Another important influence on the efficiency of heat pumps is the temperature of the heat source; the higher the temperature, the higher the efficiency of the heat pump. In this case the standard also offers a variety of temperature levels, categorized by the input temperature at the evaporator. Seasonal variations are minimized using ground or water heat sources. For brine there are three temperature categories: −5 °C (B-5), 0 °C (B0), and +5 °C (B5). For water as the heat source, two categories are defined: 10 °C (W10) and 15 °C (W15). The largest temperature span is available when using outside air as a heat source, where 12 °C (A12), 7 °C (A7), 2 °C (A2), −7 °C (A-7), and −15 °C (A-15) are defined as source temperatures.

The building sector is dominated by existing buildings. The yearly volume of new buildings is less than 1% of the volume of existing buildings. It is an undisputed goal for Europe to drastically reduce the total energy consumption and to increase the share of renewable energy. The construction of new buildings in Turkey is much higher compared to Europe due to the urban transformation plan to reinforce buildings against earthquakes. To ensure energy consumption reduction, existing buildings cannot be avoided. In fact, it is easier to utilize new technologies in new constructions. The optimal solution for each building will need a case by case development of the appropriate concept.

Whether the additional heat obtained from the ground, water or air is renewable or not, has always been discussed. This issue has been solved after the Renewable

Energy Sources Directive (2009/28/EC, RES Directive, Article 2) and Energy Performance of Buildings Directive (2010/31/EU, Article 2). The European Commission has added this extra energy from the ground, water and air into the definition of renewable energy sources. The important point that should be considered here is the fact that the energy (fossil fuel or electricity) to operate the heat pump is not renewable. The third clause of the 2009/28/EC, RES Directive states that this energy should be deducted from the total transferred energy.

In the end we are heading for "nearly Zero Energy Buildings", the so called nZEB's. Refurbishments down to that level need concepts for the building itself, i.e. the insulation and the tightness, and for the building services systems, i.e. for the heating, cooling and ventilation. The optimal solution for each building will need a case by case development of the appropriate concept. Heat pumps are a very promising component for the heat supply. The achievable seasonal coefficient of performance (SCOP) is the determining factor. It depends on the temperature levels on the heat source side and on the thermal system side, on construction and running parameters of the heat pump itself. On the heat source side we discuss different sources such as outside air, that has the advantage of being available everywhere but the disadvantage of low temperatures in winter. This then makes combinations with solar energy interesting. On the heat pump itself, developments of capacity controlled compressors are very promising to improve the SCOP. All these new heat pump concepts need appropriate and agreed testing procedures.

Energy performance characterization for heat pumps is gradually migrating from EER and COP nominal performance EN 14511 to seasonal performance EN 14825. The industry has implemented ESEER (European Seasonal Efficiency Ratio), certified since 2007; recently published regulations talk about seasonal coefficient of performance (SCOP) in the heating mode, and its equivalents SEER and SEPR in the cooling mode. To achieve nearly Zero Energy Buildings in the case of old buildings, heat pumps will be especially effective.

The possible solution to work towards eco-sustainable products with high energy efficiency and integration with renewable energy is the replacement of boilers for heating and hot sanitary water production with heat pumps. The advantages are obvious:

- Possibility of exploitation of renewable energies (air, water, geothermal);
- High efficiency (COP ever higher, 4–5);
- Versatility of use: seasonal employment winter/summer;

But by using traditional heat pumps we face, at times, a number of problems. While functioning satisfactorily in summer, have some problems, well known, in the winter. Almost always installations with standard heat pumps are complemented by traditional energy sources (boilers, electric heaters) to reach the limits of power and temperature.

Geothermal heat, as one of the renewable energy sources is based on the process of transferring heat to the ground from Sun and core of the Earth. The heating or cooling processes using this transferred energy are called Ground Source Heat Pumps (GSHPs). Since Turkey has a very high geothermal energy potential, heat pumps are basically used in this sector for Turkey. Heating and cooling processes are both carried out in heat pumps. In this regard, GHPs have had the largest growth since 1995, almost 59% or 9.7 annually in the United States and Europe. The utilization of GHPs in residential buildings is new in Turkey, although they have been in use for years in developed countries.

4.2 Cogeneration

Cogeneration in energy applications, for example, combined heat-power generation systems (CHP, Combined Heat and Power), are systems where steam and electricity are produced together. In these systems, waste heat is evaluated and energy efficiency is increased and energy is used more efficiently compared to conventional systems. Since energy is produced where it is consumed, it eliminates losses in transmission and distribution lines and provides uninterrupted and high quality electricity supply without being affected by the network (Pravadalıoğlu et al. 2011).

This system is the production of electricity and heat energy together; electricity, hot water and steam are produced simultaneously and supplied to the user (Fig. 15). In thermal power plants that produce only electricity, the efficiency is around 30–40%, whereas with cogeneration up to 80–90% high yields are achieved. Cogeneration systems are the most efficient way to convert fuel into electricity and heat.

1 m^3 of natural gas has an energy value of around 10.64 kWh and roughly corresponds to 10 kWh of electrical energy. Considering that the total cycle efficiency of

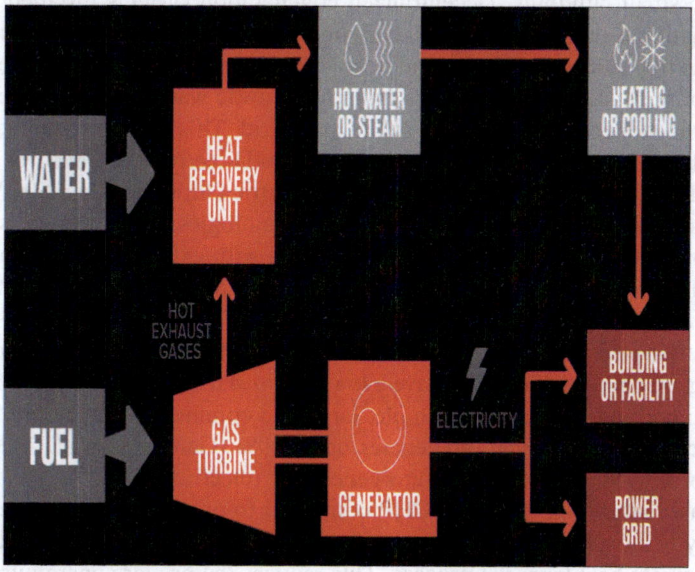

Fig. 15 Combined heat and power (https://centraxgt.com/)

the cogeneration plant is 90%, electricity and heat energies corresponding to 9 kWh electric energy are obtained from 1 m^3 of natural gas.

There are several examples of applications in the world. The EU aimed at the promotion of high-efficiency energy production facilities by issuing the Cogeneration (CHP) Directive on 11.02.2004, for its significant benefits such as, providing energy savings, preventing network losses and contributing to reducing the emissions of greenhouse gases. Combined heat-power generation is rapidly expanding in Europe and as of 2010, accounts for 18% of the total heat and electricity production. With this method, electricity and heat production reaches up to 51% of the total production in some EU countries such as Denmark (Wit 2011). In the European Union, especially to encourage the use of cogeneration units these systems are supported by facilitating connection to the network, simplifying regulations and providing financial support.

Advantages of cogeneration systems (Pravadalıoğlu et al. 2011);

- It provides uninterrupted, high quality and continuous electricity in the enterprises.
- Cogeneration plants have lower greenhouse gas emissions and bring energy savings.
- It is possible to buy and sell electricity from the network via a single port.
- It has the option of working with natural gas, fuel oil, biogas and biofuels.
- This system is ideal for industrial plants, hospitals, hotels, shopping centers, dormitories, schools and universities.
- The cost of the system is met within a short period of time if all of the electricity, hot water and steam are used.

Cogeneration can be applied at different scales, to large industrial facilities as well as to small buildings. Decentralized cogeneration systems installed in buildings are gaining more interest and diffusing nore and more into the community. The aim is to produce electricity and heat near the consumption point, meeting the demand of the end-users for heating and hot domestic water, while producing electricity for local needs and/or to sell to the main grid. This concept is called "micro-cogeneration" (micro-CHP), defined according to Directive 2004/8/EC as the local combined production of heat and electricity (electric output less than 50 kW).

We can collect the available micro-CHP Technologies under four classes: internal combustion engines (ICEs), Stirling engines with external combustion cycles, Organic Rankine Cycles (ORC) and fuel cells (PEMFC and SOFCs). The large majority of commercial micro-CHP units currently installed are based on ICEs, Stirling engines and ORCs. The ratio electrical to thermal energy depends on the technology. Combustion devices have a lower ratio and are more suitable to applications with high heating loads (example, large and commercial buildings). Fuel cell systems can reach electrical efficiencies up to 60% (Gandıglıo et al. 2020), being perfect for self-consumption and auto-generation (electricity).

The main advantage of micro-CHP compared to producing heat and electricity separately is the high primary energy efficiency (80% or more) achieved by recovering the waste heat from the process. This reduces the primary energy use and the emission of greenhouse gases (GHG). CHP is not a zero-emission solution because the vast majority run on natural gas. On the other hand, the increase in the energy efficiency means lower operating costs as well as low initial investments for the users. A micro-CHP system has the advantages of distributed electricity generation helping the grid stability by generating electricity near the consumption point, at the time of demand. It also improves the security of electricity supply (The Benefits of Micro-CHP: A Delta-ee report produced on behalf of COGEN Europe 2015).

4.3 Renewable Energy

Renewable energy is the energy that makes use of the continuous natural processes for its production and renews itself in a shorter time than the depletion rate of the resources it uses for production. The types of renewable energy include geothermal energy, wind energy, solar energy, hydroelectric, hydrogen, wave and biomass energy (Fig. 16).

Solar energy is the result of the transformation of the sun's rays into heat and electricity with the help of solar panels. Solar energy, which is among natural energy sources, can be obtained without any harm to the environment. Wind energy is the type of energy obtained from the pressure generated by the differences in the incidence angles of sun rays arriving at the earth surface and the winds generated by the rotation of the earth. Wind energy obtained from wind turbines installed in dense winds are among renewable energy sources.

Hydroelectric energy is produced by transmitting the kinetic energy generated by the flow of water to the turbines through channels. Geothermal energy is the type of energy obtained directly or indirectly where geothermal resources exist. This type of energy, which serves different purposes such as heating, cooling, electricity

Fig. 16 Renewable energy (Özkara 2018)

production and mineral production, also helps the tourism sector with the help of spas (Özkara 2018).

Hydrogen energy is a source of energy generated by the processing and conversion of hydrogen gas, which is present as compounds in nature. Although it is not a natural energy source, it is among the sustainable and alternative energy carriers. Wave energy is the type of energy obtained from the fluctuations in the sea and the pressure generated by the waves. Biomass energy is the type of energy obtained as a result of the use of biomass waste by incineration or through different processes (Özkara 2018).

Advantages of renewable energy (Özkara 2018);

- Environmentally important because it reduces fossil fuel use,
- It has great importance in the development of domestic resources,
- It reduces the dependence on foreign sources,
- It complies with international agreements,
- Provides new employment and reduces unemployment,
- It enables the use of electricity to geographic areas where it is difficult to deliver electricity.

Figure 17 shows the annual addition of renewable energy capacity between 2012–2018 by technology and their total.

According to Fig. 17, solar PV energy is increasing rapidly. By the end of 2018, total renewable energy was 181 GW. Figure 18 shows the estimated renewable energy share of global electricity generation.

As shown in Fig. 18, the estimated renewable energy share of global electricity generation is 26.2%; 15.8% of this energy comes from hydro power. Figure 19 shows

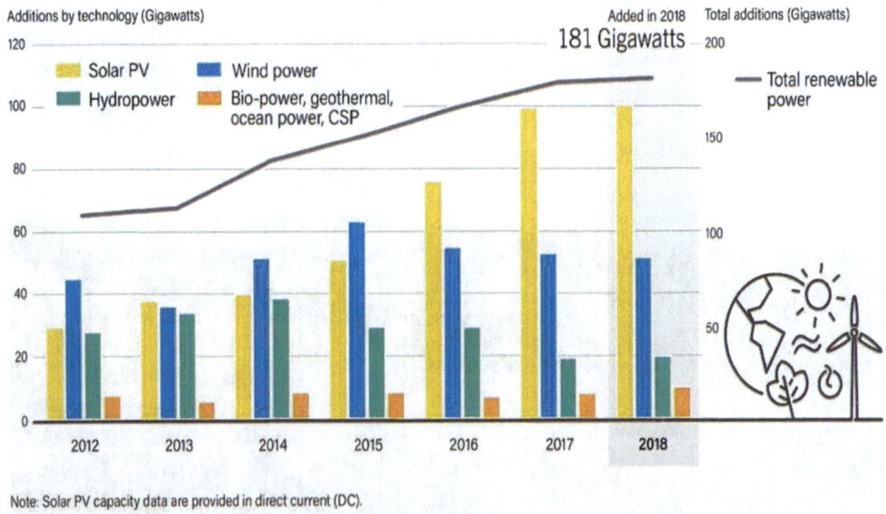

Fig. 17 Annual additions of renewable power capacity, by technology and the total, 2012–2018 (Renewables 2019)

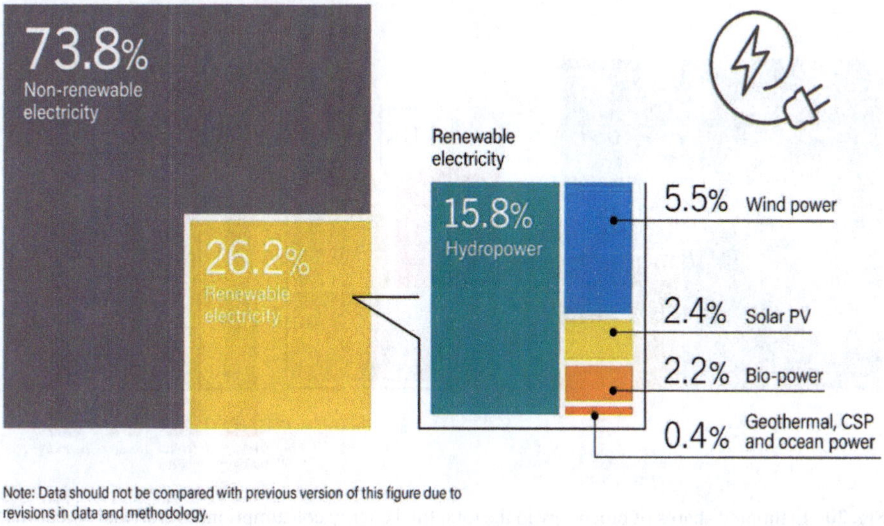

Fig. 18 Estimated renewable energy share of global electricity production, end of 2018 (Renewables 2019)

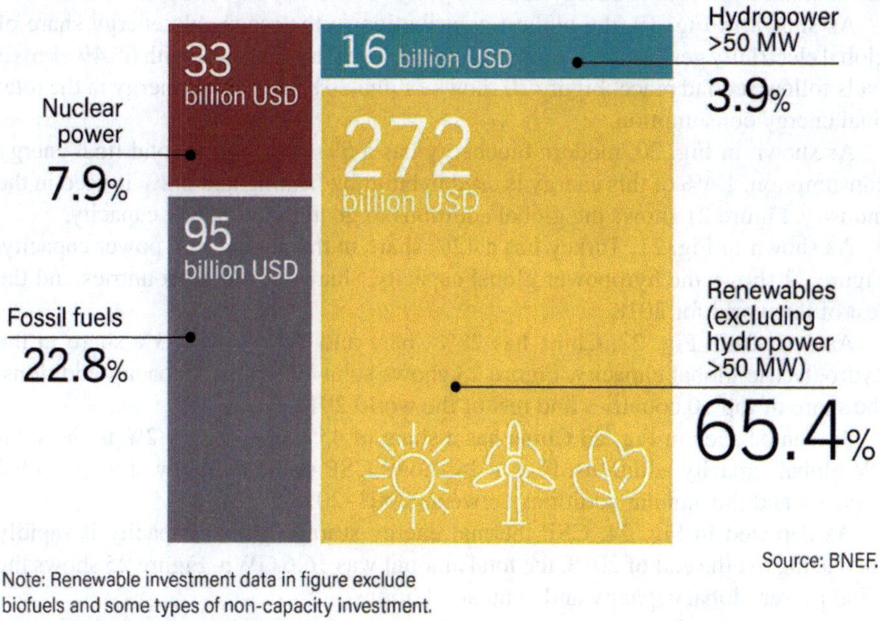

Fig. 19 Estimated Renewable Energy Share of Global Electricity Investments, End 2018 (Renewables 2019)

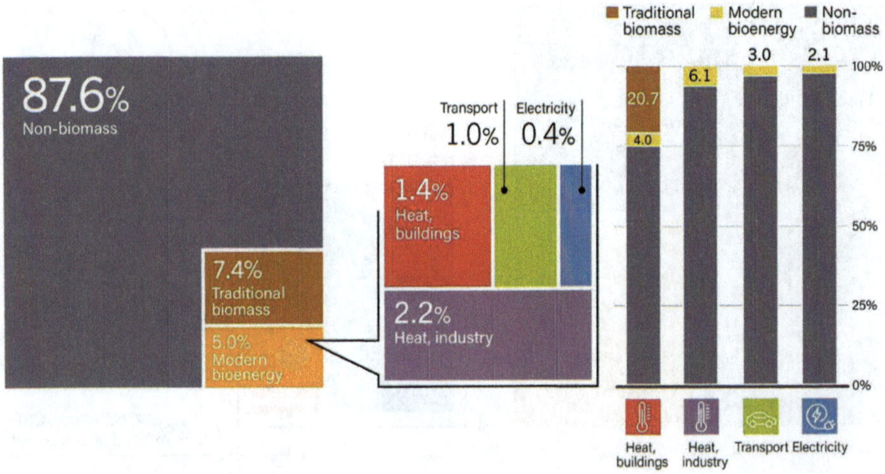

Fig. 20 Estimated shares of bioenergy in the total final energy consumption, overall and by end-use sector, 2017 (Renewables 2019)

the estimated renewable energy share of global electricity investments, end 2018.

As shown in Fig. 19, the highest contribution to the renewable energy share of global electricity generation is provided by renewable hydropower with 65.4%. Fossil fuels follow second place. Figure 20 shows estimated shares of bioenergy in the total final energy consumption.

As shown in Fig. 20, modern bioenergy has a 5% share in the total final energy consumption. 1.4% of this energy is used in building heating and 2.2% is used in the industry. Figure 21 shows the global additions to geothermal power capacity.

As shown in Fig. 21, Turkey has a 42% share in the geothermal power capacity. Figure 22 shows the hyropower global capacity, shares of top 10 countries and the rest of the world for 2018.

As seen from Fig. 22, China has 28% share and Turkey only 3% share in the hydroelectric global capacity. Figure 23 shows solar PV global capacity additions, the share of top 10 countries and rest of the world 2018.

As can be seen in Fig. 23 China has a share of 45% and Turkey 2% in the solar PV global capacity additions. Figure 24 shows CSP thermal energy storage global capacity and the annular additions between, 2008–2018.

As depicted in Fig. 24, CSP thermal energy storage global capacity is rapidly increasing. At the end of 2018, the total amount was 16.6 GWh. Figure 25 shows the wind power global capacity and annular additions.

As seen in Fig. 25, wind power global capacity is rapidly increasing. At the end of 2018, the total amount was 591 GW. Figure 26 shows the solar water heating collectors global capacity.

Figure 26 indicates that, solar water heating collectors global capacity is also rapidly increasing. At the end of 2018, the total amount was 480 GW. Figure 27 shows the solar water heating collectors additions.

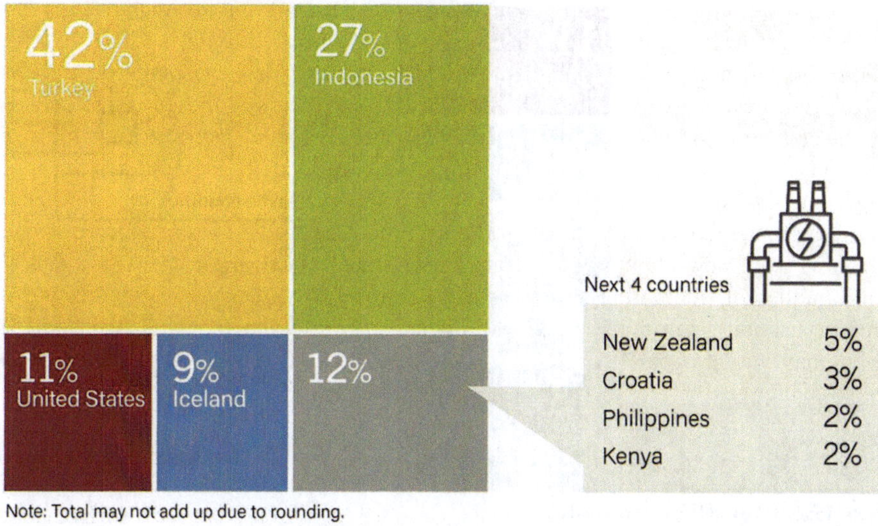

Fig. 21 Geotermal power capacity global additions, share by country 2018 (Renewables 2019)

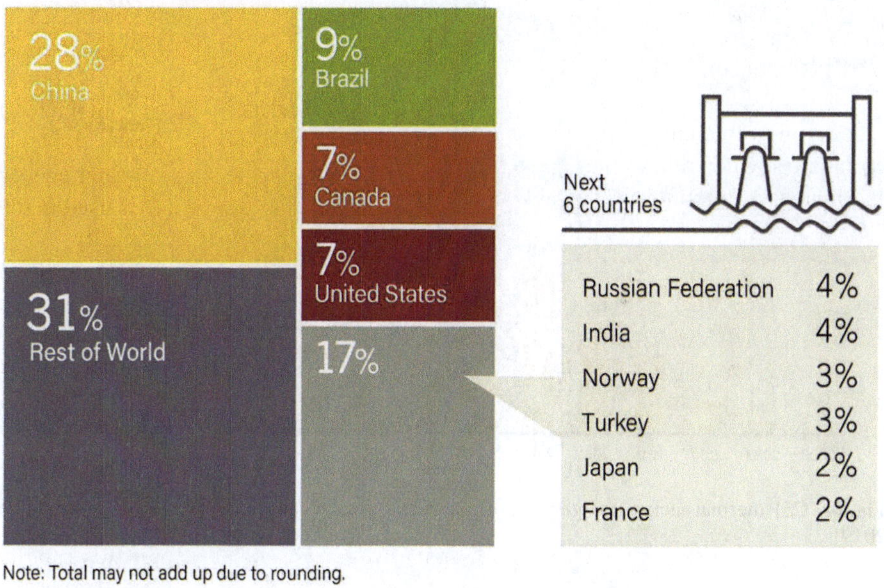

Fig. 22 Hyropower global capacity, shares of top 10 countries and rest of world 2018 (Renewables 2019)

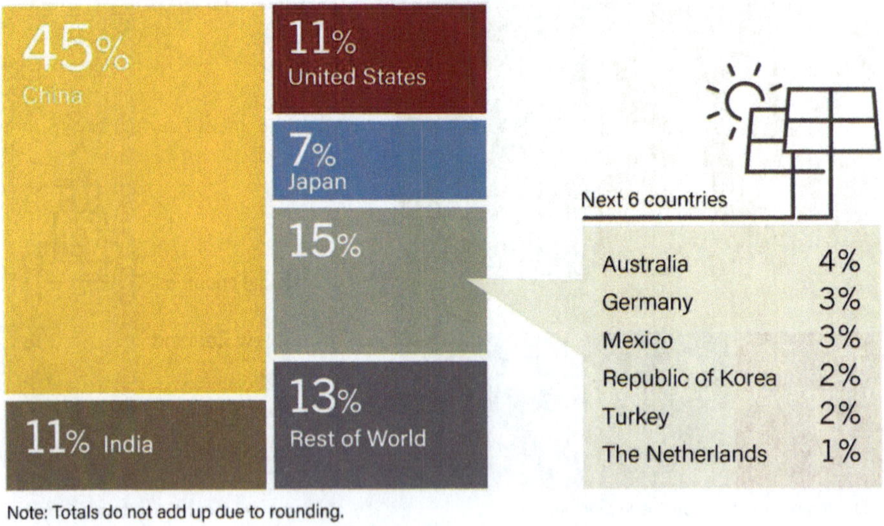

Fig. 23 Solar PV global capacity additions, share of top 10 countries and rest of world 2018 (Renewables 2019)

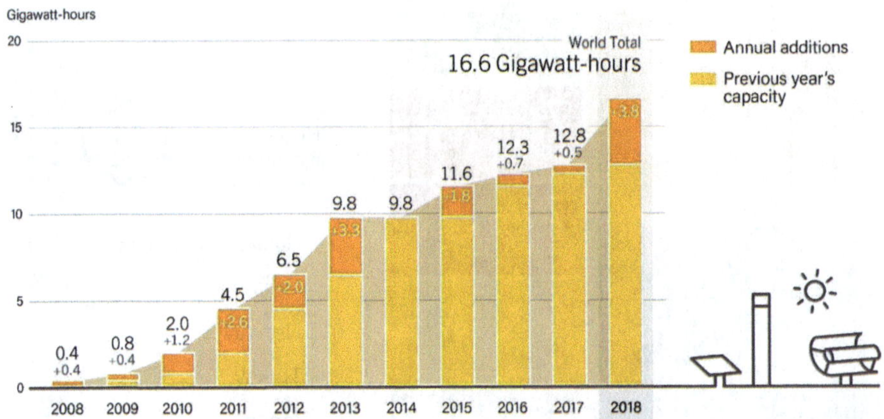

Fig. 24 CSP thermal energy storage global capacity and annular additions, 2008–2018 (Renewables 2019)

From Fig. 27 we can follow the solar water heating collectors for glazed, unglazed and evacuated tube types for the top 20 countries with the highest rate. The first place is occupied by China followed by Turkey in second place.

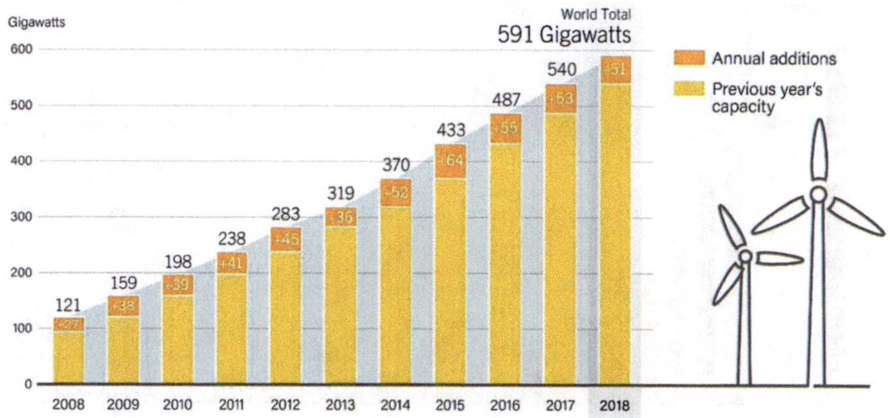

Fig. 25 Wind power global capacity and annular additions, 2008–2018 (Renewables 2019)

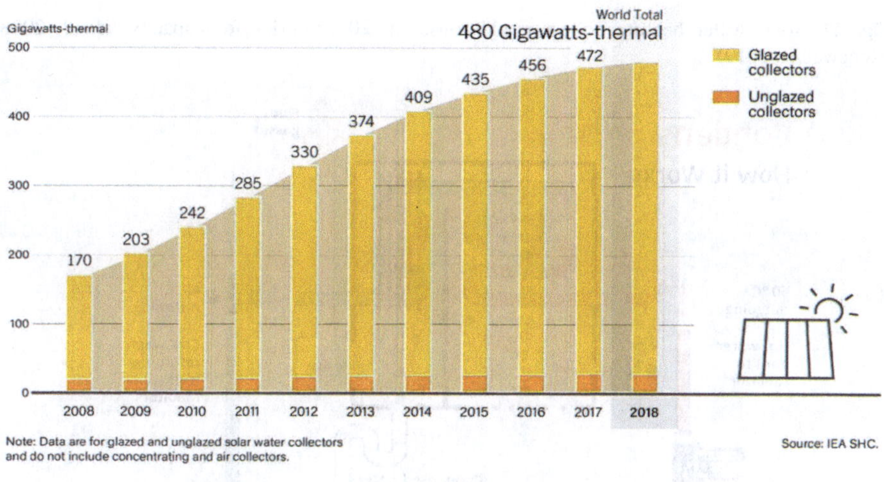

Fig. 26 Solar water heating collectors global capacity, 2008–2018 (Renewables 2019)

4.4 Condensing Boiler Systems

Condensation is the event that a gaseous substance is transformed into the liquid state by giving heat to its surroundings. Condensing combi boilers are the devices that increase the total efficiency by recovering the energy in the water vapor of the flue gas to be discharged through the chimney by means of a special heat exchanger (Fig. 28).

The efficiency of condensing combi boiler systems is higher than that of conventional combi boilers. This results in lower fuel consumption and lower flue gas

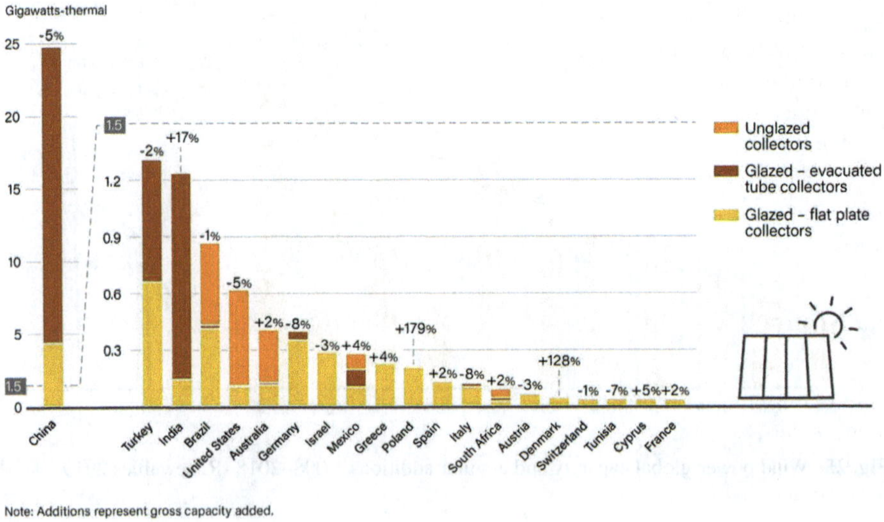

Fig. 27 Solar water heating collector additions, top 20 countries for capacity added, 2008 (Renewables 2019)

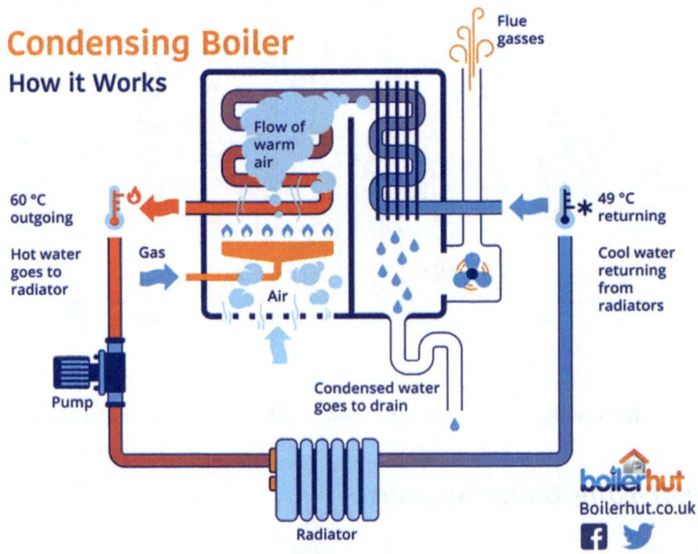

Fig. 28 Condensing boiler systems (http://boilerhut.co.uk/)

temperatures. It adjusts the flame height according to demand and operates modulated and continuously without turning on and off. With the help of the fan modulation, high combustion efficiency is obtained depending on the amount of gas consumed and a much quieter operation is achieved. It is more environmentally friendly with lower emissions (CO and NO_x). In case of operation below 50 °C; efficiency increases up to 79% in normal combi boilers and up to 109% in condensing combi boilers, thus saving 30% (Kaya 2012).

5 Smart Technologies and IoT Applications in Buildings

Smart buildings can be defined as developments where the building system management is carried out in a controlled manner within the scope of the systems within the building in order to maximize energy efficiency. These buildings provide high performance while keeping energy consumption to a minimum.

Ventilation systems, air conditioners and heating systems together with different equipment operate harmoniously so that energy efficiency is kept under control automatically. In this way, while keeping the production at the highest level, more production opportunities can be achieved with this centrally controlled system. Therefore, while it is used in factories for different production lines in various sectors, it is also preferred for different building structures for both residential and commercial use.

Controlling the complete structure within the scope of a general center, with building automation systems, energy management and control systems, together with monitoring systems brings great advantage to smart buildings. Hence, electrical and electronic systems can be used at desired levels, heating and cooling systems can become passive at certain times adapting to environmental factors and with the aid of the existing automation system, interaction with the tenants can be done in a way that is fully compatible with automated machines. Thanks to industrial automation systems which have made great progress with Industry 4.0, intelligent building applications hold an important place today (https://proente.com/akilli-bina-nedir/).

Advantages of smart building systems:

- Provides security.
- Provides comfort.
- Saves time and energy.
- Makes life easier for people experiencing physical or mental discomfort.
- Reduces responsibilities.

Internet of Things (IoT), provides the interconnection of smart objects within a widely spread large computing environment (Miorandi et al. 2012). The term Internet of Things refers to this internet-based structure which facilitates the exchange of information and data between numerous objects that are smart within themselves (Miorandi et al. 2012).

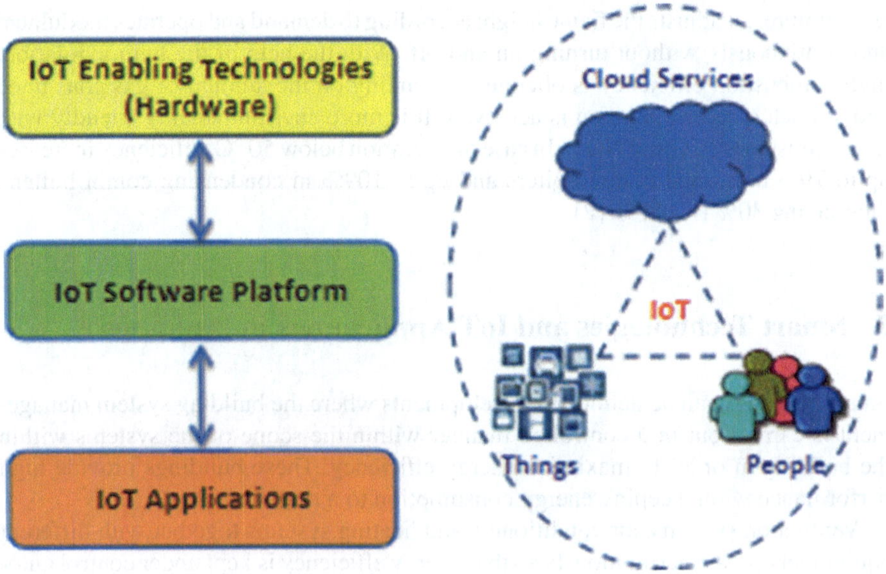

Fig. 29 The IoT architecture model (Khajenasiri et al. 2017)

The Internet of Things, IoT, is a term that reflects the growing number of intelligent, connected products and emphasizes new opportunities they can represent. What makes these smart, interconnected products fundamentally different is not particularly the Internet, but the change in the nature of objects. It is the broad capabilities of the intelligent, connected products and the big data that enable us to enter the new age of competition (Fig. 29). By analyzing this information, a completely different phenomenon is created which gives the devices the ability to decide on their own (Porter and Heppelmann 2015).

Zero energy building Technologies are considered as an important tool towards the reduction of greenhouse gas emissions. However, realization of comfort conditions without degrading them is possible with the correct design of mechanical and air conditioning systems. In the improvement of indoor air quality, the use of Internet of Things, IoT-based systems, help to reduce the total energy consumption of the building. Data from users and information collected from the environment are collated and processed in the presence of an effective communication network. Here, the fact that people living in the neighborhood are moving and moving constantly complicates the control event (Heperkan 2018). PID, proportional-integral-differential control systems make it easy to optimize comfort conditions. IoT greatly increases access to the network structure and computing capacity.

Digital Control Systems (DDC) and BIM technologies are also used in intelligent building applications. These technologies are very useful in the design, operation, control, management and monitoring of air-conditioning and mechanical systems

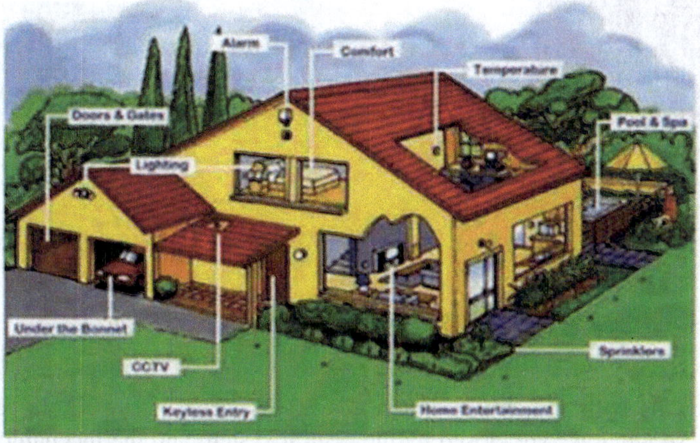

Fig. 30 Automation in the houses (Jourdan 2010)

of intelligent buildings. In a standard indoor monitoring scenario, the DDC system commands the installation valves and fresh air dampers to regulate heating/cooling by evaluating inputs such as temperature, humidity, indoor air quality. When the design parameters are selected correctly, it helps to use energy efficiently and to reduce energy costs (Heperkan 2019). Because DDC systems use electronic sensors and drives instead of pneumatics, they can save more energy as a result of a more precise control. DDC systems can be used effectively in many areas. These include residential automation (Fig. 30), air conditioning systems of commercial buildings and building automation and industrial applications (PLC, SCADA, etc.) (Heperkan 2018; Jourdan 2010).

BIM, "building information modelling" can be defined as production and management of physical and functional properties of spaces in a digital environment. Building information models are usually specific information found in files supplied in special formats. Using this information, it is easier to analyze and make decisions on the design, construction and operation of buildings. Nowadays, BIM software is used intensively by private and public institutions and companies that design, build and operate systems related to water, waste, electricity, gas, communication, road, bridge, port, tunnel, etc. Since 2007, it is being used in public buildings in Norway, in tenders over 2 million euros in Finland and in large projects in the United States. It has also been required as a prerequisite for public projects since 2012 in the Netherlands, 2014 in Hong Kong, 2016 in South Korea and the UK (Porter and Heppelmann 2015; Heperkan 2019).

Frankfurt Commerzbank Headquarters Building (Fig. 31) is an example of a smart building. The Norman Foster building, the tallest building in Europe at the time of its construction, is based on a natural ventilation system. The windows in the inner wall of the double-skinned facade and the inner courtyard windows were designed to be controlled by the users using the central building management system or the controls

Fig. 31 Commerzbank headquarters building (Yilmaz 2006)

mounted on the walls. When undesirable conditions occur in the workplaces or living spaces, these windows are closed by the central system and the mechanical HVAC system is automatically activated. The lighting in the office spaces is automatically adjusted according to the amount of daylight and occupancy of the space. Corridor and office lighting are activated by motion sensors. This building, which includes automation systems integrated with passive systems, saves 25–30% energy compared to traditional buildings. (Yilmaz 2006).

6 Case Study: Grand Bazaar, Water Source Heat Pump

The Grand Bazaar (Turkish: Kapalıçarşı, meaning "Covered Bazaar") is located inside the ancient city walls of Istanbul and is one of the largest and oldest covered markets in the world, with 61 covered streets and over 2,000 shops which attract between 250,000 to 400,000 visitors daily. It lies on the slope of the third hill of Istanbul, between the ancient Forum of Constantine and of Theodosius. There are two main buildings called the "Bedesten" one of which is dated as far as the Byzantine era (Cevahir). The other (Sandal) was built in the Ottoman reign. The complex has around 45 000 square meters of covered area. Analysis of the brickwork shows that most of the structure originates from the second half of the 15th century. It is well known for its jewelry, hand-painted ceramics, carpets, embroideries, spices and antique shops. Many of the streets in the bazaar are grouped by type of goods, with special areas for leather, gold jewelry and the like.

Today the Grand Bazaar (Fig. 32) is a thriving complex, employing 26,000 people. The heating and cooling of the shops are done with split type air conditioners at the present. The view of the roof is very disturbing and does not suit with the historical structure of the sight (Fig. 33). It is dominated with external units of the air conditioners, water storage tanks, antennas, etc. (Sevindir and Temel 2014).

A restoration project started in 2012 to renew its infrastructure, heating and lighting systems. This study is related to the HVAC design to centralize the heating and

Fig. 32 Overall view of the Grand Bazaar

Fig. 33 External units of the split air conditioners reflect a bad view

cooling systems. The variety of the different trades (jewelers, carpet shops, restaurants, etc.) requiring different indoor conditions in the complex makes the application quite difficult. The dominant loads in the Grand Bazaar are the internal loads from occupants and lighting (Sevindir and Temel 2014).

6.1 HVAC System for the Grand Bazaar

A fan coil system with a water cooled chiller, a fan coil system with an air cooled chiller, a VRV system, cooling towers or a water source heat pump system could be used for the Grand Bazaar. Water cooled chillers require cooling towers, that are difficult to install on the roof of a historical building. The chillers themselves are huge and heavy structures themselves. Due to the ancient texture of the Bazaar and the architectural restrictions imposed, together with the billing difficulties for heating and cooling among the shops, a water source heat pump system was adopted. The system solves the modern HVAC requirement of the historical site without disturbing its ambiance.

Two different water source heat pump systems have been considered for this study. The first alternative utilizes cooling towers to cool the common circuit in the summer and a boiler to heat it in the winter (Fig. 34).

The second alternative cools the common circuit utilizing sea water from the Marmara Sea (Fig. 35). A small boiler is still needed to heat the common line when required. This can easily be achieved by a condensing gas boiler, since natural gas is already available at the site of the bazaar. Water is extracted from a natural source at a depth which corresponds to year-round typical chill water temperatures. Seawater flows countercurrent to the intermediate closed cycle fresh chill water loop from the buildings to dump the heat into the sea. The intermediate fresh chill water loop never comes in direct contact with the seawater which removes concerns of contamination or silting. The chill water loop is treated freshwater which can pass safely through the serviced building central air systems without harm to existing systems or air handling units/coils. The blue and red lines under ground (marked as 3) represent the chill water loop (treated fresh water) which extracts the heat from the serviced client buildings (marked as 4 and 6) in a typical central air conditioning system through the existing air handling coils/units. The red pipeline (marked as 5) coming out of

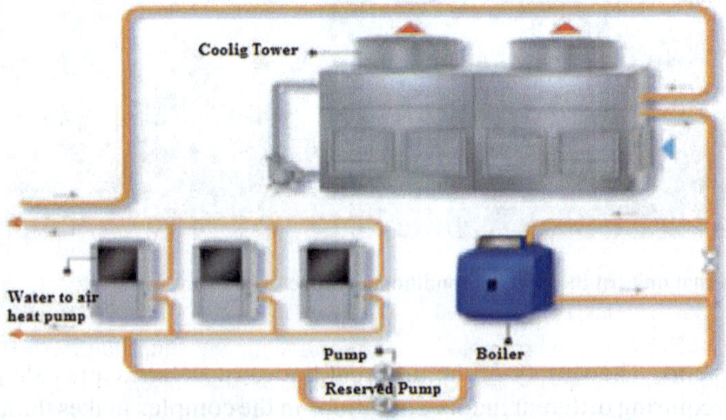

Fig. 34 A water source heat pump system with cooling towers and a boiler

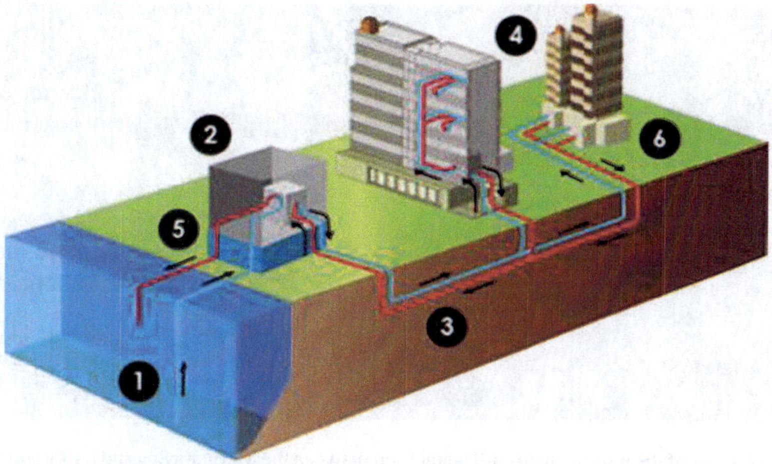

Fig. 35 Alternative solution utilizing sea water

the heat exchange building (marked as 2) is the seawater discharge which has been only modestly raised in temperature from the deep water temperature (although still at a lower temperature than the surface water) which means there is little if any environmental impact.

Our system uses two main compact plate heat exchangers that are connected through a closed circuit piping system installed underground. One heat exchanger is located at sea level and uses sea water from the Marmara Sea to exchange heat with the fluid in the connection pipes of the intermediate closed circuit fresh chill water loop. Instead of using this fresh water directly in the shops of the bazaar, a second heat exchanger is placed in a mechanical room located underground near the Grand Bazaar (Fig. 36). The figure shows a satellite image of the old city located on the ancient peninsula of Istanbul surrounded by the Byzantine era city walls. The red line represents the route of the intermediate closed circuit fresh chill water loop. This heat exchanger serves as the source for the water distribution between the shops. It also serves as an isolation between the intermediate loop and the main water lines looping the bazaar. Internal units are utilized to extract or dump heat from the shops. The two Bedestens have been treated separately since they are arranged in the form of street sales booths. A VRV system is utilized with water cooled condensers with limited fresh air support.

An alternative could have been a system utilizing cooling towers to replace the intermediate chill water loop and the heat exchanger at the Marmara Sea shore, dumping the heat to the sea. In this case, the only water loop would have been the one circulating through the bazaar area. The system would still be a water source heat pump system, but the generated heat at the internal units would be removed by the 6 cooling towers required for the project as depicted in Fig. 37. The six towers disturb the historic view of the bazaar roof top. Besides, the water vapor fumes coming out of the cooling towers continuously would create an unpleasant environment and

Fig. 36 Layout of the intermediate chill water loop between the Marmara Sea and the Grand Bazaar (red line)

Fig. 37 Alternative solution utilizing cooling towers

possibly increase the humidity of the area, creating complaints from the touristic facilities in the neighborhood.

6.2 *Load Calculations*

Following the architectural works in the Grand Bazaar, the plans of the ancient site were established. Heating and cooling loads of the shops were determined using Carrier's Hourly Analysis Program (HAP) (https://www.carrier.com/commercial/en/us/software/hvac-system-design/hourly-analysis-program/). Internal loads were considered using occupancy scenarios. Daily as well as seasonal weather conditions were adopted into the analysis. The alleys between the shops were unconditioned regions, however due to the structure, were not subject to the same outdoor conditions of the prevailing season.

HAP estimates cooling and heating design loads for commercial buildings in order to determine the required sizes for HVAC system components. Ultimately, the program provides information needed to select and specify equipment. Specifically, the program performs the following tasks (Sevindir and Temel 2014);

- Calculates design cooling and heating loads for spaces, zones and coils in the HVAC system
- Determines required airflow rates for spaces, zones and the system
- Sizes cooling and heating coils.
- Sizes air circulation fans.
- Sizes chillers and boilers.

We calculated the heating and cooling loads of the Grand Bazaar that generally stays open from 08:00 to 19:00 and is used intensively during this time period. ASHRAE design weather conditions for İstanbul International Atatürk Airport were used for the analysis. Summer design dry bulb and wet bulb temperatures for the area are 30 °C and 21.1 °C, respectively. Winter design dry bulb temperature is −3.3 °C and the relative humidity is 50%.

In this analysis we modeled the heat transfer of each shop separately. Peak loads and required airflow rates were determined for each shop. Characteristics of these spaces were taken from architectural plans. One common wall construction was used for all exterior walls. The construction consisted of 600 mm stone. The exterior surface absorption was in the "dark" category. The overall U-value was 2.174 W/m^{2}K. The overall weight was 1560 kg/m^2. (Sevindir and Temel 2014).

One uniform horizontal roof construction was used for all shops. The roof construction consisted of 400 mm stone. The exterior surface absorption was in the "dark" category. The overall U-value was 3.028 W/m^{2}K. The overall weight was 1040 kg/m^2. A 800 × 800 mm skylight was used for all shops. The overall U-value was 2.4 W/m^{2}K. One common floor type was used for all floors. The floor type was "slab floor on grade". The total floor U-value was 1.750 W/m^{2}K. Recessed, unvented lighting fixtures were used for all shops. A lighting density of 15 W/m^2 was assumed. (Sevindir and Temel 2014). The fixture ballast multiplier was 1.00. For the shops in the Grand Bazaar, a design day lighting level of 100% from 08:00 through 19:00 was used.

An occupancy density of 1 m^2/person was used. For all shops, an "office work" activity level was used (71.8 W/person sensible, 60.1 W/person latent). For shops in the Grand Bazaar, we assumed a design day occupancy level of 10% from 08:00 through 10:00, 50% from 10:00 through 11:00, and 100% from 11:00 through 19:00. Electrical equipment was being used in some shops. An electrical equipment density of 15 W/m^2 was used. For shops in the Grand Bazaar, we assumed a design day electrical equipment levels of 10% from 08:00 through 10:00, 50% from 10:00 through 11:00, and 100% from 11:00 through 19:00. An infiltration rate of 1 ACH was taken for all shops in the design cooling and heating load calculations. Infiltration occured at all times.

The data given above was used to perform the system design calculations for the Water Source Heat Pump (WSHP) system. A sample output for a shop is given in Table 1 and the Air System Design Load Summary Report is shown in Table 2.

The peak heating load of around 3410 kW was reached during the heating period and a peak cooling load of 4456 kW in June at 16:00. When the loads are examined dynamically, it can be observed that, among the shops in the Bazaar, some require cooling while others need heating. This is valid also in the cold season due to the nature of the trade in the shops.

Table 1 Sample load sheet for a shop in the Grand Bazaar

2748	Design cooling			Design heating		
	Cooling data at Jun 1900			Heating data at DES HTG		
	Cooling OA DB/WB 27.7 °C/20.6 °C			Heating OA DB/WB −3.3 °C/−5.7 °C		
	Occupied T-STAT 23.9 °C			Occupied T-STAT 21.1 °C		
Zone loads	Details	Sensible (W)	Latent (W)	Details	Sensible (W)	Latent (W)
Window and skylight solar loads	21 m^2	6429	–	21 m^2	–	–
Wall transmission	0 m^2	0	–	0 m^2	0	–
Roof transmission	330 m^2	13,347	–	330 m^2	24,443	–
Window transmission	0 m^2	0	–	0 m^2	0	–
Skylight transmission	21 m^2	149	–	21 m^2	1239	–
Door loads	0 m^2	0	–	0 m^2	0	–
Floor transmission	351 m^2	0	–	351 m^2	0	–
Partitions	0 m^2	0	–	0 m^2	0	–
Ceiling	0 m^2	0	–	0 m^2	0	–
Overhead lighting	5271 W	4316	–	0	0	–
Task lighting	1950 W	1754	–	0	0	–
Electric equipment	5271 W	4731	–	0	0	–
People	351	17,998	21,119	0	0	0
Infiltration	–	0	0	–	0	0
Miscellaneous	–	0	0	–	0	0
Safety factor	0%/0%	0	0	0%	0	0
Total zone loads	**–**	**48,723**	**21,119**	**–**	**25,682**	**0**

Table 2 Air system design load summary report

Air System Information

Air System Name .... **KAPALI ÇARŞI** Number of zones .... !
Equipment Class .... **TERM** Floor Area **17250**
Air System Type .... **WSHP** Location .... **Istanbul, Turk**

Sizing Calculation Information

Zone and Space Sizing Method:

Zone L/s .... **Sum of space airflow rates** Calculation Months .... **Jan to D**
Space L/s .... **Individual peak space loads** Sizing Data .... **Calculat**

	Design cooling			Design heating		
	Cooling data at Jun 1600			Heating data at DES HTG		
	Cooling OA DB/WB 29.4 °C/21.1 °C			Heating OA DB/WB −3.3 °C/−5.7 °C		
Zone loads	Details	Sensible (W)	Latent (W)	Details	Sensible (W)	Latent (W)
Window and skylight solar loads	867 m^2	345,419	–	867 m^2	–	–
Wall transmission	0 m^2	0	–	0 m^2	0	–
Roof transmission	13,194 m^2	412,284	–	13,194 m^2	976,424	–
Window transmission	0 m^2	0	–	0 m^2	0	–
Skylight transmission	867 m^2	8295	–	867 m^2	50,876	–
Door loads	0 m^2	0	–	0 m^2	0	–
Floor transmission	13,960 m^2	0	–	13,960 m^2	0	–
Partitions	0 m^2	0	–	0 m^2	0	–
Ceiling	0 m^2	0	–	0 m^2	0	–
Overhead lighting	258,754 W	206,137	–	0	0	–
Task lighting	121,170 W	107,490	–	0	0	–
Electric equipment	260,759 W	230,812	–	0	0	–
People	16,253	801,600	1,048,553	0	0	0
Infiltration	–	0	0	–	0	0
Miscellaneous	–	0	0	–	0	0
Safety factor	0%/0%	0	0	0%	0	0

(continued)

Table 2 (continued)

	Design cooling			Design heating		
	Cooling data at Jun 1600			Heating data at DES HTG		
	Cooling OA DB/WB 29.4 °C/21.1 °C			Heating OA DB/WB −3.3 °C/−5.7 °C		
Zone loads	Details	Sensible (W)	Latent (W)	Details	Sensible (W)	Latent (W)
Total zone loads	**–**	**2,112,037**	**1,048,553**	**–**	**1,027,299**	**0**
Zone conditioning	–	2,462,162	1,048,553	–	1,015,648	0
Plenum wall load	0%	0	–	0	0	–
Plenum roof load	0%	0	–	0	0	–
Plenum lighting load	0%	0	–	0	0	–
Exhaust fan load	0 L/s	0	–	0 L/s	0	–
Ventilation load	71,128 L/s	380,559	154,627	71,128 L/s	2,077,218	0
Ventilation fan load	0 L/s	0	–	0 L/s	0	–
Space fan coil fans	–	0	–	–	0	–
Duct heat gain/loss	0%	0	–	0%	0	–
Total system loads	**–**	**2,842,720**	**1,203,180**	**–**	**3,092,866**	**0**
Terminal unit cooling	–	2,842,720	1,204,165	–	0	0
Terminal unit heating	–	0	–	–	3,092,866	–
Total conditioning	**–**	**2,842,720**	**1,204,165**	**–**	**3,092,866**	**0**
Key	**Positive values are clg loads**			**Positive values are htg loads**		
	Negative values are htg loads			**Negative values are clg loads**		

This study suggests that a solution utilizing water source heat pumps is energy efficient, environmental friendly and practical to use for the occupantants.

References

Climate Change (İklim Değişikliği). https://www.wwf.org.tr/ne_yapiyoruz/iklim_degisikligiveenerji/iklim_degisikligi/. Accessed 25 Sept 2019

United Nations Climate Change Conference: COP24. https://www.semtrio.com/cop24-birlesmis-milletler-iklim-degisikligi-konferansi

Heperkan H, Baloğlu F, Karahan A (1994a) Modern heating systems. Inst Eng 2(16):53–55

Heperkan H, Baloğlu F, Karahan A (1994b) Modern heating systems. Inst Eng 2(17):36–39

MMO (2018) Energy outlook of Turkey (Türkiye'nin Enerji Görünümü). Yayın No: MMO/691, TMMOB Makine Mühendisleri Odası. www.mmo.org.tr/sites/default/files/EnerjiGorunumu2018_1.pdf

Goals and activities of institutions related to the ministry of energy and natural resources (Enerji ve Tabii Kaynaklar Bakanlığı ile Bağlı, ilgili ve ilişkili Kuruluşlarının Amaç ve Faaliyetleri) (2016). www.enerji.gov.tr. Last Accessed 19 March 2018

Koç A, Yağlı H, Koç Y, Uğurlu İ (2018) General evaluation of the energy outlook in Turkey and the world (Dünyada ve Türkiye'de Enerji Görünümünün Genel Değerlendirilmesi). Eng Mach 59(692):86–114 (2018)

Electricity (2019). www.enerji.gov.tr. Last Accessed 20 March 2020

ETKB-EİGM. https://www.enerji.gov.tr/tr-TR/EIGM-Raporları. Last Accessed 15 Nov 2018

Electricity consumption distribution according to sectors for Turkey (Türkiye Elektrik Tüketiminin Sektörlere Göre Dağılımı). http://www.teias.gov.tr. Last Accessed 15 Nov 2018

International Energy Agency, World Energy Outlook (2018). IEA WEQ

Energy Performance of Buildings Directive (Binalarda Enerji Performansı Yönetmeliği) (2019). www.mevzuat.gov.tr. Accessed 04 Oct 2019

Erhorn H, Erhorn-Kluttig H (2012) The path towards 2020: nearly zero-energy buildings. REHVA Europ HVAC J 49(3):12–15

UK (2017) National energy efficiency action plan and annual report. https://ec.europa.eu/energy/sites/ener/files/documents/uk_neeap_2017.pdf. Accessed 27 Sept 2019

Wittchen KB, Thomsen KE (2012) Introducing cost-optimal levels for energy requirements. REHVA Europ HVAC J 49(3):25–29

Energy Certificate (Enerji Kimlik Belgesi). http://www.enerjikimlikbelgesi.com/. Accessed 28 Sept 2019

Korkmaz E (2019) Energy performance in buildings (Binalarda Enerji Performansı) (BEP). http://www.atermit.com/Sayfalar/bep-tr. Accessed 28 Sept 2019

Voss K, Musall E (2012) Net zero energy buildings: international projects of carbon neutrality in buildings. Energy Build 43(4):971–979

DIN V 18599:2007

Hourly-Analysis-Program (HAP). https://www.carrier.com/commercial/en/us/software/hvac-system-design/hourly-analysis-program/. Accessed 28 Sept 2019

https://energyplus.net/. Accessed 28 Sept 2019

What are Heat Pumps (Isı Pompası Nedir?). https://midori.com.tr/isi-pompasi/isi-pompasi-nedir.html. Accessed 29 Sept 2019

Pravadalıoğlu S, Komisyonu EMOİŞE, Şti ÜTEML (2011) Local energy production-cogeneration systems (Yerinde Enerji Üretimi-Kojenerasyon Sistemleri). EMO

Wit JD (2011) Mini and micro cogeneration. Danish Gas Technology Centre, ICCI

https://centraxgt.com/. Accessed 28 Sept 2019

Gandıglıo M, Ferrero D, Lanzını A, Santarellı M (2020) Fuel cell cogeneration for building sector: European status. REHVA Europ HVAC J 57(1):21–26. Şubat

The Benefits of Micro-CHP: A Delta-ee report produced on behalf of COGEN Europe (2015). http://www.cogeneurope

Özkara G (2018) What are renewable energy sources? (Yenilenebilir Enerji Kaynakları Nelerdir?) https://www.enerjiportali.com/yenilenebilir-enerji-kaynaklari-nelerdir/. Accessed 3 Oct 2019

Renewables 2019 Global Status Report (2019). https://www.ren21.net/wp-content/uploads/2019/05/gsr_2019_full_report_en.pdf. Accessed 05 Oct 2019
Kaya D (2012) Why condensing boilers? (Neden Yoğuşmalı Kombi?) http://www.dogalgaz.com.tr/yayin/214/6028.html. Accessed 05 Oct 2019
Combi Boiler System. http://boilerhut.co.uk/. Accessed 06 Oct 2019
What is a Smart Building? (Akıllı Bina Nedir?) https://proente.com/akilli-bina-nedir/. Accessed 06 Oct 2019
Miorandi D, Sicari S, De Pellegrini F, Chlamtac I (2012) Internet of things: vision, applications and research challenges. Ad Hoc Netw 10(7):1497–516
Khajenasiri I, Estebsari A, Verhelst M, Gielen G (2017) A review on Internet of Things solutions for intelligent energy control in buildings for smart city applications. Energy Procedia 111:770–779
Porter ME, Heppelmann JE (2015) How do smart products change competition (Akıllı Bağlantılı Ürünler Rekabeti Nasıl Dönüştürüyor). Harvard Bus Rev
Heperkan H (2018) Digitalization in HVAC systems: smart buildings (İklimlendirme Sistemlerinde Dijitalleşme: Akıllı Binalar). Termoklima 116:12–14. http://termoklimadergisi.com/img/9fubbOs9.pdf. Accessed 21 Oct 2019
Heperkan H (2019) Heating technologies in renovated buildings (Yenilenen Binalar İçin Isıtma Teknolojileri). Termoklima 133:12–16. http://termoklimadergisi.com/img/Yeh1ZIjg.pdf
Jourdan G (2010) DDC controls and software for the HVACR Lab. www.hvacexcellence.org/nhetc/archive/docs/2010nhetcdocs/ddccontrols.pdf
Yilmaz Z (2006) Smart buildings and renewable energy (Akıllı Binalar ve Yenilenebilir Enerji). J Inst Eng, Tesisat Mühendisliği Dergisi 91:7–15
Sevindir MK, Temel S (2014) Heat pump application in the HVAC system of the Sandal Bedesten in the covered Bazaar (Kapalıçarşı Sandal Bedesten'in İklimlendirilmesinde Isı Pompası Uygulaması). J Inst Eng, Tesisat Mühendisliği Dergisi 142:42–49

Energy System Analysis, Simulation and Modelling Practices in Turkey

Egemen Sulukan, Tanay Sıdkı Uyar, Doğancan Beşikci, Doğuş Özkan and Alperen Sarı

Abstract Humankind first began to use simple machines instead of arm force. Subsequently, more advanced and sophisticated energy technologies introduced to our lives since the industrial revolution. Relatively advanced machines have begun to replace the simple mechanical systems in order to meet the various needs of humanity. Continuous improvements in related technologies have triggered the emergence of more compact, modular and more efficient vehicles over time. The rapid advances in technology naturally manifested itself also in energy technologies. Drivers in a national economy, particularly environmental problems caused by energy production and consumption, have been an area of interest in recent years. Energy technologies include increasing productivity, producing relatively more energy, or working towards an upward trend. In this perspective, this chapter aims to give an overview of the recent analysis of an energy system and modelling studies conducted by the energy joint workgroup at Marmara University and Naval Academy of National Defence University, Istanbul, Turkey.

Keywords Energy system analysis · Energy modelling · Turkey

1 Introduction

Humankind first began to use simple machines instead of arm force. Subsequently, more advanced and sophisticated energy technologies introduced to our lives since the industrial revolution. Relatively advanced machines have begun to replace the simple mechanical systems in order to meet the various needs of humanity. Continuous improvements in related technologies have triggered the emergence of more compact, modular and more efficient vehicles over time. The rapid advances in technology naturally manifested itself also in energy technologies.

E. Sulukan (✉) · D. Özkan
Mechanical Engineering Department, Turkish Naval Academy, National Defence University, İstanbul, Turkey
e-mail: esulukan@dho.edu.tr

T. S. Uyar · D. Beşikci · A. Sarı
Mechanical Engineering Department, Engineering Faculty, Marmara University, İstanbul, Turkey

T. S. Uyar (ed.), *Accelerating the Transition to a 100% Renewable Energy Era*,
Lecture Notes in Energy 74, https://doi.org/10.1007/978-3-030-40738-4_21

Drivers in a national economy, particularly environmental problems caused by energy production and consumption, have been an area of interest in recent years. Energy technologies include increasing productivity, producing relatively more energy, or working towards an upward trend.

In this perspective, this chapter aims to give an overview of the recent analysis of an energy system and modelling studies conducted by the energy joint workgroup at Marmara University and Naval Academy of National Defence University, Istanbul, Turkey.

1.1 Setting up the Reference Energy System to Be the Base of MARKAL Energy Model of Turkey

Energy modelling activities at Marmara University has been started with this study. The study emphasized that; as soon as the MARKAL Model of Turkey will be prepared, the phase of local and national decision support will be backed to join global energy resource management.

The initial step in energy system analysis and modelling studies is shaping the reference energy system (RES) of the analyzed environment. For this, the aim of this analysis has been determined to develop and make operational the RES and MARKAL Database to be used for the MARKAL Energy Decision Support System of Turkey. Related to this aim, resource technologies and energy carriers that are in use and can be used, conversion-process technologies and demands have been decided, their relations and situations in the MARKAL hierarchy have been determined, and a detailed network of energy from source to end-use, namely a RES has been created. In this phase, the statistical data of the network was unidentified, because of the missing data-statistical values i.e. technical efficiency, investment costs, fixed/variable operation and maintenance costs, transmission efficiency, the lifetime of the technologies, amount of demand; and therefore the energy balance could not be obtained.

After this step, the system was simplified and a secondary RES has been created in a more generalized level, then this network was specified by the relevant data and energy balance has been obtained; namely, the Base scenario has been constructed. Results of the Base case have been compared with the future projections in official development plans and it was seen that the reference energy system and the database are correct, with other alternative scenarios and proven that the model was operational. As a result of this study Reference Energy System of Turkey is constructed in two ways: simplified and detailed, the database has been constituted as an operational tool (Koç 2005).

1.2 *Analysis of the Effects of Greenhouse Gas Emission Reduction Strategies on Turkish Energy and Economy Systems*

Additional to the efforts given under the first study; the aim of this study has been determined as "to analyze the effects of different scenarios, particularly how to decrease greenhouse gas emissions on the Turkish energy and economy systems, by using Turkey MARKAL Energy Decision Support System and ANSWER software (Vardar 2005)."

2 Establishing Energy Efficient Utilization and Cost-Effective Energy Technologies Selection Strategies for Turkey Using MARKAL Family of Models

After above mentioned two studies; a more comprehensive study came out to bring up an operational improved model for Turkey, by using Standard MARKAL energy modeling platform, used by more than 40 countries around the world, and it has been established in the analysis period of years 2000–2025.

In this study; energy carriers, energy technologies, and demands are created in a database, their interplays and relations in MARKAL structure have been defined and a general network of energy from source to end-use has been constructed. Technical, economic and environmental effects were analyzed on the base scenario for the stated future cases by creating alternative scenarios (Sulukan 2010).

2.1 *Establishing Mitigation Strategies for Energy-Related Emissions for Turkey Using the MARKAL Family of Models*

This study has the aim to make a cost-effective contribution to the Turkish energy subsector about emission-related with energy in Turkey and integrate it into the world solution efforts.

The Standard MARKAL model which has been used in this work can connect the whole Turkish energy system ingredients like energy demands, subsidies, investment costs, capital requirements, greenhouse gas emissions, conversion, resource and process technologies in the same modeling platform. The model makes it easy to reach the right economic and environmental decisions by finding the best solution that meets the energy demand at the lowest cost by selecting energy supply and technologies.

The main priority for the creation of greenhouse gas emission reduction alternative is the use of relatively inexpensive types of non-polluting energy, especially in electricity generation, and the connection of electricity generated from renewable energy sources to the grid. It may be preferable to use electricity generated from renewable sources, especially for small consumption of non-mobile, residential, commercial, industrial or similar sectors. According to the results of this study, 50% of renewable energy use in electricity generation should be targeted in the first stage (Sağlam 2010).

2.2 A Preliminary Study for Post-Kyoto Period for Turkey by MARKAL Model

In this analysis, the RES of Turkey that has been developed by MARKAL is analyzed with the two alternative scenarios for reducing greenhouse gas emissions (GGE). The scenarios are theoretically requesting that they forecast considerable mitigations in GGE. Of course, this has seriously affected not only the prevention of economic development but also the large investment requirements for renewable energy sources.

It is unlikely that policymakers for whom economic development is of great importance will realize the significance of the scenarios focusing on the combat against climate change and its adverse effects. However, the scenarios described in this analysis demonstrate MARKAL's flexibility and suitability for developing energy analyzes for decision-makers (Sulukan et al. 2010a).

2.3 Determining Optimum Energy Strategies for Turkey by MARKAL Model

A comprehensive energy database has been developed in this study with the priority of technology selection strategies. After developing the RES for the energy system of Turkey, respective alternative scenarios have been offered to be applied against the developed Base Scenario. This process provides us with analyzing the possible effects for further and in-depth Turkey-MARKAL studies:

- The course of actions to increase the efficiency of thermal power plants by selecting their energy production, consumption and service levels in order to decrease greenhouse gas emissions,
- Analyses of future power plant growth plan to forecast the annual amount of investment and electricity capacity,
- Analyzing the possible effects of the transmission of electricity generation contributions of hydraulic, wind, solar and wave energy sources to the national grid.

- Possible options to maximize the combined heat and power contribution in sectors, especially power generation and industrial sub-sectors,
- Analyzing the possible outcomes of future nuclear power plants installed in the national energy system,
- Possible options to maximize the utilization of domestic renewable energy potentials in electricity production,
- Determining the greenhouse gas emission (GHG) mitigation scenarios to develop a sustainable energy action plan,
- Natural gas import fees possible instability is analyzed by the economic effects (Sulukan et al. 2010b).

2.4 Greenhouse Gas Emission Assessment from Maritime Transportation for Turkey by MARKAL Model

Maritime transportation is one of the anthropogenic sources of greenhouse gas (GHG) emissions which constitutes approximately 4% of the global total. Although this problem is currently being discussed at the international maritime organization and the United Nations Framework Convention on Climate Change (UNFCCC), policy effort on greenhouse gas emissions from maritime transport is insufficient yet. Shipping is the most environmentally and friendly mode of transport among other options in terms of GHG emissions.

However, it is predicted that the growth of emissions from ships will boost by 150–200% by 2050 if no precaution is taken, while it is estimated that emissions from ships will increase by 150–200% by 2050 if no action is taken. In this respect, the GHG emission inventory of Turkey is analyzed in order to determine the present state and future projection options of the maritime transportation sector as well as other sectors in the country. This study is carried out via MARKAL and its user-friendly Windows OS based interface ANSWER. MARKAL is a family of models that have been used in a variety of energy systems since the 1980s. The system is used by over 80 institutions in 40 countries.

In this study, the maritime transportation sector is designated as the main focus and the GHG emission inventory of Turkey has been identified to the model's database in the Base Scenario.

A basic GHG projection of Turkey is obtained by this study under the main assumptions. The maritime section of this inventory may give some remarkable insights in case of continuation with the Base Scenario. These issues will address the political, economic, environmental and technological measures as a part of the ongoing accession process to the European Union (Çelebi 2011).

2.5 *Deliberating Lower-Cost Emission Reduction Options*

In this study, the base scenario of Turkey MARKAL model indicates the current position that is carried out by the present policies which mainly back up the use of fossil fuel despite their huge emissions. In the base scenario, energy consumption and emission data for the base year were entered and analyzed in an integrated manner with the capacities and emission coefficients of the technologies. Even though the use of renewable energy, the most effective solution method to reduce emissions, is at a low level in Turkey, it is expected to be added to the national power system as long as economically viable. Owing to this model, we can test that not only in which renewable energy is invested in by the investors, but also determine whether we need nuclear energy to generate sustainable energy by applying alternative scenarios against the base scenario (Sağlam et al. 2013).

2.6 *Developing Reference Energy System for Agricultural Sector and Modeling Biomass-Based Biofuel Production*

Searching alternative energy sources and efficiently utilization of energy issues gains importance by gradual depletion of fossil fuel reserves. When the environmental effects are also considered, renewable energy sources are preferred in first order among other alternatives. Apart from solar and wind energy, biofuel production from biomass is specified as a significant work area in terms of relatively low cost and low greenhouse gas emissions; foreign dependency reducing effect and contribution to agricultural development.

With this perspective; firstly, the reference energy system is developed for the agricultural sector, then MARKAL energy-economy-ecology integrated model is generated including irrigation and all agricultural demands of the country from biofuels by biodiesel production from canola and bioethanol from sugar beet. MARKAL model generator is currently being improved by the Energy Technology System Analysis Programme (ETSAP); which is already a continuing multinational program under the International Energy Agency (IEA). MARKAL model is also used for scenario analyses in World Energy Outlook reports which are annually published by IEA. The model basically consists of six columns, where domestic canola and sugar beet are used as resources, biomass-based power plants, and biofuel production processes are conversion and process technologies. Electricity, biodiesel, bioethanol, (bio) hydrogen, biogas, and glycerin production is modeled as final energy carriers; these products are identified as demand technologies, which are electricity and biodiesel utilized by irrigation pumps, tractor, and agricultural transporters.

The generated energy meets the main demands of agricultural irrigation; land preparation and agricultural transportation and agricultural fertilizers, agricultural chemicals and liquid carbon dioxide consist of other products of the system as the other demands in this area (Şahbaz and Sulukan 2015).

2.7 Reference Energy System for Land Transportation in Turkey

Increasing energy consumption often adversely affects the transport sectors in developing countries. Integrating energy planning for the transport sector in these countries can diversify the options to change its suboptimal organization and reduce energy consumption.

The MARKAL family of models show a range of practical decision support tools for energy planning analyzes, as well as their relationship to environmental impacts. In the past, the model was a simple optimization program used only by researchers, but today it has turned into a very complicated package with many potential applications, from scientific research to analysis of energy/environmental policy and planning of questions.

Users who are not familiar with programming or optimization theory can use the model framework more easily through the Windows-based interface ANSWER. The MARKAL family is one of the most common decision support tools used to translate global commitments to reduce greenhouse gas emissions into projects and actions. The economic and ecological benefits of these individual activities and the additional benefits arising from the opportunities for cooperation should be assessed. The MARKAL model family makes informed decision-making with a flexible, well-understood, proven, verifiable and evolving methodology.

In this paper, a preliminary analysis is performed on the relevant statistical data of land transportation in Turkey. Then, the reference energy system of land transportation for Turkey is developed. According to statistics and RES, data obtained from the Turkish Statistical Institute (TurkStat) is entered into MARKAL-Answer software and the Base Scenario is prepared. In this way, all data is seen as a summary table. This will hopefully establish a foundation for further analyses in the transport sector in Turkey (Şener and Sulukan 2016).

2.8 Energy–Economy–Ecology–Engineering (4E) Integrated Approach for GHG Inventories

Energy is a very effective factor for the economic and social improvement of a country. Nonetheless, energy resources, energy conversion/processes or demand technologies are complicated and need to use an algorithm to optimize the energy system. Because that is the best way to get energy-efficient and clean solutions that may provide sustainability and mitigate climate change effects. It is possible for countries to achieve the purpose of sustainable development in terms of economic and environmental constraints through realistic and long-term strategic plans that combine and optimize their global and unique conditions. When performing a comprehensive analysis with realistic parameters, the current situation and the potential for the future can be taken into account. MARKet ALLocation (MARKAL) is a model manufacturer that is currently being developed and is currently being used by universities, government agencies, non-governmental organizations, energy services, and consulting companies. The Energy Technology Systems Analysis Program (ETSAP) is the implementation agreement of the International Energy Agency (IEA), which acts as a consortium of member country teams. It also invites teams that actively cooperate to build a consistent multi-country analytical capability related to energy, economy, environment, and engineering (4E).

This study sums up the progress of a MARKAL model for Turkey and the beginning predictions proceeding from the BAU predictions from 2005 to 2025. If the Turkish MARKAL model is analyzed in more detail, the basic demands to analyze for an effective energy policy road map of Turkey can be created. These can be achieved by creating and running additional alternative scenarios against the BAU scenario as follows:

- Supporting the financial decision mechanisms for renewable energy system investments,
- Accelerate the integration of renewable energy among the heating/cooling and transportation sector for meeting national targets,
- Improve alternative plans for increasing efficiency in thermal power capacity or technologies used in end-use sectors that influence the energy generation, consumption, and GHG emission levels,
- Creating potential candidate generator examinations to acquire yearly investment levels and power load rates of electricity,
- Boosting alternative capabilities of hydro, wind, solar and wave power sources of the national energy system is analyzed in terms of possible effects,
- Benefit from combined heat and power facilities also called co-generation across all sectors, particularly power plants and industrial sub-sectors,
- Analyzing upcoming nuclear power plants' effects in the energy system,
- Analyzing CO_2 reduction scenarios to determine a road map in terms of ecological aspects.

Currently utilized conventional energy resources manage Turkey's energy grid and the present energy mix involves no externalities for fossil fuels, as the energy technologies have progressed to be mainly based on fossil fuels since the industrial revolution. Demand technologies rely almost exclusively on the use of fossil fuels which bring about ecological pollution. However, management is the most significant component for deciding a useful path for decreasing the consumption of polluting harmful energy resources to reduce global warming effects and sustain society. There is a consensus among decision-makers, engineers, local authorities and non-governmental organizations on the choice of energy and technology to support environmental planning by experts with energy modeling capabilities. Additionally, it should identify relevant aspects of an energy system such as energy economy, environmental impacts, energy production and ethical responsibilities (Sulukan et al. 2017a).

2.9 Reference Energy System Development for Turkish Residential Sector

Another energy analysis and modelling of demand technologies via ANSWER-MARKAL software have been carried out by this study which focused on the Turkish residential sector. Some basic assumptions were made during data entry and model building processes. The energy decision support model can be complicated to the extent that the available data allows. Though, an initial phase of modelling is constituting the base scenario that belongs to demands technologies in the residential sector of Turkey has been completed. The next step will be to run the base scenario model and obtain the results of the BAU scenario. The next step after running the base scenario is establishing various alternative scenarios such as one with a fully renewable energy source and another with fossil fuel use set to zero, and so on.

Additionally, actions to achieve energy-efficiency and energy-saving ought to be able to bring into the model as input. The primary aim of this study is to have a projection of the next five decades and to help forward-looking alternative scenarios that will be progressed over time. These results will be used to minimize the total cost of the system by providing technical support to the decision-makers in the residential sector (Mutluel and Sulukan 2017) (Fig. 1).

2.10 Technical Efficiency Improvement Scenario Analysis for Conversion Technologies in Turkey

The results of the technical efficiency improvement option in especially conversion technologies can be analyzed through the technical efficiency improvement scenario. It is analyzed if there is a 1% increase in the technical efficiency of currently active energy production. When there is a 1% increase in technical efficiencies, the influences of the scenario occur in the efficiencies and investment costs of conversion technologies, beginning from 2015. For a 1% progress in energy efficiency, the investment costs of conversion technologies are estimated to be an additional cost of $2/kW. The investment costs of the TECH-1 scenario are calculated by adding $2/kW for each technology. After the main data are entered into the model assuming basic assumptions, this model runs against the BAU with the TECH1 Scenario. As a result, fossil fuel imports and export levels are the same in the Base and TECH-1 scenarios. According to the TECH-1 scenario, fossil fuel exportation halts during the 2010–2025 period. Final energy and use of final energy levels are the same in these two scenarios. Mining activities increased 55% and total primary energy decreased 6% in TECH1 Scenario.

Domestic fuel margin steps up 3.43% on average while annual total costs of TECH-1 diminish 3.77% on average. Fuel and net import cost diminished by 15.61% and 18.09%, on average. While the undiscounted total system cost diminished by 8.10%, the total discounted system cost diminished by 1.6% as can be seen in the

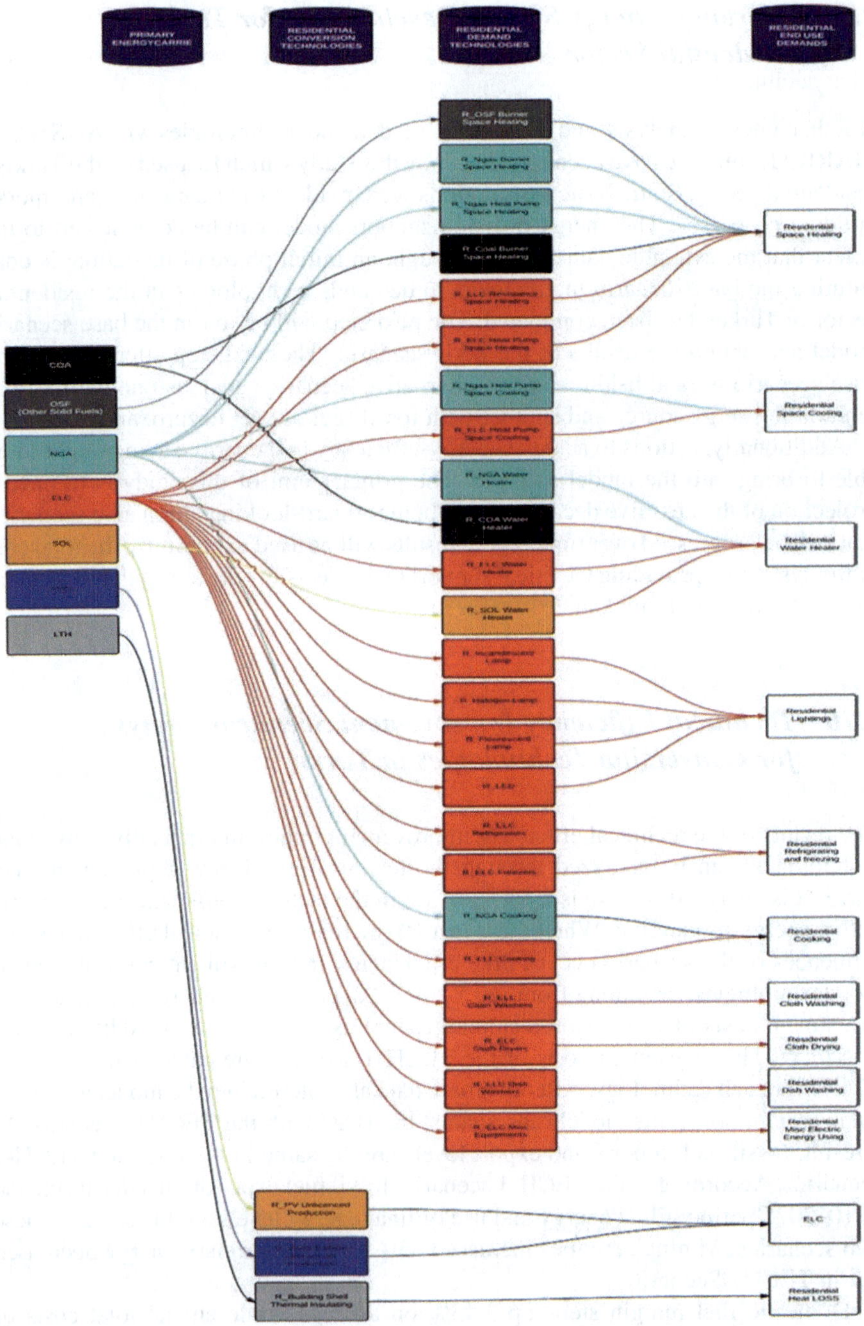

Fig. 1 Reference energy system of the residential sector in Turkey (Mutluel and Sulukan 2017)

TECH1 Scenario in this cost summary. Domestic fuel costs step up as the system prefers more domestic fuel rather than imports. Thus, import costs additionally somewhat decline. A small development has a positive effect on the national economy by creating a chain effect on the whole system (Sulukan et al. 2017b).

2.11 Analysis of Demand-Side Management Option with Cogeneration Implementations in Turkish Energy System by MARKAL Model

As stated in the official reports previously published, the main motivation of this study was the optimistic assumption of the demand side to activate the cogeneration potential and to manage it in a more efficient manner. Demand levels are assumed to decrease by 15% in industry and by 10% in the residential sector to show the possible effects of improving the use of expanding cogeneration, especially in industry. In addition to demand reduction precautions in these sectors, it is expected that cogeneration type facilities will be used to grow the share of cogeneration in total electricity production, which is expected to reach 16% in 2025 from 4% at the end of 2005. Thus, 16% of total electricity production would originate from combined heat and power in 2025.

If you compared with base scenario demand levels, the demand level is decreased by 13.9% in 2020 and by 25.9% in 2025. The total efficiency of combined heat and power generation is taken as 85%, whose 55% is for heat and 30% is for electricity production. It is also accepted that all new combined heat and power plant capacities will be ignited by natural gas.

Emission mitigations and cost-effectiveness of alternative scenarios were evaluated by comparing scenarios. Some economic and ecological values were calculated such as total economic cost, cost increase, net energy import cost, net energy added value, total CO_2 emission, emission increase and cost reduction (%) through alternative scenarios and Base Scenario values. While these modeling studies, the DSM scenario is considered as a cost-effective and climate-friendly application. At the rate of 43%, CO_2 reduction is achieved with this scenario. As a result of such measures, implementation will bring an additional cost to the country's economy in the first stage, but it will bring advantages both economically and environmentally in the long term. This idea will strengthen the position of our country in the international community in terms of determining the sustainable development target together with the IPCC process (Sulukan et al. 2017c).

2.12 A Native Energy Decision Model for Turkey

Turkey's Reference Energy System has been created over time and constantly improved throughout the study period involving the present and predicted resources, fuels, and technologies. Lastly, 28 energy carriers, 29 resource technologies, 11 process and 20 conversion technologies, 139 end-user (demand) technologies, 30 demands in five sectors, 11 emission components, 11 tax/subsidies, and 12 global items (i.e., annual discount rate, GDP in first year, fraction of year for season, time of day) have been determined in suitable positions according to MARKAL hierarchy and characterized with value by using ANSWER user interface (Fig. 2).

The menu-driven user interface of ANSWER makes it easy to adjust the data input and structuring of the MARKAL model. Demonstration cases with default data are also provided. This interface not only organizes file processing and execution but also includes result processing menus for both table and graphical display. However, the components and structure of the RES can be developed depending on economic, technological or environmental requirements for further analysis with future requirements (Sulukan et al. 2017d).

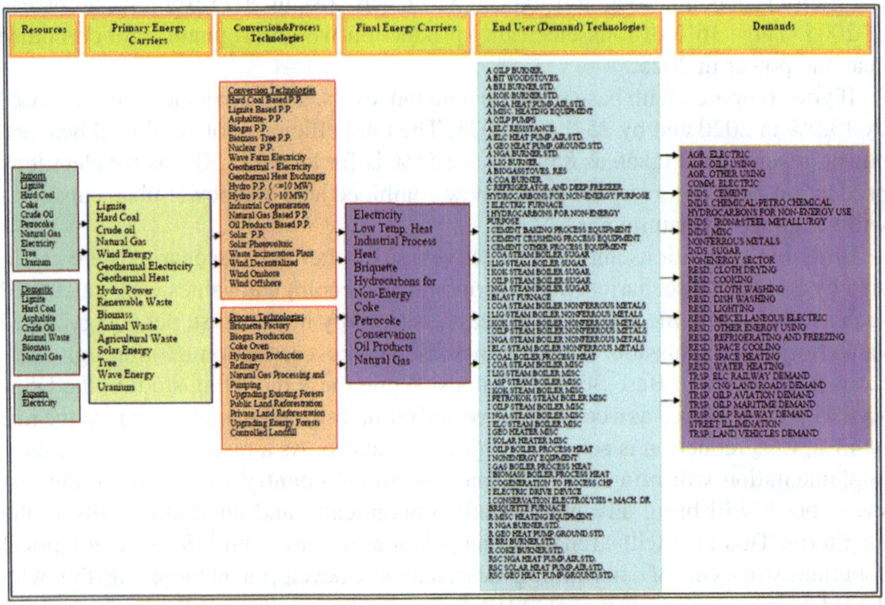

Fig. 2 Generalized RES of Turkey energy system (Sulukan et al. 2017)

2.13 Analyzing Cost-Effective Renewable Energy Contribution Options for Turkey

The scenario which is aimed to meet the energy demand in Turkey via using all renewable energy sources is named Renewable Scenario (REN100). Possible renewable energy potentials have become the upper limit on the activities in the model. All costs are given in US dollars in 2005. In particular, the estimation of demand and associated capacity has been one of the assumptions that it is proportional to Turkey's economic growth estimated by the World Bank. It is assumed that the total hydro-energy potential in the country is renewable by applying an upper limit in the scenario, ignoring the size or capacity of the dams.

Second, storage, hydrogen production, and forestry which are renewable process technologies have been determined to limit the maximum level of renewable use with existing technologies. So, allowing the model to be used even more than the reserves of the country and determining the fixed fossil fuel boundaries on which the technologies are based, turns towards a positive result as a model output. In this case, the model gives an "optimal" result, in the end, it is possible to observe a significant increase in the use of a renewable but on the contrary, there is a decline in fossil fuel use. The overall demand in the system is optimized and the overall system cost of energy is drastically reduced due to excessive exports of fossil fuel products and the generation of abundant renewable electricity. After this progression, if it is restricted to use one of the fossil fuels (e.g., coal) entirely in the energy system by setting the bounds "zero" on fossil-based and the end-use (demand) technologies, the energy carriers used in the model preferred by the end-use technologies give rise to "infeasible" result.

This means that unless there are subsidies, taxes, certain limitations or innovations in end-use technologies, the marketing system does not change direction. In order to prepare a more realistic scenario that will yield a viable result with positive input, each assumption must be carefully selected and applied one by one, showing a clear response of the model for each trial. As an illustration, setting a limit on the burden imposed by reference fossil energy technologies would be a good starting point for the reducing options, as long as the model gives optimal results, by setting it to zero at times other than the reference year. Reserves are selected as the upper limits determined at the resource level. The bounds on activities and capacities of renewable technologies are set free to allow the system to use them independently. Limits on the capacities of demand technologies that use electricity have been allowed, for example, to use renewable electricity instead of diesel in railway transport. In this case, while the whole system stops importing fossil fuels including crude oil and NGA, the increase in domestic lignite and coal usage sets up. The system does not produce any briquettes. The fact that hard coal and crude oil are the essential components of the country's energy mix until the market possesses the transportation technologies using hydrogen and alternative fuel production as hydrogen is concluded by the model.

As progress, hydrogen fuel turned out from renewable electricity and hydrogen using in end-use technologies has been introduced. It is suggested that the wave power plant will generate electricity by 2020. All potential energy resources are ready for use. Good security points are provided for domestic mining resources. It does not allow the model value to be entered to indicate wind and solar energy security. The system runs under these conditions and gives optimum results.

The model notes that the use of fossil resources has declined significantly, and that no nuclear energy is necessary to sustain the energy system with respect to conversion technologies. It is clear that renewable use is extremely high when it is a model that is allowed to use all the renewable electricity potential, regardless of whether the energy type is expensive or cheap or has other characteristics. Therefore, in REN100 Scenario, it is not possible to generate electricity from any of the asphaltites, natural gas, petroleum products, lignite and hard coal plants since the country's renewable potential is used. The model refrains from using fossil energy in electricity generation. In this case, renewable energy sources provide total electricity production. As an important indicator of dependence on foreign resources, in the REN100 Scenario; total fossil energy carrier imports decreased by 100% and total domestic fossil energy carrier supply decreased by 94.4%.

To illustrate the impact of end-use technologies on all energy systems, it should be noted that different changes in demand technology investment costs can be analyzed under different scenarios. According to the BAU results, the country's total primary energy rose from PJ 3861.6 in 2005 to PJ 7705.5 in 2025, up by 128.9% during the 20-year analysis period under the main economic assumptions. In the REN100 Scenario, this increase is seen as 90.1%. The use of renewable energy changes this view by a 97.2% increase in the BAU Scenario and provides 100% renewable electricity generation during the scenario with 892.3% in the REN100 Scenario. All these parameters can be modified or applied by the user like other scenarios on the model, relying on the leading indicators of the economy. Carbon dioxide emission levels are also affected due to increasing renewable energy use and decreasing fossil fuel use among the sources used as energy carriers. The BAU scenario foresees that CO_2 emissions will increase from 236.8 Mt in 2005 to 479.9 Mt in 2025, resulting in an increase of 131.6% in the total analysis period. The change in Carbon dioxide emissions decreased by 64.4% in the REN100 Scenario. The use of imported fossil fuels has led to an increase in the level of security by 6.3% (Sağlam et al. 2017a).

2.14 *An Alternative Carbon Dioxide Emission Estimation for Turkey*

Estimates were made using the CO_2 emission rate methodology according to the calculation method based on fuel use described in the IPCC rules (Sağlam 2010). Approach of the fuel analysis made up of calculations grounded on aggregate of results of multiplications of the quantity of fuel consumption and the fuel characteristic information containing heat content, carbon content oxidation factor, CO_2's molecular weight (44), C's molecular weight (12), conversion factor from kg to tons (1/907.2) in base scenarios of transportation, residential, agricultural and industrial sectors. While CO_2 emission is calculated, default fuel-specific emission factors (kT/PJ or kilotons/petajoule) multiplied with the quantity of fuel type on volume/mass/energy basis (PJ or petajoule) and addition the results (kT/year) for all type of fuels and this method called as a generalized approach. In this model, greenhouse gas emissions are predicted from all sources contains imports and exports, presuming they are consumed as any kind of fuels such as gas, solid or liquid and forestry effect over the system is assessed in the alternative scenarios. An example of estimating carbon dioxide emissions from sample fuel consumption in technology is compared with the results of fuel analysis and generalized approaches. The MARKAL model allows using different methods to calculate the total quantity of emissions produced from all sources (sources and technologies) in each period. However, the cross-check of methodologies is significant to confirm the correctness of the CO_2 emission estimate.

The base scenario also named as business as usual or BAU is consists of a series of information about Turkish energy and emission to develop policies on how to mitigate carbon dioxide emissions. The base scenario is also used to assess the relationship among the economic growth, energy consumption and GHG emissions of Turkey by comparing with other scenarios. Therefore, we create and run various scenarios with a base scenario and add detailed and appropriate energy sources on a sectoral basis, conducting detailed analyzes in terms of economic composition change, carbon density and energy density change. In this study, the period between 2005 and 2025 was analyzed.

The model indicates carbon dioxide emission projection as five-year periods in which TUIK information is added to the results of the base scenario. Here, especially to predict demand, domestic fossil fuel reserves and the consequent capacity, as envisaged by the World Bank, it is assumed that every sector in Turkey is directly proportional to the 3.3% economic growth. Estimates of long-term carbon dioxide emissions from energy-related use increased the rate of 183.49% from 183.59 kT to 479.95 kT between the years 1990 and 2025. The fulfilled period of carbon dioxide projections which is related to energy provides clues about the economic growth rate of a country. We foresee that in 2010, 293979.36 kT of CO_2 emissions will be released and if we compare it with the TUIK data, we can see the similarity between these values. TUIK is declared 299.11 kT CO_2 in 2009. This means one may be impressed with this study which made in 2005 and foresee the economic growth of Turkey in energy demand between 2005 and 2010 is not much more than 3.3%.

The carbon dioxide ratio of the electricity produced in Turkish power plants was calculated. From 2000 to 2025 the increase in the emission rate was 153.21%. The increase in carbon emissions from electricity generation is lower than the increase in emissions from other activities in the country, as shown in Fig. 3.

If there isn't comprehensive information associated with the process, conversion or combustion in plants or other end-use technologies such as trucks, steam boilers,

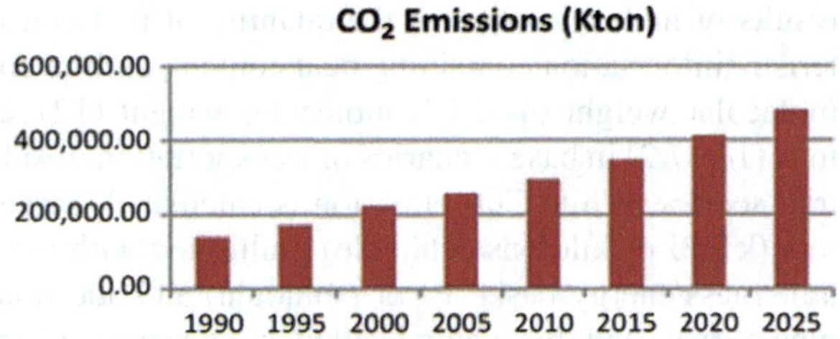

Fig. 3 CO_2 emission projections between 1990 and 2025 (Sağlam et al. 2017b)

stoves, etc., the quantity of the CO_2 by using default carbon factors can be calculated from the combustion of the resources at the beginning (generalized approach). Emission coefficient/activity (ENV_ACT) has been set to zero in the Scenarios, while the carbon dioxide emission factor is being used as an emission coefficient/resource activity (ENV_SEP) in this study. This is more accurate than what I have achieved in my previous studies. In the model, the carbon coefficients in the BAU scenario are adjusted according to the consumption given in the balances and emission stocks from each activity in 2000 and 2005. The worth of carbon dioxide coefficients acquired for each technology in 2005 was used in the period between 2010 and 2025. In the GEMIS and TCR scenarios, carbon dioxide emissions are predicted in resource technologies, which means before burning in conversion or end-use technologies. Intergovernmental Global Emission Model of Integrated Systems (GEMIS) and The Climate Registry (TCR) General Reporting Protocol provide all GHG emission factors for all sectors and technologies. In the GEMIS and TCR scenarios, the results are 4.9% and 5.1% higher than the BAU 'results in general terms, respectively. Different results are obtained with Fuel Analysis Approach for sample products extracted from the RES system and imported to RES system.

As a result; the operational data of technologies are the most significant part of trustworthy statistics, some of which are based on assumptions or estimates. Even Though it requires hard work to improve emission coefficients for fuel combustion technologies used in Turkey, they should be calculated for each of the fuels used in the Turkish sectors according to the measurement and scientific methods. Because they must be trustworthy both in tracking the actual CO_2 emission and in creating the emission inventories (Sağlam et al. 2017b).

2.15 An Analysis of Centennial Wind Power Targets of Turkey

Wind power has become one of the cost-effective options for improving the energy mix, while the wind farm installations reach higher shares by new capacity additions in Turkey, similar to many other countries, in order to support the sustainable development targets. Various methodologies are implemented and applied for planning and optimizing the energy systems of countries, taking into consideration the energy, economy, and ecology aspects as a whole. The objective of this paper is to demonstrate the implications of the "2023 energy targets" policy in terms of wind power. A futuristic scenario was tailored and applied to a comprehensive energy model for the energy system of Turkey.

The official wind energy vision and goal of Turkey were analyzed by applying a scenario based on 20,000 MW wind power capacity installation by the year 2023. The results indicated that, if applied, this action will increase the share of wind power utilization to the level of 16% in the total renewables while the installed wind capacity will reach 18%. Finally, the total cost of this strategic target which doubles the installed capacity and electricity generation by wind power, including the respective investment costs and subsidization results, was calculated as $24.78 billion on the analysis horizon (Sulukan 2018).

2.16 *A Model-Based Analysis of End-Use Energy Efficiency for Çanakkale, Turkey*

The energy-saving potential can increase the share of renewable energy in the energy mix and accelerate to become more economical by opening to new markets. This paper primarily focuses on the importance of energy efficiency in the transition of 100% renewable energy in cities.

In this study; a detailed energy network, namely Reference Energy System is developed for Çanakkale to identify the components and the interrelations between the energy supply, demand, and energy technologies as a holistic approach. This framework also aims to analyze end-use efficiency in electricity consumption in Çanakkale and will allow us to make inferences for large-scale situations from a broader perspective. It is crucial to increase end-use efficiency in order to meet or even reduce the rising energy demand and then plan a sustainable, environmentally friendly energy system (Bakirci et al. 2018).

2.17 *Urban Scaled Reference Energy System Development with a Sectoral Focus*

This study aims to analyze and determine the potential of the marble sector, which is a significant part of the energy model developed for Burdur, Turkey. Burdur, the selected city as the analysis domain, is located in the West Mediterranean region of Turkey.

Burdur province makes most of its revenue from the marble sector, because of the rich marble veins and region-specific marble types. Other than the marble, Burdur also makes income from livestock breeding, agriculture, and dairy products. As a result, high capacity marble quarries also attract local and foreign companies to

Burdur province and stimulate the establishment of various marble processing plants. The marble sector in Burdur is analyzed in the energy system structure perspective as two main segments; quarrying and processing of the marble. The entire equipment and machinery inventory were determined from a middle-sized marble quarry and processing plant. Reference Energy System is based on the year 2016, extending to 2031 by Answer-TIMES energy-economy-ecology model generator. The energy demand of the sector in the years 2021–2026–2031 was calculated and the Burdur-TIMES model is designed and developed on the urban scale, based on rich energy technology and sector-based data (Beşikçi et al. 2018a).

2.18 Reference Energy System Analysis of a Generic Ship

With the ever increase of population and technology development, energy consumption and demand have increased in diminishing sources worldwide. Therefore, energy system analysis has been an important topic to control energy consumption against limited sources in the last decade. In light of climate change, governments and organizations have been released new environmental regulations on energy consumption and emissions in the maritime sector to reduce the effects of climate change.

In this study, energy consumption characteristics and energy demand segments of a generic ship were evaluated by the energy system analysis approach and a RES is developed. On this basis, data-driven and technology-rich RES will be available to help prospective analyses on a base scenario. Furthermore, the respective energy model of a generic ship will be ready to be developed and analyzed, taking into account technical, economic and ecological constraints (Sulukan et al. 2018).

2.19 An Analysis of Centennial Solar Power Target of Burdur Province

Solar photovoltaic has recently become a significant energy conversion technology in all of the renewable alternatives. Like other countries, rising solar share in the renewable market supports sustainable development targets in Turkey. To achieve this development, Answer-TIMES optimizes possibly the best solutions among the other energy modelling methods due to the energy-economy-ecology aspects.

In this paper, 2023 renewable energy targets Turkey is discussed. In this perspective, 5000 MW installed capacity target of Turkey has been implemented to Burdur province by the share of solar PV installed capacity. Results show that, after 47.35 MW addition to current solar power plant installations, total electricity generation from solar PV peaks to 49% and triples the solar-based electricity production (Beşikçi et al. 2018b).

2.20 Calculating the Levelized Cost of Electricity by an Urban Scaled Simulation Approach

Countries are working to define and track pathways decreasing CO_2 emissions since the Kyoto Protocol has been signed in 1992. Paris Agreement in 2015 fortified renewable energy strongly; and as a result, the US and Europe had to step ahead in renewable energy roadmaps. In the renewable energy transition process, the cost of annually generated energy has great importance. On the other hand, as a small representative of a country, cities have more than one energy generation plant. Various fuel-based plants and energy generation technologies (from renewables to fossil-fuelled plants) make the calculation of the unit cost of energy generation more complex.

Manisa is an important city in the Aegean Region of Turkey, rich in energy generation technologies from renewable to lignite plants presents with a wide opportunity for assessment of energy production costs. However, alternative energy cost calculation methods exist in the literature such as ESA and LCOE, with different limitations and assumptions. In this study, a code on the MATLAB environment is prepared to simulate the Manisa electricity generation grid to determine the amount of electricity production for each power plant. The outputs of the simulation are used in the cost calculation process on the MS Excel sheets using a modified version of the LCOE methodology for the base year 2016.

The same simulation has been applied to the EnergyPLAN environment as an additional study to verify the results of the MATLAB code and provide a basis for discussion on the amount of energy and respective equivalent CO_2 emissions by these two platforms (Köker et al. 2019).

2.21 Transition to a Low Carbon Future in Maritime Fleet for Climate Change

Climate change is a recent important issue for transportation sub-sectors and the environmentalists have been working to combat climate change for decades. As 90% of the world-trade carried out by maritime transport; ships play a crucial role in the transport and trade vehicles. Approximately, 2.5% of the global GHG are caused by the ships and it has been ever-increasing, depending on the expanding maritime transport demand. Reducing the maritime-based GHG is a global challenge task, determined by the International Maritime Organization (IMO).

Nowadays; a number of measures to reduce carbon footprint and respective course of actions are recently being discussed and developed globally. IMO encourages shipping industry to minimize the carbon footprint by fostering energy-efficient onboard technologies, as the global top decision-maker. In this study, the current technology configuration on a chemical tanker ship has been modelled by LEAP (Long-range

Energy Alternatives Planning System), one of the widely used energy decision support tools. Then we focused on reducing greenhouse gas emissions with an alternative scenario according to IMO regulations and future technology developments. The results show that reducing the ship based GHG is possible and that is sure to give us a more secure future, in a cleaner atmosphere (Sarı et al. 2019).

2.22 An Urban Techno-Economic Hydrogen Penetration Scenario Analysis for Burdur, Turkey

Turkey's existing power generation infrastructure and the mixture of energy are analyzed, and two alternative scenarios based on hydrogen are implemented on Burdur's TIMES model in this paper in order to have an estimate of any environmental or economic outcomes for the years of 2016–2031. Building an advanced RES, determination of how energy carriers and the related technologies are connected and illustration of the existing energy system of Burdur was provided. Then, relevant data, which involves technologies of fuel cell driven land vehicles that are combined with demand-side of land transportation, was used to enounce this structure.

In this paper, hydrogen's applicability as an alternative bearer of energy in the fuel mixture for producing electricity in Burdur was examined to attain a continual economic growth, to enhance safety of energy through reducing relevant environmental emissions and to demonstrate the potential impacts of implementation of hydrogen supply chain and various fuel cell end-use technologies from the viewpoint of an energy modeling on city level.

The city of Burdur has been chosen as the target city to resolve hydrogen technologies' demand for land transport passengers. After putting hydrogen cars in practice in 2020; calculation showed that a total of 43.44 kT CO_2 emission can be ruled out by only 0.09 PJ of hydrogen car activity in Burdur, focusing on the 8% of the base scenario's total emission in the time period of the analysis. Lastly, hydrogen has been considered a secure and clean alternative in order to expand the mixture of energy on Burdur's existing energy generation system (Beşikçi et al. 2019).

3 Conclusion

Energy system analysis studies at Marmara University, Turkey via reference energy system concept firstly appeared in 2005. The Reference Energy System of Turkey was established and the effects of GHG reduction strategies on Turkish energy and economy systems analyzed through MARKAL which is a comprehensive energy decision support tool, have been used to create the energy database of Turkey for the period 2000–2025.

Since 2010, a number of studies have been carried out jointly in this field at Marmara University and National Defence University. Analyses of different energy systems have been done, but these studies are not only limited to regional energy analysis; also, for sectors, vehicles, buildings and so on. Besides these efforts, data mining has been used in the energy sector aiming to make a positive contribution to the energy profile of the country by increasing energy efficiency (Zorlu et al. 2019).

In the studies, the designated energy model has been created through reference energy system approach and decision support tools have been utilized to maximize the synergy between Energy–Economy–Ecology–Engineering (4E) to achieve a cleaner and more reliable future.

Disclaimer
The views and conclusions contained herein are those of the authors and should not be interpreted as necessarily representing the official policies or endorsements, either expressed or implied of Turkish Armed Forces, Turkish Naval Forces, National Defence University, Marmara University, and any affiliated organization or government.

References

Bakirci S, Razavi SP, Sulukan E, Uyar TS (2018) A model based analysis on end-use energy efficiency for Çanakkale, Turkey. In: 8th international 100% renewable energy conference (IRENEC 2018), İstanbul, pp 45–50

Beşikçi D, Sulukan E, Uyar TS (2018a) Urban scaled reference energy system development with a sectoral focus In: 8th international 100% renewable energy conference (IRENEC 2018), İstanbul, pp 40–45

Beşikçi D, Sulukan E, Uyar TS (2018b) An analysis of centennial solar power target of Burdur province. In: Solar conference and exhibition, İstanbul

Beşikçi D, Sulukan E, Uyar TS (2019) An urban techno-economic hydrogen penetration scenario analysis for Burdur, Turkey. Int J Hydrog Energy, In Press

Koç E (2005) Setting up the reference energy system to be the base of MARKAL energy model of Turkey. MSc thesis, Marmara University, İstanbul

Köker U, Koruca HI, Sulukan E, Uyar TS (2019) Calculating the levelized cost of electricity by an urban scaled simulation approach. In: 9th international 100% renewable energy conference, İstanbul

Mutluel F, Sulukan E (2017) Reference energy system development for Turkish residential sector. In: Uyar T (ed) Towards 100% renewable energy. Springer Proceedings in Energy. Springer, Cham

Sağlam M (2010) Establishing mitigation strategies for energy related emissions for Turkey using the MARKAL family of models. PhD Thesis, Marmara University, İstanbul

Sağlam M, Sulukan E, Uyar TS (2013) Deliberating lower-cost emission reduction options. In: 3rd international 100% renewable energy conference (IRENEC 2013), İstanbul, pp 370–381

Sağlam M, Sulukan E, Uyar TS (2017a) Analyzing cost-effective renewable energy contribution options for Turkey. In: Uyar T (ed) Towards 100% renewable energy. Springer Proceedings in Energy. Springer, Cham

Sağlam M, Sulukan E, Uyar TS (2017b) An alternative carbon dioxide emission estimation for Turkey. In: Uyar T (ed) Towards 100% renewable energy. Springer Proceedings in Energy. Springer, Cham

Şahbaz M, Sulukan E (2015) Developing reference energy system for agricultural sector and modeling biomass based biofuel production. In: National environmental congress, vol 2(1). Afyonkarahisar, p 15

Sarı A, Sulukan E, Özkan D, Uyar TS (2019) Transition to a low carbon future in maritime fleet for climate change. In: 9th international 100% renewable energy conference, İstanbul

Şener R, Sulukan E (2016) Reference energy system for land transportation in Turkey. In: 1st international conference on engineering technology and applied sciences. Afyon Kocatepe University, Turkey

Sulukan E (2010) Establishing energy efficient utilization and cost-effective energy technologies selection strategies for Turkey using MARKAL family of models. PhD Thesis, Marmara University, İstanbul

Sulukan E (2018) An analysis of centennial wind power targets of Turkey. Turk J Electr Eng Comput Sci 26(5):2726–2736

Sulukan E, Sağlam M, Uyar TS, Kırlıdoğ M (2010a) A preliminary study for post-Kyoto period for Turkey by MARKAL model. In: 5th international Ege energy symposium and exhibition (IEESE–5), Denizli, Turkey

Sulukan E, Sağlam M, Uyar TS, Kırlıdoğ M (2010b) Determining optimum energy strategies for Turkey by Markal model. J Nav Sci Eng 6(1):27–38

Sulukan E, Sağlam M, Çelebi, UB, Uyar TS (2011) Greenhouse gas emission assessment from maritime transportation for Turkey by MARKAL model. In: 1st international symposium on naval architecture and maritime (INT-NAM2011), Istanbul, pp 85–91

Sulukan E, Sağlam M, Uyar TS (2017a) Energy–Economy–Ecology–Engineering (4E) Integrated approach for GHG inventories. In: Erşahin S, Kapur S, Akça E, Namlı A, Erdoğan H (eds) Carbon management, technologies, and trends in Mediterranean ecosystems. The Anthropocene: Politik—Economics—Society—Science, vol 15. Springer, Cham

Sulukan E, Sağlam M, Uyar TS (2017b) Technical efficiency improvement scenario analysis for conversion technologies in Turkey. In: Uyar T (ed) Towards 100% renewable energy. Springer Proceedings in Energy. Springer, Cham

Sulukan E, Sağlam M, Uyar TS (2017c) Analysis of demand-side management option with cogeneration implementations in Turkish energy system by MARKAL model. In Uyar T (ed) Towards 100% renewable energy. Springer Proceedings in Energy. Springer, Cham

Sulukan E, Sağlam M, Uyar TS (2017d) A native energy decision model for Turkey. In: Uyar T (ed) Towards 100% renewable energy. Springer Proceedings in Energy. Springer, Cham

Sulukan E, Özkan D, Sarı A (2018) Reference energy system analysis of a generic ship. J Clean Energy Technol 6(5):371–376

Vardar YE (2005) Analysis of the effects of greenhouse gas emission reduction strategies on Turkish energy and economy systems. MSc Thesis, Marmara University, İstanbul

Zorlu SK, Önaçan MBK, ve Sulukan E (2019) General overview of data mining applications for electric power industry (Turkish). In: International congress of energy, economy and security (ENSCON '19), İstanbul

Residential Island Nano-grid for 100% Renewable Clean Energy

John O. Borland

Abstract Using smart IoT devices for home automation as well as for energy usage monitoring and control, 100% renewable clean energy using only sun light and heat with multiple-storage for all daily energy needs on sunny and partly cloudy days has been achieved for 336 days, 92% of the time in 2018 & 2019. With the smart home energy ecosystem, Grid-buy electricity was used as back-up on rainy and cloudy days accounting for 8% of annual total energy source. Rooftop solar-PV accounted for 58% of the annual energy source followed by battery storage discharge at 27% then hot and cold thermal storage at 7%. Island Nano-Grid mode of operation ensures lowest daily cost of electricity and resilience 24/7 for home safety and security.

Keywords Island Nano-grid · Zero grid buy · Multi-storage · Lifestyle behavioral changes · Solar-PV and resilience

1 Introduction

The total global carbon emissions for 2018 reached an all-time record of 37B tons up +2.5% from 2017 levels reported by Freedman (2018). Figure 1 shows China accounted for 27.7% of the global carbon emissions at 10.3B tons up +4.7% from 2017. Second was the US at 14.5% with 5.4B tons up +2.5% followed by Europe at 9.4% with 3.5B tons down –0.7% then India at 7.0% with 2.6B tons up +6.3%. The Earth's level of CO_2 also continues to raise with the May 11, 2019 reading reaching an all new record high of 415.26 PPM as reported by Miller and Rice (Miller and Rice 2019). Therefore, to fight climate change (global warming) and reduce carbon emissions requires each person to do his or her share starting today. The quickest way to have a significant impact in reducing carbon emissions caused by energy generation is to power your home or place of residence with 100% Renewable Clean Energy (RCE). Installing rooftop solar-PV with multiple storage can reduce grid-buy energy that produce CO_2 by 40–100% depending on solar-PV + storage system sizing and available utility programs (Borland 2019). Also, important is the net cost

J. O. Borland (✉)
J.O.B. Technologies, Honolulu, HI, USA
e-mail: johnoborland@aol.com

T. S. Uyar (ed.), *Accelerating the Transition to a 100% Renewable Energy Era*, Lecture Notes in Energy 74, https://doi.org/10.1007/978-3-030-40738-4_22

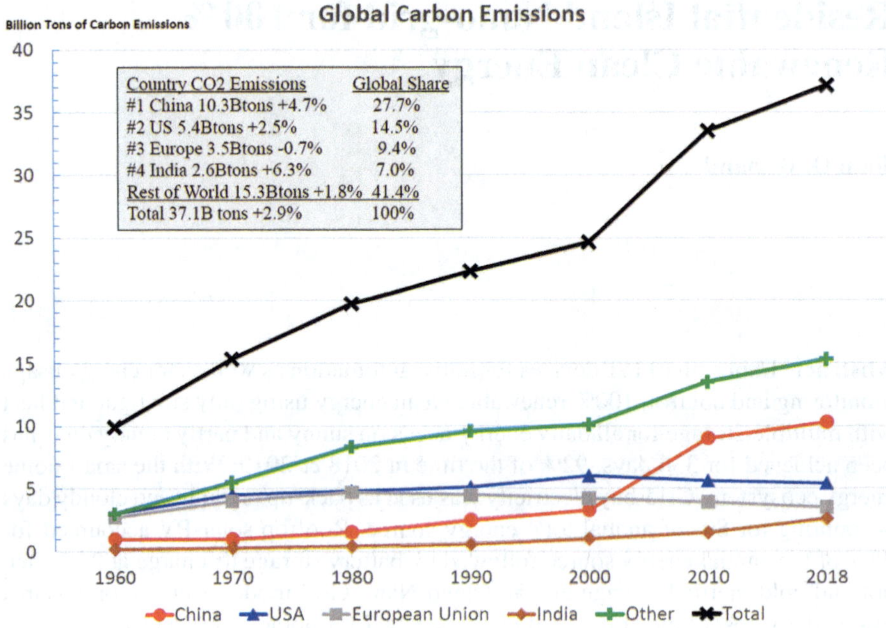

Fig. 1 Global carbon emissions from 1960 to 2018

of system (COS) equipment including installation which can vary from $3/W up to $7/W and after state and federal tax credits vary from $1.35/W to $4.90/W for a payback return on investment (ROI) of 3 years to 13 years so it is important to get the best deal when getting quotes for the rooftop solar and storage system.

2 Net Energy Metering and 1st Solar Wave

Net Energy Metering (NEM) is usually the initial program offered by most utility companies where the excess daytime PV generation can be exported back to the grid at a credit rate of full retail. In Hawaii the NEM program from Hawaii Electric (HECO) for residential rooftop solar-PV was so successful with a retail credit rate of >34 ¢/kWh from 2012 to 2014, it resulted in the high penetration level where 1 out of 3 homes have rooftop solar-PV which is called the 1st Solar Wave. Many of these rooftop solar-PV systems were oversized by 30–100% to guarantee zero-NEM (actually negative NEM) and achieve the minimum $25/month utility bill for grid connection and by using the grid as a storage battery for the excess daytime PV energy exported back to the grid to use for any overnight and rainy day energy demand. As shown in Fig. 2, case study for a residential home in Aiea, HI from Jan 2018 to Jan 2019 the yearly NEM was net export to the grid of 3.88 MWh/year or 323 kWh/month (10.8 kWh/day) achieving zero-NEM also called Net Zero Energy

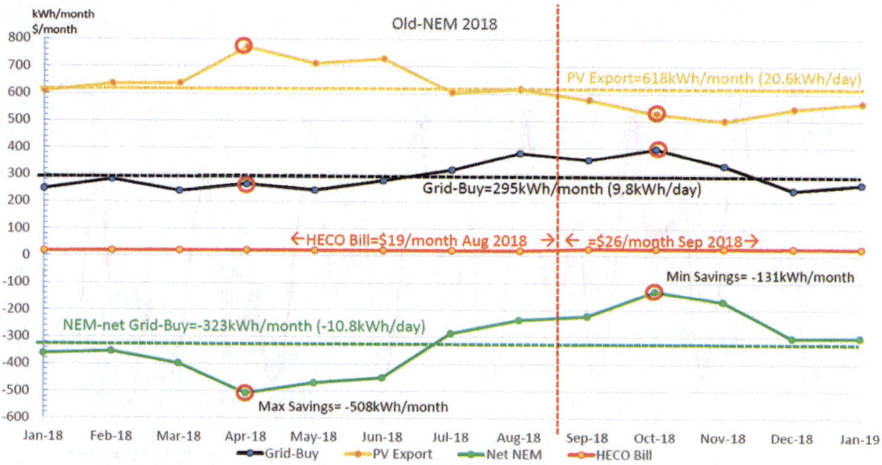

Fig. 2 Zero-NEM (negative-NEM) of −323 kWh/month for residential home in Aiea, HI in 2018

home due to oversizing of the rooftop solar-PV system. The yearly average grid-buy was 295 kWh/month (9.8 kWh/day) and solar-PV export was 618 kWh/month (20.6 kWh/day). April 2018 saw the greatest export back to the grid at 503 kWh/month and Oct 2018 was the least at 131 kWh/month. The generous NEM program also resulted in more excess energy usage by rooftop solar-PV system homes rather than energy conservation because of full retail credit from export allowed the use of as much electricity from the grid and still only pay the minimum $25/month bill. This export of excess rooftop solar-PV energy back to the grid created another problem called the "Duck Curve" as illustrated in Fig. 3 for HECO a one week result from 2013 to 2017 (www.HECO.com). The mid-daytime utility energy demand drops below the overnight low requiring a steep 400 MW mid to late afternoon ramp by the generators to meet the evening peak demand especially on Sunday. Instability to the grid also occurred resulting in the utility (HECO) ending the original NEM program in 2016. The current program called Customer Grid Supply plus (CGS+) reduced PV export credit from full retail to the wholesale rate of 10 ¢/kWh. Another program with no PV export back to the grid called Customer Self-Supply (CSS) requires both battery storage and thermal storage (Borland and Tanaka 2018). Time-of-use (TOU) electricity rates with highest rates (>35 ¢/kWh) during peak energy demand from late afternoon (5 p.m.) to early night time (10 p.m.) and lowest electricity rates (13 ¢/kWh) during daytime peak solar-PV generation periods from 9 a.m. to 5 p.m. was offered by HECO starting in February 2017 as shown in Fig. 4 (Borland and Tanaka 2018). San Diego Gas & Electric (SDG&E) also shifted their TOU rates in Dec 2017 from peak rate between 10 a.m. and 5 p.m. at 53 ¢/kWh to 4 p.m. and 9 p.m. at 54 ¢/kWh to coincide to the evening peak demand time period. This reduces the value of having excess solar-PV with daytime export back to the grid and contributed in a 84% decline in residential rooftop solar-PV for Oahu, Hawaii going from a rooftop

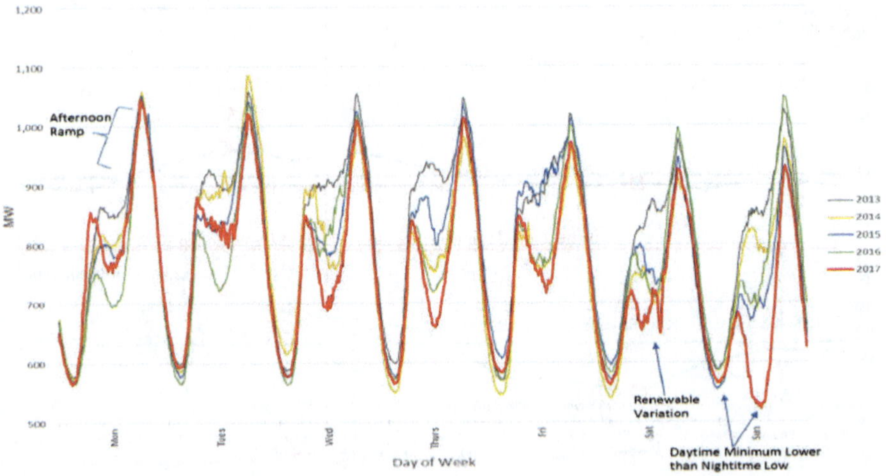

Fig. 3 Hawaiian electric duck curve due to mid-day rooftop solar PV backfeed to the grid (www.HECO.com)

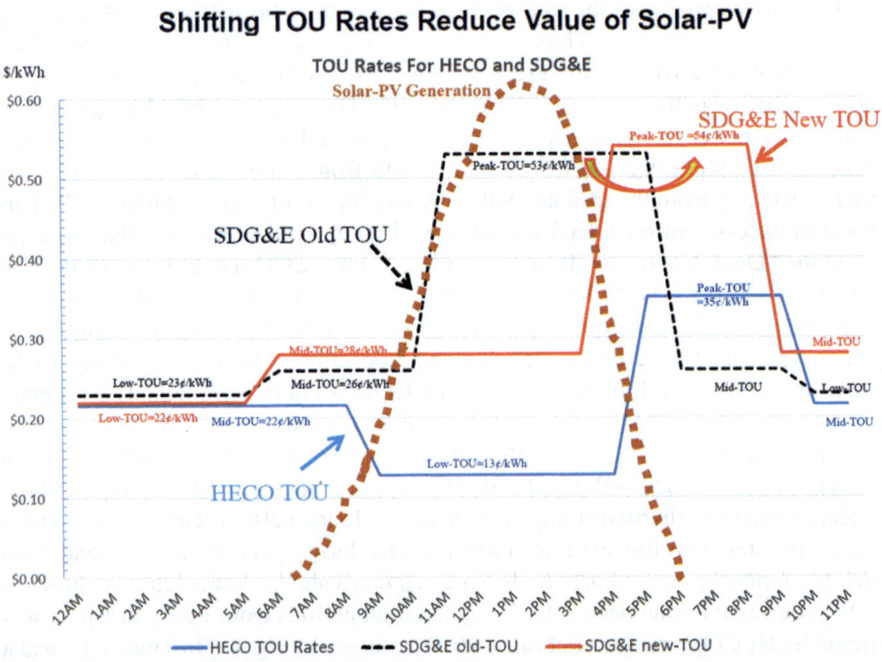

Fig. 4 TOU rates from SDG&E and Hawaiian electric with peak rates during peak demand

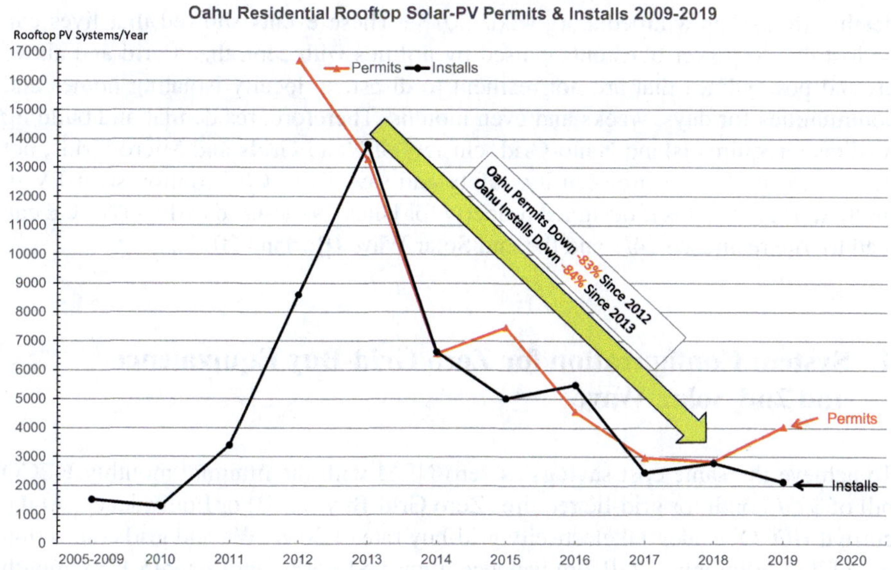

Fig. 5 Decline in rooftop solar-PV permits on Oahu, HI by −84%

solar boom in 2012 with >16,700 permits to solar bust by 2017 with <3,000 permits as shown in Fig. 5 (Borland 2019; Borland et al. 2018).

3 Island Nano-grid for Resilience

Utility companies with centralized grid are using solar-PV farms for daytime energy demand and to charge battery storage farms for discharge to meet evening peak energy demand. The battery storage farm night time electrical discharge will eliminate the Duck Curve by diverting the excess daytime PV generation to battery storage to first off-set the evening peak demand periods and if the storage capacity is large enough to extend to morning peak demand as now being implemented by several island utility companies in Hawaii to meet the state 2045 target date for 100% renewable energy. However, lessons learned from natural disasters such as hurricane and earthquake as well as man-made wildfire disasters: (1) Puerto Rico hurricane Maria in September 2017 caused months of no power and costing nearly 3,000 lives (Borland and Tanaka 2018), (2) California wildfires in November 2018 caused by overhead utility power lines had many reports of people describing not being able to see at night to escape their homes after losing power from PG&E, highlighting safety issues and importance of resilience with 85 lives lost and (3) Hokkaido Island Japan earthquake in September 2018 causing 2 days of island wide power blackout to all 2.9 M customers and many hospitals on emergency backup power had to turn away emergency patients (41 deaths) (https://en.wikipedia.org/wiki/ 2018). These events showed that lives can be lost due to power blackouts caused by utilities with centralized grid and above ground power lines that are not resilient to disasters, locally isolating homes and communities for days, weeks and even months. Therefore, residential and building resiliency requires Island Nano-Grid, clusters of Nano-Grids and Micro-Grids, not centralized grid to ensure security, safety and save lives. CSS rooftop solar-PV + multi-storage systems if designed properly for Island Nano-Grid and 100% RCE can lead to true resilience 24/7 and the 2nd Solar Wave (Borland 2017).

4 System Configuration for Zero Grid-Buy Equivalence and 2nd Solar Wave

To achieve the same cost savings as zero-NEM with the minimal monthly HECO bill of $25/month for grid-tie requires Zero Grid-Buy (ZGB) or Equivalence. At the current HECO residential electricity grid-buy rate of 30 ¢/kWh and grid connection fee of $11.50/month, ZGB-Equivalence allows $13.50/month or <45 kWh/month grid-buy with CSS for rainy/cloudy days and still realize $25/month minimum bill. On average, only 25% of a household daily energy usage occurs during daylight solar generation hours between 9 a.m. and 5 p.m. leaving the remaining 75% usage in the evening and overnight hours (www.HECO.com). Costs and energy usage simulations show that the night time grid-buy usage needs to be reduced to <3.3 kWh/night. A CSS residential case study for Aiea, HI in Fig. 6 shows the HECO monthly utility electricity bill dating back to Jan 2012 for per-solar and post-solar starting in June 2016 for solar phase 1, 2, 3 & 4. For the pre-solar years of Jan 2012–May 2016 the grid-buy monthly average was 1250 kWh/month ($396/month) with a peak of >1500 kWh/month (>$500/month) for July and Dec 2013, Mar 2014 and Apr 2016 with a daily cost of electricity (D-COE) of $13.73/day and rate of 34.5 ¢/kWh. For post-solar they achieved ZGB-Equivalence utility monthly bill of $25/month for a total of 3 months in 2017 with solar phase 2 & 3, 5 months in 2018 and 7 months in 2019 with solar phase 4. For all of 2018 the average HECO grid-buy was 103.3kWh/month or $30.92/month for a 92.4% reduction in both carbon emissions and costs (Borland 2019). 2019 improved to 97.5kWh/month or $29.92/month for a 92.5% reduction. On average this equates to ~2.3 rainy/cloudy days per month.

5 Internet of Things (IoT) and Smart Devices with Data Analytics

The key to achieving ZGB-Equivalence is having the right data analytics for daily energy usage monitoring and being renewable energy conservation minded with interactive assets that can be controlled to make the necessary lifestyle behavioral

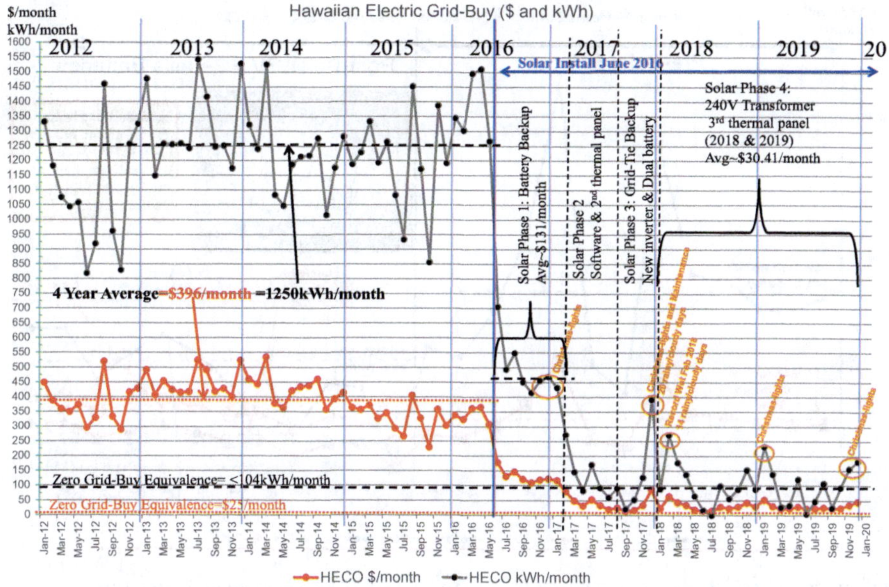

Fig. 6 Monthly HECO residential utility bill from Jan 2012 to Dec 2019 for pre-solar and solar phase 1–4

changes to achieve 100% RCE with only Solar Energy + Multi-storage. Integration of IoT home automation devices and Smart Home/Building Energy Management System to maximize solar-PV generation, maximize solar thermal generation, optimize electrical battery storage/discharge and maximize hot & cold thermal storage is critical for high efficient communication to balance RCE. By just adding solar + storage (2-Tesla batteries) the monthly HECO grid-buy electricity usage for residential case study in Kunia, HI was reduced by only 44.6% going from an average of 1291 kWh/month or $357/month (27.7 ¢/kWh) for 2017 to 714 kWh/month or $226/month (31.6 ¢/kWh) for 2018, a savings of 577 kWh/month ($131/month) as shown in Fig. 7. This CSS system is a solar + storage lease program with a fixed power purchase agreement (PPA) of $185/month for 20 years so the net monthly cost of electricity is $226 + $185 for a total of $411/month which is 15% higher than pre-solar of $357/month, therefore this case study example which achieved 44% RCE actually costs more with a negative payback costing $13 K more over the 20 year PPA so poor economic choice (Borland 2019). This example shows the importance of achieving ZGB-equivalence especially when you add the monthly costs of a solar loan or PPA to realize at best a monthly savings of $147/month, $1,764/year and $35,280 over the 20 year PPA of $185/month. The new monthly total electricity costs would be $25/month (HECO) + $185/month (PPA) for a total of $210/month. Using IoT and smart devices for energy usage monitoring as shown in Fig. 8a for peak summer time daily energy usage and Fig. 8b for winter time daily energy usage with one minute resolution we can identify the root cause of each household appliance energy

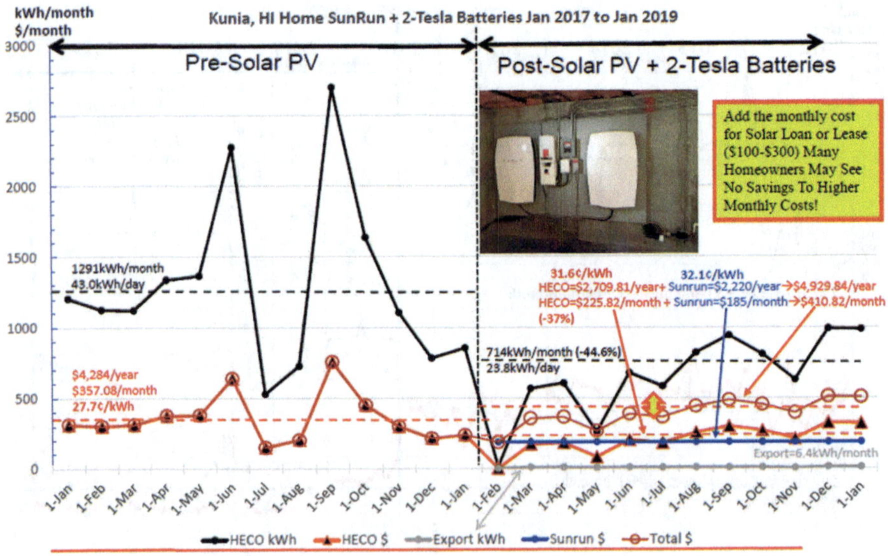

Fig. 7 Pre and post comparison for CSS solar-PV + battery storage system with a PPA

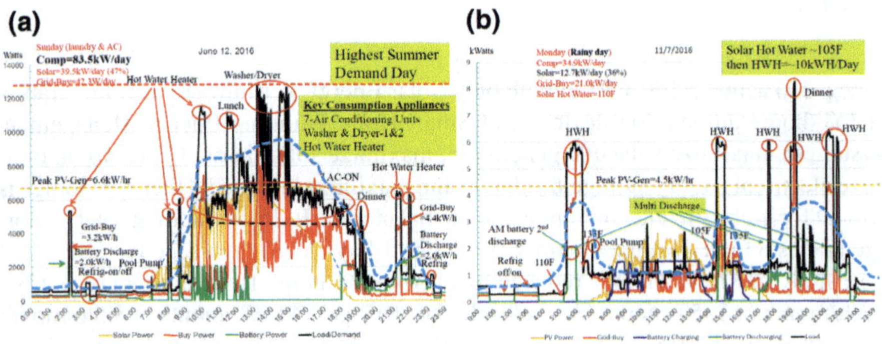

Fig. 8 Comparison of **a** peak summer day to **b** winter day energy usage

demand spike so we can make the required lifestyle behavioral changes to reduce grid-buy and increase the use of RCE. The high resolution data analysis shows that the morning and evening energy peaks are actually comprised of a number of energy spikes dominated by morning and evening electrical hot water heating at 4.5 kW and cooking (stove/oven at 4.5 kW, toaster oven at 1.0 kW and microwave oven at 1.0 kW). The mid-day weekend energy spikes is from the clothes dryer at 5.5 kW and multiple portable air conditioners varying between 2.5 and 5.5 kW. Ranking the top 4 energy consuming household appliances as shown in Fig. 9, summertime air conditioner ranks as #1 at 18–45 kWh/day with 37–60% energy usage followed by

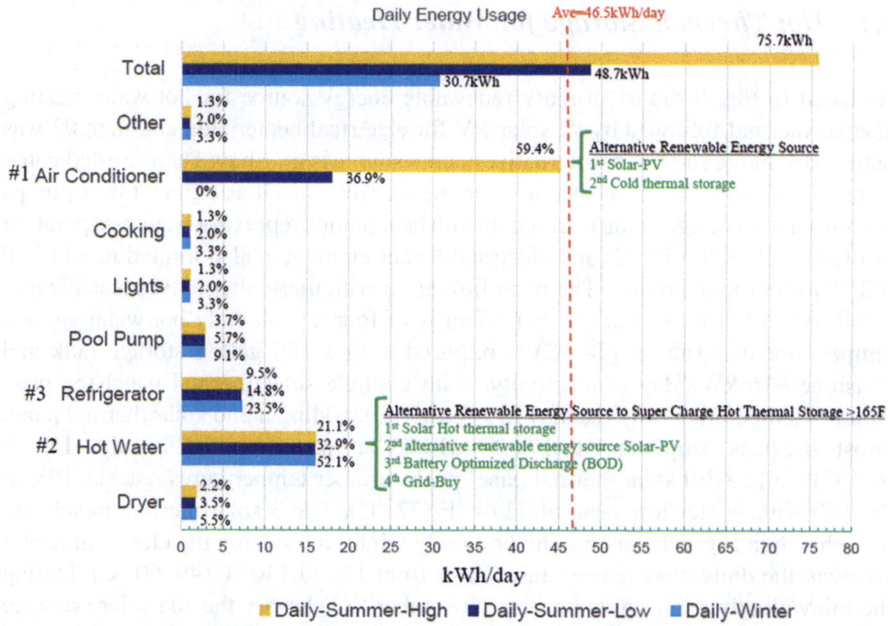

Fig. 9 Comparison of summer high and low to winter daily energy usage

#2 hot water heating at 16 kWh/day with 21–33% energy usage then #3 refrigerator/freezer at 7.2 kWh/day with 9–15% energy usage and pool pump at 3 kWh/day with 4–6% energy usage (Borland and Tanaka 2018). Winter time air conditioner drops off the list and the new #4 is clothes dryer at 1.7 kWh/day with 5.5% of energy usage and hot water heating increases to 52%.

6 Using Sun Light and Heat for 100% Residential Energy Needs

To fully use all the available RCE resources that the sun can provide, we used sun light with solar-PV panels for electricity generation and sun heat with solar thermal panels for heat generation to achieve 100% RCE. The key is to use multiple-storage for the daily light and heat energy from the sun. This is also called renewable thermal technologies for hot water heating and for space heating or cooling.

6.1 Hot Thermal Storage for Water Heating

As listed in Fig. 9 the #1 primary renewable energy source for hot water heating is solar thermal followed by #2 solar-PV for electrical heater element, then #3 was battery discharge and last #4 alternative energy source is grid-buy. There are dedicated solar-PV hot water systems available on the market as well as efficient heat pumps for hot water heating. Using refrigerant with heat pumps report hot water temperature limited to <131 °F (55 °C) and electrical heater element is also limited to <145 °F (63 °C) water temperature. The main limitation with these alternatives is achieving 1 full day of hot water storage. For a family of four a peak daily hot water storage temperature of >165 °F (74 °C) is required with a 120 gallon storage tank and consume ~16 kWh/day of electricity. With a single solar thermal panel the peak summer temperature only reaches 157 °F (69 °C), adding a 2nd solar thermal panel boosts the peak temperature to 178 °F (81 °C) but the winter high is only ~145 °F (63 °C) while a 3rd solar thermal panel peak summer temperatures reaches 198 °F (92 °C) with winter low peak of ~170 °F (77 °C). The 3 solar thermal panels are also shown in Fig. 10 and on rainy or cloudy winter days using the electrical heater element, the daily peak temperature varies from 120 to 140 °F (49–60 °C). During the rainy/cloudy winter months from November to February the sun solar radiance

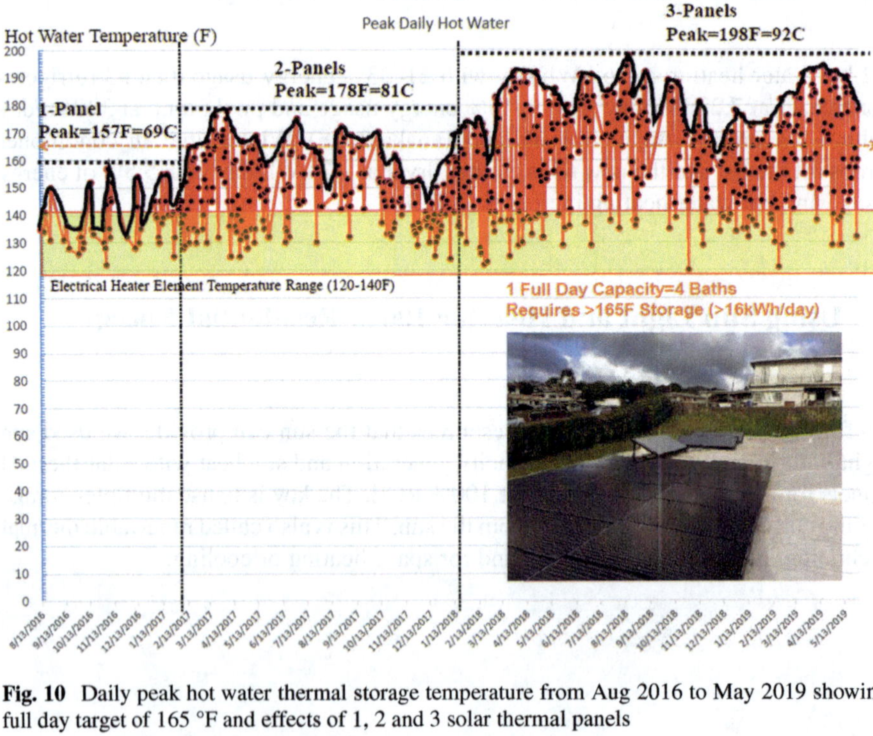

Fig. 10 Daily peak hot water thermal storage temperature from Aug 2016 to May 2019 showing full day target of 165 °F and effects of 1, 2 and 3 solar thermal panels

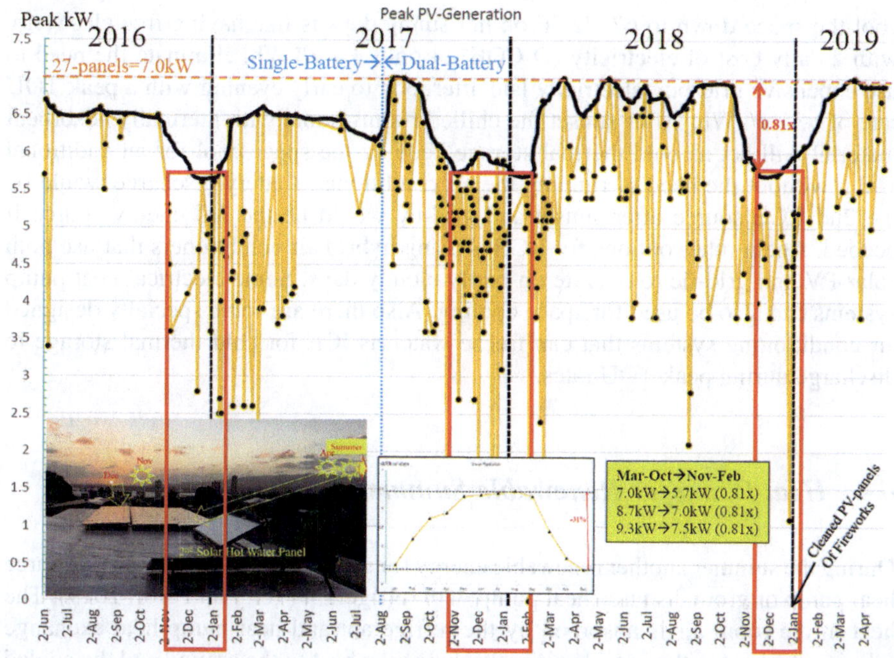

Fig. 11 Daily solar-PV generation peak from June 2016 to April 2019

is reduced by ~31% as shown in the insert in Fig. 11. The actual daily peak solar-PV generation for the 7 kW system during winter was only ~5.6 kW so a 20% reduction from the 7.0 kW spring/summer peak solar-PV generation. If a winter time minimal solar-PV generation of 7.0 kW is the target then the system size should be 20% higher (8.7 kW).

6.2 *Hot Thermal Storage for Winter Space Heating*

With a winter time hot water thermal storage daily high peak temperature of 170 °F (77 °C), this RCE source would work for hot water space heating with low-temperature radiators (Terry and Stori 2019).

6.3 *Cold Thermal Storage for Space Cooling with Air Conditioning*

For room or home space cooling with air conditioner, the primary energy source is daytime solar-PV generation. Keeping the air conditioner on during the day to cool the space down to 67–72 °F on hot sunny days is like having free electricity with a daily cost of electricity (D-COE) of only 3.4 ¢/kWh, eliminate the need to use expensive grid-buy electricity late afternoon to early evening with a peak TOU rate of >35 ¢/kWh. After sunset the chilled rooms/home with thermally adsorbent material will act as cold thermal storage keeping the space cool for an additional 1–2 h without the need of running the air conditioner. Battery discharge would be the 2nd RCE source after sunset and grid-by would be the 3rd energy source if needed. Again other options for RCE is using hybrid air conditioners that use both solar-PV and grid-tie to operate on partly cloudy days. Some electrical heat pump systems can also be used for space cooling. Also there are some specially designed air conditioning systems that can freeze water as ICE for cold thermal storage to discharge during peak TOU rates.

6.4 Heat Pump for Renewable Summer Time Space Cooling

During the summer another renewable energy for space cooling is to use an air source heat pump or ground source heat pump with refrigerant (Terry and Stori 2019). The heat in the room air is absorbed by the refrigerant and heat pump heat exchange releases the heat to the outside air or discharge the heat to the ground and the cooled down refrigerant returns back to the room to absorb more heat from the indoor air.

6.5 Heat Pump for Renewable Winter Time Spacing Heating

During winter another renewable energy for space cooling is to use an air source heat pump or ground source heat pump with refrigerant (Terry and Stori 2019). The opposite process to summer spacing heating, heat from the air or ground during winter time is absorbed by the refrigerant and the heat pump heat exchange releases the heat into the room.

7 Customer Self-supply Results with Solar + Multi-storage

By successfully implementing CSS with Solar + Multi-Storage and lifestyle behavioral changes to maximize and achieve 100% RCE on sunny and partly cloudy days, the 2012–2019 residential energy usage is shown in Fig. 12. As mentioned earlier in Sect. 4, pre-solar 4 year grid-buy electricity average from 2012 to 2015 was 14.5 MWh/year (39.7 kWh/day) while for 2019 grid-buy was reduced to 1.17 MWh/year (3.2 kWh/day) or 8.0% of the energy, solar-PV generation was 8.4 MWh/year (23.1 kWh/day) or 58% energy usage, battery discharge was 3.9 MWh/year (10.8 kWh/day) or 27% energy usage leaving thermal storage at

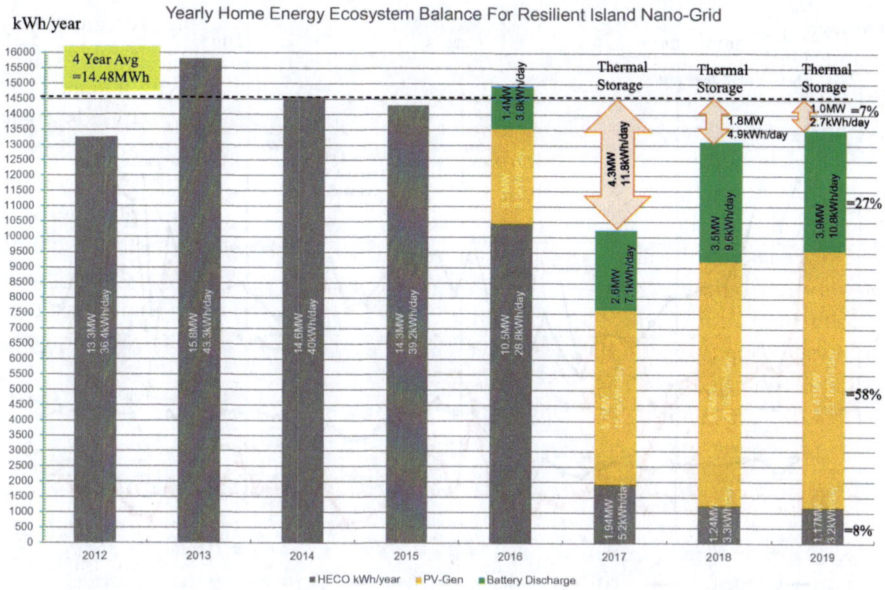

Fig. 12 Breakdown of multiple home energy sources from 2012 to 2019

1.0 MWh/year (2.7 kWh/day) or 7% energy usage. The detailed monthly breakdown each of these multiple energy source is shown in Fig. 13 for 2016–2019. When the single 10 kWh Li-ion storage battery was used between June 2016 and Aug 2017 the average daily discharge was ~7.0 kWh and with the dual battery (20 kWh) the daily average discharge increased to ~12.0 kWh. Li-ion battery discharge is limited to 75% of capacity to maximize battery life to 10–15 years based on a charge decay level of >60%. Zero Grid-Buy Equivalence requires <3 kWh/day grid-buy and this was usually achieved each month with <9 rainy/cloudy days per month.

8 Daily Cost of Electricity Analysis for Multiple Energy Sources

The Daily Cost of Electricity (D-COE) can be calculated from the lifetime cost of electricity (L-COE) for each energy source as shown in Fig. 14. The rooftop solar-PV L-COE of 9.8 ¢/kWh is based on a $3.20/W COS and 20 year product lifecycle. Add the Hawaii 35% tax credit and Federal 30% tax credit reduces L-COE to 3.4 ¢/kWh and with 58% (23.1 kWh/day) energy source D-COE from PV is 78.5 ¢/day. Li-ion battery system product lifecycle and warranty is 10 years for an L-COE of 37.4 ¢/kWh and after tax credits drops to 14.3 ¢/kWh with 27% (10.8 kWh/day) energy source for a D-COE of $1.54/day. Hot thermal storage L-COE after tax credit is 6.8 ¢/kWh with 7.0% (2.7 kWh/day) energy source for a D-COE of 18.4 ¢/day.

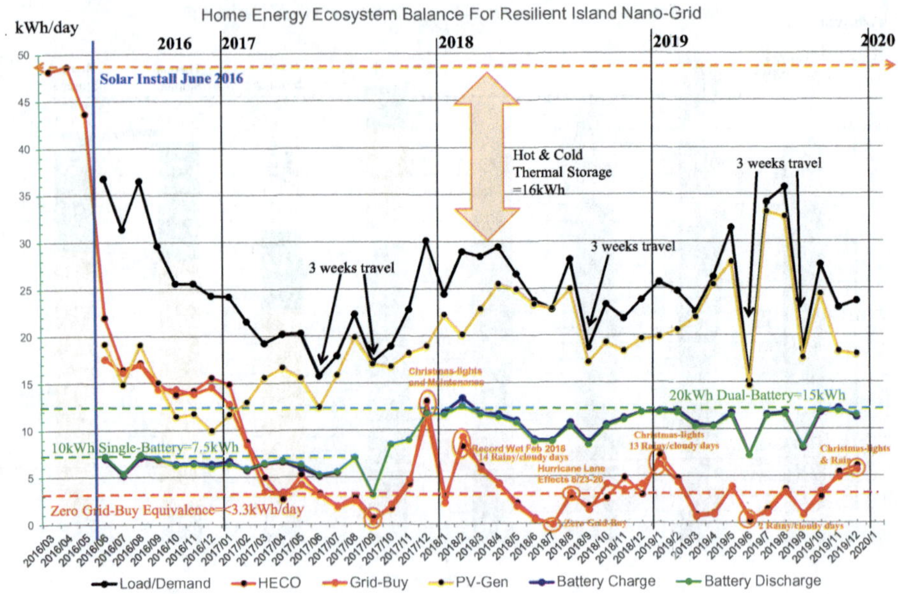

Fig. 13 Detail breakdown in multiple home energy sources from 2016 to 2019

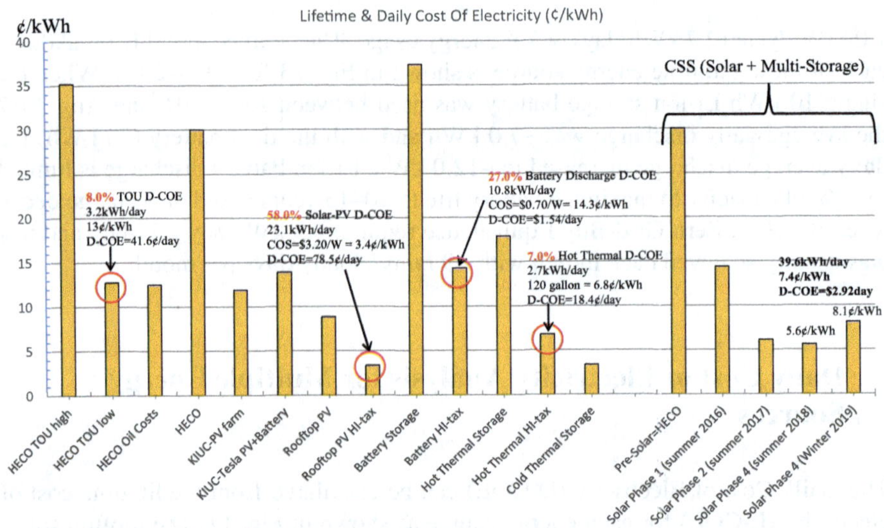

Fig. 14 Daily cost of electricity analysis for multiple energy sources

Grid-Buy during low TOU L-COE is 13 ¢/kWh with 8.0% (3.2 kWh/day) energy source for a D-COE of 41.6 ¢/day. Adding D-COE for PV + Battery + Hot Thermal + Grid-Buy gives a total D-COE of \$2.92/day and L-COE of 7.4 ¢/kWh.

9 Pay Back Return-on-Investment (ROI) Based on Cost-of-System (COS) and Yearly Savings

Figure 15 shows the payback ROI analysis for a 7 kW rooftop solar-PV + multi-storage (10 kWh battery and 80 gallon thermal hot water) system costing \$29,500 pre-tax or \$13,500 post-tax credit. Based on a yearly HECO bill of \$5,384 for 2013 the breakeven point for 50% savings is 16 years if no tax credit and 5.0 years for full tax credit. Increasing the savings to 94.5% reduces ROI to 2.7 years and a 25 year savings of \$99,800. Fig. 16 shows the pre-solar grid-buy from 2012 to 2015 averages 14.5 MWh/year at a cost of \$4,754.50/year and after half a year of solar energy in 2016 grid-buy was reduced 10.5 MWh/year for a 52.3% savings of \$2,266.50. 2017 was the first full year of solar energy with 100% RCE realized 88.3% of the time reducing grid-buy to 1.94 MWh/year (\$553/year) and savings of \$4,201.50. 2018 was the 2nd full year with solar (phase 4) reducing grid-buy to 1.24 MWh/year (\$371/year) and a savings of \$4,383.50 or 92.4% achieving 100% RCE for 336 days. For 2019, grid-buy was reduced to 1.17 MWh/year (359/year) and a savings of 4,395.50 or 92.5%. Note that if 100% RCE was achieved for 365 days a year (everyday), with the HECO

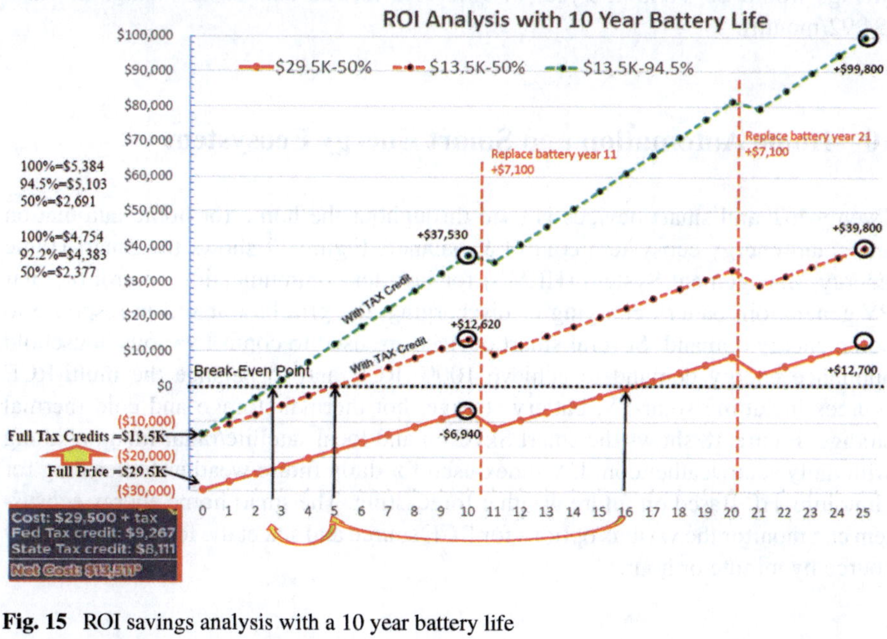

Fig. 15 ROI savings analysis with a 10 year battery life

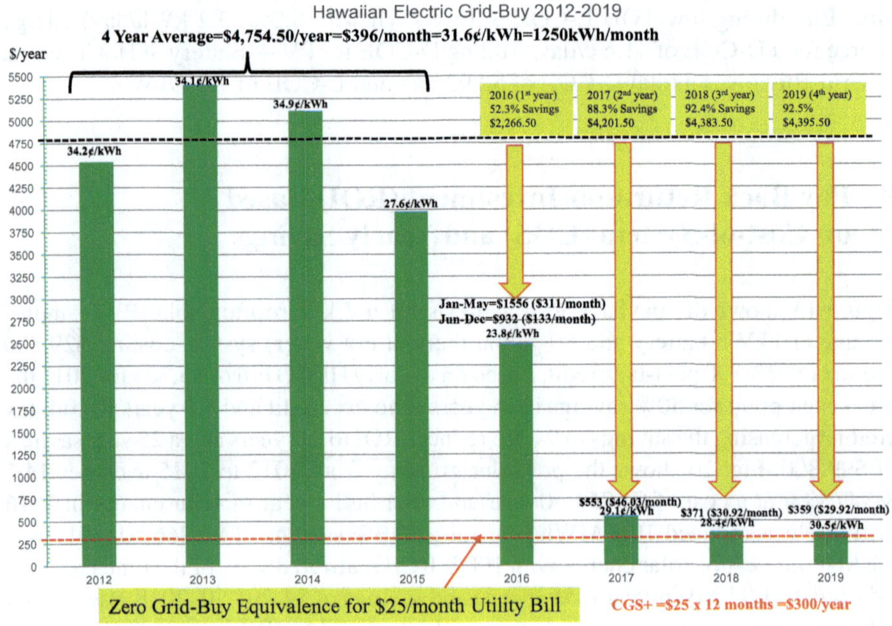

Fig. 16 Actual cost savings for CSS rooftop solar-PV + multi-storage residence in Aiea

minimum monthly residential bill of $25/month or $300/year the maximum yearly savings would be $4,454.50/year of 93.7% or just an additional savings of 1.2% ($4.92/month).

10 Home Automation and Smart Energy Ecosystem

Twenty IoT and smart devices is used throughout the home for home automation and smart energy ecosystem control & balance. Figure 17 shows the Smart Home Energy Management System (HEMS) for inverter communication control of solar PV-generation, battery charging or discharging and grid-buy or sell in response to home energy demand. Several smart devices are used to control various household appliance energy demand to achieve 100% RCE and to balance the multi-RCE sources including solar-PV, battery storage, hot thermal storage and cold thermal storage. Figure 18 shows the smart SkyCam and local satellite/radar imaging along with daily accuweather.com UV index used for daily future weather forecasting for Honolulu, HI. Based on future weather forecasting. The smart home energy ecosystem can monitor the various options for RCE source and select the lowest cost energy source by minute or hour.

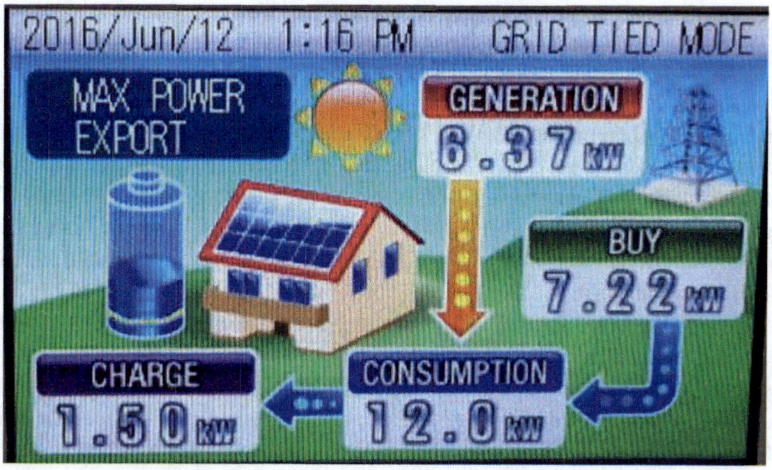

Fig. 17 Smart home energy management system showing inverter communication control of solar PV-generation, battery charging/discharging and grid-buy in response to demand consumption

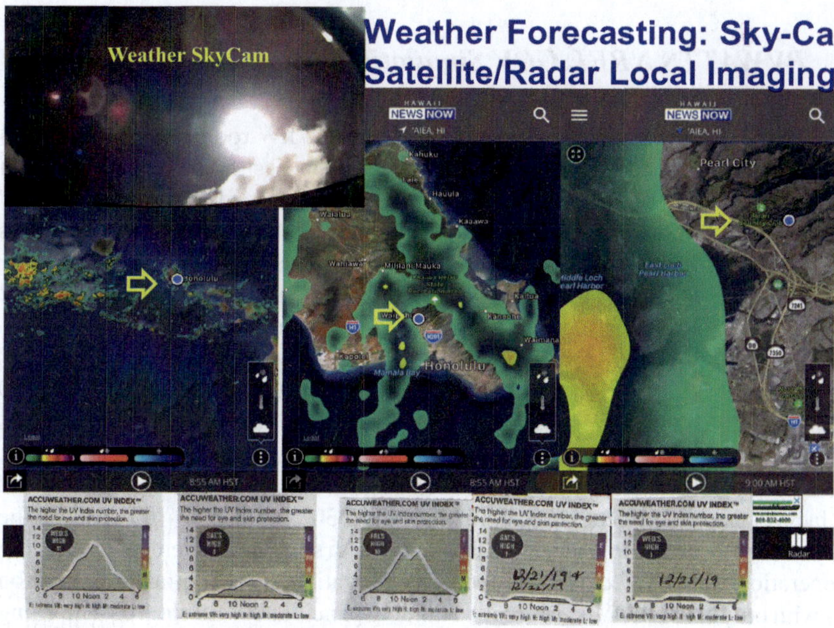

Fig. 18 Future weather forecasting using smart SkyCam with local satellite/radar weather imaging and daily accuweather.com UV index

11 Solar Panel Orientation to Improve Morning and Evening PV Generation

Borland and Singh reported PVWATTS.NREL.GOV simulation and experimental solar irradiation sensor results to achieve box-like solar-PV generation profiles by shifting the mid-day peak by 4 h to the left in the morning to 8 a.m. and 4.5 h to the right in the afternoon to 4:30 p.m. without the added costs of using trackers by optimizing tilt angles of East, West and South facing mono-facial or bifacial panels (Borland and Singh 2018). They determined the optimum panel tilt angle to be between 45 and 60° for Aiea, HI due to nearby mountain ranges. Benefits for residential rooftop solar-PV customers are: (1) eliminate peak mid-day export of excess rooftop PV-generation back to the grid reducing the Duck Curve effects, (2) reduce morning grid-buy demand, (3) reduce early evening grid-buy demand and the high TOU rates, (4) with battery storage delay the start of evening battery discharge extending overnight battery life and (5) improve morning battery charging for 2nd discharge further reducing morning grid-buy demand.

11.1 PVWATTS.NREL.GOV Simulation Results

The PV.WATTS simulations results for the 27 panel 7.0 kW rooftop solar-PV with 3 stringers based on panel tilt angle 10–90° and direction East-South-West are shown in Fig. 19. The control case (180-10 (3x)) which is flat roof and all 3 stringers at 10° tilt facing south has a maximum PV power of 5.1 kW when the actual field data is 7.0 kW so only 73% of expected value. Case 1 (E30 + S10 + W30) is for stringer 1 East at 30° tilt + stringer 2 South at 10° tilt + stringer 3 West at 30° tilt. Case 2 (E45 + S10 + W45) is for stringer 1 East at 45° tilt + stringer 2 South at 10° tilt + stringer 3 West at 45° tilt. Case 3 (E90 + S10 + W90) is for stringer 1 East at 90° tilt + stringer 2 South at 10° tilt + stringer 3 West at 90° tilt. Lastly Case 4 (E90 + W90) is for 1.5 × stringer 1 East at 90° tilt + 1.5 × stringer 3 West at 90° tilt which would simulate a Bifacial panel deployment by using back to back placement of East/West facing mono-facial panels. The results show that combining East and West facing panels/stringers going from 10° tilt to 30° tilt then 45° tilt reduces the peak mid-day PV generation level by 19% with only a slight improvement in morning PV generation and no effect on afternoon PV generation. For the bifacial simulation case 4 with only East and West vertically standing panels at 90° tilt showed a morning peak shift of 5 h to 7:30 a.m. with a drop in peak PV generation level of 20% with an afternoon 2nd peak at 4:30 p.m. and drop in peak PV generation level of 29% and the mid-day PV generation valley level was 67% lower than the 7:30 a.m. peak level. Case 3 shows that by optimizing the number of panels in stringer 2 at 10° tilt we can engineer the mid-day valley PV generation level increasing it up to 70% of the morning peak level and with more optimization, actually create a box-like PV generation profile similar to a tracker.

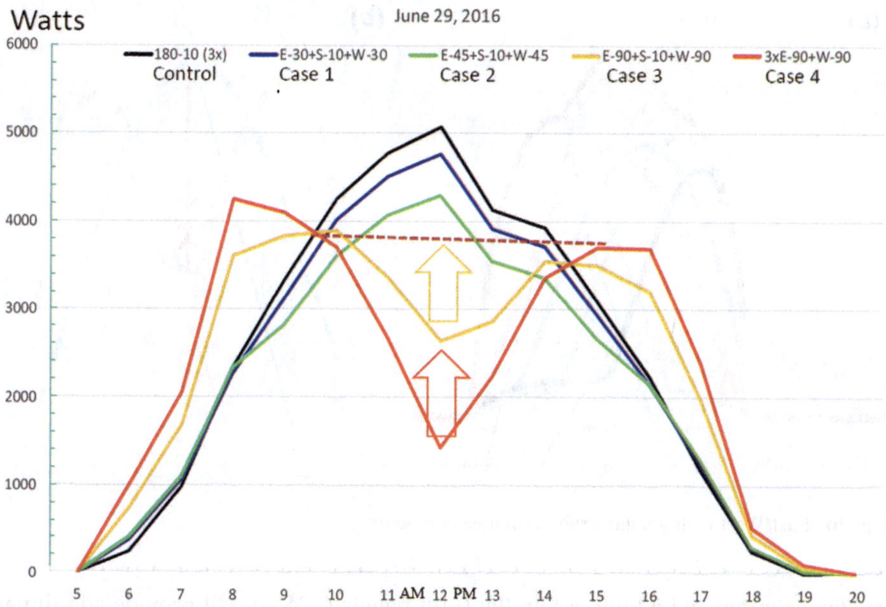

Fig. 19 PV.WATTS simulations results for 27 panel rooftop solar-PV with 3 stringers based on panel tilt angle 10°–90° and direction East–South–West

11.2 Solar Irradiance Meter Experimental Results

Since solar-PV panel hardware at these various East/West facing orientation and different tilt angles were not available for actual testing we used 6 solar irradiance meters to simulate each stringer of 9-panels and conducted the various different field data measurements in Utah during the month of September 2017 where the daytime sky are usually clear and free from the numerous trade wind clouds in Hawaii. Figure 20a shows the irradiance measurement results for case 1 & 2 at 0 tilt showing typical solar irradiance day light curve with a peak at ~900 W/m^2 With a 22.5° tilt facing East sensor the peak shifts to the left with an afternoon shadowing tail now visible. 45° tilt facing East sensor increases the peak shift to the left with a slight increase in irradiance level up to ~950. Going to 90° tilt facing East continued to move the peak to earlier morning but not the steep vertical portion of the morning profile which is already vertical after 45° tilt but now the peak irradiation level drops to ~850 and the shadowing tail portion of the curve starts at mid-day as expected. 45° tilt facing West is a mirror of the 45° tilt facing East but most interesting is the combined 45 E + 45 W irradiance curves which shows a triple-hump with a mid-day peak at >1250. Figure 20b shows the results for 0 tilt, combined 60° tilt East and West facing irradiance and combined 75° tilt East and West facing irradiance. Notice we can now change the mid-day valley irradiance shape to be flat and at a higher value. So somewhere between 60° E&W tilt and 45° E&W tilt will be a box-like

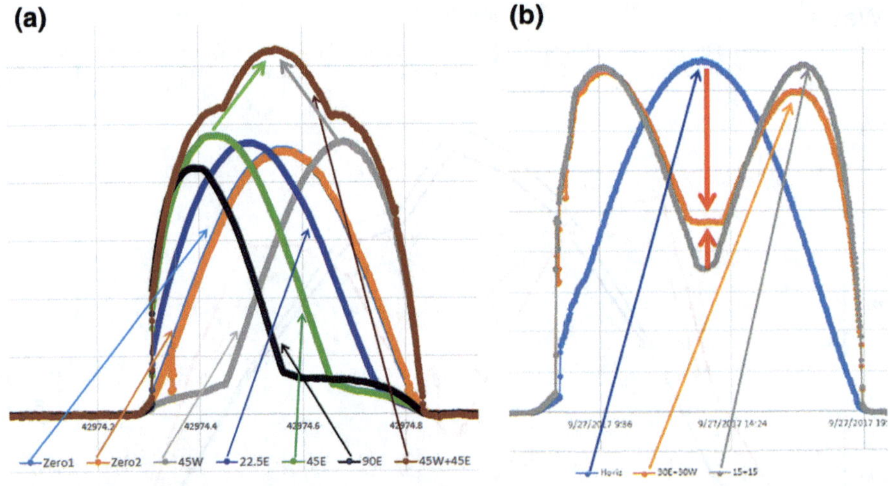

Fig. 20 East/West facing solar irradiation measurements

irradiance curve and adding a few flat 0 tilt panels (~25%) will provide additional engineering control for box-like PV generation.

12 Home Electric Vehicle (EV) Charging

Adding overnight home electric vehicle (EV) charging will increase your monthly electric bill by 60–80% on average based on two case studies (Borland 2019). The daily grid-buy electricity curves over a 2 week period showing the effects of overnight EV charging for a Tesla-75 from Jan 28, 2019 to Feb 10, 2019 shown in Fig. 21 with level 2 charging of about 7.5 kW. The residence average grid-buy before adding EV was 985 kWh/month and monthly bill of $272/month (27.6 ¢/kWh). After installing a level-2 EV-charger for a Tesla-75 the monthly grid-buy jumped 56% (+552 kWh) to 1536 kWh/month and monthly bill increased by 73% (+$199) to $471/month (30.6 ¢/kWh). For comparison Fig. 22 shows the results for a residential Kailua, HI NEM customer adding EV charging level 1 for Chev Bolt in Nov 2017. Pre-solar results for 2003 shows the yearly average HECO grid-buy of 388 kWh/month. Rooftop solar-PV under the NEM program was added in 2010 and for the 3 years 2014, 2015 and 2017 under NEM the grid-buy averaged 215 kWh/month, PV export averaged 272 kWh/month for a net NEM of −57 kWh/month (zero-NEM) with a credit at full retail rate for a $24/month HECO bill. Note that every year the Jan HECO bill was the highest (>350 kWh/month grid-buy) due to Christmas lights and extra holiday cooking. After adding EV home charging in Nov 2017 the monthly HECO bill for 2018 jumped 80% from $25/month to $45/month with a new NEM average of 106 kWh/month due to an increased daily grid-buy of 5.4 kWh/day (+163 kWh/month)

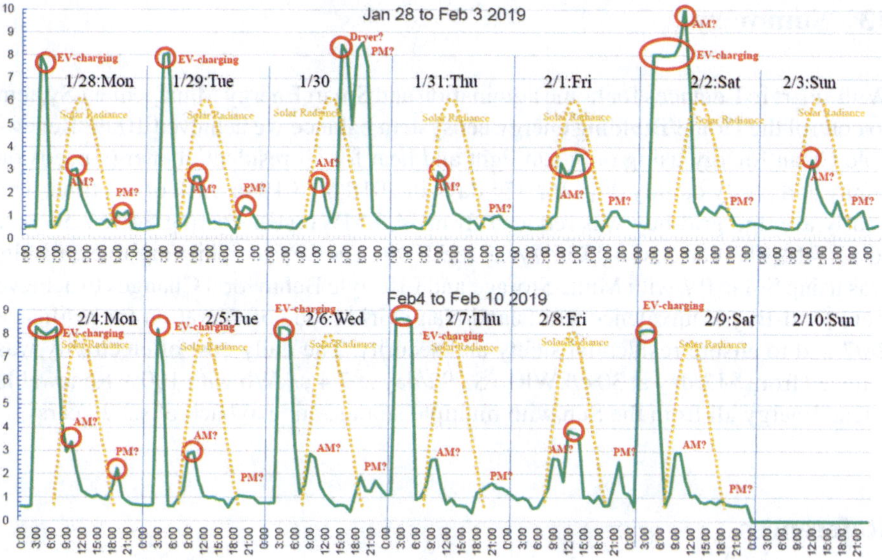

Fig. 21 Daily grid-buy electricity curves over a 2 week period showing the effects of overnight EV charging for a Tesla-75 using level 2 charging

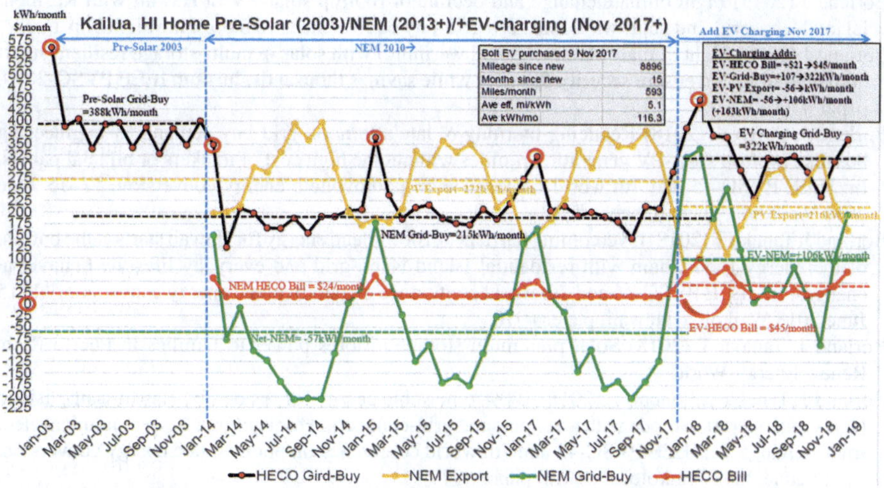

Fig. 22 NEM customer adding EV charging level 1 for Chev Bolt in Nov 2017

for driving an average of 21.7 miles per day at 4 miles/kWh. EV charging stations use utility grid electricity which is from fossil fuels so not carbon neutral but starting in late 2020 several EV cars manufacturers including Lightyear, Toyota, Hyundai and Joylong will offer PV-powered EVs using Si or III-V high efficiency solar cells for daytime battery charging for a driving range of 25–50 miles (Tsagas 2019).

13 Summary

With smart IoT devices for home automation and Smart Energy Management System to control the Home/Building energy ecosystem balance we achieved 100% Renewable Clean Energy using only sun light and heat for all residential energy needs on sunny and partly cloudy days for 336 days in 2018 & 2019 or 92% of the time. The yearly average grid-buy was reduced from 14.5 MWh/year costing $4,757.50/year to 1.17 MWh/year costing $359/year for 2019. The key to achieving these results was using Solar-PV with Multi-Storage and Lifestyle Behavioral Changes to achieve Zero Grid-Buy Equivalence and Island Nano-Grid mode of operation for resilience 24/7 and to ensure residential safety and security. The daily cost of electricity was reduced from $12/day at 30 ¢/kWh to $2.92/day at 7.4 ¢/kWh with 100% Renewable Clean Energy all from the Sun with multiple storage and payback of <3.2 years.

References

Borland J (2017) Be proactive not reactive to the next island blackout: a case study for achieving 100% renewable energy from the sun + multiple storage for hawaii residences with <3 years payback & off-grid operation. Wiliki-o-Hawaii 53(10):1

Borland J (2019) Fight climate change and decline of rooftop solar-PV in Hawaii with Resilient Island Nano-grid 2nd solar wave and zero grid-buy equivalence. Wiliki-o-Hawaii 5(2):p 1

Borland J (2019) Fight climate change (global warming) with solar + multi-storage resilient smart home/building and ensure security & safety while savings thousands. In: 46th IEEE PVSC-2019, paper #100

Borland J, Singh A (2018) Reducing morning & late afternoon grid-buy demand by engineering box-like rooftop solar-PV generation profiles without the high cost of trackers or bifacial panels. In: IEEE PVSC-45 and 7th world conference on photovoltaic energy conversion, 11–15 June 2018 Waikoloa, Hawaii, paper #737

Borland J, Tanaka T (2018) Overcoming barriers to 100% clean energy for Hawaii starts at the bottom of the energy food Chain with residential Island Nano-grid and everyday lifestyle behavioral changes. In: IEEE PVSC-45 and 7th world conference on photovoltaic energy conversion, 11–15 June 2018 Waikoloa, Hawaii, paper #970

Borland J, Tanaka T (2018) Solar plus multi-storage restores power to families in Puerto Rico. Renew Energy World

Borland J, Tanaka T, Poncho C (2018) 100% renewable energy for residential Hawaii using Island Nano-grid operation mode with no export and grid-connection for back-up: best economical electricity savings. In: IEEE PVSC-45 and 7th world conference on photovoltaic energy conversion, 11–15 June 2018 Waikoloa, Hawaii, paper #977

Freedman A (2018) The countries that pushed carbon emissions to record levels. Axios News, Dec 6

HECO report on 25% average daytime energy usage. www.HECO.com website

https://en.wikipedia.org/wiki/2018_Hokkaido_Eastern_Iburi_earthquake

Miller R, Rice D (2019) CO_2 level highest in human history. USA Today, 14 May, p 1

Terry G, Stori V (2019) Community campaigns for renewable heating and cooling technologies four case studies. Clean Energy State Alliance, June 2019. www.cesa.org

Tsagas I (2019) World's first solar-powered passenger car on roads next year. PV Mag Int, 14 Feb 2019

Solar Chimneys: Technology and Their Role for Transition to 100% Renewable Energy

Dogan Eryener

Abstract The need for widespread use of renewable energy has created intense interest in developing various renewable energy technologies that can be used in different applications. Solar chimney is a system that can be used for cooling or heating the buildings, and power production. Moreover, in recent years solar chimneys were widely used for air cleaning, urban planning and sea water desalination. One of the reasons behind these various purposes is the simple working principle of solar chimneys. Due to its simplicity, it can play a key role in transitioning to 100% renewable energy. However, solar chimneys have weaknesses, and many of these issues need new approaches and innovative solutions. It is not a mature technology yet, either for power production or building energy efficient projects. Standardization, integration, hybridization, and finally developing the key components are the main subjects of solar chimney research and development. However, this unique system has great potential for harnessing solar energy in the near future in various applications.

Keywords Solar chimney · Solar chimney power plant · Solar updraft tower · Natural ventilation

1 Introduction

Global warming and pursuit of sustainable development have induced engineers to search for new ways of power production from renewable energy, and some "old" methods and techniques, like solar chimney, a hundred years idea, become popular again. Actually, solar chimney is a very ancient method, which has been used for passive ventilation of buildings since thousands of years ago. The mechanism is very simple, therefore many architects and building engineers paid close attention to it in recent years. Many new sustainable, low energy buildings use solar chimney for ventilation, heating and cooling purposes. Solar chimneys can do it in a natural, more sustainable way, without consuming electricity like mechanical building

D. Eryener (✉)
Mechanical Engineering Department, Trakya University, Edirne, Turkey
e-mail: deryener@trakya.edu.tr

T. S. Uyar (ed.), *Accelerating the Transition to a 100% Renewable Energy Era*, Lecture Notes in Energy 74, https://doi.org/10.1007/978-3-030-40738-4_23

HVAC systems. Nowadays, solar chimney, the thousand year old method for building ventilation has become very common and popular technology for production of renewable energy. Also, solar chimneys not only considered for production of renewable electricity and building ventilation, but also for other different purposes such as urban planning, city air purifying, solar drying and desalination, and freshwater production. In this regard, solar chimney is a very unique concept, which is very different from other technologies.

Despite the simple and very well-known working mechanism of solar chimneys, which makes them a technology of choice for many engineers and architects in their sustainable and renewable energy projects, there are lots of aspects yet to be developed to make solar chimneys fit for either renewable energy production or its use in the buildings' climate control. Therefore, this very ancient method, based on rising hot air, has some uncertainties for power production and building heating or cooling applications. However, solar chimney still offers a lot of potential in various application opportunities, so it can be stated that solar chimneys will play an important role in 100% renewable energy transition.

2 Background

Solar powered chimney systems have been used for centuries, from ancient Minoan to Middle East architecture. Usually solar chimneys were accoupled with other natural ventilation systems, such as wind towers or scoops. Naturally, it was very common in hot climates, especially when combined with wind catchers. Some of these structures stood the test of time and are still standing today, like famous windcatchers in the city of Yazd in Iran. In general, a tall chimney, which is integrated in the building structure, creates a temperature difference within building by absorbing solar energy. This temperature difference leads to a displacement of ventilation across the building, and the fresh air comes from a wind tower, ground or windows. This method not only supplies fresh air, it also provides a passive cooling for building, today considered as a part of a new definition which called ventilative cooling, and some works are ongoing for standardization in modern buildings. Solar chimneys also can be used for passive heating of the buildings (Zhai et al. 2011). In recent years, many extensive studies and works on solar chimneys have been written, also many buildings integrated solar chimneys were constructed.

Solar heated air and solar chimney brought another brilliant idea a century ago: Using solar chimneys for power production. Spanish colonel Isidoro Cabanyes offered this system in his study (Breeze 2014). Actually, Cabanyes was not the first person who thought about producing power from rising hot air, the idea goes back to much earlier times: Leonardo Da Vinci described it five hundred years ago in his works (Mackay 2015), however it was not powered by solar energy. Later, in early 20th century, many patents and concepts were published, which described systems that used hot air to produce power, some of them were heating air by solar energy. In the 1970s, the first and most important attempt to build a solar chimney power plant was made by German engineer Jörg Schlaich, who can be considered to be the father of solar chimney power plant today. Schlaich and his team designed the first big solar chimney power plant and built it in Manzanares, Spain, in 1982 (Haaf et al. 1983; Haaf 1984). Today, all the studies, knowledge, correlations, mathematical models, CFD works are based on this unique experimental solar chimney tower, which actually operated for only 9 years until it collapsed. Schlaich and his team built a 200 m solar chimney and called it solar updraft tower, so they also renamed the solar chimney as solar updraft tower for modern era. Today, this renaming sometimes creates confusion in literature, because some researchers still prefer to use name "solar chimney", adding "power plant", however "solar chimney" has a broader meaning. So, solar updraft tower doesn't represent only a modern solar chimney, it also eliminates the other uses of solar chimneys.

The first big demonstration of solar updraft tower built in Manzanares, shown in Fig. 1, had a 196 m chimney with 48,000 m^2 transparent solar air heating collector, 10 m diameter tower, one turbine, with the capacity of 50 kW, was serving to produce electricity from hot air. Actually, the main purpose of the experimental unit was not power production, but testing the performance of different collectors, therefore the tower was designed to last for two years, however it lasted nine. Results showed that the thermal efficiency of transparent greenhouse collectors was low, however it was caused by the heat transfer mechanism, not the collector's design. Later in 2014, metal transpired solar collectors were proposed instead of greenhouse-like collectors. They provided many advantages such as durability, 3 times higher efficiency, low area usage and low cost, but it should be noted that new applications also need further detailed studies (Eryener et al. 2017). Despite the fact that the purpose was

Fig. 1 First large-scale solar updraft tower demonstration, Manzanares, Spain, 1982

not the testing of power production, Schlaich's research team obtained the power production data which showed that the power output of Manzanares tower was very low, 50 kW maximum for 46,000 m^2. This circumstance also affected similar projects and developments. Because the efficiency of the system was very low compared to other solar power technologies, no big tower projects were carried out in the 20 years following Manzanares project, neither for research nor commercial purposes. Therefore, solar updraft tower was never considered to be a serious alternative to other renewable energy production systems until 2000s. Twenty years of research on Manzanares tower showed that large scale solar updraft towers can be a valid alternative for renewable energy production. In 2010, 28 years after Manzanares project, first large solar updraft tower test unit was built by Chinese government in Wuhai, Inner Mongolia, for supplying clean renewable energy as an alternative to coal based electric production (Zhou and Xu 2016). Fifty meters high tower was built smaller than initially designed because of the close proximity to the airport, however, the obtained results were never shared with the scientific community in details. On the other hand, in 2000s large scale tower projects were proposed by various companies in different countries; different project with towers varying from 750 to 1300 m in height with significantly large greenhouse collectors were designed in Spain, Namibia, Australia and USA. Some of them were officially approved by governments, but the projects were never realized until today. These large scale towers projected in the range of 40–400 MW, and also promising different purposed such as touristic activities, agricultural, observation wildlife and 1000 m tall tower is an attractive point for governments and investors. On the other hand, in 2000s some studies claimed that compared to coal fired power plants 1000 m high solar updraft towers can have lower LCOE (levelized cost of energy) under certain economic parameters (Schlaich et al. 2005).

3 Advantages and Disadvantages of Solar Chimney

3.1 Advantages of Solar Chimney Use in Buildings

- Solar chimneys provide energy efficiency in a building. During hot summer months, a solar chimney operates as an effective passive cooling device. This function reduces dependency on mechanical air conditioners.
- Solar chimney reduces operational cost related to fresh air ventilation. Some recent projects can achieve up to 42% electricity saving from ventilation.
- Solar chimneys are cost effective solutions for buildings. The chimney can be constructed by using available building materials, and since it works with thermal buoyancy there are no additional operational costs.
- Solar chimneys are problem-free systems, they work independently, no switching on–off or pre-settings required. It also requires little effort to maintain (e.g. cleaning).

- Solar chimneys don't utilize conventional energy for building heating, cooling and ventilation, which makes them more sustainable. Also, solar chimneys don't produce noise and carbon dioxide emissions.
- Solar chimneys can be integrated with new systems for space heating and cooling applications.

3.2 *Advantages of Solar Chimney Use in Electricity Production*

- Solar chimney power plant is simple in terms of technological level. It doesn't require complex manufacturing processes as photovoltaic panels or concentrated solar power systems. Necessary materials are available in almost any country, except the turbine (i.e. wind) technology and grid converters.
- It works with direct and diffuse solar irradiation, which is an advantage compared to other solar power technologies that only work with direct solar irradiation.
- It needs relatively low temperature difference between inlet and outlet of within system boundaries, which means that solar chimney power plants can work with additional heat sources such as ground, waste or geothermal, not only solar energy. Hence, the system can be combined with other sources and operate day and night.
- Heat storage is possible in solar chimney power plants. A wide range of materials, from phase changing materials to stones, can be used under the collector roof for storing solar radiation during daytime.
- Operational and maintenance costs are very low compared to concentrated solar power systems. Also, there is no need for cooling water and surface cleaning.

3.3 *Disadvantages of Solar Chimney Use in Buildings*

- Since solar chimney operates in a natural way, which depends on solar radiation, overventilation or underventilation can occur frequently. Therefore, it might need mechanical control systems depending on application, especially in modern multi-storey office buildings.
- It needs a proper architectural design. Solar chimney placement depends on the exact location of residential areas of buildings, therefore a detailed flow analysis before implementation is necessary.
- There are no solar chimney standards for buildings, so it must be a tailormade solution.
- Solar chimney height, which effects the performance significantly, can be potentially restricted because of the building codes.

3.4 *Disadvantages of Solar Chimney Use in Electricity Production*

- The investment cost of the construction of a commercial plant is high due to its large-scale: The area collector is about 3,800 ha, and the tower is 1,000 m high and 130 m in diameter.
- Its efficiency is low compared to other solar power technologies, such as photovoltaics and concentrated power systems. The electricity production efficiency is about 1% and this is very low for an electricity production system. Therefore, the decision of building a solar updraft tower depends more on other financial parameters, such as levelized electricity cost or interest rate.
- The chimney height is an important parameter for increasing the capacity factor, however there are technological constraints and atmospheric and geophysical factors that can impose certain limitations on constructing a very high tower.
- There are environmental concerns. For a typical 200 MW power output, 3,800 ha field would be covered with solar collectors. Despite the fact that solar updraft towers are sustainable atmospheric devices (Richter et al. 2017), construction and operation of a large scale solar updraft tower may have a negative impact on the local environment and lives of different plants and animal species.

4 Solar Chimney Basics

4.1 *Working Principle of Solar Chimney in Buildings*

The chimney effect in a building, also known as stack effect, is a thermal buoyancy-driven air movement which is related to air density difference. In general, air density is a function of local pressure, temperature and humidity, however the main driver behind the stack effect is the temperature difference between indoor of structure and outdoor conditions. The higher temperature difference and the height of the structure, where the air passes, creates more buoyancy force, and thus the stack effect. Therefore, a solar heated chimney structure in a building can create strong stack effect, and associated natural displacement ventilation, which helps cooling the indoor conditions.

Technically, the driving mechanism for solar chimneys is the same as the stack effect, therefore, classical stack effect equation can be applied for solar chimneys, except related to solar radiation. Assuming temperature and barometric pressure are constant over the height of interest, the stack pressure decreases linearly as the separation above the reference point increases. For a single column of air, the stack pressure can be calculated as (ASHRAE 2003)

$$P_s = P_r - \rho g H \tag{1}$$

where

P_s stack pressure, Pa
P_r stack pressure at reference height, Pa
g gravitational acceleration, 9.81 m/s^2
ρ indoor or outdoor air density, kg/m^3
H height above reference plane, m

Neglecting vertical density gradients, the stack pressure difference for a horizontal leak at any vertical location is given by

$$\Delta P_s = (\rho_o - \rho_i)g(H_{NPL} - H) \tag{2}$$

or

$$\Delta P_s = \rho_o\left(\frac{T_o - T_i}{T_i}\right)g(H_{NPL} - H) \tag{3}$$

where

T_o outdoor temperature, K
T_i indoor temperature, K
ρ_o outdoor air density, kg/m^3
ρ_i indoor air density, kg/m^3
H_{NPL} height of neutral pressure level above reference plane

The effectiveness of solar chimney is influenced by:

- The solar radiation.
- The height of the solar chimney.
- The temperature difference between the indoor and the top of the chimney.
- Pressure differences outside the building.
- Area of chimney and building openings.

4.2 *Working Principle of Solar Chimney Power Plant*

A conventional solar chimney power plant or solar updraft tower has a simple operation principle as shown in Fig. 2a. The ambient air, which enters from the open perimeter of a large transparent collector, is heated via "solar greenhouse effect" and flows into a centrally positioned chimney with horizontal axis turbines to produce electricity by using the air flow to turn turbines. Recent studies show that metal transpired solar collectors can be used instead of classical transparent collector as a modern approach to solar chimney power plants. In this case, air enters through the entire surface instead of the perimeter in, as seen in Fig. 2a, b. In conventional solar chimney power plant, outdoor air enters from open perimeter of the transparent roof

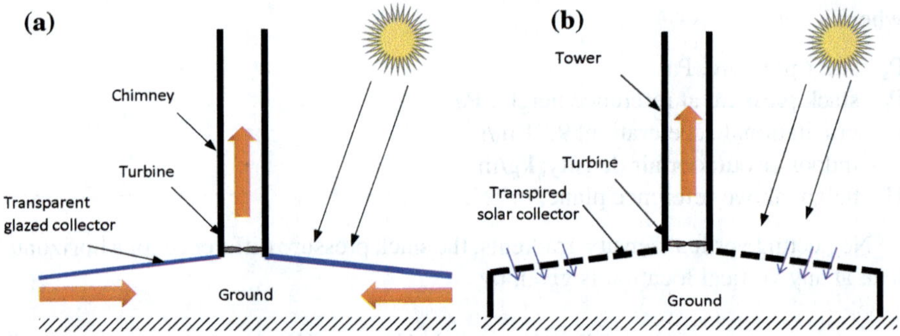

Fig. 2 Schematic of a conventional and transpired solar collector updraft tower respectively

and is heated mostly by warmer ground under the roof. Metal transpired collectors change the air flow method, air enters through millions of perforations along the metal surface of the solar collector and is heated directly by absorber, not above the ground. Consequently, the function and definition of solar collector is changed for solar chimney power plants with the new concept. However, these two different working principles change only the thermal efficiency, not the operational principle of the turbine, which is located at the bottom of tower, and power production. Hence, classical equations for solar chimneys can be used for both concepts.

In a solar chimney power plant, driving potential of rising hot air is calculated as similar to the stack pressure. It therefore can be given as (Schlaich et al. 2005):

$$\Delta P_{pot} = g \int_{0}^{H_c} (\rho_o - \rho) dz \tag{4}$$

where

ρ_o outdoor air density, kg/m^3
ρ chimney density, kg/m^3
H_c chimney height, m
g gravity, 9.81 m/s^2

The potential pressure difference can be subdivided into a static and a dynamic component, neglecting friction losses:

$$\Delta P_{pot} = \Delta P_{st} + \Delta P_{dyn} \tag{5}$$

where the maximum power from a solar chimney power plant can be obtained as

$$P_{max} = \frac{2}{3} V_c A_c \Delta P_{pot} \tag{6}$$

where

V_c Air velocity in chimney, kg/m^3
A_c chimney cross-section, kg/m^3

Solar chimney efficiency is given as

$$\eta = \frac{gH_c}{cT_o} \tag{7}$$

As seen from Eq. (7), solar chimney efficiency is strongly connected with chimney height. This is the fundamental reason why solar chimney should be as high as 1,000 m for large scale power production. Power production is also related to air flow; however, classical transparent collectors have low efficiencies which results in larger areas needed for 200–300 MW systems.

On the other hand, recent approach to tower shape became one of the current issues of solar chimney works. Studies showed that diffuser shaped towers or divergent chimneys may increase the power output significantly compared to classical constant cross sectional tower. For example, some studies claim that it is possible to obtain 20 times more power with a divergent tower for classical Manzanares test unit, because of the pressure recovery at the exit of tower, which also changes the potential pressure equations as

$$\Delta P_{pot} = g(\rho_o - \rho)H_c - \frac{1}{2}\rho(v_o^2 - v_i^2) \tag{8}$$

where

v_o air velocity at chimney outlet, m/s
v_i air velocity at chimney inlet, m/s

5 Global Potential for Solar Chimney

5.1 *Energy Efficient Buildings*

According to international energy agency report for construction sector, buildings construction and operations accounted for 40% of global final energy use and 36% of energy-related carbon dioxide emissions in 2018 (The Global Alliance for Buildings and Construction 2018). Therefore, new building codes and regulations become more important during transition to 100% renewable energy in buildings. It is obvious that new building design considerations must include more sustainable and renewable energy strategies, because it significantly affects human comfort, energy consumption and carbon dioxide emissions in buildings. Among new building designs, passive buildings are the most efficient and effective way to reduce operational energy consumptions for space heating, cooling, ventilation and lighting in a building. Also, a building with passive design can reduce some structural problems of conventional buildings such as solar gain, insufficient building envelope, indoor ventilation and use of daylight. Today, the need for new energy efficient and sustainable building standards and guides based on passive building design is increased. On the other hand, the number of new modern energy efficient and sustainable buildings is increasing fast. Most designs include the passive building design considerations and concepts with renewable energy such as solar energy. High energy performance of building concepts, nearly zero-energy building and net zero energy buildings are based on passive house design considerations and renewable energy use. The supplementary energy needs for building requirements are provided by renewable sources. In this regard, solar chimney system for space heating, cooling and ventilation can contribute to energy savings in buildings.

The ventilation in buildings is becoming very important with tight building envelopes, however, mechanical ventilation is not desirable because of energy consumption, noise and also other operational issues. Natural ventilation is an important way to reduce or sometimes entirely eliminate problems associated with mechanical ventilation. Also, it helps to ventilate building which has no mechanical ventilation systems. A solar chimney is a very good way for supporting of a natural ventilation with substantial energy savings. However, solar chimneys are still not fully recognized in modern construction sector. Despite the fact that solar chimneys are more suitable for hot climates and are mostly used in southern countries, there are very good examples of solar chimneys which have been operating in cold climates for many years, and results show that there is an important energy saving potential during ventilation not only in sunny days but also in windy days. Also, during sunny days, solar chimneys can save a lot of energy by air preheating.

It is obvious that natural stack ventilation strategies can provide a weak stack effect and ventilation, because temperature difference inside the building might not be sufficient. In this regard, solar chimney offers more promising outcomes because it provides higher indoor-outdoor temperature difference by solar radiation. An important benefit of a solar chimney is the reduced demand for space cooling. Studies show that solar chimneys can reduce cooling energy consumption by 10–20% and indoor temperature can be lower as much as 4–5 °C.

Apart from ventilation and passive cooling, solar chimneys can also provide space heating in winter period. As seen in Fig. 3, the new energy efficient building of York University in Toronto is a characteristic example which shows the multi-use of solar chimney in a building. Fifteen-meter-high iconic design of solar chimney not only increases the visual impact of the building, but also provides natural ventilation, passive cooling and heating as mentioned above. This is a very good example of integration of a solar chimney in a building structure. Solar chimneys in buildings can be also combined with earth-air heat exchangers, phase changing materials and photovoltaics.

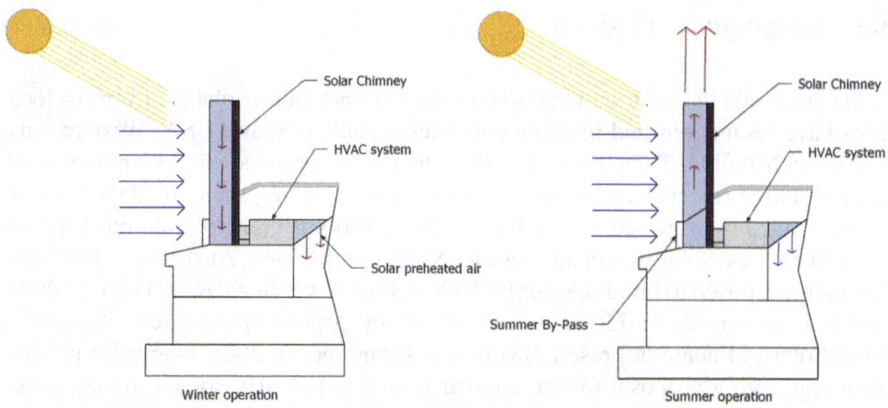

Fig. 3 Solar chimney application for cooling, ventilation and heating modes

5.2 *Renewable Electricity Production*

Although the principle of a solar updraft tower has been known for long, it has not been applied in practice on a large scale yet. Solar chimney power plants are not commercial systems, which are bankable and ready for market like CSP and PV power plants. Therefore, there is not existing available map which shows the potential of solar chimney power plants as in PV and CSP systems, and also there is no certified software available to calculate the power output of a solar chimney for electricity production. At this moment, the situation of large scale solar chimney power plans is similar to first modern dams used for electricity production from hydro power. Solar chimney power plants need some key developments for the next phase, however, the potential of solar chimney power plants for different regions has been extensively studied in the literature, especially some regional studies, as in China, Nigeria or Australia. Despite the absence of map of potential locations of solar chimney power plants, studies show that solar chimney power plants can be used in any geographic location, because the system works with low temperature difference, however, strong solar radiation obviously improves the performance. On the other hand, most of the barren lands, mainly including the Sahara Desert, the Arabian Peninsula Deserts, the Australian Desert, the Kalahari Desert, the Atacama Desert, the deserts in Iran and the surrounding regions, the North American Desert, the Taklimakan Desert, and the Gobi, where solar radiation is high, are suitable for large scale solar chimney power plants (Zhou and Xu 2016). Also, high altitudes with mountains or hills are another potential area. Despite lower solar radiation it is possible to get performance as high as 85% of the locations with the highest solar radiation (Bilgen and Rheault 2015).

5.3 Large Scale Projects

Some proposals to construct large scale solar chimney power plants in various locations have been submitted in many countries including Spain, USA, Western Australia, and Namibia. Though some of them had been approved on government level, none of these projects have progressed beyond the planning stage. In 2001, an Australian company proposed to build a 200 MW, 1000 m high solar chimney power plant known as Solar Tower in western New South Wales, Australia. Also, same company proposed to build the similar 1000 m high tower in Arizona, USA. In 2006, a proposal to construct a 750 m high 40 MW chimney power plant titled Ciudad Real Torre Solar in Fuente el Fresco, Spain, was submitted. In 2008, Namibian government approved a proposal for the construction of a 400 MW solar chimney called the 'Greentower'. The tower is planned to be 1.5 km high and 280 m in diameter, and the base will consist of 37 km^2 of greenhouse in which cash crops can be grown. In 2011, another Australian Company "Hyperion Energy", was reported to be planning to build a 1 km high solar updraft tower near Meekatharra to supply power to Mid-West mining projects. The high cost, difficulty and possible dangers of constructing such a super high building seem to be the main factors that precludes solar chimney power plants from commercialization. Additionally, the efficiency of solar chimney power plants is significantly low in comparison to other solar power technologies, therefore hybrid uses and combinations are proposed for solar chimneys, from compost to photovoltaic integrations in relevant literature.

6 Research and Development for the Transition Toward a Sustainable Future

6.1 Solar Chimney Development Towards a Zero Emission, Energy Efficient Buildings

According to international energy agency report on building sector, the global building floor area is expected to double by 2050, as well as energy demand and related GHG emissions for construction of the buildings.

In the last decade the number of construction regulations has increased significantly all over the world. Now over 60 countries apply either optional or compulsory building energy regulations. However, so far only 85 countries are in the process of developing building certification guidelines. In order to achieve sustainable environment comprehensive and integrated design approach to construction is required. Solar chimneys are an integral part of holistic building design and construction. Nowadays, solar chimney applications in today's ambitious low energy buildings are very popular and often become the main structure of the architectural design of the buildings. We can safely assume that there are no modern low energy buildings without integrated solar chimneys. However, now the main challenge of designing a solar chimney is to optimize and standardize its performance with the lowest cost. On the other hand, new standardization projects related to natural ventilation applications in buildings have officially been started under Technical Committees (TC) in the European Committee for Standardization (CEN/TC 156) and the International Organization for Standardization (ISO/TC 205) since 2017.

Even though the benefits of natural ventilation associated with solar chimney application are widely acknowledged, its use by designers or architects strongly depends on the following challenges:

- The acceptable modeling of solar chimney for natural ventilation, and especially flow design for cross ventilation with solar chimneys.
- The suitable design for cooling in summer to prevent overheating risks for buildings.
- The adequate estimation of the expected benefit of solar chimney for thermal comfort and ventilation requirements, as well as energy performance when using a solar chimney in buildings.

While the first issue can be handled by the use of airflow models and CFD analyses, the second and third problems require new calculations in building regulations and standards.

6.2 *Solar Chimneys in Urban Environment*

Large scale power production with solar chimney integrated cities are popular subjects in relevant literature. Some futuristic solar chimney concepts in the last decade were proposed for different cities, mostly for sustainable city development and eliminating air pollution. Most of the proposed works are directly associated with city life, and create new habitat. Since these projects often seem futuristic and can significantly change the city landscape, it can be stated that these proposals are not realistic from many standpoints and can hardly be realized. However, many solar chimney concepts are designed very well from artistic and architectural standpoint.

An architectural studio Heri and Salli (2016) sketched a solar updraft tower for Vienna's urban space, called Energy Land Tower, which is integrated with the city of Vienna and creates a new vertical and horizontal habitat for the city, as shown in Fig. 4. Around a tube-like concrete core as updraft shaft of the so called energy-land-tower develops a 1000 m high habitable landscape. By means of an orthogonal three-dimensional structure following grid like urban planning, they begin by spreading out and opening up a space, that can be used afterwards. The grid formulates opening up connecting and public zones that are available. Solar updraft tower equipped with solar cells are able to move according to the light and if necessary, provide shade or protect from the wind. Excess rainwater is saved and used for irrigation. When required it's possible to create proper rain, mist or other microclimate areas.

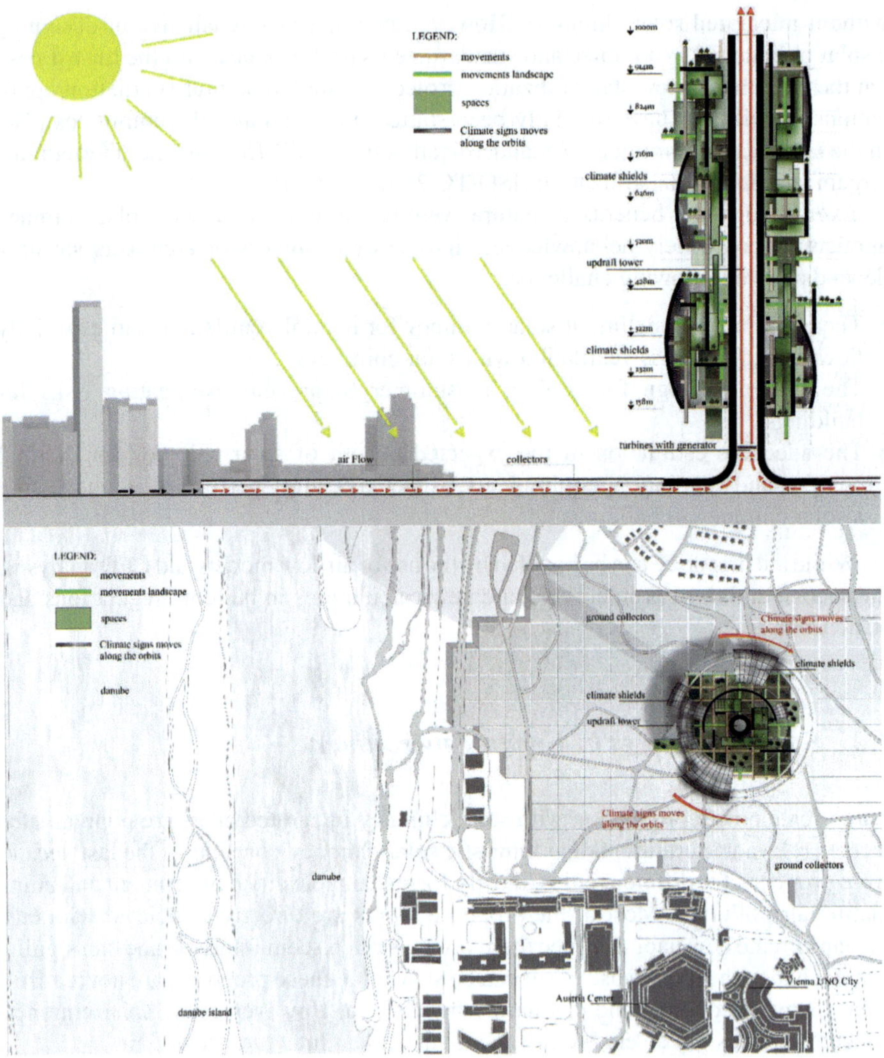

Fig. 4 A solar updraft a compressed habitat, Vienna

Another project was proposed recently for the smart city project in Graz, Austria (Smart Future Graz "Urban building blocks" self sufficient in energy. Energy Innovation Austria 4/2013). The whole project includes research, testing and demonstration of new technology components: new photovoltaic modules, solar cooling, urban generation of solar generated electricity, solar updraft power plant, integrated technologies of facades, a small block heating power station (biogas), smart heat grids and new strategies and concepts for sustainable urban mobility including e-mobility. A solar chimney project "A Proposal to Create an Energy-Producing Megatall for

Kunming" was submitted by Thomas Kraubitz for China (Kraubizt 2012). The concept combines a 750 m high solar updraft tower and sustainable high-rise building which has 30 MW power production capacity. As seen in Fig. 5, urban using of solar chimney power plant which is completely integrated into the city environment to provide energy and balancing the heat island with air cleaning option, is sketched by Except architectural and innovation company (Bosschaert 2009). A tower in a city could hardly be the size of one in a field, so its generative power would be diminished. However, it could provide three distinct advantages alongside electricity generation. First, the tower could be fitted with particulate, carbon and other air filters. The air rushing through the chimney would thus be cleaned resulting in urban air quality improvement, while at the same time generating some electricity. Systems like these would be very suitable for highly polluted cities. Second, in cold climates, the heat dissipation of the urban environment can be buffered. The trapped heat will aid the power generation, but will also reduce the temperature gradient between interiors and exteriors, resulting in energy savings for building heating and higher insulation efficiencies. Third, in hot climates, a second layer with a semi-transparent PV membrane could be installed under the transparent roof collector. This would partially block out the sun, causing the temperature gradient to drop. Then there will be two layers generating convection, possibly increasing the efficiency of the tower. The top layer would ensure the heat is not trapped in the bottom layer, thus preventing the heating up of the city ambient. The constant air pull of the solar updraft tower will partially combat the heat island effect.

On the other hand, one experimental 60 m high tower was built in Xian, China, in the end of 2018, to clean the air by using chimney method (Cao 2018). The tower

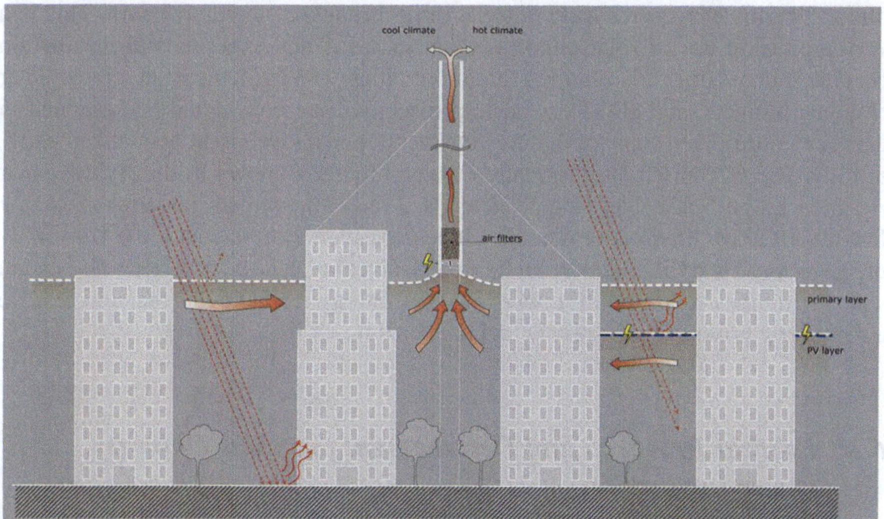

Fig. 5 A proposal for urban use of solar updraft towers

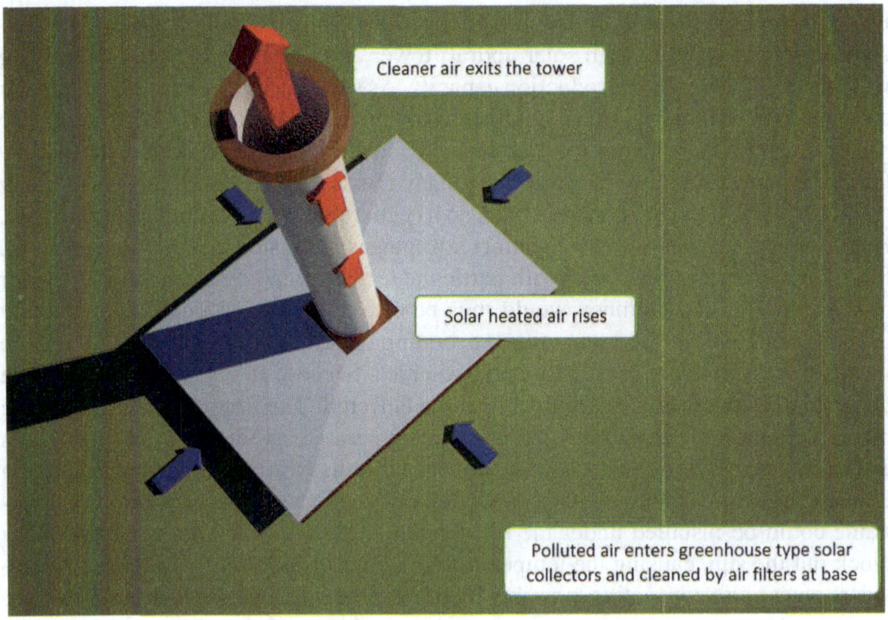

Fig. 6 A solar updraft tower for air cleaning in cities

works through greenhouse covering about half the size of a soccer field around the base of the tower. As seen in Fig. 6, air is sucked in and heated up by solar energy; hot air then rises through the tower and passes through multiple layers of cleaning filters. The tower requires little power to run throughout daylight hours. This is a very promising use of solar chimney method and if this experimental installation becomes successful, it is obvious that more cities will implement integrated solar chimney projects, and also large scale projects will be considered and planned in the near future. Even some of the projects mentioned above might be realized partly or fully. The 60 m high air purification tower, officially known as the Hybrid Solar Assisted Large Scale Cleaning System, is a scientific project undertaken by the Institute of Earth Environment at the Chinese Academy of Sciences, the University of Minnesota Twin Cities, and Xian Jiaotong University. It is the first cleaning system, which combines solar updraft tower with new energy technologies that can remove atmospheric pollutants effectively and prevent smog.

6.3 *Further Research and Development for Solar Chimney Power Integration*

Solar chimney power plant and associated technologies need a series of common and coordinated research and development phases until the first large scale commercial tower operation is possible. According to some projections, first commercial 100 MW solar chimney power plant can start operation between 2030–2040. It means that there is a transition/maturity period for solar chimneys. The most important objectives for solar chimney power plant research and development are:

- Certified design and power estimation software for solar chimney.
- Turbine design and production specifically for solar chimneys.
- New chimney and collector designs.
- The preparation of solar chimney power potential map.
- Power electronics for grid connection.
- Environmental works, structural works, economic analysis.
- Hybrid systems.

At this stage, especially during transition period, hybrid solar chimney power plants become more important than stand-alone projects, because solar chimney power plant is a versatile system in general and it can be combined with different energy systems for different purposes. Hybrid concept is important to solar chimneys, because the performance of the solar chimneys significantly increases with integration of other energy systems such as photovoltaics, waste heat, biomass, etc. Additionally, the hybrid approach would reduce the transition period for stand-alone solar chimney project, and would provide many advantages for developing solar chimney systems, as well as more data for the future scientific studies. Moreover, 100 years old concept can be operated for the transition period of 100% renewable energy. Some important examples and possibilities of hybridization show that solar chimneys combined with other renewable energy technologies can be a good solution for transition period.

The experimental studies show that photovoltaic integrated transpired solar collectors have a great potential for wide use of solar chimney power plants (Eryener and Kuscu 2017). Solar chimney with photovoltaic panel is a promising combination which provides many advantages, such as cooling the PV panels and increasing their efficiency via air flow of solar chimney, benefit from the waste heat of PV panels, reduced dimensions of solar chimneys, increased land usage efficiency, optimization opportunity for both systems (Fig. 7).

Using compost waste heat in solar chimney projects is another opportunity for implementation of hybrid solar chimney power plant (Anderson 2016). Compost has a high heat capacity and the space under the solar chimney collector is very suitable for storage and use of waste heat as the source of additional heat energy in solar chimney when the solar radiation is not sufficient for production of electricity. As a hybrid compost solar chimney power plant addressing sustainable renewable energy needs, system can also be integrated with photovoltaic panels. Experimental and numerical results show that 100 m high hybrid compost solar chimney with photovoltaic panels can potentially produce over 100 MW for 24 acres solar collector field (Fig. 8).

The method which generates power by downward movement of the air inside the tower is called the Downdraft Energy Tower. The downdraft technique, in which the atmospheric air is quickly moved downwards from the top of a chimney or tower using various methods such as wind catcher or humidifying techniques, is dated back

Fig. 7 A hybrid solar updraft tower experimental test unit in Trakya University, Edirne, Turkey

to the 4th millennium BC. Normally, both tower systems operate separately. Recent studies show that downdraft mechanism can be used with solar chimneys (Hybrid energy tower 2018). Thus, smaller tower dimensions and greater power output are achievable in the hybrid downdraft and updraft system compared to conventional solar power systems. When there is no solar radiation, system can be operated with only wind power, and conversely, hybrid system can produce energy using only solar radiation when there is no wind. Also, it can be combined with photovoltaics and waste heat/geothermal sources as mentioned above (Fig. 9).

Integration of geothermal sources into solar chimneys is one of the most important innovations. In general, for production of electricity, the recommended temperature of geothermal resources is minimum 100 °C. Lower temperatures are used for space heating such as greenhouses or building heating. However, geothermal sources below 100 °C can be used in solar chimneys, because solar chimneys can produce energy with low temperature systems, and their large heat transfer surfaces are very suitable for geothermal sources, hence additional renewable electricity production can be obtained from geothermal sources via solar chimney power plants (Fig. 10). It also provides a continuous energy production in a solar chimney. A geothermal powered solar enhanced natural draft dry cooling tower can produce 20 times more renewable electricity than a stand-alone solar chimney of the same size (Zue et al. 2012).

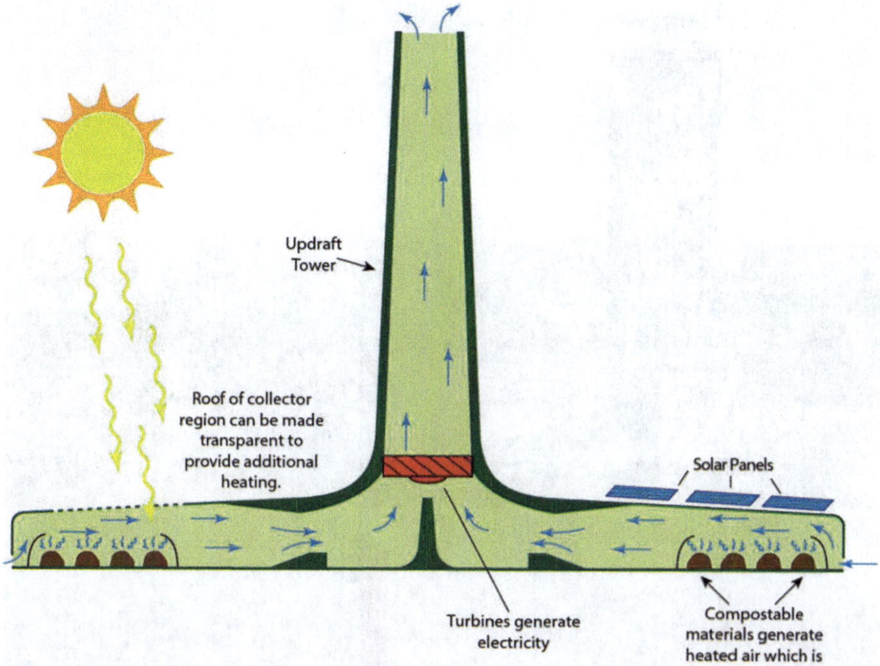

Fig. 8 Hybrid compost waste heat solar chimney power plant

7 Conclusion

Solar chimneys are unique systems that can be used for various purposes, from air cleaning to renewable energy production. Solar chimneys are very well known and have been used for a long time, and therefore cannot be defined as a new technology, however, its performance is still unknown in many applications; one of the reasons is low efficiency compared to other renewable energies, therefore, despite simplicity being its most important advantage, solar chimneys need comprehensive research and development for 100% renewable energy transition. Besides, studies show that large scale solar chimneys can be functional and operational for building ventilation and air pollution elimination in cities, as well as urban planning and large scale renewable energy production in next decade. Therefore, this unique system can potentially be used for a very wide variety of projects.

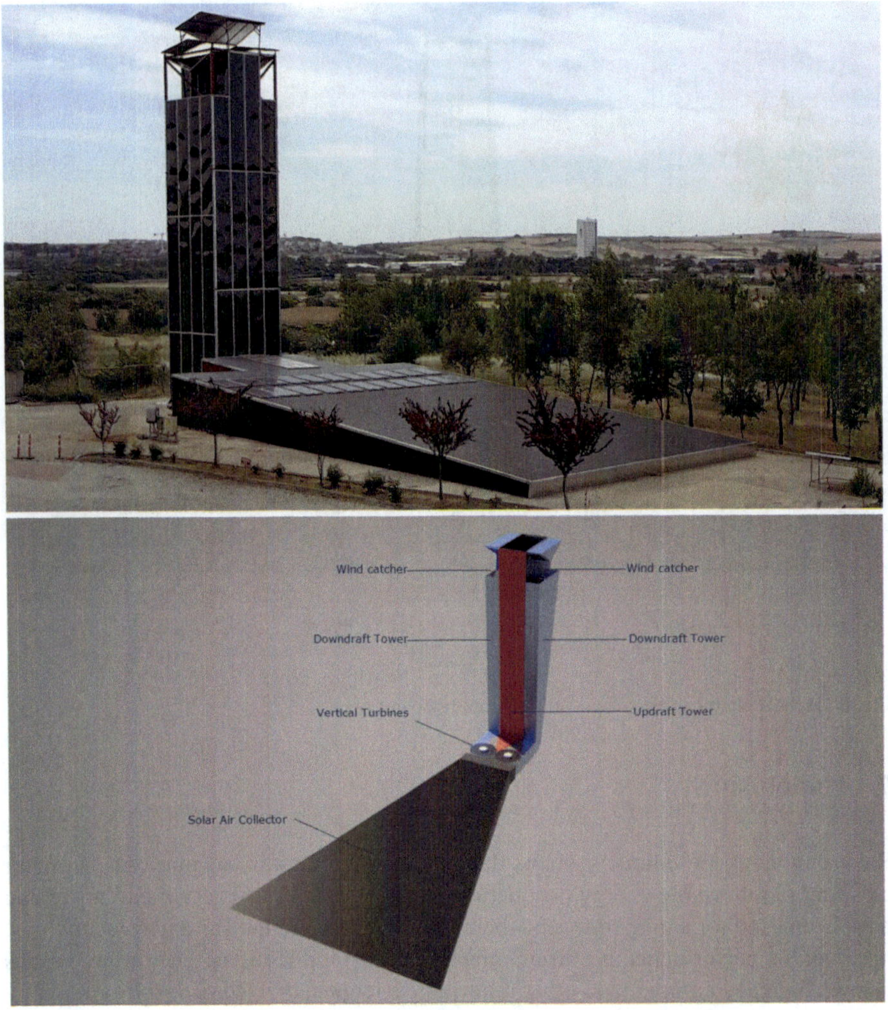

Fig. 9 A hybrid solar updraft tower test unit in Trakya University, Edirne, which combines updraft and downdraft mechanisms in a single structure

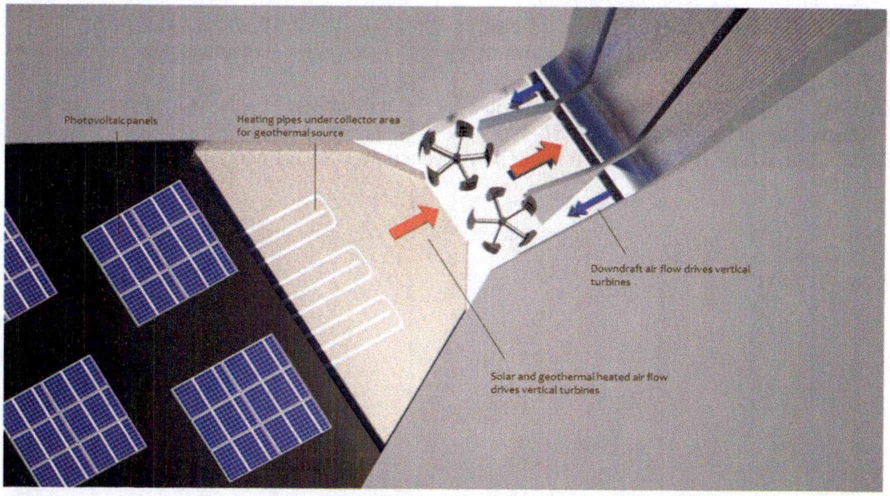

Fig. 10 A solar updraft tower combined with geothermal, photovoltaic panels and downdraft towers

References

Anderson K et al (2016) Design of a compost waste heat to energy solar chimney power plant. J Civ Eng Res 6(3):47–54

ASHRAE (2003) Ventilation and infiltration. Handbook—HVAC applications

Bilgen E, Rheault J (2015) Solar chimney power plants for high latitudes. Sol Energy 79(5):449–458

Bosschaert T (2009) Improving vertical updraft towers for renewable energy. http://www.except.nl/en/articles

Breeze P (2014) Solar power. Power generation technologies, 2nd ed. Newnes, pp 259–286

Cao Q et al (2018) Urban-scale SALSCS, Part I: Experimental evaluation and numerical modeling of a demonstration unit. Aerosol Air Qual Res 18:2865–2878

Eryener D, Kuscu H (2017) Hybrid transpired solar updraft tower. Sol Energy 159:561–571

Eryener D, Hollick J, Kuscu H (2017) Thermal performance of a transpired solar collector updraft tower. Energy Convers Manag 142:286–295

Haaf W (1984) Solar chimneys, Part II: Preliminary test results from the Manzanares pilot plant. Int J. Solar Energy 2:141–161

Haaf W, Friedrich K, Mayr G, Schlaich J (1983) Solar chimneys, Part I: Principle and construction of the pilot plant in Manzanares. Int J Solar Energy 2:3–20

Heri&Salli (2016) Energy land tower. http://www.heriundsalli.com/en/energie-land-turm/

Hybrid energy tower (2018). http://en.hybridenergytower.com

Kraubizt T (2012) A proposal to create an energy producing Megatall For Kunming China. CTBUH J 4:30–35

Mackay ME (2015) Solar energy: an introduction. Oxford University Press, Oxford

Richter R, Tingzhen M, Davies D, Wei L, Caillol S (2017) Removal of non-CO_2 greenhouse gases by large-scale atmospheric solar photocatalysis. Prog Energy Combust Sci 60:68–96

Schlaich J, Bergermann R, Schiel W, Weinrebe G (2005) Design of commercial solar updraft tower systems—utilization of solar induced convective flows for power generation. J Sol Energy Eng 127:117–124

Smart Future Graz "Urban building blocks" self sufficient in energy. Energy Innovation Austria 4/2013

The Global Alliance for Buildings and Construction (2018) The 2018 global status report
Zue Z et al (2012) Solar enhanced natural draft dry cooling tower for geothermal power applications. Sol Energy 86:2686–2694
Zhai XQ, Song ZP, Wang RZ (2011) A review for the applications of solar chimneys in buildings. Renew Sustain Energy Rev 15:3757–3767
Zhou X, Xu Y (2016) Solar updraft tower power generation. Solar Energy 128:95–125

Towards a Low-Carbon Energy World: Some Pilot Projects in China, Europe and the United States

Wolfgang Palz

1 Zhangjiakou: China's First-Ever National Renewable Demonstration Zone

On the pathway to a low-carbon future the State Council of China has designated Zhangjiakou, a region neighbouring the capital Beijing with 4.4 million people, to become China's first Renewable Energy (RE) Demonstration Zone. The project is, among others, supported by China National RE Centre, IRENA in Bonn, and the German Ministry of the Environment—they presented an official project analysis in Nov. 2019 (irena.org/publication).

The Winter Olympics in 2022 will be held in Beijing with the mountains of Zhiangjiakou to serve for the skiing part of them. The decision by the central government of China to develop the RE Demonstration Zone of Zhiangjiakou is actually a follow up of the objective to make these Olympics "low-carbon". The area with the skiing facilities will be powered 100% by Renewable electricity at the time of the Olympics.

Zhangjiakou is a pioneer of wind power development that it started already in 1993. By 2017 it had a wind power capacity of 8.7 GW and nearly 3 GW of PV. Installations increased further continuously since then. But with the rapid growth, transmission of electricity on the local grids had emerged as a major problem. Between 2012 and 2013 wind curtailment increased from 20 to 30% and 1 TWh of green electricity was wasted. Since then local and regional grid capacities were enhanced. One also has begun using excessive wind power for heating purposes. In 2017 curtailment dropped to 7%, in line with the national goal of 5% by 2020.

Following the targets set out by Zhangjiakou City the RE sources will account by 2020 for 55% of electricity consumption and 30% for total final energy consumption. By 2030 those consumption figures are planned to increase to 80% and 50% respectively.

W. Palz (✉)
Brussels, Belgium
e-mail: w@palz.be

T. S. Uyar (ed.), *Accelerating the Transition to a 100% Renewable Energy Era*, Lecture Notes in Energy 74, https://doi.org/10.1007/978-3-030-40738-4_24

China has a sound basis to achieve those ambitious targets. The country has in recent years become the world's leader for most of the RE markets. China's domestic markets are the world's first ones for

- Wind Power
- Solar PV
- Hydropower
- Bio-power
- Solar Heat Collectors
- Biogas
- Wood Pellets.

Zhangjiakou's energy structure is currently much linked to coal. It must be fundamentally changed on the way to a low-carbon scenario. Coal mining and production had a key role in the region's energy activities. In 2012 the phasing out of coal mining began aiming at near-zero coal mining by 2020. By then nearly all the coal mining and production companies will have to be closed. So much to the mining sector.

As far as consumption is concerned, coal is still significant in the region for electricity and heat generation through CHP and heating plants. The region has 6.6 GW of coal-fired CHP plants and 3,500 km distribution grids of heat generated from coal.

Coke coal is employed for iron and steel making, too. To get this sector under control it is planned to shut down a major steel factory, Xuan Steel, or to upgrade the use of blast furnaces to electric-arc furnaces.

In the residential sector and coal's overwhelming role there for space heating it is envisaged to increase the use of individual heat pumps as well as large heat pumps in district heating systems. Important is China's new attempt to improve the energy performance of buildings and to introduce near-zero heat demand standards. Since 2016 building retrofitting is already under way in China. Eventually thermal energy demand in buildings could be reduced down to 30 kWh/m^{2}/year, the international standard for passive buildings.

Ultimately the projection of a RE-2050-Electricity Scenario anticipates that industry would be fully electrified. The 30 GW of solar PV and 40 GW of wind power would be fully exploited. Enabling technologies would include heat storage and seasonal storage for heat and electricity. Hydrogen generated from RE sources is considered potentially a major player: by 2050—that is only 30 years from now—some 93% of all energy demand would be met by the RE in the form of electricity, heat, and renewable hydrogen.

The author is member of the expert advisory Committee for the Zhangjiakou Demonstration Zone. Picture from 2019, 11. The author in the centre of the front row, with the Vice Mayor of Zhangjiakou, 3rd from the right

2 EU: The Renewables in the European Union

The objective of the EU is to make Europe the first climate-neutral continent by 2050.

Since the year 2000 already, the domestic markets of the Renewables keep increasing there all the time. Currently in 2019 the EU is second after China—and just before the United States—in terms of total wind power and solar PV capacity installed. In the coming years a further increase of the Renewable power markets in Europe is on the way. This is in particular true for solar PV that is gaining the lead over all of them.

In the building sector, the EU has in place the EPBD, the Energy Performance of Buildings Directive. It demands that all new buildings on the continent be "nearly zero energy buildings NZEB". The term means that buildings consume nothing else than energy of the renewable type, namely PV. The so-called +energy buildings that are exclusively equipped with Renewable energies, too, produce even more than the energy they consume. The EU Directive is binding for all new public buildings since 2019 and generally for all new buildings after 2021.

In the transport sector the EU has another Directive in place. It is called the FQD, the Fuel Quality Directive. It demands that the Greenhouse Gas GHG Intensity of all transport fuels be reduced by 6% in 2020. It is actually a huge incentive for blending the petrol with bio-fuels (bio-alcohols or vegetable oils) that are neutral in carbon emissions. The reduction can be achieved, too, by switching to electric cars or cars employing hydrogen from Renewable power. But the blending with bio-fuels is by far the cheapest of all options. Petrol in the EU is today typically blended with up to 5% of bio-alcohol. In the United States it is already 10%, the E10.

There are a great number of cities and towns, regions, and islands in the EU that have adopted the target to reduce carbon emissions to zero by 2050.

Here is the example of Paris, the capital of France, Europe's biggest city and one of the most beautiful.

The author's main residence is located in the centre of Paris.

The city of Paris has set out for 2050 the objective to become neutral in carbon emissions and a 100% supplied with Renewable energies. 80% of the Renewables will have to be imported from neighbouring areas. The world of agricultural and economic associations will play its role. All wastes will have to be recycled by that time; none of it will be allowed for burying. Energy consumption is to be cut by half in 2050. Existing buildings are to be renovated, a million dwellings and half a million shops and offices. The energetic renovation of buildings will entail a better quality of life and a reduction of residence taxes. 20% of all roofs will be equipped with PV. Public transport will be increased. Paris will have less individual cars and those on the streets will be less pollutant. Biking will be further encouraged—it is already very much today: by 2020 all streets of Paris are ready for biking.

Paris has an Agency for the Climate giving advice and financial aid. Recently 36,000 dwellings were renovated, 50,000 m^2 of building integrated PV was deployed. Energy consumption of public lighting was reduced by 30%. 240 municipal schools were renovated. 1 bn € were spent on building renovation. The city has a solidarity fund for dwellings. After 2024 diesel cars are no more allowed on the streets in Paris. By 2020 more streets will be appeased and made greener. "Eco-quarters" are being created that are neutral in carbon emissions. A first one is "Saint Vincent de Paul".

Paris has a "Plan Climate Energy" that includes 500 individual measures of action. For 2030 the action plan calls for 50% less GHG emissions and 45% of RE consumption as well as 35% less energy consumption.

The city has a Charter "Paris Climate Action" for companies to become members. Paris has volunteers of the climate.

Each year an appraisal of achievements is established.

3 America's First Solar Town: Babcock Ranch in Southern Florida

The information that follows was provided by my American friend Neville Williams. He was one of the first promoters of PV in Asia and wrote books on solar energy, too; he lives now in Florida. When we last met, it was in Nepal in a remote village. Now he said: "(we) have seen such incredible changes in RE. I remember the early solar days in China, visiting a primitive PV factory, and installing our little solar home systems."

The Babcock Ranch in Florida, USA, is a new town built from scratch on a green field on the coast by private initiative. Today it is provided by 100% Renewable energy, all being PV. After a decade of planning it has now several hundreds of residents. The town has a huge central 75 MW PV array and 10 MW of battery storage. The local utility Florida Power and Light FPL has taken the lead on the solar projects. FPL and other utilities are building thousands more MWs of solar farms all over the State. They know it is the cheapest form of energy there is today and they want to keep their distribution monopoly, save on their national gas costs, and not pass along any of the savings from the sun to customers. They all fought solar for years, then monopolised it. Not stupid! says Neville.

Nevertheless, in the new town of Babcock Ranch where all buildings are new constructions there are also a lot of small players to install PV on private residences…

MIX
Papier aus verantwortungsvollen Quellen
Paper from responsible sources
FSC® C105338

Printed by Books on Demand, Germany